AF893471

Methods in Molecular Biology

Series Editor
John M. Walker
School of Life and Medical Sciences,
University of Hertfordshire, Hatfield,
Hertfordshire AL10 9AB, UK

For further volumes:
http://www.springer.com/series/7651

Cancer Systems Biology

Methods and Protocols

Edited by

Louise von Stechow

NNF Center for Protein Research, University of Copenhagen
Copenhagen, Denmark

Humana Press

Editor
Louise von Stechow
NNF Center for Protein Research
University of Copenhagen
Copenhagen, Denmark

ISSN 1064-3745 ISSN 1940-6029 (electronic)
Methods in Molecular Biology
ISBN 978-1-4939-7492-4 ISBN 978-1-4939-7493-1 (eBook)
https://doi.org/10.1007/978-1-4939-7493-1

Library of Congress Control Number: 2017961108

© Springer Science+Business Media LLC 2018
Chapters 5, 6 and 7 are licensed under the terms of the Creative Commons Attribution 4.0 International License (http://creativecommons.org/licenses/by/4.0/). For further details see license information in the chapters.
This work is subject to copyright. All rights are reserved by the Publisher, whether the whole or part of the material is concerned, specifically the rights of translation, reprinting, reuse of illustrations, recitation, broadcasting, reproduction on microfilms or in any other physical way, and transmission or information storage and retrieval, electronic adaptation, computer software, or by similar or dissimilar methodology now known or hereafter developed.
The use of general descriptive names, registered names, trademarks, service marks, etc. in this publication does not imply, even in the absence of a specific statement, that such names are exempt from the relevant protective laws and regulations and therefore free for general use.
The publisher, the authors and the editors are safe to assume that the advice and information in this book are believed to be true and accurate at the date of publication. Neither the publisher nor the authors or the editors give a warranty, express or implied, with respect to the material contained herein or for any errors or omissions that may have been made. The publisher remains neutral with regard to jurisdictional claims in published maps and institutional affiliations.

Printed on acid-free paper

This Humana Press imprint is published by Springer Nature
The registered company is Springer Science+Business Media, LLC
The registered company address is: 233 Spring Street, New York, NY 10013, U.S.A.

Preface

Cancer is a highly complex disease that is often characterized by vast changes in the genetic and epigenetic landscape. Those changes result in altered protein expression levels in tumors compared to healthy tissues. Moreover, posttranscriptional alterations lead to deregulation of signaling processes, and altered metabolic pathways can produce aberrant metabolic signatures in cancer cells.

A wealth of high-throughput information has emerged over the last decade, including global measurements of genes, proteins, and metabolites, as well as many other molecular species. Those studies provide a glimpse of the molecular makeup of cancer cells on various levels. In order to classify tumor types and predict clinical outcomes of cancer, researchers often employ sophisticated computational tools to extract cancer-specific events from the excessive amounts of data that have been compiled.

This volume on "Cancer Systems Biology" comprises protocols, which describe systems biology methodologies and computational tools, offering a variety of ways to analyze different types of high-throughput cancer data. Those include for example network- and pathway-based analyses. Other chapters cover descriptive and predictive mathematical models used to analyze complex cancer phenotypes and responses to anticancer drugs.

A number of chapters give an overview of data types available in large-scale data repositories, describe state-of-the-art computational methods used, and highlight key trends in the field of cancer systems biology.

Copenhagen, Denmark *Louise von Stechow*

Contents

Part IV Systems Biology of Metastasis and Tumor/ Microenvironment Interactions

Part V Modeling Drug Responses in Cancer Cells

Contributors

TERO AITTOKALLIO • *Institute for Molecular Medicine Finland (FIMM), University of Helsinki, Helsinki, Finland; Department of Mathematics and Statistics, University of Turku, Turku, Finland*

ASH A. ALIZADEH • *Division of Oncology, Department of Medicine, Stanford Cancer Institute, Stanford University, Stanford, CA, USA; Division of Hematology, Department of Medicine, Stanford Cancer Institute, Stanford University, Stanford, CA, USA; Stanford Cancer Institute, Stanford University, Stanford, CA, USA; Institute for Stem Cell Biology and Regenerative Medicine, Stanford University, Stanford, CA, USA*

MIRIAM R. AURE • *Department of Cancer Genetics, Institute for Cancer Research, The Norwegian Radium Hospital, Oslo University Hospital, Oslo, Norway*

HAMIDREZA BADRI • *Industrial and Systems Engineering, University of Minnesota, Minneapolis, MN, USA*

TONE F. BATHEN • *Department of Circulation and Medical Imaging - MR Center, Faculty of Medicine and Health Sciences, NTNU - Norwegian University of Science and Technology, Trondheim, Norway*

JAMES CAMPBELL • *CRUK-Centre Core Bioinformatics Facility, Department of Data Science, The Institute of Cancer Research, London, UK*

BINBIN CHEN • *Department of Genetics, Stanford University School of Medicine, Stanford, CA, USA*

JUERGEN COX • *Computational Systems Biochemistry Group, Max-Planck Institute of Biochemistry, Martinsried, Germany*

REBECCA J. CRITCHLEY-THORNE • *Cernostics, Inc., Pittsburgh, PA, USA*

PEDRO CUTILLAS • *Barts Cancer Institute, Queen Mary University of London, London, UK*

AARON DEWARD • *Cernostics, Inc., Pittsburgh, PA, USA*

STEPHANIE L. ELDRIDGE • *Institute for Computational and Engineering Sciences, The University of Texas at Austin, Austin, TX, USA; Biomedical Engineering, The University of Texas at Austin, Austin, TX, USA*

OLIVIER ELEMENTO • *Department of Physiology and Biophysics, Institute for Precision Medicine, Institute for Computational Biomedicine, Weill Cornell Medical College, New York, NY, USA*

LESLIE R. EUCEDA • *Department of Circulation and Medical Imaging - MR Center, Faculty of Medicine and Health Sciences, NTNU - Norwegian University of Science and Technology, Trondheim, Norway*

ERNEST FRAENKEL • *Computational and Systems Biology, Massachusetts Institute of Technology, Cambridge, MA, USA; Department of Biological Engineering, Massachusetts Institute of Technology, Cambridge, MA, USA*

GURO F. GISKEØDEGÅRD • *Department of Circulation and Medical Imaging - MR Center, Faculty of Medicine and Health Sciences, NTNU - Norwegian University of Science and Technology, Trondheim, Norway; St. Olavs Hospital, Trondheim University Hospital, Trondheim, Norway*

CHRISTOPHE GOMEZ • *Aix Marseille Université, CNRS, Centrale Marseille, Marseille, France*

NIKLAS HARTUNG • *Department of Clinical Pharmacy and Biochemistry, Freie Universität Berlin, Berlin, Germany; Institute of Mathematics, Universität Potsdam, Potsdam, Germany*

TONJE H. HAUKAAS • *Department of Circulation and Medical Imaging - MR Center, Faculty of Medicine and Health Sciences, NTNU - Norwegian University of Science and Technology, Trondheim, Norway*

LIYE HE • *Institute for Molecular Medicine Finland (FIMM), University of Helsinki, Helsinki, Finland*

DAVID A. HORMUTH • *Institute for Computational and Engineering Sciences, The University of Texas at Austin, Austin, TX, USA*

HANNAH JOHNSON • *Department of Biological Engineering, Massachusetts Institute of Technology, Cambridge, MA, USA; Signalling Laboratory, The Babraham Institute, Cambridge, UK*

AMANDA J. KEDAIGLE • *Computational and Systems Biology, Massachusetts Institute of Technology, Cambridge, MA, USA*

MICHAEL S. KHODADOUST • *Division of Oncology, Department of Medicine, Stanford Cancer Institute, Stanford University, Stanford, CA, USA; Division of Hematology, Department of Medicine, Stanford Cancer Institute, Stanford University, Stanford, CA, USA; Stanford Cancer Institute, Stanford University, Stanford, CA, USA*

VESSELA N. KRISTENSEN • *Department of Clinical Molecular Biology (EpiGen), Division of Medicine, Akershus University Hospital, Lørenskog, Norway; Department of Cancer Genetics, Institute for Cancer Research, The Norwegian Radium Hospital, Oslo University Hospital, Oslo, Norway*

EVGENY KULESSKIY • *Institute for Molecular Medicine Finland (FIMM), University of Helsinki, Helsinki, Finland*

KEVIN LEDER • *Industrial and Systems Engineering, University of Minnesota, Minneapolis, MN, USA*

CHIH LONG LIU • *Division of Oncology, Department of Medicine, Stanford Cancer Institute, Stanford University, Stanford, CA, USA*

CHRISTOPHER J. LORD • *The Breast Cancer Now Toby Robins Breast Cancer Research Centre and CRUK Gene Function Laboratory, The Institute of Cancer Research, London, UK*

NEEL S. MADHUKAR • *Department of Physiology and Biophysics, Institute for Precision Medicine, Institute for Computational Biomedicine, Weill Cornell Medical College, New York, NY, USA*

MICHAEL I. MIGA • *Biomedical Engineering, Vanderbilt University, Nashville, TN, USA; Department of Radiology, Vanderbilt University, Nashville, TN, USA; Department of Radiological Sciences, Vanderbilt University, Nashville, TN, USA; Diagnostic Medicine, The University of Texas at Austin, Austin, TX, USA*

AARON M. NEWMAN • *Division of Oncology, Department of Medicine, Stanford Cancer Institute, Stanford University, Stanford, CA, USA; Institute for Stem Cell Biology and Regenerative Medicine, Stanford University, Stanford, CA, USA*

HENG PAN • *Department of Physiology and Biophysics, Institute for Precision Medicine, Institute for Computational Biomedicine, Weill Cornell Medical College, New York, NY, USA*

COLM J. RYAN • *Systems Biology Ireland, University College Dublin, Dublin 4, Ireland*

JANI SAARELA • *Institute for Molecular Medicine Finland (FIMM), University of Helsinki, Helsinki, Finland*

JULIO SAEZ-RODRIGUEZ • *Joint Research Center for Computational Biomedicine (JRC-COMBINE), Faculty of Medicine, RWTH Aachen University, Aachen, Germany; European Molecular Biology Laboratory - European Bioinformatics Institute (EMBL-EBI), Cambridge, UK*

MARY E. SEHL • *Division of Hematology-Oncology, Department of Medicine, David Geffen School of Medicine, University of California, Los Angeles, CA, USA; Department of Biomathematics, David Geffen School of Medicine, University of California, Los Angeles, CA, USA*

ANDLIENA TAHIRI • *Department of Clinical Molecular Biology (EpiGen), Division of Medicine, Akershus University Hospital, Lørenskog, Norway*

HUA TAN • *Department of Radiology, Wake Forest School of Medicine, Center for Bioinformatics & Systems Biology, Winston-Salem, NC, USA*

JING TANG • *Institute for Molecular Medicine Finland (FIMM), University of Helsinki, Helsinki, Finland; Department of Mathematics and Statistics, University of Turku, Turku, Finland; Institute of Biomedicine, University of Helsinki, Helsinki, Finland*

LAURA TURUNEN • *Institute for Molecular Medicine Finland (FIMM), University of Helsinki, Helsinki, Finland*

STEFKA TYANOVA • *Computational Systems Biochemistry Group, Max-Planck Institute of Biochemistry, Martinsried, Germany*

JARED A. WEIS • *Department of Biomedical Engineering, Wake Forest School of Medicine, Winston-Salem, NC, USA; Comprehensive Cancer Center, Wake Forest Baptist Medical Center, Winston-Salem, NC, USA*

KRISTER WENNERBERG • *Institute for Molecular Medicine Finland (FIMM), University of Helsinki, Helsinki, Finland*

FOREST M. WHITE • *Department of Biological Engineering, Massachusetts Institute of Technology, Cambridge, MA, USA; Koch Institute for Integrative Cancer Research, Massachusetts Institute of Technology, Cambridge, MA, USA*

MAX S. WICHA • *Department of Internal Medicine, University of Michigan, Ann Arbor, MI, USA*

JAKOB WIRBEL • *Joint Research Center for Computational Biomedicine (JRC-COMBINE), Faculty of Medicine, RWTH Aachen University, Aachen, Germany; Institute for Pharmacy and Molecular Biotechnology (IPMB), University of Heidelberg, Heidelberg, Germany*

THOMAS E. YANKEELOV • *Institute for Computational and Engineering Sciences, The University of Texas at Austin, Austin, TX, USA; Biomedical Engineering, The University of Texas at Austin, Austin, TX, USA; Diagnostic Medicine, The University of Texas at Austin, Austin, TX, USA; Livestrong Cancer Institutes, The University of Texas at Austin, Austin, TX, USA*

XIAOBO ZHOU • *Department of Radiology, Wake Forest School of Medicine, Center for Bioinformatics & Systems Biology, Winston-Salem, NC, USA*

JUNFENG ZHU • *Industrial and Systems Engineering, University of Minnesota, Minneapolis, MN, USA*

Part I

Systems Biology of the Cancer Genetic and Epigenetic Landscape

Chapter 1

Detection of Combinatorial Mutational Patterns in Human Cancer Genomes by Exclusivity Analysis

Hua Tan and Xiaobo Zhou

Abstract

Cancer genes may tend to mutate in a co-mutational or mutually exclusive manner in a tumor sample of a specific cancer, which constitute two known combinatorial mutational patterns for a given gene set. Previous studies have established that genes functioning in different signaling pathways can mutate in the same sample, i.e., a tumor from one patient, while genes operating in the same pathway are rarely mutated in the same cancer genome. Therefore, reliable identification of combinatorial mutational patterns of candidate cancer genes has important ramifications in inferring signaling network modules in a particular cancer type. While algorithms for discovering mutated driver pathways based on mutual exclusivity of mutations in cancer genes have been proposed, a systematic pipeline for identifying both co-mutational and mutually exclusive patterns with rational significance estimation is still lacking. Here, we describe a reliable framework with detailed procedures to simultaneously explore both combinatorial mutational patterns from public cross-sectional gene mutation data.

Key words Cancer genomics, Co-mutation, Mutual exclusivity, Signaling pathway, Hypergeometric test

1 Introduction

Genetic aberrations and deleterious environment exposure orchestrate to govern the development of various human diseases including cancer [1–7]. In particular, somatic driver mutations accumulating in the human genome are largely recognized as the culprit of human cancer initiation/progression [1, 2, 4]. While numerous somatic mutations can be detected in a single tumor, the mutations are distributed across the genome in a cancer-specific and sample-specific manner [4, 5, 8, 9]. The cancer-specific property refers to the scenario that mutational pattern varies between different cancer tissue types, e.g., liver, lung and breast cancers; while the sample-specific sense corresponds to the mutational variety between different patient samples with the same cancer type. For a specified cancer type, some genes are altered commonly across patient samples, while others exemplify apparent sample-specificity

Louise von Stechow (ed.), *Cancer Systems Biology: Methods and Protocols*, Methods in Molecular Biology, vol. 1711,
https://doi.org/10.1007/978-1-4939-7493-1_1,

© Springer Science+Business Media, LLC 2018

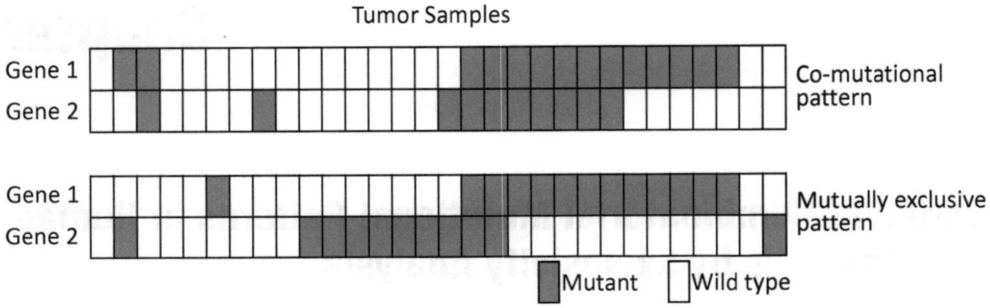

Fig. 1 Schematic representation of two combinatorial mutational patterns studied in this protocol: the co-mutational pattern (*upper panel*) refers to the scenario that a set of genes tends to mutate simultaneously in a tumor sample, whereas the mutually exclusive pattern (*lower panel*) represents the opposite scenario: genes in a given set tend to avoid mutating simultaneously in any one tumor sample

[5]. Previous experimental and statistical analyses have consistently revealed two combinatorial mutational patterns for a given set of genes, termed co-mutational and mutually exclusive patterns [5, 8–10]. As shown in Fig. 1, the co-mutational pattern occurs when a set of genes tend to mutate simultaneously in a single tumor, while the mutually exclusive pattern refers to the scenario in which one and only one of a set of genes is likely to be altered in a tumor.

Mutually exclusive genes are likely to function in the same signaling pathway, whereas co-mutational genes are likely to take effect in different pathways [11]. Combinatorial patterns of genes can be leveraged to infer signaling networks implicated in human cancer development and progression. Indeed, many efforts have been devoted to de novo discover novel driver pathways based on mutual exclusivity of gene mutations [11–13]. Therefore, it has essential biological relevance to identify gene pairs or gene sets with significant combinatorial mutational patterns.

Previous work proposed a statistical method to deal with this question and nominated a number of gene sets with significant combinatorial patterns [10]. However, this analysis was performed on a batch of very limited cell line data. The analysis thus lacks an elaborate procedure to preprocess data from a giant mutation database which consists of a large number of clinical samples of various cancer types (e.g., the recently released Catalog of Somatic Mutations in Cancer—COSMIC [14] and the Cancer Genome Atlas—TCGA, https://tcga-data.nci.nih.gov/tcga/). In addition, the analysis by Yeang et al. adopted different hypothesis tests to estimate the significance levels of the two combinatorial mutational patterns, which tend to yield a too conservative *p*-value for the co-mutational pattern [10].

To address these issues, we here describe a systematic and reliable pipeline to identify both combinatorial mutational patterns in cancer genomes. Here, somatic mutations exclude the synonymous point mutations which will not change an amino acid (marked as "coding silent" in COSMIC). Those mutations typically have little, if any, impact on the biological function of corresponding proteins and are uninformative for signaling pathway inference [11, 15]. Other types of mutations, such as missense and nonsense point mutations, small insertions and deletions, frame shifts, gene fusions, and translocations, etc., could be counted as effective mutations when performing exclusivity analysis. Furthermore, the mutations should be detected based on genome-wide or exome-wide screening efforts ensuring that all protein-coding genes were covered, to minimize the statistical bias induced by incomplete sample coverage.

The step-by-step procedure for data acquisition and criteria of data quality control, as well as the specific formulae used to calculate the likelihood ratio (LR) and significance level (p-value), are elucidated in the following sections. Figure 2 illustrates the overall procedures of this pipeline for mutational pattern determination.

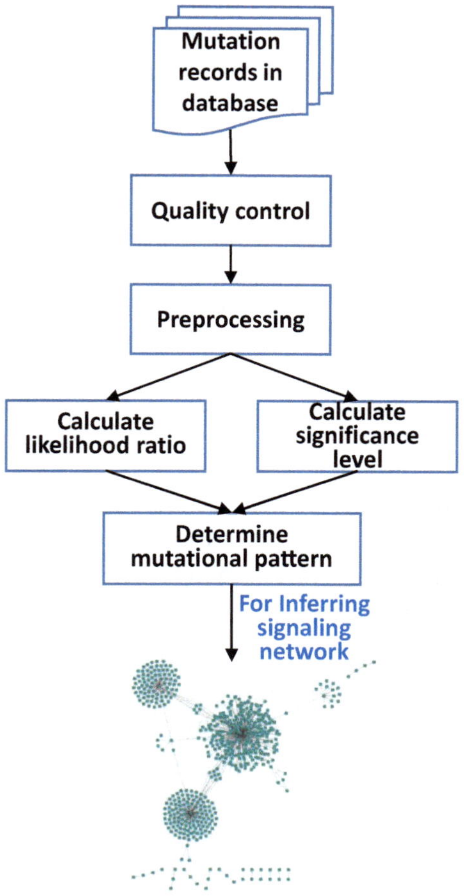

Fig. 2 Schematic of the overall pipeline proposed in this protocol. The specific steps of text processing, computation, and visualization are provided in Subheading 3

This pipeline has been shown to be highly effective and efficient in identifying mutational patterns of gene pairs in cancer mutation data from COSMIC v68, as described in our earlier publication [5]. We recently applied this pipeline to analyze the data from the latest COSMIC release (version 76), which has been threefold expanded since the release of v68, and well recapitulated and significantly improved the previous results (data not shown). However, our previous efforts were mainly devoted to the biological discoveries instead of technical details of the analysis. In this protocol, we address this gap by providing extensive practical details and highlighting alterative solutions when encountering problems in the users' particular applications.

2 Materials

The pipeline proposed in this protocol has been successfully tested in the Catalog of Somatic Mutations in Cancer (COSMIC, release v68 and v76). Therefore, the procedures described in the below section will be mainly based on the COSMIC database. However, it is noteworthy that this protocol is applicable to other databases such as TCGA that contain information of both gene mutation and associated patient sample IDs. Synonymous mutations and mutations that are not from a genome-wide or exome-wide study should be excluded prior to further analysis. All the text processing, subsequent computation, and visualization can be implemented in Matlab (The Math Works, Inc.), as described previously [5], or in R [16], another popular language for data analysis and graphics.

3 Methods

3.1 Data Quality Control and Preprocessing of COSMIC Mutation Entries

1. Extract mutations of a designated cancer type from the mixed mutation records in COSMIC by the keyword "Primary site" (*see* **Note 1**).
2. Remove synonymous mutations by the keyword "Substitution-coding silent" (*see* **Note 1**).
3. Remove mutation records that are not from a genome-wide study by the keyword "genome-wide screen" (*see* **Note 1**).
4. Generate a gene mutation pattern matrix based on the mutations and sample IDs. The rows and columns of the matrix refer to samples and genes, respectively. The entry at row i and column j of the matrix refers to the number of mutations occurring on gene j in tumor sample i. Figure 3 highlights an example showing the 9th tumor sample has a mutation on gene 2 by marking the coordinate (9,2) (*see* **Note 2**).

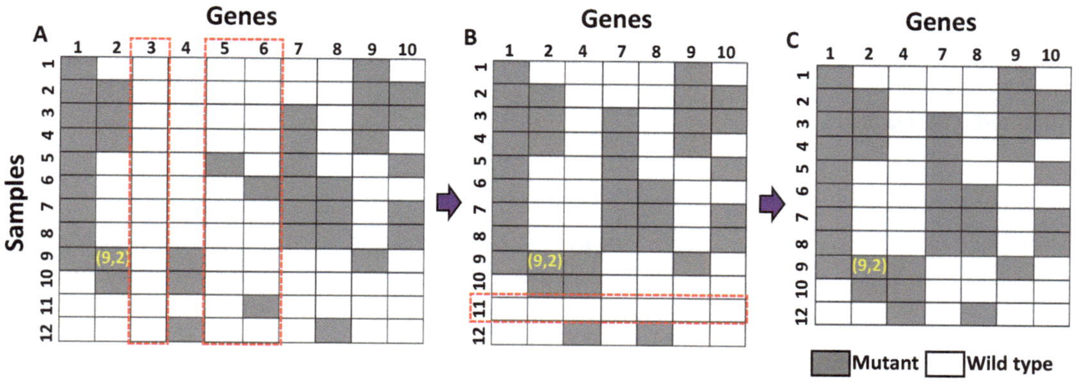

Fig. 3 Schematic depicting the mutation pattern matrix and entry filtering criteria. (**a**) A mutation pattern matrix is generated to represent the mutation profiles of the tumor samples across all genes. A *gray grid* indicates the corresponding sample has at least one mutation on the gene specified by the column ID. (**b**) Columns 3, 5, and 6 are deleted since the associated genes are mutated in only a small fraction of samples (the threshold of fraction can be prescribed). (**c**) Row 11 is deleted as the corresponding sample has no mutation in the remaining genes after the processing in (**b**)

3.2 Calculation of Likelihood of Co-occurrence of Mutant Genes

1. Exclude genes (columns) that do not exceed a prescribed threshold of sample coverage. As shown in Fig. 3a, b, if the percentage of nonzero entries in a column is lower than a threshold, then delete this column (*see* **Note 3**).
2. Remove samples (rows) that do not harbor any mutations across the remaining genes, as shown in Fig. 3b, c.
3. Calculate the likelihood ratio LR_{comb} of co-occurrence for each gene pair by the formula (1):

$$LR_{comb} = \frac{P(g_1 = 1, g_2 = 1)}{P(g_1 = 1)P(g_2 = 1)} \tag{1}$$

where $P(g_1 = 1)$ and $P(g_1 = 1, g_2 = 1)$ correspond to the percentage of samples in which a single gene or both the genes are mutated, respectively.

4. Determine the threshold of the likelihood ratio (LR_{comb}) for pattern categorization based on a mixture Gaussian distribution fitting model using an Expectation-Maximization algorithm [17]. Specifically, suppose m_1, m_2 are the means of the low and high components of all LR_{comb}'s, and δ_1, δ_2 their standard deviations respectively. Then the thresholds for the co-mutational pattern (lower bound) and exclusive pattern (upper bound) are calculated as $\theta_1 = m_2 - \delta_2/2$ and $\theta_2 = m_1 + \delta_1/2$, respectively (*see* **Note 4**).

3.3 Calculation of Significance of Combinatorial Mutational Patterns

1. Calculate the significance level of the co-mutational pattern P_{co} by the hypergeometric test as the formula (2):

$$P_{co} = \sum_{k=n_{12}}^{n_2} \binom{n_1}{k} \binom{n-n_1}{n_2-k} \Big/ \binom{n}{n_2} \tag{2}$$

where n, n_1, n_2, n_{12} represent the numbers of total samples, samples harboring gene 1 mutation, samples harboring gene 2 mutation, and samples harboring both gene 1 and gene 2 mutations, respectively.

2. Calculate the significance level of the mutually exclusive pattern P_{excl} by formula (3):

$$P_{excl} = \sum_{k=0}^{n_{12}} \binom{n_1}{k} \binom{n-n_1}{n_2-k} \Big/ \binom{n}{n_2} \tag{3}$$

where n, n_1, n_2, n_{12} are defined as in the formula of P_{co} above.

4 Interpretation of the Results

Both co-mutational and mutually exclusive patterns of gene pairs have biological meaning implicated in signaling network inference. For the co-mutational pattern, genes are likely to function in different signaling pathways and exert synergistic impact on tumor progression. Therefore, multiple oncogenic pathways driving the tumorigenesis for a particular cancer type could be identified by analyzing these co-mutational patterns. For the exclusive pattern, more insights could be obtained. In particular, affiliation of genes to a signaling pathway can be inferred from a list of gene pairs with exclusive pattern. For example, if A-B, B-C, and C-A are all exclusive gene pairs, then it is reasonable to conclude that genes A, B, and C are likely to operate in the same signaling pathway for the particular cancer type in question. When the whole list of gene pairs is visualized in one graph, as shown in the bottom of Fig. 2 (by Cytoscape [18]), a signaling network can emerge. However, this preliminary signaling network subjects to modification based on prior knowledge about gene-gene/protein-protein interactions and experimental evidence for real applications (*see* **Note 5**). To conclude, the exclusive patterns can be used to infer a specific cancer-associated signaling pathway, while the co-mutational patterns can assist in exploring whether multiple oncogenic pathways were involved, as described in our previous work [5] and references therein.

5 Notes

1. If combinatorial mutational pattern analysis needs to be conducted on tumor subtypes instead of tissue types, the keyword "Site subtype" can be used to further divide mutations into smaller groups of subtypes. For TCGA data, since genomic and epigenomic data are deposited separately for different cancer types, the mutation data for a cancer type of interest can be downloaded directly from the TCGA web site by choosing level 2 MAF (Mutation Annotation Format) file. Also, the information of amino acid change corresponding to nucleotide alteration is not available in TCGA, **step 2** (Subheading 3.1, which aims at removing the coding silent mutations) can be skipped. Since all the mutation data in TCGA are based on exome-wide sequencing, therefore, it is no necessary to implement the screening procedure specified in **step 3** (Subheading 3.1).
2. When generating the gene mutation pattern matrix, the mutations counted for each gene could be restricted to particular mutation types such as missense point mutations or gene fusions, depending on the biological question to be answered and the working model hypothesis to be tested.
3. The threshold of sample coverage used to exclude the less frequently mutated genes can be adjusted according to the sample size and/or gene set size, to yield a reasonable number of combinations of genes. In our previous practices, the threshold sample coverage was set to 2–10%.
4. When determining the thresholds for mutational pattern categorization, the mixture Gaussian model based on the Expectation-Maximization algorithm sometimes can produce inconsistent outcomes over technical replicates. This is largely due to the stochastic properties implicated in the EM (Expectation Maximization) optimization procedure. A reliable alternative is to simply use $LR_{comb} = 1$ to divide candidate gene pairs into two groups, with $LR_{comb} < 1$ referring to exclusive pattern and the remainder co-mutational pattern. Then rank the pairs in each group according to *P* values. After that, select a reasonable number (e.g., 20–30) of the top-ranked significant gene pairs in respective groups.
5. Although the pipeline is applied to gene pairs, sets of genes with particular combinatorial patterns could emerge by integrating gene pairs of corresponding patterns. Thus, the pipeline introduced in this protocol can serve as a starting point for inferring signaling network modules in particular cancers.

Acknowledgments

This work was partially supported by the Beijing Normal University youth funding (105502GK and 2013YB43 to H.T.) and National Institutes of Health (1U01CA166886, 1R01LM010185, and 1U01HL111560 to X.Z.).

References

1. Hanahan D, Weinberg RA (2000) The hallmarks of cancer. Cell 100(1):57–70. https://doi.org/10.1016/S0092-8674(00)81683-9
2. Stratton MR, Campbell PJ, Futreal PA (2009) The cancer genome. Nature 458 (7239):719–724. https://doi.org/10.1038/Nature07943
3. Peng H, Tan H, Zhao W, Jin G, Sharma S, Xing F, Watabe K, Zhou X (2016) Computational systems biology in cancer brain metastasis. Front Biosci 8:169–186
4. Tan H, Bao J, Zhou X (2012) A novel missense-mutation-related feature extraction scheme for 'driver' mutation identification. Bioinformatics 28(22):2948–2955. https://doi.org/10.1093/bioinformatics/bts558
5. Tan H, Bao J, Zhou X (2015) Genome-wide mutational spectra analysis reveals significant cancer-specific heterogeneity. Sci Rep 5:12566. https://doi.org/10.1038/srep12566
6. Tan H, Li F, Singh J, Xia X, Cridebring D, Yang J, Bao J, Ma J, Zhan M, Wong STC (2012) A 3-dimentional multiscale model to simulate tumor progression in response to interactions between cancer stem cells and tumor microenvironmental factors. IEEE 6th International Conference on Systems Biology (ISB):297–303. https://doi.org/10.1109/ISB.2012.6314153
7. Tan H, Wei K, Bao J, Zhou X (2013) In silico study on multidrug resistance conferred by I223R/H275Y double mutant neuraminidase. Mol BioSyst 9(11):2764–2774. https://doi.org/10.1039/c3mb70253g
8. Vogelstein B, Papadopoulos N, Velculescu VE, Zhou S, Diaz LA Jr, Kinzler KW (2013) Cancer genome landscapes. Science 339 (6127):1546–1558. https://doi.org/10.1126/science.1235122
9. Vogelstein B, Kinzler KW (2004) Cancer genes and the pathways they control. Nat Med 10 (8):789–799. https://doi.org/10.1038/nm1087
10. Yeang CH, McCormick F, Levine A (2008) Combinatorial patterns of somatic gene mutations in cancer. FASEB J 22 (8):2605–2622. https://doi.org/10.1096/fj.08-108985
11. Ciriello G, Cerami E, Sander C, Schultz N (2012) Mutual exclusivity analysis identifies oncogenic network modules. Genome Res 22 (2):398–406. https://doi.org/10.1101/gr.125567.111
12. Vandin F, Upfal E, Raphael BJ (2012) De novo discovery of mutated driver pathways in cancer. Genome Res 22(2):375–385. https://doi.org/10.1101/gr.120477.111
13. Leiserson MD, Blokh D, Sharan R, Raphael BJ (2013) Simultaneous identification of multiple driver pathways in cancer. PLoS Comput Biol 9 (5):e1003054. https://doi.org/10.1371/journal.pcbi.1003054
14. Forbes SA, Beare D, Gunasekaran P, Leung K, Bindal N, Boutselakis H, Ding M, Bamford S, Cole C, Ward S, Kok CY, Jia M, De T, Teague JW, Stratton MR, McDermott U, Campbell PJ (2015) COSMIC: exploring the world's knowledge of somatic mutations in human cancer. Nucleic Acids Res 43(Database issue): D805–D811. https://doi.org/10.1093/nar/gku1075
15. Greenman C, Stephens P, Smith R, Dalgliesh GL, Hunter C, Bignell G, Davies H, Teague J, Butler A, Stevens C, Edkins S, O'Meara S, Vastrik I, Schmidt EE, Avis T, Barthorpe S, Bhamra G, Buck G, Choudhury B, Clements J, Cole J, Dicks E, Forbes S, Gray K, Halliday K, Harrison R, Hills K, Hinton J, Jenkinson A, Jones D, Menzies A, Mironenko T, Perry J, Raine K, Richardson D, Shepherd R, Small A, Tofts C, Varian J, Webb T, West S, Widaa S, Yates A, Cahill DP, Louis DN, Goldstraw P, Nicholson AG, Brasseur F, Looijenga L, Weber BL, Chiew YE, DeFazio A, Greaves MF, Green AR, Campbell P, Birney E, Easton DF, Chenevix-Trench G, Tan MH, Khoo SK, Teh BT, Yuen ST, Leung SY, Wooster R, Futreal PA, Stratton MR (2007) Patterns of somatic mutation in human cancer genomes. Nature 446 (7132):153–158. https://doi.org/10.1038/nature05610

16. Ihaka P, Gentleman R (1996) R: a language for data analysis and graphics. J Comput Graph Stat 5(3):299–314
17. Dempster AP, Laird NM, Rubin DB (1977) Maximum likelihood from incomplete data via EM Algorithm. J Roy Stat Soc B Met 39 (1):1–38
18. Shannon P, Markiel A, Ozier O, Baliga NS, Wang JT, Ramage D, Amin N, Schwikowski B, Ideker T (2003) Cytoscape: a software environment for integrated models of biomolecular interaction networks. Genome Res 13(11):2498–2504. https://doi.org/10.1101/gr.1239303. 13/11/2498 [pii]

Chapter 2

Discovering Altered Regulation and Signaling Through Network-based Integration of Transcriptomic, Epigenomic, and Proteomic Tumor Data

Amanda J. Kedaigle and Ernest Fraenkel

Abstract

With the extraordinary rise in available biological data, biologists and clinicians need unbiased tools for data integration in order to reach accurate, succinct conclusions. Network biology provides one such method for high-throughput data integration, but comes with its own set of algorithmic problems and needed expertise. We provide a step-by-step guide for using Omics Integrator, a software package designed for the integration of transcriptomic, epigenomic, and proteomic data. Omics Integrator can be found at http://fraenkel.mit.edu/omicsintegrator.

Key words Data integration, Network biology, Computational biology, High-throughput data

1 Introduction

As biologists gain access to increasing amounts of data, the challenges associated with interpreting those data have increased. Biologists and clinicians can obtain high-throughput information about a cell's genome, transcriptome, epigenome, and proteome with reasonable effort and constantly decreasing costs. Indeed, much of those data are freely available to scientists through resources such as The Cancer Genome Atlas [1] and ENCODE [2]. The challenge remains, however, in knowing how to interpret those rich datasets. These "omic" data can be extraordinarily valuable. However, this value can only be extracted if data are properly analyzed using methods that account for the relatively high error rate of high-throughput experiments [3], and then condensed into understandable and actionable hypotheses about the underlying biology. This process can be especially difficult, and especially rewarding, when attempting to integrate several kinds of high-throughput data. Our group and others have shown that integrating data from several

Louise von Stechow (ed.), *Cancer Systems Biology: Methods and Protocols*, Methods in Molecular Biology, vol. 1711,
https://doi.org/10.1007/978-1-4939-7493-1_2,

© Springer Science+Business Media, LLC 2018

sources can lead to novel discoveries that each assay could have missed on its own [4–6].

Network biology is a fast-growing category of methods for this type of analysis [7]. Network models provide a valuable resource for biologists looking to analyze their high-throughput data in a systems context. By mapping "hits" from high-throughput assays onto interaction networks, the mechanistic connections between the hits become obvious, and investigators can focus on pathways, or series of interactions in the cell that are related to a certain function, that may be perturbed in the system.

Network methods typically involve modeling the molecules within a cell—which can for example be DNA, mRNAs, proteins, or metabolites—as nodes in a graph. Edges between these nodes connect molecules that are functionally or physically connected [7]. For example, a protein-protein interaction network (PPI) would represent the binding of protein A to protein B by drawing an edge between the "A" node and "B" node in the network. Several publicly available databases have been created to translate experimentally discovered protein interactions into PPIs, such as iRefIndex [8], BioGRID [9], and STRING [10]. There are also databases that store interactions of proteins with other molecules, such as metabolites [11–13]. In other types of networks, the edges can represent more abstract relationships. For example, in a correlation-based network, edges between nodes might represent probable co-regulation, rather than physical interactions, based on covariance between the concentration of molecule A and molecule B [14, 15].

Mapping high-throughput hits onto networks in search of affected pathways has several advantages. Hits that are close to each other in a network might function in the same pathway. Focusing on subnetworks of functionally related nodes can produce a more tractable number of targets, rather than the potentially hundreds of individual factors identified in high-throughput experiments. In addition, this type of pathway identification reduces the chance of devoting resources to the analysis of false positives from the high-throughput screen. Although the confidence for each hit in a screen may be low, the confidence in a pathway that contains many hits is much higher. Finally, pathway analysis can help to find novel nodes that may not have appeared in a high-throughput screen. These "hidden nodes" can be false negatives in a screen, or true negatives that are nonetheless important players in the investigated biological system. Our work has shown that these hidden nodes can often be important to a system under study, despite the lack of direct experimental evidence [4, 16, 17]. Using the PPI to discover these pathways de novo, rather than relying on predetermined pathway databases like KEGG [18], expands our ability to find novel information, and avoids biasing the results toward well-studied pathways.

However, network analysis is not as simple as just mapping high-throughput assay hits onto PPIs and finding all possible connections through them. Because of the large and highly connected nature of most biological networks [7], this "brute force" method results in extremely dense, uninterpretable "hairballs" rather than clear pathways [16]. Moreover, combining several types of experimental assays into a unified analysis can be complex. For example, experiments assessing changes in mRNA levels and protein levels are often not well correlated [19]. It is not trivial to map them onto one protein or RNA interaction network together. This chapter will walk you through the use of Omics Integrator, a software package that proposes a solution to these problems [17].

Omics Integrator is a new software tool designed to help biologists analyze and synthesize several kinds of high-throughput omics data, and reduce it to a few important, high-confidence pathways. Omics Integrator is designed for ease of use by biologists with basic computer skills (comfort with using the Unix command line is helpful). Omics Integrator first uses transcriptomic and epigenomic data to reconstruct transcriptional regulatory networks, and then integrates those with proteomic data by mapping them onto a protein interaction network [17]. It uses two modules—Garnet and Forest, which are designed to run sequentially, but can also be run individually. Garnet mines transcriptomic and epigenomic information in order to predict transcription factors that may be responsible for gene expression changes in the studied system. Forest maps these transcription factors and protein-level experimental information onto a PPI. Forest then implements the Prize-Collecting Steiner Forest algorithm [16] to predict high-confidence low-density protein interaction pathways that are important to the studied system (*see* Fig. 1).

2 Materials

2.1 Finding Transcriptional Regulators with Garnet

1. Transcriptomics data, i.e., differential gene expression between different conditions in your study (i.e., tumor vs. control).
2. Epigenomic data from a source such as TCGA [1], ENCODE [2], Roadmap [20], Omics Integrator example data, or experimentally derived epigenomic data (in a BED formatted file).
3. Transcription factor sequence binding motif predictions, from a source such as TRANSFAC [21], and/or Neph et al. [22]. Omics Integrator provides a file derived from the TRANSFAC database.

2.2 Network Integration with Forrest

1. Prize-collecting Steiner tree algorithm executable (msgsteiner can be downloaded from http://areeweb.polito.it/ricerca/cmp/code/bpsteiner).

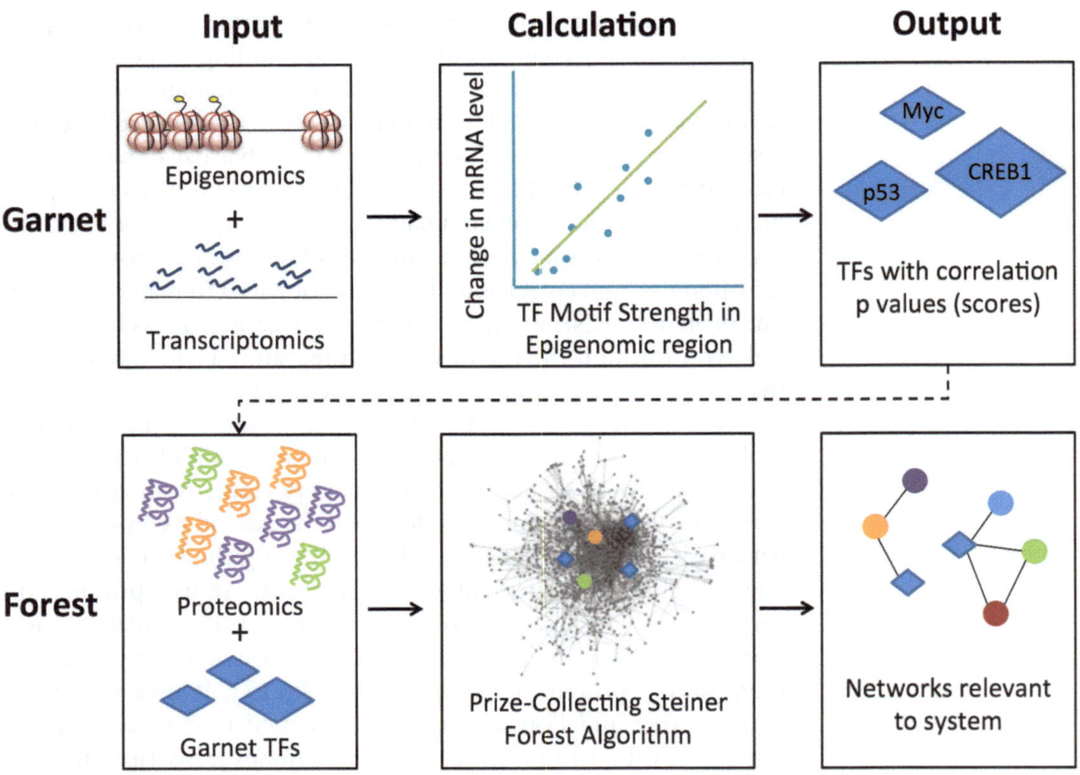

Fig. 1 Outline of the Omics Integrator workflow. Epigenomic data (open chromatin regions or histone marks) and transcriptomic data are used to predict influential transcription factors (TFs). Transcription factors and proteomic data are then mapped onto an interactome, and the Prize Collecting Steiner Forest algorithm is used to produce small pathways and subnetworks predicted to be relevant to the experimental system

2. Interactome file indicating all known interactions between proteins. Omics Integrator provides an interactome for mouse and human proteins derived from iRefIndex [8].
3. Input prize file, indicating the proteins you would like to include in the final solution (*see* **Note 1**).
4. (Optional) Output from Garnet to include transcription factors implicated by transcriptomic data in the final solution.
5. Cytoscape [25] to visualize the final network solution.

3 Methods

3.1 Installation of Omics Integrator

1. You can run Omics Integrator as a web tool on our website: http://fraenkel.mit.edu/omicsintegrator/or install it on your own computer using the instructions at https://github.com/fraenkel-lab/OmicsIntegrator. You should make sure you have all dependencies (*see* **Note 2**) installed and that you have the most updated version of Omics Integrator from our GitHub page (*see* **Note 3**).

3.2 Finding Transcriptional Regulators with Garnet

Garnet uses differentially expressed genes from your transcriptomic assays (i.e., RNA-seq) to predict transcription factors (TFs) that are likely to be responsible for the altered gene expression. It uses epigenomic data to find regions of the genome to look for differential TF binding. For example, this could be ATAC-seq data that points out accessible regions of the genome in your cell type. The algorithm will search for transcription factor binding motifs within regions implicated by your epigenomic data. The strength of these motifs is then correlated with the magnitude of change of nearby differentially expressed genes to give each TF a score.

1. Obtain epigenomic data for cell lines related to your samples from one of the sources listed under Subheading 2.1. Alternatively, if you have epigenomic data for your own samples, you can use this as well. These data can be in the form of histone marks ChIP-seq, or DNase-seq or ATAC-seq, all of which indicate accessible chromatin regions where a TF might be bound. Collect these data in a BED-formatted file.
2. Go to the Galaxy webserver [23] (*see* **Note 4**) to extract the DNA sequences for your epigenomic regions. Upload your BED file to Galaxy under the "Get Data" tool, specify which genome you are using, and then use the "Fetch Alignments/ Sequences">"Extract Genomic DNA" tool to download a FASTA-formatted file.
3. Format your experimentally derived gene expression data in a tab-delimited file with two columns. The first should be the name of the gene, and the second should be the log-fold-change of that gene in the study conditions (i.e., tumor vs. control). We recommend only including genes with a statistically significant change in expression (*see* **Note 5**).
4. Create the Garnet configuration file. For an example configuration file, see the README on the Omics Integrator GitHub page, or the comment on the top of scripts/garnet.py. Your configuration file should be formatted similarly, but you should replace the paths to the *bedfile*, *fastafile*, and *expressionFile* with the paths to the files you created in **steps 1–3** in Subheading 3.2. Make sure the annotation files referenced by *genefile*, *xreffile*, and *genome* are using the correct genome for your sample (files for mm9 and hg19 are provided with Omics Integrator).
5. You can change the parameters to your liking (Table 1).
6. Run Garnet on the command line by navigating to the directory with garnet.py and running *python garnet.py yourconfigfile.cfg*. You can also add a *--outdir directoryname* flag if you would like to put the output from garnet into a different directory.

Table 1
An explanation of the parameters used by Garnet

windowsize	This parameter determines the maximum distance in nucleotides from a gene TSS to a TF binding motif to consider them related. Higher values will find more TFs, but their binding may be farther away from the gene, and thus, less likely to be directly related to expression. Values usually range from 2000 to 20,000
pvalThresh	The *p* value of a correlation measures how likely you are to get this correlation value if the events were not correlated. This threshold determines which transcription factors will be passed to Forest. Only those whose correlation with expression falls below the provided threshold will be included. Recommended values range from 0.01 to 0.05. Leave this value blank to use a *q* value threshold rather than a *p* value
qvalThresh	A *q* value is a False Discovery Rate adjusted *p* value. This measurement will result in fewer false positives. This threshold determines which transcription factors will be passed to Forest. Only those whose correlation with expression falls below the provided threshold will be included. Recommended values range from 0.01 to 0.05. Leave this value blank if a *p* value threshold is sufficient. (If you are going on to run Forest, a *p* value is generally sufficient since the network nature of Forest make false positives less likely to appear in a final network)

Garnet will run through several steps, informing you on the command line where it is in the process. These steps include:

- Mapping the genes to nearby epigenetic regions.
- Scanning those regions for TF binding motifs.
- Building a matrix of gene expression changes and binding motif scores for each TF.
- Running a regression to check the correlation of TF binding score with differential gene expression.

Garnet will print results into several tab-delimited files. These files are described in the README file on the Omics Integrator GitHub page. The file that ends in regression_results.tsv shows all TFs, clustered by similar binding sites, along with their *p*- and *q*-values from the regression. The file that ends in FOREST_INPUT.tsv contains only significant results and will be used in future steps.

3.3 Network Integration with Forest

Forest integrates proteomic data and the output from Garnet into a network. After mapping the data onto a provided interactome network, it uses the prize-collecting Steiner tree algorithm (solved by the msgsteiner code that you downloaded and installed) to find an optimal set of subnetworks. These subnetworks can then be analyzed for pathway context.

1. If you are not using the default interactome provided with Omics Integrator, prepare your input interactome file. An interactome file (or "edge file") contains the large network of all known connections between nodes. The file should be

formatted in three tab-delimited columns. Each line should have the form "interactor1 (tab) interacter2 (tab) weight." The third column contains an edge weight, between 0 and 0.99, usually representing the confidence in the validity of that edge. Optionally, you can include a fourth column indicating whether that edge is directed ("D") or undirected ("U"). The current default interactome for human or mouse tissue is derived from iRefIndex (version 13) [8] and scored with the MIScore system [24]. You can find it in the data folder, called iref_mitab_miscore_2013_interactome.txt. You should create your own interactome file if you are not running your experiments in mouse or human cell models, or if you have a more updated interactome for your experiments.

2. Prepare your input prize file. This file contains significant features from your proteomic data (*see* **Note 6**). It should have two tab-delimited columns: the protein name (matching the interactome file exactly), and the protein prize. You should assign higher prizes to proteins for which you have stronger evidence that they should be in the final network.
3. Prepare your configuration file. This file contains input parameters for your run of Forest (Table 2). An example can be found in the example/a549 folder, called tgfb_forest.cfg. At a minimum this file must contain values for the parameters *w*, *b*,

Table 2
An explanation of the parameters used by Forest

w	This parameter influences the number of separate trees detected, which can aid in identifying functionally distinct processes. Higher values of *w* lead to more trees in the optimal forest, while lower values force most prizes to be found in the same tree. Values usually range from 1 to 10. *See* Tuncbag et al. [14] for a more detailed explanation
b	This parameter linearly scales the prizes, thereby changing the relative weighting of edge weights and node prizes. Higher values lead to larger trees, including some low-confidence edges, while lower values force networks to be small and use only high confidence edges, and lead to the possible exclusion of some prize terminals. Values usually range from 1 to 20
D	This parameter sets the maximum depth from the dummy node, or root of the tree, to the leaf nodes. Higher values lead to long pathways, while lower values lead to shorter disparate pathways. Values usually range from 5 to 15
μ	This parameter controls negative prizes in Forest. Negative prizes are explained in detail in Section 3.4.2. The default value is zero, and if you want to use negative prizes, values usually range from 0.0001 to 0.1
garnetBeta	This parameter controls the relative weighting of TF scores derived from Garnet and prize values on proteomic nodes. Higher values will encourage the inclusion of more TF nodes in the network, while lower values force networks to include only the most significant or pathway-relevant TF nodes. Typically, the value for this parameter is set to the median value of the proteomic prizes divided by the median value of the TF scores

and *D*. If you are including results from Garnet, you will also need a *garnetBeta* parameter. *See* Subheading 3.4.1 for more information.

4. You can now run forest with the command *python forest.py –p yourprizefile.txt –e youredgefile.txt –c yourconfigfile.txt --garnet yourgarnetoutput_FOREST_INPUT.tsv*. You can also add a *--outlabel yourexperimentname* flag to give your output files a prefix and a *--outpath directoryname* flag if you would like to put the output from forest into a different directory. You may need to add a *--msgpath directoryname* flag to indicate where you installed the msgsteiner code during the installation step. There are several other optional flags you can add to this command if wanted (*see* **Note 7**).
5. Forest will run through several steps, informing you on the command line where it is in the process. These steps include:
   - reading in your input files.
   - running the msgsteiner optimization.
   - writing the output files.

   Output files are described in the README file on the Omics Integrator GitHub page (*see* **Note 8**).
6. To visualize the network output, open the Forest output files in Cytoscape [25]. Open Cytoscape and import a network. The Forest output files that end in .sif have been formatted for this purpose. The file ending in optimalForest.sif contains only those edges used in the optimal Steiner forest, while augmentedForest.sif contains all edges in the interactome between the nodes in the final forest, and is recommended for final analysis. You can then import tables to annotate those networks; the nodeattributes.tsv file and the edgeattibutes.tsv file, to view information about the nodes and edges in the network, such as the edge weights and the node prizes. Node attributes also include the node prize type: TF, proteomic, or blank to indicate a hidden node which had no input prize but was chosen by the algorithm to connect prize nodes. Cytoscape has many useful visualization tools that you can use to better represent these values and types [25, 26] (*see* **Note 9**).

3.4 Network Quality Control

1. We recommend checking the robustness and specificity of your networks. You can do this by adding flags to the forest.py command. Add *--noisyEdges 10* to test robustness of your network to noise in the edge weights. This command will add Gaussian noise to the edgeweights, re-run Forest ten (or your input number of) times, and then merge the results into output files with noisyEdges in the filenames. Add *--randomTerminals 10* to test specificity of your network to your input terminals.

This command will randomly redistribute your prizes among the interactome, keeping the degree distribution of your original prizes, re-run Forest ten times, and then merge the results into output files with randomTerminals in the filenames. Both of these flags will increase the runtime of forest significantly (*see* **Note 10**).

2. Forest results include an attribute representing the fraction of optimal forests containing each node, which indicates how often that node appeared in the various forest runs with noise or random inputs. A robust network will have high FractionOfOptimalForestsContaining values for most nodes in noisyEdges run, and nodes that are specific to your input data will have low FractionOfOptimalForestsContaining values after randomTerminals runs. These metrics can be especially useful ways to judge the importance of hidden nodes to your system.

3.4.1 Choosing Parameters for Forest

The resulting network from this data integration algorithm is highly dependent on several parameters. These include w, b, D, μ, and *garnetBeta* (Table 2).

We recommend running Forest over a range of these values to find the best set for your system. To see an example of a script for testing parameters, *see* OmicsIntegrator/example/GBM/GBM_case_study.py. Once you have several resulting networks, we recommend choosing the best result by

1. Choosing a set of parameters that maximizes the fraction of input prize nodes that are included in the final network and are robust to noise (as judged by the noisyEdges runs).
2. Some parameters will lead to networks with large "hubs," that is, one hidden protein in the middle connected to several prize nodes with few interactions between these "spokes." These hubs are usually not informative or very specific to one system. We recommend choosing parameters that minimize this by measuring the average degree of hidden nodes in your network (i.e., the number of edges connecting to those nodes in the interactome) compared to the average degree of prize nodes. A good parameter set will minimize the distance between these metrics. Figure 2 shows an example of this analysis using the data in the example/a549 folder (*see* Fig. 2).
3. Once conditions 1 and 2 are satisfied, we prefer larger networks, as those provide the most opportunities for novel discoveries of hidden nodes and pathways enriched in the subnetworks.

3.4.2 Negative Prizes in Forest

One of the more innovative aspects of Omics Integrator is its ability to incorporate negative evidence. There are two settings in which

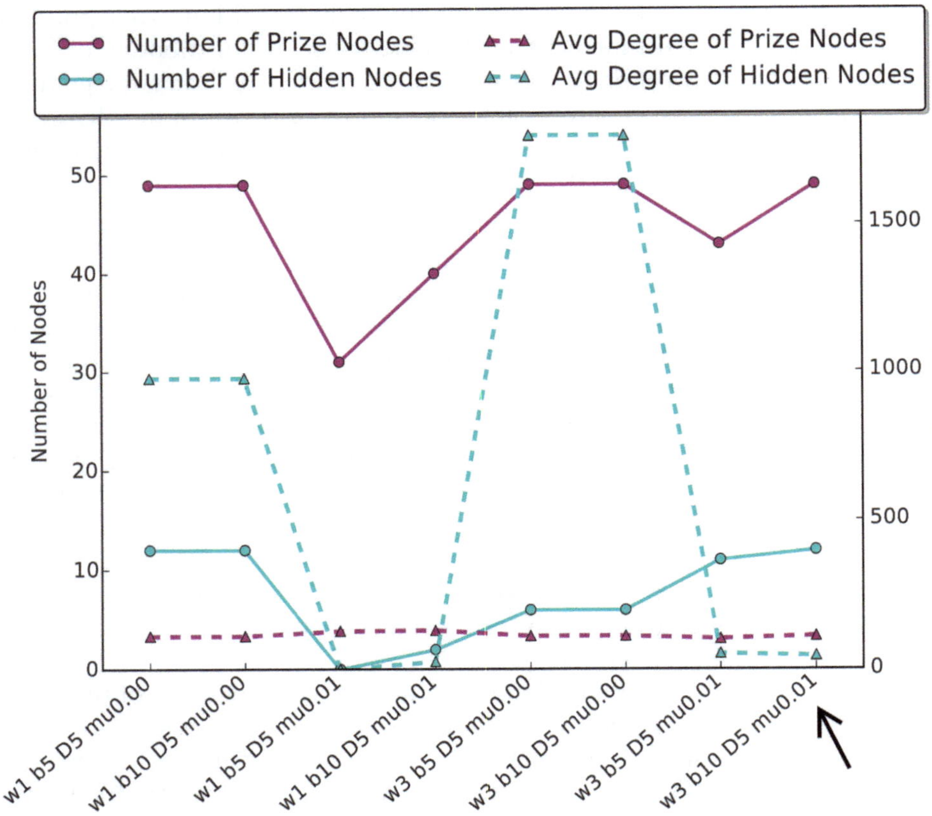

Fig. 2 An analysis of several parameter sets when running Forest on the sample A549 data provided with Omics Integrator. A good parameter set will minimize the difference between the average degree of prize nodes and hidden nodes, and will include a large number of prize nodes. A good choice of a parameter set is highlighted by the *black arrow*. The A549 dataset reflects phosphoproteomic changes in a lung cancer cell line when stimulated with TGF-beta. The *black arrow* highlights a network that includes relevant nodes such as EGFR, while networks with large average degree of hidden nodes are mostly comprised of a hub centered on ubiquitin-C, which connects to most prize nodes in the interactome, but is not specific to the lung cancer cell system

negative prizes can be useful. First, if you have reason to believe certain nodes should *not* show up in your optimal network you can assign a negative prize to a node and include it in the input prizes file along with positive prizes. Second, negative prizes can be used to avoid bias toward "hub nodes."

We have found that in many cases, certain nodes are overrepresented in network integration solutions because they have a high "degree," or number of edges connecting to that node, in the interactome. This could be because they bind with low specificity, e.g., chaperone proteins, or because they are highly studied proteins, causing more of their interactions to be discovered and represented in the literature. Because the optimal solution to the PCSF problem has the lowest cost method of connecting nodes, it will tend to use these nodes regardless of the input data. Simply removing these nodes from the network is not desirable, as there are settings in which they are relevant. To prevent hubs from being over-represented in all networks, Forest adds a penalty to nodes based on their degree. This penalty discourages solutions that include hubs but still allows them to be present when indicated by the data. This has been shown to improve accuracy in certain networks [17]. A positive number of the parameter μ will cause all nodes in the interactome to incur a penalty of μ **degree*.

4 Notes

1. Problems in running Omics integrator can originate from spaces in node names, or mismatched node names. Input files to Garnet and Forest should have no spaces in the protein and gene names. In addition, all node names in the input files should match those in the interactome exactly. Forest will try to catch this error by letting you know if a large percentage of your input nodes were not found in the interactome. The provided iRefIndex interactome uses Official Gene Symbols for protein nodes, so when using this interactome, input files should also use this nomenclature.
2. Currently, Omics Integrator requires Python 2.6 or 2.7, with the python packages numpy, scipy, matplotlib, and Networkx. You will need Cytoscape (http://www.cytoscape.org) [25, 26] for viewing network results. Any updates will be reflected in the "System Requirements" section on our GitHub page (*see* **Note 3**).
3. GitHub is an online hosting service for repositories of code. It lets the community contribute to improvements of open source projects like Omics Integrator, and keeps track of changes made and bugs reported. The latest version of Omics Integrator can be found on its GitHub page: https://github.com/fraenkel-lab/OmicsIntegrator. A new version of Omics Integrator, using Python version 3, is under development at https://github.com/fraenkel-lab/OmicsIntegrator2.
4. Galaxy is an online platform for computational biologists. In addition to the Extract Genomic Sequences tool described here, Galaxy provides several tools and workflows for analyzing biological data [23].
5. Genes used in Garnet should be significantly differentially expressed according to your transcriptomic data. For example, RNA-seq data can be analyzed with tools such as DEseq [27] or CuffDiff [28]. Genes that these tools report as differential with a p value less than 0.05 should be used as the input to Garnet.

6. Similar to transcriptomic data, your proteomics data will indicate which proteins should be used as the input to Forest. A review of tools for differential proteomics can be found here [29]. Many of these tools will provide a metric for determining statistical significance of differential expression of proteins, such as a *p* value. We generally use all proteins with a (modified) *p* value of less than 0.05. Prizes for the proteins are then the absolute value of the log of the fold change of protein expression. Be sure to use the absolute value, to avoid assigning a negative prize to downregulated proteins, which would encourage the algorithm to leave that node out of the networks, rather than including it.
7. There are several other flags available for advanced users, which change the behavior of forest.py. For example, you can change the group of nodes Forest uses to root each resulting tree (by default, this is all nodes which have been assigned a positive prize). There is a knockout option for doing an in silico knockout experiment by removing a protein from the interactome. For details on these and other flags, run *python forest.py -h* or read our GitHub repository page.
8. Many problems can lead to the final Forest output being empty (i.e., not containing any nodes). Check the output file ending in "info.txt" for some statistics of the run. One common problem, once formatting and input protein name problems have been ruled out, is a mu parameter set too high or other Forest parameters that lead to an empty optimal solution. Try changing your parameter values.
9. Cytoscape is a popular open source software for visualizing and analyzing networks [25, 26]. It is highly flexible and there are several available plug-ins for extending its use [30]. Omics Integrator can output results formatted for import into Cytoscape versions 2.8 or 3 by the use of a flag for forest.py (it defaults to version 3). Once the networks and node and edge attributes are imported into Cytoscape, you can use options in Cytoscape to create informative figures of your results. For example, we often use the Style tab to change the color of a node to represent its prize, the shape of a node to represent its Terminal Type (TF vs. proteomic vs. hidden node), and the edge width to represent its confidence. We recommend playing around with Styles and Layouts to best display your network.
10. Depending on your input data and run setup, a run of Omics Integrator can take a few hours. We recommend running in a screen session (https://www.gnu.org/software/screen/) or tmux (https://tmux.github.io/), which will allow the program to run continuously in the background, or on a computer that

is set not to turn off or interrupt the run. You can also run Omics Integrator on a cloud server. However, if the run is taking more than a day, you should cancel the run and look for errors. In particular, try running Forest without or with a smaller input to noisyEdges or randomTerminals, as these options can lead to large memory and time consumption. High values for the D parameter can also increase runtime.

Acknowledgments

This work was supported by grants from National Institute of Health (R01-NS089076, T32-GM008334, and U01-CA184898). We thank Tobias Ehrenberger and Renan Escalante-Chong for helpful comments on the manuscript.

References

1. Tomczak K, Czerwińska P, Wiznerowicz M (2015) The Cancer Genome Atlas (TCGA): an immeasurable source of knowledge. Contemp Oncol (Pozn) 19:A68–A77. https://doi.org/10.5114/wo.2014.47136
2. Encode Consortium (2013) An integrated encyclopedia of DNA elements in the human genome. Nature 489:57–74. https://doi.org/10.1038/nature11247
3. Malo N, Hanley JA, Cerquozzi S et al (2006) Statistical practice in high-throughput screening data analysis. Nat Biotechnol 24:167–175. https://doi.org/10.1038/nbt1186
4. Huang S-SC, Fraenkel E (2009) Integrating proteomic, transcriptional, and interactome data reveals hidden components of signaling and regulatory networks. Sci Signal 2:ra40. https://doi.org/10.1126/scisignal.2000350
5. Ideker T, Thorsson V, Ranish JA et al (2001) Integrated genomic and proteomic analyses of a systematically perturbed metabolic network. Science 292:929–934. https://doi.org/10.1126/science.292.5518.929
6. Huang SSC, Clarke DC, Gosline SJC et al (2013) Linking proteomic and transcriptional data through the interactome and epigenome reveals a map of oncogene-induced signaling. PLoS Comput Biol 9(2):e1002887. https://doi.org/10.1371/journal.pcbi.1002887
7. Barabási A-L, Oltvai ZN (2004) Network biology: understanding the cell's functional organization. Nat Rev Genet 5:101–113. https://doi.org/10.1038/nrg1272
8. Razick S, Magklaras G, Donaldson IM (2008) iRefIndex: a consolidated protein interaction database with provenance. BMC Bioinformatics 9:405. https://doi.org/10.1186/1471-2105-9-405
9. Tyers M, Breitkreutz A, Stark C et al (2006) BioGRID: a general repository for interaction datasets. Nucleic Acids Res 34:D535–D539. https://doi.org/10.1093/nar/gkj109
10. Szklarczyk D, Franceschini A, Wyder S et al (2015) STRING v10: protein-protein interaction networks, integrated over the tree of life. Nucleic Acids Res 43:D447–D452. https://doi.org/10.1093/nar/gku1003
11. Wishart DS, Jewison T, Guo AC et al (2013) HMDB 3.0—the human metabolome database in 2013. Nucleic Acids Res 41(Database issue): D801–D807. https://doi.org/10.1093/nar/gks1065
12. Thiele I, Swainston N, Fleming RMT et al (2013) A community-driven global reconstruction of human metabolism. Nat Biotechnol 31: 419–425. https://doi.org/10.1038/nbt.2488
13. Kuhn M, Szklarczyk D, Pletscher-Frankild S et al (2014) STITCH 4: integration of protein-chemical interactions with user data. Nucleic Acids Res 42(Database issue): D401–D407. https://doi.org/10.1093/nar/gkt1207
14. Valcárcel B, Würtz P, al Basatena NKS et al (2011) A differential network approach to exploring differences between biological states: an application to prediabetes. PLoS One 6(9): e24702. https://doi.org/10.1371/journal.pone.0024702
15. Kotze HL, Armitage EG, Sharkey KJ et al (2013) A novel untargeted metabolomics

correlation-based network analysis incorporating human metabolic reconstructions. BMC Syst Biol 7:107. https://doi.org/10.1186/1752-0509-7-107

16. Tuncbag N, Braunstein A, Pagnani A et al (2013) Simultaneous reconstruction of multiple signaling pathways via the prize-collecting steiner forest problem. J Comput Biol 20:124–136. https://doi.org/10.1089/cmb.2012.0092
17. Tuncbag N, Gosline SJ, Kedaigle AJ et al (2016) Network-based interpretation of diverse high-throughput datasets through the Omics Integrator software package. PLoS Comput Biol 12(4):e1004879
18. Aoki-Kinoshita KF, Kanehisa M (2007) Gene annotation and pathway mapping in KEGG. Methods Mol Biol 396:71–91. https://doi.org/10.1007/978-1-59745-515-2_6
19. Maier T, Güell M, Serrano L (2009) Correlation of mRNA and protein in complex biological samples. FEBS Lett 583:3966–3973. https://doi.org/10.1016/j.febslet.2009.10.036
20. Bernstein BE, Stamatoyannopoulos JA, Costello JF et al (2010) The NIH Roadmap Epigenomics Mapping Consortium. Nat Biotechnol 28:1045–1048. https://doi.org/10.1038/nbt1010-1045
21. Matys V, Kel-Margoulis OV, Fricke E et al (2006) TRANSFAC and its module TRANSCompel: transcriptional gene regulation in eukaryotes. Nucleic Acids Res 34:D108–D110. https://doi.org/10.1093/nar/gkj143
22. Neph S, Vierstra J, Stergachis AB et al (2012) An expansive human regulatory lexicon encoded in transcription factor footprints. Nature 489:83–90. https://doi.org/10.1038/nature11212
23. Blankenberg D, Von Kuster G, Coraor N et al (2010) Galaxy: a web-based genome analysis tool for experimentalists. Curr Protoc Mol Biol. https://doi.org/10.1002/0471142727.mb1910s89
24. Villaveces JM, Jiménez RC, Porras P et al (2015) Merging and scoring molecular interactions utilising existing community standards: tools, use-cases and a case study. Database 2015:bau131. https://doi.org/10.1093/database/bau131
25. Shannon P, Markiel A, Ozier O et al (2003) Cytoscape: a software environment for integrated models of biomolecular interaction networks. Genome Res 13:2498–2504. https://doi.org/10.1101/gr.1239303
26. Smoot ME, Ono K, Ruscheinski J et al (2011) Cytoscape 2.8: new features for data integration and network visualization. Bioinformatics 27:431–432. https://doi.org/10.1093/bioinformatics/btq675
27. Love MI, Anders S, Huber W (2014) Moderated estimation of fold change and dispersion for RNA-seq data with DESeq2. Genome Biol. https://doi.org/10.1186/s13059-014-0550-8
28. Trapnell C, Hendrickson DG, Sauvageau M et al (2013) Differential analysis of gene regulation at transcript resolution with RNA-seq. Nat Biotechnol 31:46–53. https://doi.org/10.1038/nbt.2450
29. Bantscheff M, Lemeer S, Savitski MM, Kuster B (2012) Quantitative mass spectrometry in proteomics: critical review update from 2007 to the present. Anal Bioanal Chem 404:939–965. https://doi.org/10.1007/s00216-012-6203-4
30. Saito R, Smoot ME, Ono K et al (2012) A travel guide to Cytoscape plugins. Nat Methods 9:1069–1076. https://doi.org/10.1038/nmeth.2212

Chapter 3

Analyzing DNA Methylation Patterns During Tumor Evolution

Heng Pan and Olivier Elemento

Abstract

Epigenetic modifications play a key role in cellular development and tumorigenesis. Recent large-scale genomic studies have shown that mutations in players of the epigenetic machinery and concomitant perturbation of epigenomic patterning are frequent events in tumors. Among epigenetic marks, DNA methylation is one of the best studied. Hyper- and hypo-methylation events of specific regulatory elements (such as promoters and enhancers) are sometimes thought to be correlated with expression of nearby genes. High-throughput bisulfite converted sequencing is currently the technology of choice for studying DNA methylation in base-pair resolution and on whole-genome scale. Such broad and high-resolution coverage investigations of the epigenome provide unprecedented opportunities to analyze DNA methylation patterns, which are correlated with tumorigenesis, tumor evolution, and tumor progression. However, few computational pipelines are available to the public to perform systematic DNA methylation analysis. Utilizing open source tools, we here describe a comprehensive computational methodology to thoroughly analyze DNA methylation patterns during tumor evolution based on bisulfite converted sequencing data, including intra-tumor methylation heterogeneity.

Key words DNA methylation, ERRBS, DMRs, Intra-tumor methylation heterogeneity

1 Introduction

Epigenetic modification plays a key role in the regulation of all DNA-based processes including transcription, DNA repair, and replication, which are fundamental to tumorigenesis [1]. Recent large-scale genomic studies have shown that mutations in the epigenetic machinery and concomitant perturbation of epigenomic patterning are frequent events in tumors, such as B-cell lymphomas, leukemia, and prostate cancers [2–5]. DNA methylation is one of the best-studied epigenetic markers. DNA methylation is characterized by the attachment of a methyl group to carbon 5 of cytosines, principally in the context of CpG dinucleotides. Hyper- or hypo-methylation of genomic regions (for example promoters or enhancers) can lead to repression or activation of the expression of nearby genes. Several examples of promoter methylation levels that are inversely correlated with gene expression levels have been

Louise von Stechow (ed.), *Cancer Systems Biology: Methods and Protocols*, Methods in Molecular Biology, vol. 1711,
https://doi.org/10.1007/978-1-4939-7493-1_3, © Springer Science+Business Media, LLC 2018

identified [6]. A number of tumor suppressor genes are silenced by promoter hypermethylation [7]. Thus, identifying differentially methylated cytosines (DMCs) and differentially methylated regions (DMRs), especially those perturbed in tumors, has become a central objective in cancer methylome analysis.

The advent of high-throughput DNA sequencing technologies has provided new opportunities to study DNA methylation, allowing for fast, single-base resolution scans in targeted or enriched regions or at whole-genome scale. Large-scale sequencing projects have generated hundreds of methylation profiles of tumors of different origins and at different stages. In turn, markers based on methylation usually provide important information about cellular phenotypes in healthy and diseased tissues. In many cases, assessing methylation profiles enabled improved patient stratification over other approaches based on transcriptomics or mutation profiles [4, 8].

Enhanced reduced representation bisulfite sequencing (ERRBS) is a powerful high-throughput sequencing platform, which can provide high sequencing depth and coverage of millions of CpGs in the human genome [9]. To date, few computational pipelines for analyzing bisulfite sequencing exist, even though such data are increasingly widely used. Here, we describe a comprehensive DNA methylation profiling analysis based on ERRBS data. This method is also applicable to reduced representation bisulfite sequencing (RRBS) data or whole-genome bisulfite sequencing (WGBS) data [10].

Tumors evolve following a Darwinian process, in which cells continuously acquire mutations that alter their fitness. The fittest cancer cells may divide faster, and will be more likely to survive inhibitory signals from the microenvironment than other less fit cells. Those cells are therefore more likely to expand in abundance within a tumor. Initiation of anti-cancer treatment can alter the fitness landscapes within tumors and frequently leads to selection and growth of cells that have acquired resistance mutations. Because every tumor potentially evolves along a different trajectory as a result of distinct environments and exposures to treatment, tumor evolution introduces individual features into each tumor [11]. While the contribution of genetic mechanisms to tumor evolution is well documented, the contribution of epigenetic mechanisms to tumor evolution has only recently begun to be studied [4, 12]. In this chapter, we specifically focus our analyses on capturing and analyzing DNA methylation patterns during tumor evolution.

Following generation of an ERRBS dataset, a typical analysis workflow consists of first identifying DMCs and DMRs, followed by correlating those regions to biological relevant genes and pathways. Other analyses relevant to tumor evolution may include quantifying intra-tumor methylation heterogeneity (MH). Indeed, compared to the genetic code, DNA methylation is more flexible and cells within a cell population may have distinct methylation patterns. Thus a tumor cell population may harbor varying levels intra-tumor MH. Such heterogeneity is emerging as a powerful predictor of tumor evolution, progression, and relapse [4].

The analysis of DNA methylation patterns during tumor evolution requires ERRBS samples from different tumor development stages, or diagnosis-relapse sample pairs from several patients. ERRBS samples from normal healthy (which can be used as baseline) can add an additional layer of information to the analysis. The computational analysis scheme is outlined in Fig. 1. Sample preparation and high-through sequencing are described by Akalin et al. [9]. Sequencing reads from every stage of tumor evolution are mapped independently to a bisulfite converted reference genome, generating separate DNA methylation profiles. DNA methylation status for every single CpG is determined in each sample, separately. Next, DMC and DMR calling is performed between the normal sample and each tumor stage or between any two tumor stages. Intra-tumor MH analysis is performed as a separate analysis on the same samples. Finally, identified DMCs/DMRs and MH hotspots are analyzed for enriched gene functions, in order to unravel pathways relevant to tumor evolution.

2 Materials

2.1 Software

1. ERRBS data quality control: FastQC (available at http://www.bioinformatics.babraham.ac.uk/projects/fastqc/) [13].
2. Adaptive quality and adapter trimming: Trim Galore (available at http://www.bioinformatics.babraham.ac.uk/projects/trim_galore/) [14].
3. Bisulfite converted sequence reads mapping and cytosine methylation states calling: Bismark (available at http://www.bioinformatics.babraham.ac.uk/projects/bismark/) [15].
4. DMCs and DMRs calling: RRBSseeqer (available at http://icb.med.cornell.edu/wiki/index.php/Elementolab/) [4].
5. Region annotations: ChIPseeqer (available at http://icb.med.cornell.edu/wiki/index.php/Elementolab/) [16].
6. Pathway analysis: iPAGE (available at http://icb.med.cornell.edu/wiki/index.php/Elementolab/) [17].
7. ERRBS data analyzing tools: Errbs-tools including methylCall_from_Bismark.py, regionMethyl.R and regionMH.R (available at https://github.com/SpursHeng90/errbs-tools/) [4].

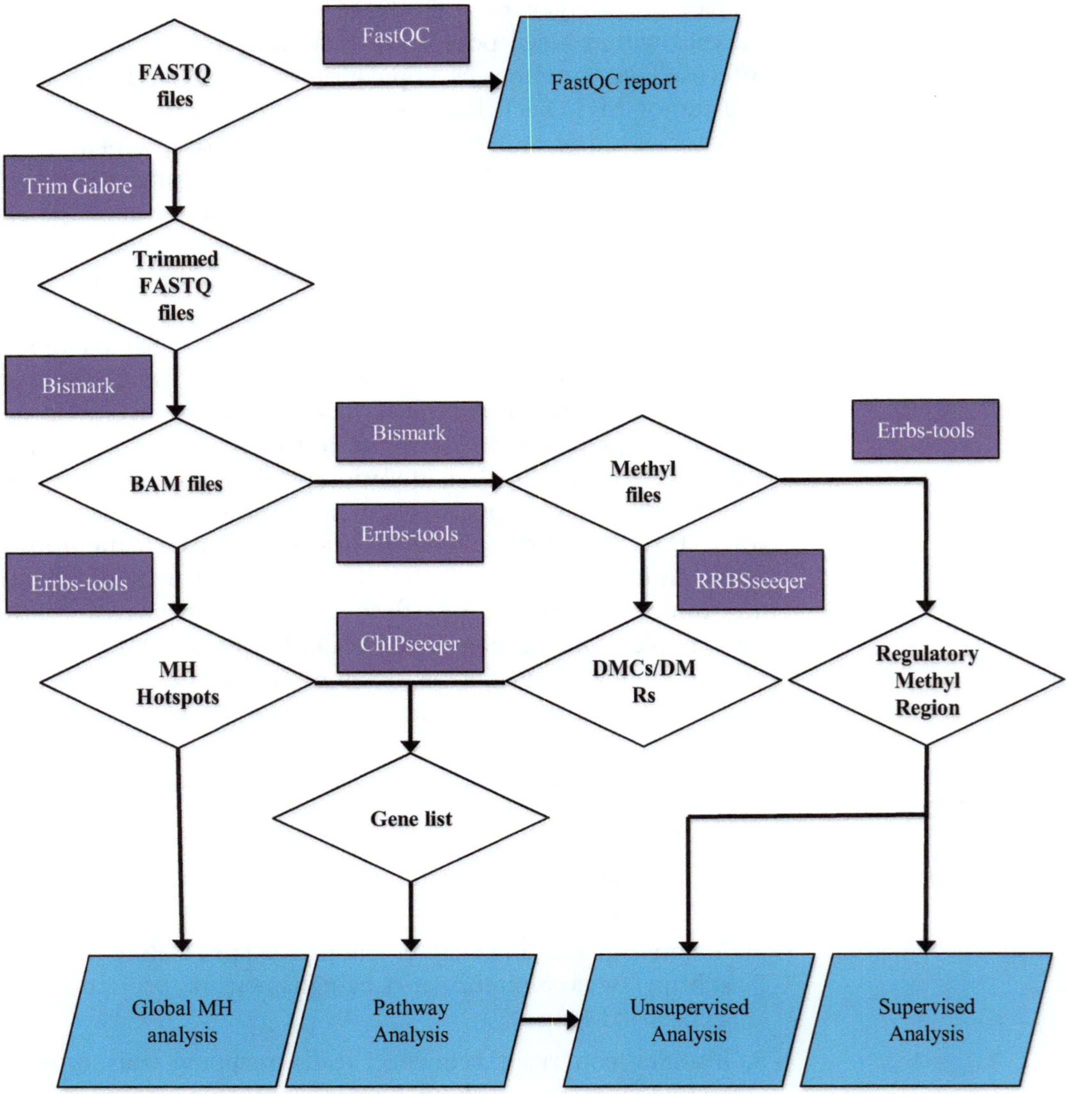

Fig. 1 Schematic of ERRBS data analysis pipelines. This comprehensive computational pipelines start from the ERRBS FASTQ files. The first step is to use FastQC to perform quality check of ERRBS data and make sure the data quality is good enough to make downstream analysis. Second, Trim Galore is used to remove adapter contaminations. Third, Bismark is used to map reads to bisulfite converted genomes and call methyl files, which indicates methylation status for each CpG site in genomes. Next, many computational tools including Errbs-tools, ChIPseeqer and RRBSseeqer are used to perform downstream analysis. MH hotspots can be used to perform global MH analysis and link MH to tumor evolution and disease progression. Individual DMCs/DMRs can be annotated to nearest genes and such gene lists can be used to perform pathway analysis. Methylation levels for regulatory regions are good inputs for both supervised and unsupervised types of downstream analysis

2.2 Input Files

1. For **fastqc** in FastQC: FASTQ format files of normal or tumor samples are most common inputs, BAM or SAM format files are also acceptable.
2. For **trim_galore** in Trim Galore: FASTQ format files of healthy tissue or tumor samples are required.
3. For **bismark_genome_preparation** in Bismark: FASTQ/ FASTA format files of genome reference are required.
4. For **bismark** in Bismark: FASTQ format files processed with **trim_galore** are required. FASTA format files are also acceptable but not recommended since the quality values are missing for such types of data.
5. For **bismark_methylation_extractor** in Bismark: BAM files from **bismark** are used as inputs.
6. For **methylCall_from_Bismark.py** in Errbs-tools: **CpG_OT_-sample.RRBS_trimmed.1bp.fq_bismark.txt** and **CpG_OB_-sample.RRBS_trimmed.1bp.fq_bismark.txt** from **bismark_m ethylation_extractor** are used as input files. Reads in file set 1 (labeled with OT) reflect methylation levels of CpGs in the forward strand. Reads in file set 2 (labeled with OB) contain m ethylation information of CpGs in the reverse strand.
7. For **epicore2calls.pl** in RRBSseeqer: Methyl files from **methyl-Call_from_Bismark.py** are used as input files (*see* Table 1).
8. For **RRBSseeqer_CG** in RRBSseeqer: output files from **epi-core2calls.pl** are used as inputs.
9. For **RRBSidentifyUpDownDMR.pl** in RRBSseeqer: output files with DMCs information from **RRBSseeqer_CG** are used as input files.
10. For **ChIPseeqerAnnotate**, **mergeCSAnnotateGenesCol-umns.pl**, **make_PAGE_input.pl** and **page.pl** in ChIPseeqer: files with DMR information from **RRBSidentifyUp-DownDMR.pl** are used as inputs. Each tool uses the output files from the previous one for those four sequential tools.
11. For **regionMethyl.R** in Errbs-tools: two kinds of input files are required. One is the Methyl file from **methylCall_from_Bis-mark.py** (*see* Table 1), the other one is RDS format file including genomic region annotations in GRanges or GRangesList objects [18, 19]. RDS is a special R based format, which can store a single R object.
12. For **regionMH.R** in Errbs-tools: three types of input files are required. The first type is the Methyl file from **methylCall_-from_Bismark.py** (*see* Table 1). The second one is the BAM file from **bismark**, which is a binary format for storing sequence data. BAM format is a more space-saving format as compared to SAM format data. The last one is the RDS file format including the genomic region annotations in GRanges or GRangesList objects [18, 19].

Table 1
RRBSeeqer input files example

chrBase	chr	Base	Strand	Coverage	freqC	freqT
chr1.10542	chr1	10542	F	587	99.83	0.17
chr1.10636	chr1	10636	F	57	85.96	14.04
chr1.10617	chr1	10617	F	58	100	0
chr1.10589	chr1	10589	F	58	100	0
chr1.10631	chr1	10631	F	56	100	0
chr1.10638	chr1	10638	F	57	85.96	14.04
chr1.10609	chr1	10609	F	58	98.28	1.72
chr1.10620	chr1	10620	F	59	93.22	6.78
chr1.10525	chr1	10525	F	609	95.24	4.76
chr1.10497	chr1	10497	F	606	97.85	2.15
chr1.10633	chr1	10633	F	58	89.66	10.34
chr1.133181	chr1	133181	R	118	61.02	38.98
chr1.133218	chr1	133218	R	117	54.7	45.3
chr1.133180	chr1	133180	F	131	40.46	59.54
chr1.133165	chr1	133165	F	136	88.24	11.76
chr1.135028	chr1	135028	F	168	88.1	11.9
chr1.135203	chr1	135203	R	77	87.01	12.99
chr1.135208	chr1	135208	R	77	90.91	9.09
chr1.135173	chr1	135173	R	78	87.18	12.82
chr1.134999	chr1	134999	F	170	31.18	68.82
chr1.135191	chr1	135191	R	76	94.74	5.26
chr1.135179	chr1	135179	R	79	67.09	32.91
chr1.135031	chr1	135031	F	168	90.48	9.52
chr1.135218	chr1	135218	R	71	78.87	21.13
chr1.136911	chr1	136911	F	103	92.23	7.77
chr1.136913	chr1	136913	F	101	94.06	5.94
chr1.136895	chr1	136895	F	104	67.31	32.69
chr1.136876	chr1	136876	F	104	95.19	4.81
chr1.136925	chr1	136925	F	103	0.97	99.03
chr1.137120	chr1	137120	F	29	96.55	3.45
chr1.137157	chr1	137157	F	29	100	0

(continued)

Table 1
(continued)

chrBase	chr	Base	Strand	Coverage	freqC	freqT
chr1.137169	chr1	137169	F	28	0	100
chr1.139059	chr1	139059	F	13	0	100
chr1.139029	chr1	139029	F	13	61.54	38.46
chr1.139073	chr1	139073	F	13	61.54	38.46
chr1.237094	chr1	237094	F	99	7.07	92.93
chr1.249382	chr1	249382	R	29	0	100
chr1.249429	chr1	249429	R	29	93.1	6.9
chr1.531247	chr1	531247	R	29	100	0
chr1.531265	chr1	531265	R	29	100	0

3 Methods

3.1 Pre-alignment Quality Control and Data Cleaning Processes

Similar to other high-throughput sequencing technologies, ERRBS is prone to systematic errors and artifacts such as PCR duplicates, GC-content shifts, and adapter contamination. In addition to these common problems, ERRBS data can suffer more critical problems such as erroneous methylation status of cytosine introduced by end-repair and low bisulfite conversion rates. Thus, it is always a good idea to perform a simple quality control analysis to avoid any biases that may affect subsequent analyses. ERRBS generates output in FASTQ format, like most high-throughput sequencing assays. We have successfully used a publicly available tool -FastQC- to perform short read quality control (*see* Fig. 1). Several features are very import to pay attention to. Those include per base sequence quality, per sequence quality scores, per base sequences content, and adapter content. Details about how to interpret the FastQC results are available in Andrews et al. (*see***Note 1**) [13].

Tools such as FastQC may reveal a variety of artifacts including adapter contamination. Adapter contamination is one of the most important technical issues for next-generation sequencing data, in that it may affect read mapping, leading to low mapping efficiencies and may even result in incorrect mapping and/or unreliable methylation calling in ERRBS. Moreover, positions filled in during end-repair can introduce artificial methylation readouts in ERRBS. The restriction endonuclease MspI selects relatively small fragment sizes (usually between 40 and 220 bp, but with quite a few MspI-MspI fragments even shorter than 40 bp). This can become a problem especially for sequencing reads with longer lengths. If the read length is longer than the MspI-MspI fragment size, there is a higher chance that the sequencing read would contain the adapter sequence on the 3′ end. To address this problem, adapters from longer reads need to be trimmed. We have successfully used Trim Galore to perform adapter trimming in our pipelines (*see* Fig. 1).

Altogether the pre-alignment process follows these three steps:

1. Quality check for original ERRBS reads with FastQC:

```
$ fastqc [-o output dir] [-f fastq|bam|sam] seqfiles1 ... seqfilesN
```

-**o** parameter specifies the directory where all outputs from FastQC should be stored. -**f** parameter indicates the input file format, usually FASTQ format files. **"seqfiles1 ... seqfilesN"** indicates that multiple input FASTQ files can be analyzed.

2. Adapter trimming with Trim Galore:

```
$ trim_galore [options] <filenames>
```

To run this analysis, just specify optional parameters and indicate your input FASTQ files after the options. The most relevant options are --**rrbs** and --**adapter**. --**rrbs** specifies that the data is an MspI digested library and --**adapter** specifies that adapter sequences need to be trimmed from reads.

3. Quality check for adapter trimmed ERRBS reads with FastQC: Perform the same analysis as in **step 1**, but use trimmed FASTQ files instead.

Example:

Take one of our FASTQ files as an example (**DLBCL_1D.ERRBS.fq**), the actual commands are:

1. **$ fastqc -o. -f fastq DLBCL_1D.ERRBS.fq**
2. **$ trim_galore --rrbs --adapter TGAGATCGGAAGAGCGGTTCAGCAGGAATGCCGAGACCGATCTCGTATGC --output_dir. DLBCL_1D.ERRBS.fq**
3. **$fastqc -o. -f fastq DLBCL_1D.ERRBS_trimmed.fq**

In our examples, we assume all of the samples and files to be analyzed are placed in the current working directory. If not, please specify the exact path of the files instead of using **DLBCL_1D.ERRBS.fq** directly. In **step 2**, the sequence we used is the standard adapter for Illumina. If no sequence is supplied, **trim_galore** will attempt to auto-detect the adapter that has been used (Illumina, Small RNA, and Nextera platforms standard adapters would be used).

In general, per base sequence quality provided by FastQC should be higher than cutoff (20 in most cases). Also, average per sequence quality should be around 38 (Phred+33, 0-41 scale). The other important restriction is that there should be no overrepresented sequences or adapter contents, which can be an indication of potential adapter contamination. Ideally, your dataset should pass most quality control steps after adapter trimming (hence **step 3** above) to confirm the data quality for downstream analysis (*see* **Note 1**). Also, besides the two parameters in **step 2** we mentioned, there are several items worth consideration during quality control, the details of which are provided in Subheading 4 (*see* **Notes 2** and **3**).

3.2 ERRBS Reads Alignment

Mapping ERRBS reads to a bisulfite converted genome presents many computational challenges. Alignments should allow for mismatches, especially for potential methylation sites. Also, alignments should be unique considering the numerous possibilities combining all the methylation statuses in each read to avoid miscalling of methylation levels. Among all the publicly available mapping tools such as BSMAP, RMAP-bs, MAQ, or BS seeker, we have chosen Bismark [15] to map ERRBS reads due to a couple of substantial advantages (*see* **Note 4**) [20–23].

The alignment process requires two steps:

1. Bisulfite converted genome preparation: typically no parameter changes are required for the genome preparation process. The only thing that absolutely needs to be specified is the directory where genome references are located. Such files need to be in FASTA/FASTQ format and can be downloaded from public databases such as UCSC genome browser or Ensembl [24, 25]. A recent genome build is recommended, e.g., hg19 or GRCh38.

```
$ bismark_genome_preparation [options] <path_to_genome_folder>
```

<path_to_genome_folder> specifies the directory of genome reference. Using **--bowtie1** will create bisulfite indexes for Bowtie 1 instead of Bowtie 2 (Default).

2. ERRBS reads alignment: for this step, the path to the bisulfite converted genome directory and the adapter trimmed FASTQ files are needed and the alignment can be run with default parameters.

```
$ bismark [options] <genome_folder> {-1 <mates1> -2
<mates2> | <singles>}
```

<genome_folder> specifies the directory of bisulfite converted genome reference from the first step. When paired-end data are used as input, **-1 <mates1> -2 <mates2>** is used to indicate each FASTQ file. If single-end data are used here, **<singles>** is used to specify the ERRBS reads file in FASTQ format. Several parameters are often used in options. For example, **--bowtie1** indicates bismark will use Bowtie 1 instead of Bowtie 2. **-l** specifies the "seed length," which is the number of bases at the high quality end of the read to which the mismatches ceiling applies. Typically, the length of sequencing reads can be used for this parameter. **--multicore** sets the number of parallel instances of bismark to be run concurrently. **--output_dir** specifies that all output files are written into the specified directory in BAM format.

Example:

Using one of our trimmed FASTQ files as an example (**DLBCL_1D.ERRBS_trimmed.fq**), the actual commands are:

1. **$ bismark_genome_preparation --bowtie1 genome/**
2. **$bismark--bowtie1-l50--multicore6genome/--output_dir. DLBCL_1D.ERRBS_trimmed.fq**

In our examples, we assume that all sample files (.fq) and reference genomes are placed in the current working directory. If not, the exact path of the files needs to be specified instead of using **genome/** and **DLBCL_1D.ERRBS_trimmed.fq** directly. Additional information regarding other optional parameters in the second step (Bismark alignment), such as usage of Bowtie 1 or Bowtie 2, as well as directional or non-directional sequencing can be found in Subheading 4 (*see***Note 5**) [26, 27].

3.3 Cytosine Methylation State Calling

Once suitable ERRBS alignments are generated, the methylation level for individual sites (mostly CpG sites) can be determined. To be consistent with our alignment processes, we utilize a simple script, named **bismark_methylation_extractor** from Bismark to achieve this goal. After methylation levels are generated, we need to perform quality checks to assess the accuracy of individual CpG methylation levels. Then we convert the data into a user-friendlier format for further analysis. This process consists of the following steps:

1. Extract CpG methylation levels from BAM files: we use **bismark_methylation_extractor** from Bismark to extract CpG methylation levels from each read in the BAM files. This tool is one of the most important advantages of Bismark compared to other computational tools (*see***Note 4**).

```
$ bismark_methylation_extractor [options] <genome_folder> <filenames>
```

<genome_folder> specifies the directory of bisulfite converted genome reference (outputs from

bismark_genome_preparation). <**filenames**> specifies the BAM files from bismark alignment (outputs from **bismark** alignment). **-s** option indicates that single-end type sequence read data was used. --**multicore** sets the number of parallel instances to be run concurrently. --**output_dir** specifies the directory to which all output files are exported. Using default options, **bismark_methylation_extractor** will give six different output files. Those include two types of possible strand-specific methylation information, original top strand reads (OT) and original bottom strand reads (OB), within three different contexts (CpG, CHG, CHH). Details about those files can be found in Subheading 4 (*see* **Note 6**). Each file has methylation information of a single-cytosine group, for example, methylation information for cytosines in CpG context from OT reads. Those files are tab delimited with 1-based coordinates:

```
<seq-ID> <methylation state> <chromosome> <start posi-
tion (=end position)> <methylation call>
```

Each line in these files represents single CpG site information from a single read. The second column specifies the methylation state: "+" and "−" indicate methylated and unmethylated status, respectively. Also, the fifth column has methylation call information. "Z" and "z" indicate methylated and unmethylated CpGs. Meanwhile, "X" and "x" represent methylated and unmethylated CHGs and "H" and "h" represent methylated and unmethylated CHH context. In our analysis, we only focus on cytosines in CpG dinucleotide context and we use --**merge_non_CpG** to merge non-CpG methylation (CHG context and CHH context) results into one file. Reads in files labeled with OT reflect methylation levels of CpGs in the forward strand and reads in files labeled with OB contain methylation information of CpGs in the reverse strand.

2. Quality check for CpG methylation extraction: after we extract methylation information from each read, we need to check several features to make sure the alignment and methylation status are correct and ready for downstream analysis. All the features below can be found in report files from **bismark** alignment (*see* Table 2).

   - It is important to check that all the methylation call statuses in the OT and OB files are "Z" and "z" to make sure that all the information is collected from the CpG context.
   - One of the most important features is average conversion rate, which detects how bisulfite treatment can successfully convert unmethylated cytosines. This number should be very close to 100% (*see* Table 2).

Table 2
Bismark output statistics example

ID	#reads	Mapping efficiency	Average conversion rate	#CpGs (10×)	Average coverage	Average CpG methylation levels(%)
1	73293324	64.50%	99.8979	2848900	51.82	34.40%
2	76101498	66.00%	99.8947	2933770	50.14	31.50%
3	85482964	66.50%	99.8838	2822016	56.84	39.50%
4	80272372	66.50%	99.8138	2795046	56.97	35.90%
5	64361288	66.60%	99.7396	2625874	50.8	34.50%
6	78897537	66.90%	99.8999	2782944	56.52	33.80%
7	76009876	66.00%	99.8751	2850695	52.66	38.20%
8	75431630	65.90%	99.778	2893682	52.82	38.40%
9	76653335	65.20%	99.8868	2843240	54.8	39.00%
10	78481481	65.00%	99.8859	2808703	54.04	38.50%
11	73287618	67.60%	99.8617	2733715	52.24	37.20%
12	73847281	67.50%	99.7764	2920328	50.99	36.50%
13	81504963	66.30%	99.8747	2822336	58.31	41.90%
14	96892822	62.50%	99.8899	2795175	59.89	48.80%
15	43137414	64.20%	99.8866	1609370	48.55	30.80%
16	72217478	66.50%	99.7681	2741285	54.08	45.90%
17	72922475	66.90%	99.8376	2872522	53.3	38.20%
18	75434628	66.80%	99.8349	2839250	52.93	38.30%
19	73437411	66.30%	99.8747	2767916	52.07	39.10%
20	82936367	68.00%	99.8547	2823987	52.74	36.10%

- Samples should have acceptable mapping efficiency (uniquely mapped reads out of all the input sequenced reads after adapter trimming, should be 60% or higher), which typically decreases with increasing sequencing read length (*see* Table 2).
- Samples analyzed together should ideally have similar numbers of covered CpGs, similar average coverage across the whole dataset, and similar average CpG methylation levels in each category/group (*see* Table 2).

3. Call CpG methylation level for each CpG: after quality check, we are ready to collect and combine the methylation status of all reads overlapping with individual CpGs into an overall

methylation level. We created a Python script to automate this process.

```
$ python methylCall_from_Bismark.py [options]
<sample_name> <input_dir> <output_dir>
```

<sample_name> sets the name of healthy tissue or tumor sample to be analyzed. This script can perform methylation calls for all the samples in the targeted directory. **<input_dir>** specifies where input files are located. **<output_dir>** specifies a directory into which all output Methyl files are written (*see* Table 1). **-c** is the only parameter that the user needs to adjust here; it specifies the minimum coverage required per CpG site.

4. Data transformation for RRBSseeqer: RRBSseeqer requires special format for input data files. We use a Perl script to convert Methyl files to RRBSseeqer acceptable data formats.

```
$ perl epicore2calls.pl <input_file> | gzip >
<output_file>
```

<input_file> sets the name of the Methyl files from **step 3** (*see* Table 1). **<output_dir>** specifies that all output files are written into this directory.

Example:

Using one of our BAM files as an example (**DLBCL_1D.ERRBS_trimmed.fq_bismark.bam**), the commands are as follows:

1. **$ bismark_methylation_extractor -s --output. --merge_non_CpG --multicore 6 --genome_folder genome/ DLBCL_1D.ERRBS_trimmed.fq_bismark.bam**
2. **$ python methylCall_from_Bismark.py -c 10 DLBCL_1D bismark_output/ cpg/**
3. **$ perl epicore2calls.pl cpg.DLBCL_1D.mincov10.txt | gzip > cpg.DLBCL_1D.mincov10.txt.calls.gz**

As before, all files, directories, and samples are assumed to be in the current working directory. The full path to each file and directory needs to be specified otherwise, if the files are present in a different location. When working with non-directional ERRBS data, additional parameters are required as indicated in Subheading 4 (*see***Notes 6** and 7). In the above example, the minimum coverage per CpG was set to 10. Enough reads can support the reliability of methylation levels for CpGs. For ERRBS analysis, 10 is always used as the cutoff, which is a tradeoff value considering the available number of CpGs and sequencing cost. Please find suggestions about how to choose this parameter in Subheading 4 (*see***Note 8**).

3.4 Patient-Specific DMRs Analysis (Unsupervised)

The identification of DMCs and DMRs is an important component of DNA methylation analysis. DMCs and DMRs typically reflect local DNA methylation changes during tumor evolution, such as those occurring in tumors between diagnosis and relapse. Several methods enable discovery of consistently hyper- or hypo-methylated DMCs or DMRs across several samples. DMCs and DMRs can be defined based on groups of samples or between two samples from the same patient. We focus our analysis here on defining DMCs and DMRs on a patient-specific basis. We use our in-house tool—**RRBSseeqer**—to extract DMCs and DMRs from individual patient data. This analysis is unsupervised in that any region of the genome can be a DMR.

To investigate cancer progression and tumor evolution, one may want to compare diagnosis and relapse samples from the same cancer patient and analyze DMCs and DMRs between these sample pairs. To examine the functional role of these methylation changes, one may ask whether DMCs and DMRs are near genes belonging to specific pathways, which might be relevant for cancer biology. These types of analyses frequently include identification of gene sets and pathways, which are over-represented within the DMR-associated genes in each patient. To explore commonalities across patients, one can perform unsupervised analyses of over-represented pathways (Fig. 2a). In more detail, such analyses consist of the following steps:

1. Identify DMCs: DMCs are identified by comparing two samples, for example healthy tissue and tumor, or diagnosis and relapse from the same patient. We identify DMCs using Fisher Exact or Chi-Square Tests comparing fractions of methylated to total reads at individual CpGs. We use a default false discovery rate = 20% for this analysis. Data formatting scripts are used to create tab-delimited output files.

```
$ RRBSseeqer_CG -rrbs1 <control_file> -rrbs2 <experiment_file> | sort_column.pl | sort_column_alpnum.pl > <output_file>
```

   <control_file> after **-rrbs1** represent the baseline sample (for example diagnosis sample), **<experiment_file>** after **-rrbs2** represents the second sample to compare to baseline (for example relapse sample). These files should be in the format produced by **epicore2calls.pl** above. **<output_file>** sets the name of the output file containing an analysis of each CpG.

2. Identify DMRs: we have defined DMRs as regions containing at least five DMCs separated by less than 250bp, and whose average methylation difference (including non-DMC in the region) is more than 10%. We use a Perl script called **RRBSidentifyUpDownDMR.pl** to identify DMRs based on DMCs.

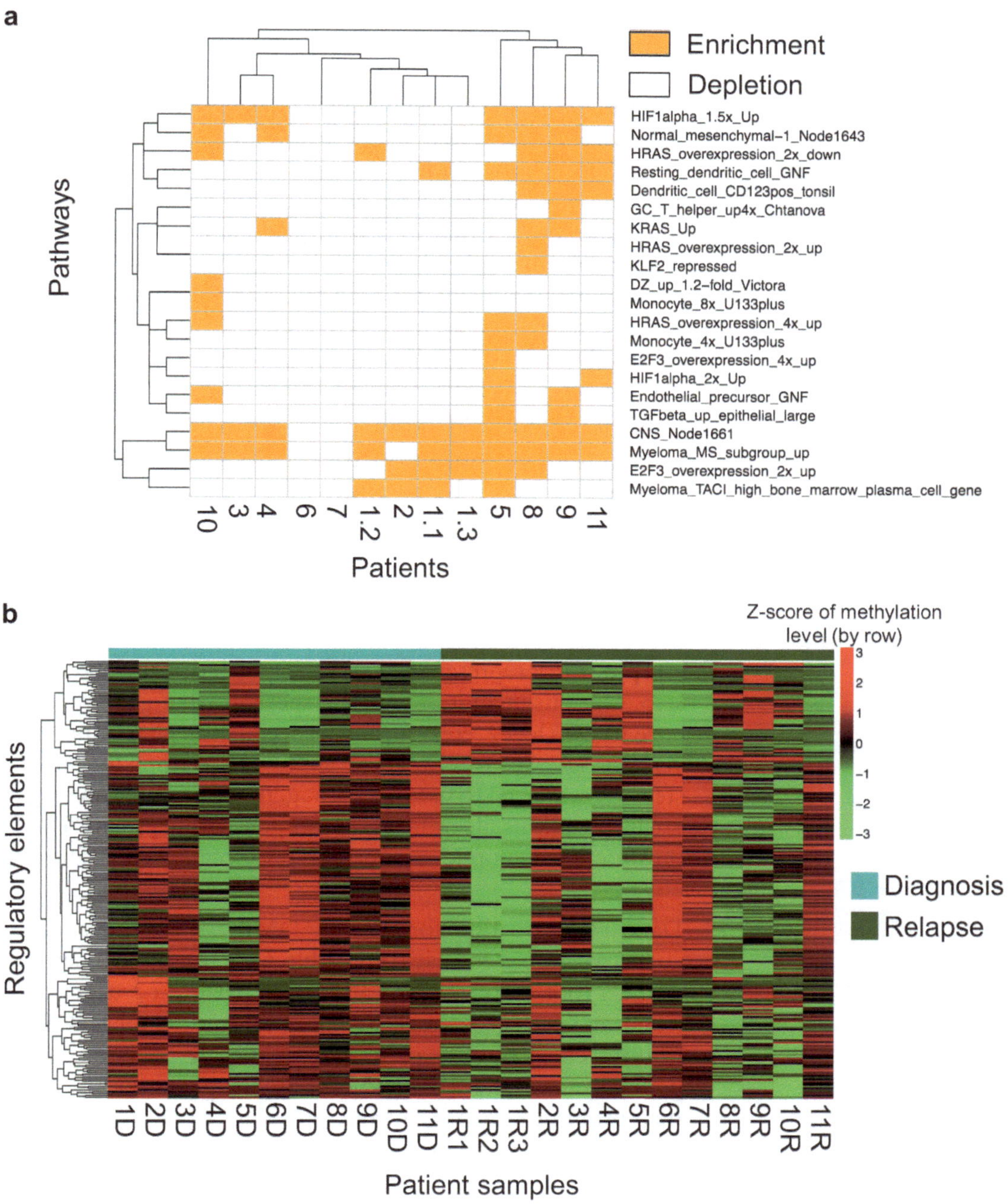

Fig. 2 Examples of DMRs identification and visualization. (**a**) Pathways overrepresented among hypermethylated genes (promoters overlapped with hypermethylation DMRs) of individual patients were illustrated here. Each row represents a single pathway and each column represents a patient pair. (**b**) Each row represents a single differentially methylated regulatory element. Each column represents single diagnosis/relapsed sample from patients. Scale bars represent z-score of methylation levels. Values were centered and scaled in row direction

```
$ perl RRBSidentifyUpDownDMR.pl --metfile=<input_file> --
outfile=<output_file> [options]
```

<**input_file**> is a tab-delimited file with CpG comparison from **RRBSseeqer_CG** (**step 1**). <**output_file**> is a file containing DMR information. DMRs are represented with a single DMR per row including chromosome, start position, end position, size, number of CpGs in DMR, and methylation difference. Additional options can be specified. -**dmax** specifies the largest distance between two DMCs (Default: 250). -**minmetdx** specifies the minimum average DNA methylation difference for DMRs (Default: 0.1). -**minnumcg** specifies the minimum number of DMCs needed to define a DMR (Default: 5).

3. Annotate DMRs with nearest genes: we use **ChIPseeqerAnnotate** from the ChIPseeqer package to annotate DMRs with the closest genes and identify genomic regions (exons, introns, promoters, intergenic regions) where gene and DMR may overlap.

```
$ ChIPseeqerAnnotate --peakfile=<input_file> [options]
```

<**input_file**> represents a DMR file produced at the previous step. Options include: --**genome** specifies what genome reference to use (hg19, etc.) and --**db** specifies the gene annotation versions (RefSeq, etc.). There are several output files in this step. We will need to use files with **.genes.annotated.txt** suffix. Each row in this file indicates if this gene overlaps with DMRs on different genomic regions like promoters, exons, introns, etc.

4. Data transformation for iPAGE: DMR-associated genes are converted into a format compatible with the iPAGE pathway analysis tool. Likewise this analysis is performed using tools from the ChIPseeqer package. First, we run the **mergeCSAnnotateGenesColumns.pl** to extract specific columns from **.genes.annotated.txt** file and retrieve the genes with peaks in their promoters/exons/introns, etc. Next, a perl script **make_PAGE_input.pl** is used to convert data into iPAGE acceptable input format.

```
$ mergeCSAnnotateGenesColumns.pl --genefile=<input_file>
--outfile=<output_file> [options]
```

<**input_file**> points to the output from **ChIPseeqerAnnotate**, specifically the file that ends with **.genes.annotated.txt.** <**output_file**> defines the output file. Options include: --**geneparts,** which specifies which gene parts overlapping with DMRs should be used for downstream analysis. P (Promoter),

I (Intron), E (Exon) etc. and combination of these (separated by commas) can be used here. --**showORF** specifies whether gene id (1) or transcript id (0) should be used in the output.

```
$ make_PAGE_input.pl --geneslist=<input_file> [options]
```

This tool creates an input file for iPAGE. Briefly, each gene is labeled as "gene of interest" (a gene near a DMR) or "background". **<input_file>** indicates the output files from **mergeCSAnnotateGenesColumns.pl**, which should be used as input for this step. The --**refgene** parameter specifies the gene data annotation used by ChIPseeqer and is used to create the background gene category.

5. Pathway analysis of DMR-related genes: given a gene profile with genes labeled either as genes of interest or as background, iPAGE is used to run pathway analysis against known pathways and gene sets. It uses mutual information to connect input gene sets and published gene sets and pathways.

```
$ page.pl -expfile=<input_file> [options]
```

<input_file> indicates the input file from last step. The --**pathways** option in iPAGE defines the database of pathways to use (Gene Ontology (GO), the Lymphoid Gene signatures and many other databases are supported [28, 29]. It is also feasible to use custom-defined pathways). The output of iPAGE indicates over- or under-representation of the input gene sets within specific gene sets or pathways with hypergeometric distribution log10 enrichment *p*-values as pathway enrichment scores.

6. Unsupervised analysis of over-represented pathways within DMR-related genes: we use an R-based package pheatmap [18, 30]. The input for this analysis is a matrix where each row represents a pathway and each column represents a single-sample pair (a diagnosis-relapse pair for example). Each entry in the matrix value is 1 if a specific pathway is significantly enriched in this patient, otherwise the value will be set to 0.

Example:

Here, we provide an example where we analyze DMCs and DMRs between diagnosis and relapse tumor sample (DLBCL). After the identification of DMCs and DMRs, following the strategy outlined above we identify tumor evolution-related DMCs/DMRs. Subsequently, we can perform downstream analysis investigating for example how those DMRs occur or disappear during tumor progression. We use **cpg.DLBCL_1D.mincov10.txt** as our < **control_file**>, **cpg.DLBCL_1R.mincov10.txt** as our <

experiment_file>. Examples of commands for calling DMCs and DMRs, and subsequently annotating them are:

1. **$ RRBSseeqer_CG -rrbs1 cpg.DLBCL_1D.mincov10.txt.calls.gz -rrbs2 cpg.DLBCL_1R.mincov10.txt.calls.gz -test chi | sort_column.pl 1 | sort_column_alpnum.pl 0 > DMC.DLBCL_1.txt**
2. **$ perl RRBSidentifyUpDownDMR.pl --metfile=DMC.DLBCL_1.txt --dmax=250 --minmetdx=0.1 --minnumcg=5 –outfile=DMR.DLBCL_1.txt**
3. **$ ChIPseeqerAnnotate --peakfile=DMR.DLBCL_1.txt --genome=hg19 --db=refSeq**
4. **$ mergeCSAnnotateGenesColumns.pl --genefile=DMR.DLBCL_1.txt.refSeq.GP.genes.annotated.txt --geneparts=P --showORF=1 --outfile=DMR.DLBCL_1.pro.txt**

   $ make_PAGE_input.pl --geneslist=DMR.DLBCL_1.pro.txt --refgene=/data/hg19/refSeq
5. **$ page.pl –expfile=DMR.DLBCL_1.pro.txt.ORF.txt --pathways=human_go_orf --cattypes=P,C,F -suffix=GO**
6. > **pheatmap(mat, ...)**

In **step 3**, **ChIPseeqerAnnotate** is used to annotate DMRs based on hg19 human genome and RefSeq gene annotations. In **step 4**, we specifically extract DMRs overlapping with promoters (defined as ±2 kb windows centered on RefSeq transcription start site). In **step 5**, we run pathway analysis against known Biological Processes (BP) in the Gene Ontology [28]. Other pathway databases such as KEGG pathways or msigDB pathways can be used in this step [31–33]. **Step 6** is different from other commands we used in this chapter. It is an R command and needs to be run in R environment. The 1-0 matrix **mat** needs to be provided to retrieve the heatmaps. Generally, several options can be used in this function to modify heatmaps in R. For example, **scale** option is a character indicating if the values should be centered and scaled in either the row or the column direction, or none. **cluster_rows** and **cluster_cols** are boolean values determining if rows/columns should be clustered.

3.5 Genomic Region-Specific DMRs Analysis (Supervised)

The analysis in Subheading 3.4 is currently limited to pairwise sample analysis. While it can be extended to more than two samples, an alternative approach is to compare the methylation levels of specific regions across two groups of samples. Groups of samples can be defined based on clinical variables such as diagnosis and relapse, chemo-resistant versus chemo-refractory, etc. Genomic regions can be defined as promoters, CpG Islands, enhancers, and binding sites for certain proteins, e.g., CTCF [34]. The proposed analysis identifies which of these predefined genomic regions are differentially methylated between the two sample groups.

We created an R script to collect methylation levels for specified regions and then perform supervised analysis between the two groups in R [18]. The analysis applies statistical testing followed by correction for multiple testing to assess differential methylation. We usually need two steps to perform this analysis:

1. We use an R script named **regionMethyl.R** in Errbs-tools to generate methylation levels for promoters of each patient. The methylation level for each region in each sample is calculated by averaging the methylation levels of all CpGs (with a threshold of a minimum number of CpGs) inside the corresponding regions. This script generates a matrix with methylation levels for all the regions across all the samples.

   ```
   $ R CMD BATCH --no-save --no-restore [options] region-
   Methyl.R regionMethyl.log
   ```

   We utilize R CMD BATCH to run the R script from the command line. The **regionMethyl.R** script can be found in Errbs-tools. **regionMethyl.log** stores the running log for the script. --**no-save** specifies that nothing will be saved in the. Rdata file. --**no-restore** specifies that R does not read the. Rdata file in the current directory. We use those two arguments since objects with identical names in the current R working space could cause bugs in the program or the outputs of the script could cause changes in the user's working space. Several arguments need to be specified in this step. --**input_dir** specifies where Methyl files should be found (*see* Table 1). --**output_dir** specifies where the output files should be created. --**regions** specifies a RDS format file that contains genomic region coordinates as GRanges or GRangesList objects [18, 19].
2. Perform supervised analysis comparing methylation levels of input regions between groups: we use paired T-tests or Wilcoxon-tests between diagnosis and relapse sample pairs. This analysis is followed by correction for multiple testing. A minimum methylation difference is also often used to ensure biological relevance of any significant change (at least 10% methylation difference). Differentially methylated promoters across all the patients can be visualized using a heatmap (Fig. 2b). As before, we use the pheatmap R package [18, 30].

   Example:

1. **$R CMD BATCH** --**no-save** --**no-restore '**--**args input_-dir=cpg/ output_dir=. regions=promoter22kb.rds regionMethyl.R regionMethyl.log**

2. > **pheatmap(mat, . . .)**

In this example the, --**regions** parameter specifies a list of promoters (defined as ± 2kb windows centered on RefSeq transcription start site) in GRanges data format [19]. Other regions such as CpG islands and enhancers can also be used here.

3.6 Intra-tumor MH Characterization

It is common to use DMRs or average DNA methylation levels within a specified region to characterize DNA methylation. However, DNA methylation is generally measured on cell populations consisting of hundreds of thousands or even millions of cells. Bisulfite converted DNA reads may contain multiple CpGs thus enabling a per-read analysis of DNA methylation patterns. In such analyses, loci with identical average DNA methylation levels can have distinct DNA methylation patterns between samples. Such distinct patterns are the result of intra-tumor MH (*see* Fig. 3a). Intra-tumor MH has been connected to gene expression levels and to clinical outcomes in several types of tumors including chronic lymphocytic leukemia (CLL) and Diffuse large B-cell lymphoma (DLBCL) [4, 12]. For example, higher MH in the promoter of certain genes was linked to lower expression of those same genes in CLL. Patients with higher global MH levels at diagnosis stage of DLBCL have a higher chance to relapse after chemotherapy [4].

The concept of epipolymorphism has been used to describe and quantify methylation heterogeneity [35]. The epipolymorphism level of a four-CpG locus (reads containing four or more contiguous CpGs) was defined as the probability that epialleles randomly sampled from the locus differ from one another. Higher epipolymorphism corresponds to higher intra-tumor MH and vice-versa. Epipolymorphism can be analyzed at individual loci based on ERRBS data. A global epipolymorphism level can also be calculated. We perform MH analysis using the following steps:

1. Epipolymorphism is calculated for each locus in a sample. The epipolymorphism level of a 4-CpG locus in the cell population is defined as the probability that epialleles randomly sampled from the locus differ from one another. More specifically, if we denote p_i as for the fraction of each DNA methylation pattern i in the cell population. The epipolymorphism equals $1 - \Sigma pi^2$. The higher the epipolymorphism, the higher the intra-tumor heterogeneity is. We created a script called **regionMH.R** to evaluate epipolymorphism levels (the script can be found in Errbs-tools). We utilize R CMD BATCH to run the R script from command line:

```
$R CMD BATCH --no-save --no-restore [options] regionMH.R
regionMH.log
```

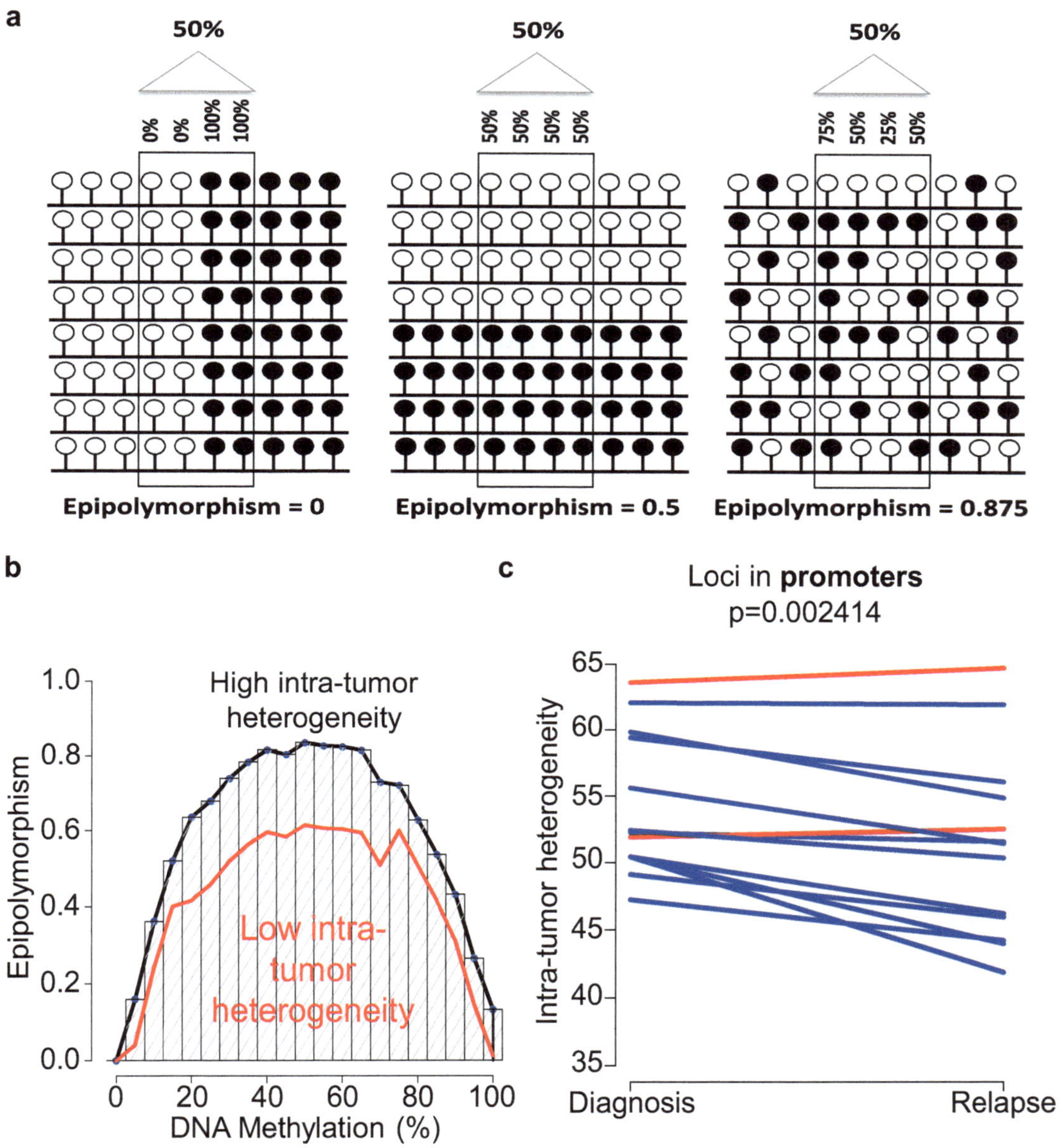

Fig. 3 Examples of intra-tumor MH analysis. (**a**) Epipolymorphism levels are dependent on DNA methylation levels. All loci are divided into different groups based on their methylation level and the median epipolymorphism of each group is calculated. Genome-wide intra-tumor MH is quantified by area under the median line. (**b**) Median epipolymorphism lines for diagnosis and relapse tumors from patient 1.1 in our cohort. Intra-tumor MH visibly decreased with tumor evolution. (**c**) Relapsed samples displayed significant lower intra-tumor MH. All the loci located in gene promoter

Several arguments need to be specified in this analysis. --**cpg_dir** sets the location of CpG Methyl files (*see* Table 1), --**bam_dir** sets the directory containing BAM files created by Bismark. Output files are written to --**output_dir**. --**regions** point to promoter regions (defined as ±2 kb windows centered on RefSeq transcription start site) in GRanges format [19].

2. Epipolymorphism is correlated with the methylation level, which means a locus has lower expected epipolymorphism values when it is fully methylated or fully unmethylated compared to the locus with 50% methylation levels) (*see* Fig. 3b). Therefore, global epipolymorphism must be normalized by methylation levels. Our global analysis divides loci into different bins depending on their methylation levels and median epipolymorphism is calculated for each bin. The area under the median line is defined as MH for each patient (*see* Fig. 3b, c). With this analysis, we can study the correlation between MH and tumor evolution.

Example:

1. **$R CMD BATCH --no-save --no-restore '--args sample=DLBCL_1 cpg_dir=cpg/ bam_dir=bam/ output_dir=. regions=promoter22kb.rds regionMH.R regionMH.log**

A growing number of studies have shown that intra-tumor MH is predictive of clinical outcome and tumor evolution. Such studies are showing that tumors with higher MH progress faster and earlier than tumors with low methylation heterogeneity. It is however worth noting that there are several methods for calculating intra-tumor MH (*see***Note 9**).

3.7 Conclusion and Outlook

Epigenetic modifications play a key role in cell development and tumorigenesis. DNA methylation is one of the best studied epigenetic modifications. High-throughput bisulfite converted sequencing technology provides great opportunities to analyze DNA methylation patterns during various physiological and pathophysiological processes. DNA methylation is relevant for cancer biology and has been link to tumor evolution. We here describe a comprehensive computational methodology to analyze DNA methylation, utilizing open source tools and our own in-house software. Our methodology starts from pre-alignment quality control and data cleaning processes, followed by data alignment, methylation state calling, and multiple downstream analyses. Following our directions, users can perform supervised and unsupervised analysis to different scales, including base pair DMCs, patient-specific DMRs, and genomic region-specific DMRs. Utilizing the above-mentioned tools to identify DNA methylation abnormalities can allow linking those to cellular development, tumor progression, and tumor evolution.

It is still unclear how DNA methylation or epigenetic modifications contribute to genetic changes and subsequently influence tumor evolution. Computationally, faster and more accurate alignment is still needed to perform larger-scale and more reliable analyses. Moreover, it will be equally important to design new algorithms to identify DMCs/DMRs with lower false discovery rate. There are still a number of unanswered questions in the field of tumor methylation analysis. For example, promoter hypermethylation could so far only be correlated to lower gene expression levels in a subset of genes. Global correlation between promoter hypermethylation and gene expression is still weak, making it difficult to draw any mechanistic conclusions from methylation patterns. Compared to successful genetic perturbation, modifying DNA methylation for specific regions is still a big challenge both in vivo and in vitro. Long-term efforts are needed for this intriguing but complex study in tumor evolution.

4 Notes

1. FastQC performs a series of quality control analyses including per base sequence quality, per sequence quality scores, per base sequences content, adapter content. Each test is flagged with a pass (green tick), warning (orange exclamation mark), or fail (red cross). The assigned status depends on how much a sample deviates from good quality benchmark samples. Examples of good and bad quality samples can be found at http://www.bioinformatics.babraham.ac.uk/projects/fastqc/ [13].
2. When running **trim_galore**, several parameters need to be considered. First, the default value of **-q/--quality** <**INT**> is 20, which means reads with low-quality ends (under 20) are trimmed. This parameter is acceptable for ERRBS analysis but can be less stringent if more reads are needed for the analysis. Second, the sequencing platform needs to be factored in. As default **trim_galore** will use ASCII+33 quality scores as Phred scores (option --**phred33**). ASCII+33 quality scores are usually used by Illumina 1.8+, which encode a Phred quality score from 0 to 41 using ASCII 33 to 74. If the sequencer did not use ASCII+33 quality scores, use --**phred64** option to specify alternative quality scores. Third, when no adapter sequence was provided, **trim_galore** will analyze the first one million sequences of the first specified file and attempt to find the first 12 or 13 bps of the following standard adapters:

   Illumina: AGATCGGAAGAGC
   Small RNA: TGGAATTCTCGG
   Nextera: CTGTCTCTTATA

   If using other adapters, it is important to provide correct adapter sequences in this step.

3. One of the most important parameters for **trim_galore** is -s/--**stringency** <**INT**>, which specifies the minimum number of required overlaps with the adapter sequence. The default value (1) is very stringent since even one overlap with the adapter sequence would be removed. If a less stringent value is used here, there is a higher chance of including too much adapter contamination into the downstream analysis, thus distorting the results. However, if one uses a very stringent cutoff (such as the default value), it is possible that some reads are mistakenly removed due to the first base being identical to adapters by chance. If sequencing data does not have enough reads after adapter trimming, CpGs coverage may be too low and downstream analyses such as DMCs/DMRs calling may be difficult.
4. There are several computational tools available, which the users can employ to solve the alignment of bisulfite converted data such as ERRBS. Before Bismark, different groups developed analysis tools for bisulfite converted data, including BSMAP, RMAP-bs, MAQ or BS seeker [20–23]. BS Seeker outperformed other mapping programs mentioned above, such as BSMAP, RMAP-bs, or MAQ, in terms of mapping efficiency, accuracy, and required CPU time [23]. Although the principles underlying BS Seeker and Bismark are similar, Bismark offers a number of advantages over BS Seeker [15]. For example, Bismark can support single-end and paired-end data, variable read length, adjustable insert size, and more adjustable mapping parameters. Bismark is much faster than BS Seeker. Also, Bismark can support non-directional library directly. The most important advantage compared to other tools is that Bismark not only does read mapping but, it also has tools for CpG methylation calling, an important feature of ERRBS type data analysis. For these reasons, Bismark is the most convenient tool available and accordingly most widely used. However, BS Seeker and Picard (https://broadinstitute.github.io/picard/) are also good alternatives.
5. When mapping reads to the genome, one should be aware of the sequencing library context. Directional sequencing libraries are common. However, if sequencing is not directional the --**non_directional** parameter should be used. Also, according to the Bismark tutorial, Bowtie 1 instead of Bowtie 2 should be used when trying to run alignment faster or when the sequencing reads are short [15]. Bowtie 1 usually performs equally well as Bowtie 2 in such condition. However, when applied to library with long fragment size (75 bp or above), Bowtie 2 is always recommended and always shows better performance.
6. In Bismark, there are four kinds of output files storing methylation status. Those are labeled with OT, OB, CTOT, and CTOB. Those files comprise reads, which are versions of the

original top strand, the original bottom strand, strands complementary to OT, and strands complementary to OB, respectively. If a library is directional, only reads that are versions of OT and OB will be sequenced. Forward strand CpG methylation information can be retrieved from OT files and reverse strand CpG information can be extracted from OB files. However, if libraries are constructed in a non-directional model, all four different strands generated will end up in the sequencing library with roughly the same likelihood. In this case, it is important to extract forward strand CpG methylation information from OT and CTOT files. OB and CTOB files collect reverse strand DNA methylation information.

7. If the applied sequencing library is non-directional, it is important to specify this when trying to use the **bismark_methylation_extractor**.

8. Extracting methylation information from low coverage CpG is typically considered not reliable due to potential sequencing errors for specific base pairs in routine ERRBS data passing process. It is recommended to remove CpGs covered by less than ten reads in the original ERRBS method paper [9]. However, decreasing this parameter is possible when analyzing WGBS data since WGBS is routinely low (10–15×). In fact. considering some people use 3× as the minimum reads in WGBS data [36], lower coverage may be acceptable.

9. Many other types of analyses can be used to characterize MH. Those include M-scores, Eloci, and PDR [8, 12, 37]. All these features have positive correlations with MH but they show differences in other respects. For example, M-scores only capture MH in each CpG site, which ignores the heterogeneity relation between adjacent CpGs. Eloci and PDR cannot consider hyper/hypo direction when they estimate MH changes. It is recommended to use different methods to study MH and explore any potential differences between methods.

References

1. Dawson MA, Kouzarides T (2012) Cancer epigenetics: from mechanism to therapy. Cell 150:12–27
2. Clozel T, Yang S, Elstrom RL, Tam W, Martin P, Kormaksson M, Banerjee S, Vasanthakumar A, Culjkovic B, Scott DW, Wyman S, Leser M, Shaknovich R, Chadburn A, Tabbo F, Godley LA, Gascoyne RD, Borden KL, Inghirami G, Leonard JP, Melnick A, Cerchietti L (2013) Mechanism-based epigenetic chemosensitization therapy of diffuse large B-cell lymphoma. Cancer Discov 3:1002–1019. https://doi.org/10.1158/2159-8290.CD-13-0117
3. Shaknovich R, Melnick A (2011) Epigenetics and B-cell lymphoma. Curr Opin Hematol 18:293–299. https://doi.org/10.1097/MOH.0b013e32834788cf
4. Pan H, Jiang Y, Boi M, Tabbò F, Redmond D, Nie K, Ladetto M, Chiappella A, Cerchietti L, Shaknovich R, Melnick AM, Inghirami GG, Tam W, Elemento O (2015) Epigenomic evolution in diffuse large B-cell lymphomas. Nat Commun 6:6921

5. Lin P-CC, Giannopoulou EG, Park K, Mosquera JM, Sboner A, Tewari AK, Garraway LA, Beltran H, Rubin MA, Elemento O (2013) Epigenomic alterations in localized and advanced prostate cancer. Neoplasia 15:373–383. https://doi.org/10.1593/neo.122146
6. Pike BL, Greiner TC, Wang X, Weisenburger DD, Hsu Y-H, Renaud G, Wolfsberg TG, Kim M, Weisenberger DJ, Siegmund KD, Ye W, Groshen S, Mehrian-Shai R, Delabie J, Chan WC, Laird PW, Hacia JG (2008) DNA methylation profiles in diffuse large B-cell lymphoma and their relationship to gene expression status. Leukemia 22:1035–1043. https://doi.org/10.1038/leu.2008.18
7. Esteller M (2002) CpG island hypermethylation and tumor suppressor genes: a booming present, a brighter future. Oncogene 21:5427–5440
8. Shaknovich R, Geng H, Johnson NA, Tsikitas L, Cerchietti L, Greally JM, Gascoyne RD, Elemento O, Melnick A (2010) DNA methylation signatures define molecular subtypes of diffuse large B-cell lymphoma. Blood 116:e81–e89
9. Akalin A, Garrett-Bakelman FE, Kormaksson M, Busuttil J, Zhang L, Khrebtukova I, Milne TA, Huang Y, Biswas D, Hess JL, Allis CD, Roeder RG, Valk PJM, Löwenberg B, Delwel R, Fernandez HF, Paietta E, Tallman MS, Schroth GP, Mason CE, Melnick A, Figueroa ME (2012) Base-pair resolution DNA methylation sequencing reveals profoundly divergent epigenetic landscapes in acute myeloid leukemia. PLoS Genet 8:e1002781. https://doi.org/10.1371/journal.pgen.1002781
10. Meissner A, Gnirke A, Bell GW, Ramsahoye B, Lander ES, Jaenisch R (2005) Reduced representation bisulfite sequencing for comparative high-resolution DNA methylation analysis. Nucleic Acids Res 33:5868–5877
11. Sidow A, Spies N (2015) Concepts in solid tumor evolution. Trends Genet 31:208–214
12. Landau DA, Clement K, Ziller MJ, Boyle P, Fan J, Gu H, Stevenson K, Sougnez C, Wang L, Li S, Kotliar D, Zhang W, Ghandi M, Garraway L, Fernandes SM, Livak KJ, Gabriel S, Gnirke A, Lander ES, Brown JR, Neuberg D, Kharchenko PV, Hacohen N, Getz G, Meissner A, Wu CJ (2014) Locally disordered methylation forms the basis of intratumor methylome variation in chronic lymphocytic leukemia. Cancer Cell 26:813–825. https://doi.org/10.1016/j.ccell.2014.10.012
13. Andrews S (2010) FastQC: a quality control tool for high throughput sequence data. http://www.bioinformatics.babraham.ac.uk/projects/fastqc/http://www.bioinformatics.babraham.ac.uk/projects/. doi: citeulike-article-id:11583827
14. Krueger F (2012) Trim Galore!. http://www.bioinformatics.babraham.ac.uk/projects/trim_galore/
15. Krueger F, Andrews SR (2011) Bismark: a flexible aligner and methylation caller for Bisulfite-Seq applications. Bioinformatics 27:1571–1572
16. Giannopoulou EG, Elemento O (2011) An integrated ChIP-seq analysis platform with customizable workflows. BMC Bioinformatics 12:277
17. Goodarzi H, Elemento O, Tavazoie S (2009) Revealing global regulatory perturbations across human cancers. Mol Cell 36:900–911. https://doi.org/10.1016/j.molcel.2009.11.016
18. R Development Core Team (2011) R Foundation for Statistical Computing, Vienna AI 3-900051-07-0. R A Lang Environ Stat Comput 55:275–286
19. Lawrence M, Huber W, Pagès H, Aboyoun P, Carlson M, Gentleman R, Morgan MT, Carey VJ (2013) Software for computing and annotating genomic ranges. PLoS Comput Biol 9:e1003118
20. Xi Y, Li W (2009) BSMAP: whole genome bisulfite sequence MAPping program. BMC Bioinformatics 10:1–9
21. Smith AD, Chung WY, Hodges E, Kendall J, Hannon G, Hicks J, Xuan Z, Zhang MQ (2009) Updates to the RMAP short-read mapping software. Bioinformatics 25:2841–2842
22. Li H, Ruan J, Durbin R (2008) Mapping short DNA sequencing reads and calling variants using mapping quality scores. Genome Res 18:1851–1858
23. Chen P-Y, Cokus SJ, Pellegrini M (2010) BS Seeker: precise mapping for bisulfite sequencing. BMC Bioinformatics 11:203
24. Kent WJ, Sugnet CW, Furey TS, Roskin KM, Pringle TH, Zahler AM, Haussler a D (2002) The Human Genome Browser at UCSC. Genome Res 12:996–1006. https://doi.org/10.1101/gr.229102
25. Aken BL, Ayling S, Barrell D, Clarke L, Curwen V, Fairley S, Fernandez-Banet J, Billis K, Garcia-Giron C, Hourlier T, Howe KL, Kahari AK, Kokocinski F, Martin FJ, Murphy DN, Nag R, Ruffier M, Schuster M, Tang YA, Vogel J-H, White S, Zadissa A, Flicek P,

Searle SMJ (2016) The Ensembl gene annotation system. Database (Oxford) 2016:baw093. https://doi.org/10.1093/database/baw093

26. Langmead B, Trapnell C, Pop M, Salzberg SL (2009) Ultrafast and memory-efficient alignment of short DNA sequences to the human genome. Genome Biol 10:1–25. https://doi.org/10.1186/gb-2009-10-3-r25. gb-2009-10-3-r25 [pii]\r
27. Langmead B, Salzberg SL (2012) Fast gapped-read alignment with Bowtie 2. Nat Methods 9:357–359
28. Ashburner M, Ball CA, Blake JA, Botstein D, Butler H, Cherry JM, Davis AP, Dolinski K, Dwight SS, Eppig JT, Harris MA, Hill DP, Issel-Tarver L, Kasarskis A, Lewis S, Matese JC, Richardson JE, Ringwald M, Rubin GM, Sherlock G (2000) Gene ontology: tool for the unification of biology. The Gene Ontology Consortium. Nat Genet 25:25–29
29. Shaffer AL, Wright G, Yang L, Powell J, Ngo V, Lamy L, Lam LT, Davis RE, Staudt LM (2006) A library of gene expression signatures to illuminate normal and pathological lymphoid biology. Immunol Rev 210:67–85. https://doi.org/10.1111/j.0105-2896.2006.00373.x
30. Kolde R (2012) Package 'pheatmap'. Bioconductor:1–6
31. Kanehisa M, Sato Y, Kawashima M, Furumichi M, Tanabe M (2016) KEGG as a reference resource for gene and protein annotation. Nucleic Acids Res 44:D457–D462
32. Ogata H, Goto S, Sato K, Fujibuchi W, Bono H, Kanehisa M (1999) KEGG: Kyoto encyclopedia of genes and genomes. Nucleic Acids Res 27:29–34
33. Subramanian A, Tamayo P, Mootha VK, Mukherjee S, Ebert BL, Gillette MA, Paulovich A, Pomeroy SL, Golub TR, Lander ES, Mesirov JP (2005) Gene set enrichment analysis: a knowledge-based approach for interpreting genome-wide expression profiles. Proc Natl Acad Sci U S A 102:15545–15550
34. Lai AY, Fatemi M, Dhasarathy A, Malone C, Sobol SE, Geigerman C, Jaye DL, Mav D, Shah R, Li L, Wade PA (2010) DNA methylation prevents CTCF-mediated silencing of the oncogene BCL6 in B cell lymphomas. J Exp Med 207:1939–1950
35. Landan G, Cohen NM, Mukamel Z, Bar A, Molchadsky A, Brosh R, Horn-Saban S, Zalcenstein DA, Goldfinger N, Zundelevich A, Gal-Yam EN, Rotter V, Tanay A (2012) Epigenetic polymorphism and the stochastic formation of differentially methylated regions in normal and cancerous tissues. Nat Genet 44:1207–1214. https://doi.org/10.1038/ng.2442
36. Eichten SR, Stuart T, Srivastava A, Lister R, Borevitz JO (2016) DNA methylation profiles of diverse Brachypodium distachyon aligns with underlying genetic diversity. Genome Res 26:1520–1531. https://doi.org/10.1101/gr.205468.116
37. Li S, Garrett-Bakelman F, Perl AE, Luger SM, Zhang C, To BL, Lewis ID, Brown AL, D'Andrea RJ, Ross ME, Levine R, Carroll M, Melnick A, Mason CE (2014) Dynamic evolution of clonal epialleles revealed by methclone. Genome Biol 15:472

Chapter 4

MicroRNA Networks in Breast Cancer Cells

Andliena Tahiri, Miriam R. Aure, and Vessela N. Kristensen

Abstract

A variety of molecular techniques can be used in order to unravel the molecular composition of cells. In particular, the microarray technology has been used to identify novel biomarkers that may be useful in the diagnosis, prognosis, or treatment of cancer. The microarray technology is ideal for biomarker discovery as it allows for the screening of a large number of molecules at once. In this review, we focus on microRNAs (miRNAs) which are key molecules in cells and regulate gene expression post-transcriptionally. miRNAs are small, single-stranded RNA molecules that bind to complementary mRNAs. Binding of miRNAs to mRNAs leads either to degradation, or translational inhibition of the target mRNA. Roughly one third of all the mRNAs are postulated to be regulated by miRNAs. miRNAs are known to be deregulated in different types of cancer, including breast cancer, and it has been demonstrated that deregulation of several miRNAs can be used as biological markers in cancer. miRNA expression can for example discriminate between normal, benign and malignant breast tissue, and between different breast cancer subtypes.

In the post-genomic era, an important task of molecular biology is to understand gene regulation in the context of biological networks. Because miRNAs have such a pronounced role in cells, it is pivotal to understand the mechanisms that underlie their control, and to identify how miRNAs influence cancer development and progression.

Key words Biomarkers, Breast cancer, Cancer, Microarrays, microRNA, Systems biology

1 microRNA Biology

1.1 microRNAs: A Historical Perspective

The central dogma in molecular biology has for a long time been "DNA makes RNA that makes protein" [1]. However, the impact of a gene on the phenotype is highly dependent on different mechanisms that allow a particular gene to be turned "on" or "off" in a particular state, in a particular cell, at a particular time. One way this type of regulation can be performed is by small RNA regulatory units called microRNAs (miRNAs). miRNAs are small, non-protein-coding RNA molecules that function as negative regulators of gene expression either by inhibiting translation or inducing degradation of messenger RNA (mRNA). Lin-4 was the first

Andliena Tahiri and Miriam R. Aure contributed equally to this work.

Louise von Stechow (ed.), *Cancer Systems Biology: Methods and Protocols*, Methods in Molecular Biology, vol. 1711,
https://doi.org/10.1007/978-1-4939-7493-1_4, © Springer Science+Business Media, LLC 2018

miRNA that was discovered in the nematode *Caenorhabditis elegans* (*C. elegans*) by Lee et al. in 1993 [2]. It took researchers 7 more years to identify another miRNA, let-7, in *C. elegans* [3]. Let-7 consisted of only 21 nucleotides, and was identified to have a significant role in nematode development. Resultantly, gene expression studies were complemented with the studies of the novel molecules regulating gene expression. Let-7 was thereafter identified in several organisms, including humans, with highly conserved sequences among different species. Today, more than 2500 mature human miRNAs have been annotated (miRBase version 21; [4]), and more than 60% of all the genes are predicted to be regulated by miRNAs [5].

After the initial description of miRNAs in *C. elegans* in 1993, it took several years before the role of miRNAs started to be fully appreciated. Over the last two decades, the number of papers published on miRNAs has exploded. However, there are still many unanswered questions regarding the detailed mechanisms by which miRNAs exert their regulatory roles.

Part of the difficulty in studying miRNA function is due to the complexity of miRNA biology. One miRNA may target several genes and one mRNA transcript has putative binding sites for various miRNAs. Thus, trying to dissect the in vivo connections between miRNAs and target mRNAs is a complex combinatorial challenge. Adding to the complexity of validating miRNA-mRNA relations is the fact that miRNA expression is tissue and time-specific, i.e., the context dependence is high.

1.2 miRNA Biogenesis and Function

The process of generating mature miRNAs in the cell consists of a series of nuclear and cytoplasmic steps (*see* Fig. 1). miRNAs are encoded either independently of protein-coding genes (intergenic) or inside introns of a host gene (intronic). Transcription occurs in the nucleus by RNA polymerase II and produces a long primary hairpin transcript called the primary miRNA (pri-miRNA). The pri-miRNA is long (>1 kb) and contains a local stem–loop structure, which is cleaved by the microprocessor complex (RNase III Drosha, in combination with DiGeorge syndrome critical region gene 8) in order to generate a precursor miRNA (pre-miRNA) [6, 7]. The pre-miRNA is exported to the cytoplasm by Exportin-5 and RAN-GTP, where it is further processed by RNase III endonuclease Dicer, to form a double-stranded miRNA duplex (~22 nt) [8]. The duplex is made up of two mature miRNA strands (named -5p and -3p depending on the 5′ and 3′ directions of the strand), and is subsequently loaded onto an Argonaut (AGO) protein to form an effector complex called the RNA-induced silencing complex (RISC) [7]. Usually, the RNA-strand with the unstable 5′-end is recruited into RISC, whereas the other strand (-3p) is released and quickly degraded. However, some studies have shown that the less abundant strand is also active in silencing, albeit usually less

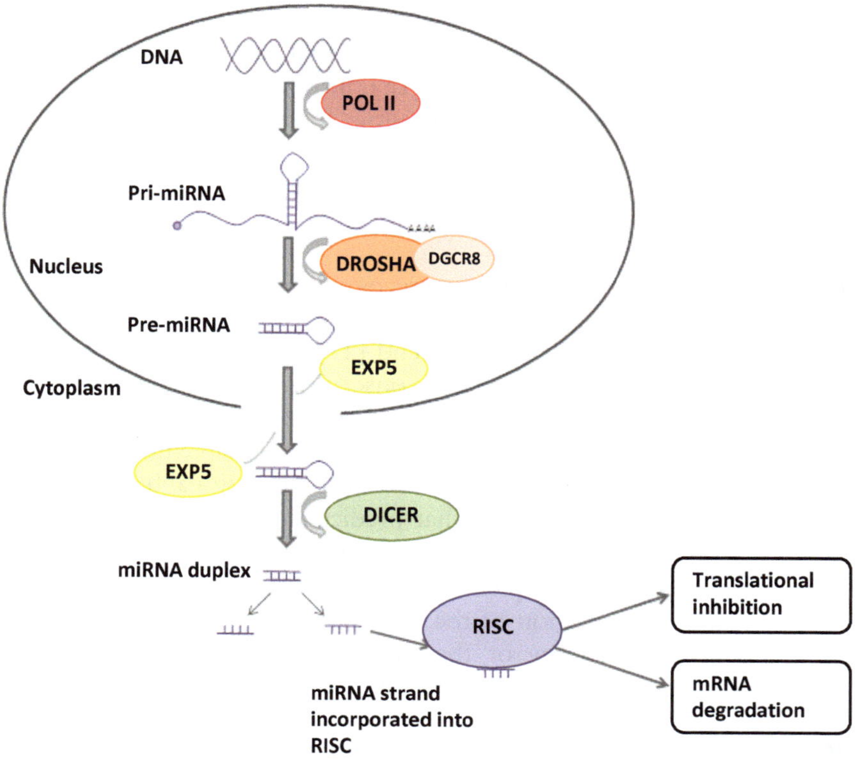

Fig. 1 The canonical miRNA biogenesis pathway and miRNA function (see the text for details). Pri-miRNA, primary miRNA; *EXP5*, Exportin 5; *POL II*, RNA polymerase II; *pre-miRNA*, precursor miRNA; *RISC*, RNA-induced silencing complex

potently than the more abundant guide strand [19]. Once the mature miRNA strand is incorporated into the RISC complex, the miRNA sequence targets mRNA through either perfect or imperfect complimentary binding to the 3′ untranslated region (UTR), the coding region or the 5′-UTR of genes [9, 10]. Binding of RISC to target mRNA can have different outcomes. Imperfect complementary binding of miRNAs to their targets inhibits translation and reduces protein expression without affecting the mRNA levels of these genes. Perfect complementary pairing between miRNA and mRNA targets the mRNA for degradation by RISC [11]. The exact mechanism of protein reduction is not fully understood, but it is likely that it occurs through both RNA degradation and translational repression pathways, with different miRNAs contributing to each pathway in different proportions [12]. mRNA degradation in mammals involves poly (A)-tail shortening (deadenylation) and other de-capping methods at the 5′-end of the mRNA strand. It is believed that miRNAs regulate a substantial portion of all protein coding genes. The complexity of mRNA regulation through miRNAs is remarkable as each miRNA can potentially regulate hundreds of genes, and one gene can be regulated by several miRNAs. Additionally, several transcription factors have been identified that can directly influence the expression of miRNAs [13–15]. miRNAs are considered important regulators of gene expression, involved in cellular processes such as development, cell proliferation, apoptosis, metabolism, cell differentiation, and stem cell division [16]; processes that are also highly involved in cancer pathogenesis.

1.3 miRNA–Target Gene Interactions and Predictions

The dominant target recognition sequence in the miRNA is termed the "seed" sequence and is located in nucleotides 2–8 in the miRNA from the 5′-end [17]. These positions in the miRNA are often evolutionary conserved. Other compensatory rules for miRNA-mRNA target recognition also exist, but all of them include some degree of sequence complementarity. miRNA target prediction is a major task in computational biology. Several in silico approaches exist that predict targets for a given miRNA and are described later in more detail. Those are based on different criteria such as complementarity to the miRNA seed region, evolutionary conservation of the miRNA recognition elements in the mRNA, free energy of the miRNA-mRNA hetero-duplex, and mRNA sequence features outside the target site [18, 19].

1.4 miRNA Function on a Cancer Systems Level

miRNAs are able to fine-tune the protein level of thousands of genes, either directly or indirectly, and thereby make fine-scaled adjustments to protein output [20]. The variety and abundance of targets offer an enormous level of combinatorial possibilities. This high level of complexity suggests that miRNAs and their targets form an intricate regulatory network intertwined with other cellular networks. It is pivotal to understand how miRNAs regulate cellular processes at the systems level, including miRNA regulation of cellular networks, metabolic processes, protein interactions, and gene regulatory networks. Studying different networks to assess the influence of miRNAs on their targets will help to identify miRNAs that have a strong influence on breast cancer development and progression.

2 miRNAs in Cancer

2.1 Breast Pathophysiology

Cancer is a complex disease involving abnormal growth of cells, invasion to surrounding tissue, and migration and invasion to distant sites. Breast cancer is the most common type of cancer and cause of cancer-related deaths in women worldwide [21].

However, the most frequently observed abnormalities in the breast are usually benign. Different benign conditions can take place in the breast. Those can be divided into three groups based on how they affect breast cancer risk [22]; (1) Non-proliferative lesions such as cysts or fibrosis considered with almost no breast cancer risk; (2) Proliferative lesions without atypia such as breast fibroadenoma or fibroadenomatosis which show excessive growth of cells in the ducts or lobules of the breast tissue, and slightly increase a woman's risk of developing breast cancer; (3) Proliferative lesions with atypia such as atypical ductal/lobular hyperplasia (ADH/ALH) show an overgrowth of cells in ducts or lobules of the breast. They have a strong effect on breast cancer risk [23–28].

Although there are many types of benign lesions in the breast, most research is focused on malignant breast tumors. Breast cancer types can be grouped based on the origin of tumor formation into invasive or noninvasive types of breast cancer. These include lobular carcinoma in situ (LCIS), ductal carcinoma in situ (DCIS), invasive ductal carcinoma (IDC), and invasive lobular carcinoma (ILC). DCIS and LCIS are considered pre-invasive cancer as they can in some cases metastasize [29]. IDC is the most common type of breast cancer which starts in a milk duct of the breast, breaks through the wall of the duct, and grows into the fatty tissue of the breast. At this point, it may metastasize to other parts of the body through the lymphatic system and bloodstream. ILC on the other hand, starts in the lobules, and like IDC, it can metastasize. There are about 5–15% of breast cancer cases that involve ILCs, and they are more difficult to detect than IDCs through physical examination, mammography, and even through gross pathologic evaluation [30].

Cancer has for a long time been viewed as a genetic disease. However, as the unraveling of the molecular biology of breast cancer progresses, it is no longer seen as a single disease, but rather as a complex disease involving many subtypes with different outcomes based on differences in the genetic makeup [31, 32]. For example, in breast cancer the expression of estrogen receptor (ER), progesterone receptor (PR), and human epidermal growth factor receptor 2 (HER2)/*neu* have implications for prognosis and therapy selection that are independent of TNM staging, which describes tumor size or depth (T), lymph node spread (N), and presence or absence of metastases (M) [33]. Ki-67 is a prognostic marker of breast cancer that has recently been applied in the clinics [34]. Ki-67 is a marker of proliferation, and higher percentage of Ki-67 in breast cancers (usually above 15%) indicates worse prognosis.

Thanks to advances in molecular biology and the use of microarray technology, breast cancer can be divided into four subtypes based on the genetic profiles, with each having a different clinical outcome [31, 32, 35–37]. These subgroups are named; (1) luminal A (ER+, HER2−), (2) luminal B (ER+, HER2+/HER2−), (3) HER2-enriched, and (4) basal-like, also often termed triple negative (ER−, PR−, HER2−). The luminal A subtype has the best prognosis compared to the other subtypes. Luminal B tumors are characterized by high proliferation activity (Ki-67 index), may be positive for HER2 expression, and have a worse prognosis than luminal A tumors [38]. However, HER2 expression in the Luminal B subtype is lower than in HER2-enriched tumors. The HER2-enriched subtype is often associated with nodal metastasis, whereas the basal-like often occurs in younger patients, is more frequently associated with visceral organ metastasis, and has a very poor prognosis [39]. It is important to note that not all triple negative tumors are identified as basal-like by gene expression, and not all basal-like tumors are triple negative [40].

2.2 Cancer Biomarkers

A biomarker is a biological molecule present in any biological material (e.g., tissue, cell, or body fluid) that can be used as a measurable indicator of normal biological processes, pathogenic processes, or response to therapy [41]. Biomarkers in cancer can be divided into three main subgroups providing different purposes in the clinic; (1) Risk assessment markers; (2) Diagnostic markers; and (3) Prognostic markers (Table 1). Medical scientists still strive to find the best biomarkers that can provide a reliable diagnosis, tell us which therapy is the best for a particular patient, or even better; tell us whether a person is at risk of getting a certain disease without reaching the diseased stage. Early cancer detection would dramatically reduce mortality associated with the disease; however, early diagnosis relies on clinically validated biomarkers with high specificity and sensitivity.

Since 1985, the TNM staging [42] has provided doctors with the basis for the prediction of survival, choice of treatment, and stratification of patients. At the same time, it has provided consistency among healthcare providers. In some cases, tumor grade, histological subtype, or patient age would be added to TNM staging when such information was important for the prediction of survival or response to therapy. Today, the findings of new

Table 1
Different types of biomarkers important in clinical settings of cancer research

Biomarker	Purpose	Example	References
1. Risk assessment and screening	• Aid in cancer prevention • Provide the earliest evidence of potential cancer in persons not yet diagnosed with the disease	Breast and ovarian cancer: BRCA1/BRCA2	[42]
2. Diagnostic	• Establish a diagnosis • Assist with staging, grading, and selection of initial therapy	TNM staging	[43]
3. Prognostic and predictive	• Estimate the aggressiveness of a condition • Predict how well a patient will respond to a specific treatment	Breast cancer: ER, PR, HER2, and Ki-67 Melanoma: BRAF (V600E)	[31, 44, 45]

molecular markers that can predict survival and efficacy of therapy provide additional important information to TNM staging, and are used in the clinics worldwide.

Next to genetic changes, post-transcriptional, posttranslational modifications, and metabolic changes play a role in cancer formation [43–45]. Better appreciation of the complexity in carcinogenesis has provided us with a number of candidate biomarkers valuable for risk assessment, screening, diagnosis, prognosis, and selection and monitoring of therapy. Nonetheless, although several markers are identified and used for diagnostic and prognostic purposes with implications in therapy treatment, histological examination is still required for diagnosis, whereas immunohistochemistry and genetic tests are utilized for treatment decisions and prognosis determination.

2.3 miRNA Function in Cancer

The first discovery of the implication of miRNAs in cancer was observed in B-cell chronic lymphocytic leukemia (CLL) in the search of tumor suppressors at chromosome 13q14, which is commonly deleted in CLL patients [46]. In this study, the authors found that miR-15a and miR-16-1 were located in this region. Since loss of this chromosome was frequent in CLL, it indicated that loss of these miRNAs also occurred, raising the question whether miRNAs could be involved in the pathogenesis of cancer. Later, the same group identified several miRNAs located in frequently deleted or amplified regions in the genome in different tumors [47].

Iorio et al. in 2005 described the first breast cancer miRNA signature which could discriminate tumors from normal tissues [48]. Subsequent studies have increased our understanding of miRNA involvement in breast cancer, and identified aberrant miRNA expression related to survival, metastasis, stage, proliferation, molecular subtype, *TP53* mutational status, hormone receptor status, and response to treatment [49–53]. The studies revealed that changes in miRNA expression profiles can serve as phenotypic signatures of specific types of cancer. Aberrant miRNA expression associated with tumorigenesis can be a result of various mechanisms. Several studies point to transcriptional deregulation, copy number aberrations, mutations, epigenetic alterations, and defects in the miRNA biogenesis machinery as contributors to miRNA deregulation in cancer [54]. Some miRNAs may be causally linked to tumorigenesis by directly modifying tumor-suppressor or oncogenic pathways. For example, the overexpression of miRNAs can inhibit tumor-suppressor genes in a pathway. Conversely, reduced miRNA expression through loss-of-function mutations could result in increased expression of oncogenes, also contributing to cancer development and progression (*see* Fig. 2).

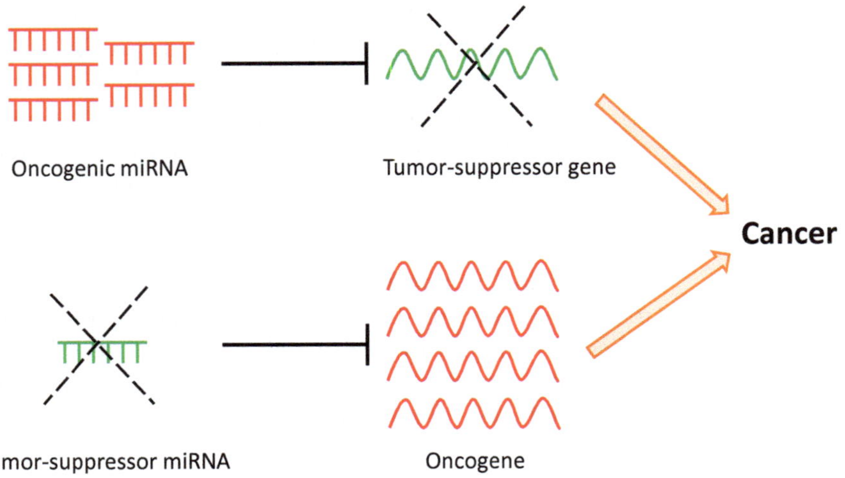

Fig. 2 miRNAs may have oncogenic or tumor-suppressive roles in cancer. Upregulation of oncogenic miRNAs results in increased repression of tumor-suppressor target genes. Conversely, downregulation of tumor-suppressor miRNAs results in decreased repression and thus increased expression of target oncogenes. Both scenarios may lead to cancer development and progression. Figure based on Lujambio and Lowe [119]

2.4 miRNAs as Cancer Biomarkers

Various studies provide evidence that miRNAs can be used as biomarkers for different purposes [55, 56]. Deregulated expression profiles of miRNAs have been discovered in a wide variety of human cancers, including breast cancer [57], colorectal cancer [58], glioma [59], lymphoma [60], and prostate cancer [61].

The survival and prognosis of a patient is highly dependent on the stage of the tumor at the time of detection. The earlier a tumor is detected, the better the prognosis is. Thus, a major clinical challenge in cancer is the identification of biomarkers that can detect cancer at an early stage. miRNAs can be reliably extracted and detected from frozen and paraffin-embedded tissues. They can moreover be found circulating freely in the blood or bound to circulating exosomes, and in different body fluids like urine, saliva, and sputum [62]. The fact that miRNAs are stable in body fluids, and that they are easily detectable through noninvasive procedures makes miRNAs attractive biomarker candidates. For example, miRNA signatures in plasma had strong diagnostic and prognostic potential detecting lung cancer before disease onset, as plasma samples were collected 1–2 years before lung cancer was detected by CT [63]. Another recent study by Cava et al. [64] showed that miRNA profiling improved breast cancer classification and could differentiate patients with breast cancer as responding or not responding to therapy, with promising results. The correct classification of breast cancer is a fundamental factor in determining the appropriate treatment, and it is now evident that miRNAs have the potential to provide new diagnostic, prognostic, and predictive biomarkers for cancer, with a great impact in the clinics. However, their use in the clinics has not been implemented yet as there still are many hurdles to overcome.

3 Techniques for Studying miRNA Networks in Cancer

3.1 The Microarray Principle

Oligonucleotide microarray is a high-throughput technique based on hybridizing labeled sample material to complementary probes that are immobilized on a solid surface. The amount of material that has hybridized to the probes is quantified by a laser that scans the array and excites the fluorescent dye attached to the labeled sample. One array contains thousands of probes, each representing a defined sequence that is complementary to an mRNA or miRNA transcript. The microarray technology is a useful tool to study the expression of thousands of miRNAs or mRNAs simultaneously. Many different platforms exist with varying probe contents and length, and labeling techniques. Figure 3 illustrate the steps of Agilent-based miRNA/mRNA expression profiling.

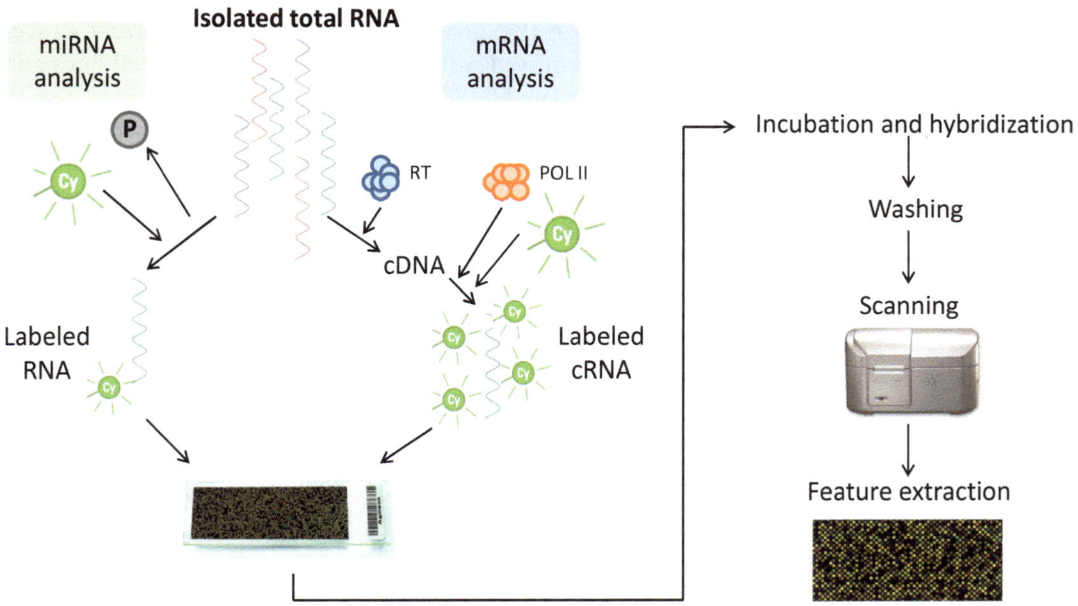

Fig. 3 miRNA and mRNA expression profiling using Agilent microarrays. RNA is labeled with a fluorescent dye (Cyanine 3; Cy) and transferred to the microarray where the sample material hybridizes to complementary probes during incubation. Then follows washing and scanning of the array, and finally feature extraction where probe hybridization intensities are quantified. The protocol deviates slightly between microRNA and mRNA analysis. For the former RNA is treated with phosphatase to remove the 3′-phosphate group (P), which is followed by labeling. For mRNA profiling the RNA is first converted to complementary DNA (cDNA) by reverse transcriptase (RT), and then the cDNA is further transcribed into complementary RNA (cRNA) by the use of RNA polymerase (POL II) where labeled cytosine residues are incorporated

3.2 Functional Experiments to Validate miRNA Targets and Their Effect on Cells

Data from functional studies of miRNAs in cell lines can be generated after identifying interesting candidates from analyses of high-throughput data. The aim is to determine whether the candidate miRNA is functionally involved in cancer-associated processes. This can be done by testing the effects of silencing or overexpression of the candidate miRNA on the viability and proliferation of cancer cells. Knockdown of potential tumor driver miRNAs can be performed using small, single-stranded anti-miRs which are miRNA inhibitors that bind to and inhibit endogenous miRNAs [65]. Conversely, the effect of candidate tumor-suppressor miRNAs can be assessed by overexpression, for example by adding miRNA mimics and measuring the effect on cell viability. In order to effectively study the functional role of miRNAs in cell lines, high-throughput screens can be performed. Leivonen et al. used libraries of either miRNA mimics or anti-miRNAs which were tested simultaneously in large scale and used to measure the effect of miRNA overexpression or knockdown, respectively [66]. miRNAs can be spotted in 96- or 384-well formats, and incubated with cells from a cell line of interest. The phenotypic end-points of such screens may measure the effects that miRNAs have on cell viability, apoptosis, and proliferation, as well as expression of marker proteins. Leivonen et al. [66] performed a high-throughput screen to identify miRNAs that were important for the growth of HER2-positive breast cancer cells. They overexpressed miRNAs in HER2-positive cell lines and assessed the effect on HER2 protein levels, proliferation (Ki67), and apoptosis (cleaved PARP). Thirty-eight miRNAs were identified that inhibited HER2 signaling and cell growth. In another study [53], miRNAs that were identified as differentially expressed between high and low proliferative tumor samples (scored by immunohistochemistry) were further functionally validated by transfecting a library of pre-miR constructs into breast cancer cell lines. The cells were lysed and the lysates printed on slides that were then stained with an antibody against Ki67 to assess the effect of the miRNAs on proliferation. Among the 123 identified differentially expressed miRNAs, 13 showed a corresponding functional effect on Ki67 protein levels [53].

The measurement of ATP using luciferase is one of the most commonly used assays for assessing cell viability in high-throughput screening applications [67]. The assay is fast and easy to use, sensitive, and also less prone to artifacts than other viability assay methods [67, 68]. However, the assay measures metabolically active cells, which cannot be translated into viable cells in all contexts. Another method that is widely used is the MTT Tetrazolium Reduction Assay. Yet, the MTT assay lacks sensitivity, is more time-consuming and more prone to variation, due to multiple experimental steps involved compared to the ATP assay [68]. Other methods that are used to measure the effect of miRNAs on cell viability or proliferation include the TUNEL assay, Trypan Blue staining assay, Tetrazolium Reduction Assays, etc. The choice of method relies on the investigators' preferences, and there are both benefits and pitfalls for each assay which have to be taken into consideration.

3.3 Databases and Tools

Different databases exist that list miRNAs, their chromosomal location, sequence and their putative target genes. For example, the miRBase database contains all published miRNA sequences and annotations [4]. The Ingenuity Pathway Analysis database (IPA, Ingenuity Systems; www.ingenuity.com) can be used to associate genes correlated to candidate miRNAs with pathways and for various gene annotation purposes. The SEEK tool [69] can be used to identify and annotate genes that are co-expressed with miRNA-correlated genes. Different computational tools are readily available for the analysis of miRNA target sites, such as miRanda [70–72], TargetScan [5, 73, 74], PicTar [75–77], and DianaMicroT-CDS [78, 79] (Table 2). Those can be used to predict potential targets of a miRNA that has been identified (for example in cancer tissue), or vice versa, identify candidate miRNAs predicted to bind to a gene of interest.

Feedback from functional validation results has greatly improved the performance of these in silico miRNA target prediction algorithms. The miRanda software was initially designed to predict miRNA target genes in *Drosophila melanogaster* [70, 71]. The algorithm searches for highly overlapping basepairs in the 3′ UTRs for identifying potential binding sites [70]. A higher score is given for sequences which are complementary to the 5′ end of the miRNA compared to the 3′ end, leading to higher prediction scores for seed regions with perfect, or nearly perfect match.

TargetScan is an algorithm developed by Lewis et al. [74], and was the first miRNA target prediction tool for the human genome, using a different search approach than miRanda. TargetScan searches for perfect complementarity in the seed region and beyond [74]. If there is complementarity outside the seed region, it will filter out the false positives more efficiently prior to prediction. Data from conservation analysis derived from orthologous 3′

Table 2
Computational algorithms for miRNA target prediction

Algorithm	Website	References
TargetScan	www.targetscan.org	[5, 71, 72]
miRanda	www.microrna.org	[68–70]
PicTar	pictar.mdc-berlin.de	[73–75]
DianaMicroT-CDS	www.microrna.gr/microT-CDS	[76, 77]

UTRs are used as input early in the process. Also, thermodynamic stability is tested to filter predicted target sites [80].

PicTar is the first algorithm for analyzing miRNAs and target mRNAs in co-expression at a specific time and place. The PicTar software fully relies on data from several species to identify common targets for miRNAs [75]. It uses conservation data from 3′ UTR as input and searches for alignment of complementary seed regions. Binding sites are tested for thermodynamic stability and each result is given a score [75, 80].

The DianaMicroT algorithm scans for larger complementarity regions and focuses on coding regions of target mRNAs [79]. It also calculates and uses the free energy of binding sites as an input for the prediction of targets. Importantly, many miRNAs share sequence composition and are thus grouped into families based on sequence homology. Members of the same miRNA family are believed to at least partly be able to target some of the same genes due to this sequence similarity [71]. There is still a gap between in silico predictions and knowledge of the in vivo relations, but further advances in molecular technology will reduce this gap.

3.4 (Epi-) Genome–Transcriptome Analysis

DNA aberrations are a hallmark of cancer genomes [81], and the phenotypic effects of such alterations are commonly investigated through the integration of genomic and transcriptomic data. Analyzing changes in DNA copy number can be used to identify aberrant cancer genes. The correlation between copy number and mRNA expression can be utilized to single out genes for which DNA aberration is manifested in the altered expression of the gene. In a similar manner, DNA copy number and methylation status can be used together with miRNA expression to identify miRNAs altered on the (epi-)genomic level with effects on the transcriptomic level. The rationale behind such integrative approaches is that recurrent alterations across tumor samples may indicate functionality through the effect on the transcription levels of the corresponding miRNAs or genes. Thus, RNA expression is used as an additional layer to the genomic or epigenetic data to further identify potential candidate genes. If a change in DNA copy number affects the expression of a miRNA, the miRNA is more likely to be under selection in the tumor and hence might be important for tumorigenesis.

Studies integrating DNA copy number and mRNA expression in breast cancer have revealed a clear dosage effect of gene copy number on gene expression [82, 83], which also holds true for miRNA expression [84]. Lahti et al. [85] divided implementations for the integrative analysis of DNA copy number and expression into four main categories of approaches; two-step approaches, correlation-based approaches, regression-based approaches, and latent variable models. In a two-step approach, tumor samples and miRNAs/genes are first grouped based on altered copy number and/or methylation levels, and then in the second step, differential expression is assessed between the different groups. Both correlation and regression-based methods can be used; this ensures a potential functional implication on the expression of the altered miRNA/gene. For example, Aure et al. [84] investigated the effect of DNA copy number and methylation alterations on miRNA expression in breast cancer. First, each miRNA in each patient was assigned to one of the two groups altered or non-altered based on copy number or methylation status. Then, Wilcoxon rank-sum tests were used to assess if the expression of a given miRNA was different in the two groups considering alterations on the copy number, the methylation level, or both. Using this approach the authors identified miRNAs whose expression was increased due to gain and/or hypomethylation. The authors further identified miRNAs, whose expression was reduced due to loss or hypermethylation of the miRNA gene. In this way, the study provided evidence of the mechanisms behind miRNA dysregulation in breast cancer. Interestingly, it was found that miRNAs from the same family (i.e., sharing seed sequence and are predicted to regulate the same target genes) were altered by different mechanisms in different patients, but with the same net effect on miRNA expression (increased or decreased), emphasizing alteration of miRNA expression in breast cancer through variable genomic changes.

Comparative studies of methods integrating copy number and expression data have shown that the different methods vary in sensitivity and specificity, as well as in their performance in small and large samples sizes [85, 86]. The objective of a study, e.g., sub-classification of tumor types or the identification of prognostic or therapeutic targets, should decide which approach should be used, together with the end-point chosen, e.g., altered genes, gene-sets, pathways, or genomic regions [87]. However, important cancer genes or miRNAs may be overlooked by such integrative approaches that require variation across samples and which focus on simultaneous changes in both, e.g., copy number and expression [85]. For example, despite an observed increase in the expression of an oncogene, the in-*cis* correlation may be low if the increased expression is caused by a mixture of amplification, mutation, or hypomethylation across the patients.

3.5 Integration of Multi-dimensional Data

The development of high-throughput technologies has made it possible to simultaneously profile the genome, epigenome, transcriptome, and proteome of biological samples such as breast tumor tissue taken from biopsies. These so-called multi-dimensional data represents several molecular levels that together can be used to characterize biological systems [88]. Uncovering the relations between the biological components of these systems—DNA copy number, methylation state, genes, mRNAs, miRNAs, and proteins—allows approaching breast cancer at a system level. Integration of multi-dimensional data from various molecular levels is required to reveal the underlying system in greater detail. Bioinformatic approaches address these challenges by representing the system as biological networks and pathways [89]. The ultimate goal of taking such a systems biology approach is to go from cancer genomes with all their aberrations to cancer models where these aberrations may be put in a system in order to identify common denominators, and ultimately provide mechanistic insight into the development and progression of cancer [90].

Due to the inherent complexity of cancer biology, a further rationale for an integrative approach is that by combining data from different levels and across patients, one may find cancer-relevant events that might not have been found if only single layers were assessed. For example, if expression of a gene is increased due to gain, activating mutations, promoter hypomethylation, or altered miRNA expression across patients, this would indicate that the gene is a candidate oncogene, even though each alteration itself may be infrequent [91]. Approaching breast cancer at the systems biology level through top-down integration of multi-level biological data is facilitated by having an outline on how to combine the available data and tools before the analysis starts, and includes several steps (*see* Fig. 4). Studies in which several "omics" levels were integrated to examine the aberrations that occur in breast cancer were previously performed [92, 93]. In the study by Curtis et al. [93] a new integrative classification system of breast tumors was identified based on both genomic and transcriptomic data [43]. The authors performed an integrated analysis of DNA copy number and gene expression, and identified novel subgroups with distinct clinical outcomes, named iClusts 1–10. These subgroups include one high-risk ER-positive 11q13/14 *cis*-acting subgroup and a favorable prognosis subgroup devoid of copy number alterations (CNA). Another study performed by Dvinge et al. [92] performed a systems-level analysis of miRNA expression in breast tumors by analyzing miRNA expression and integrating it with matched mRNA expression and CNA [92]. The authors reveal that at the whole-genome level, miRNAs behave more as fine-tuners/modulators of gene expression. This modulatory role of miRNAs was especially evident in CNA-devoid breast tumors, in which the immune response is prominent.

Molecular data are typically generated on different scales and units and must be processed prior to integration. Then, individual relations and interactions must be identified in the data, and finally put into the context of a larger system where alterations at the global scale may be identified from the more local findings. This is typically achieved by assessing the alterations found in the framework of biological pathways or networks. To fully complement a systems biology approach, findings from integration of high-

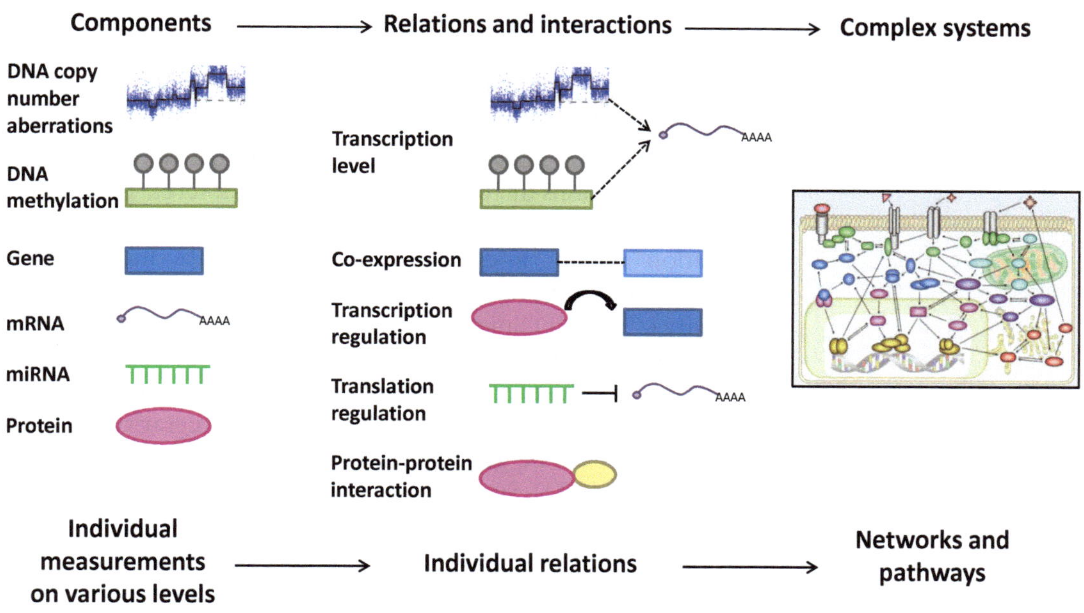

Fig. 4 Integration and analysis of multi-dimensional data. Biological components are measured across individuals and platforms, and their relations and interactions are identified. From this, complete networks and pathways are overlaid or built, and the emerging system is interrogated for alterations. Figure based on McDermott et al. [89]

throughput data must be combined with functional experiments to further evaluate and validate the findings [94].

Several large-scale projects have been launched that molecularly profile human tumors at multiple levels with the aim of integrating various data types to reveal molecular mechanisms of cancer. Some examples are The Cancer Genome Atlas (TCGA) (http://cancergenome.nih.gov/) [95], METABRIC (Molecular Taxonomy of Breast Cancer International Consortium) [93], and the International Cancer Genome Consortium (ICGC) (http://icgc.org/). These studies have provided a comprehensive picture of the great genetic diversity of breast cancers. They have moreover increased the resolution of classification suggesting the presence of additional molecular subgroups.

The most comprehensive molecular profiling of human breast tumors published to date has been done by TCGA [95]. By integrating DNA copy number, methylation data, somatic mutations, exome sequencing, mRNA arrays, miRNA sequencing, and reverse-phase protein arrays, the consortium identified four major groups of breast cancer types. To a large extent the groups recapitulate the molecular subtypes [95]. Using the expression of the 25% most variable miRNAs, the TCGA study identified seven miRNA subtypes by consensus non-negative matrix factorization clustering. These miRNA subtypes correlated with the mRNA subtypes, ER, PR and HER2 clinical status, but not with mutation status. The

TCGA study further confirmed that breast cancer is a heterogeneous disease; however, they suggested that most of the heterogeneity is found within, and not across the major subtypes.

Integrative studies aid in deciphering new subgroups and identify alterations seen across levels and patients that may ultimately lead to interruptions at the pathway or network level. In practice, the analyses and interpretation of multi-level high-throughput information remain a daunting task. Challenges include data handling, normalization and standardization, database annotation, availability of patient clinical information, and dissection of intrinsic tumor heterogeneity [90, 91]. To be able to exploit these large amounts of data, further development of computational tools and improved infrastructure is needed. From these integrative analyses, new hypotheses can be generated that require experimental testing and validation [91]. Succeeding in integrating multi-level data holds the promise of a comprehensive understanding of the alterations that are responsible for tumor initiation, maintenance, and progression. Such findings may be further translated into improved strategies for tumor sub-classification, early detection, more accurate prognostication, and a tailored therapy regime, in addition to revealing new targets for therapy.

4 miRNA Regulation in Breast Cancer

4.1 Methods to Study miRNA Regulation and Target Validation

miRNAs play an important role in the post-transcriptional regulation of gene expression. To date, the number of experimentally validated targets is low compared to the hundreds of putative targets predicted by the different in silico prediction algorithms [96]. The most common methods for the validation of miRNA targets include the transfection of reporter vector constructs or mimic miRNAs into cells, or the use of miRNA inhibitors. Those are followed by assessing the effects on mRNA (by, e.g., qRT-PCR, microarrays or sequencing) or protein levels (by, e.g., western blot) of the putative miRNA targets. The challenge entailed in these techniques lies in distinguishing direct from indirect effects [96]. Alternatively, direct methods for the validation of miRNA targets are based on the immunoprecipitation of the RISC complex together with the bound miRNA-mRNA complex. RNA isolated by crosslinking immunoprecipitation (HITS-CLIP) can then be analyzed by high-throughput sequencing [97]. Yet, also for Co-IP protocols, unspecific binding or co-isolation of secondary binders is common.

Most analyses of miRNA crosslinking to date have not included protein data. Indeed, the majority of studies modeling the regulatory impact of miRNAs have been performed on joint miRNA-mRNA expression data. While the physical interaction takes place between miRNA and mRNA, in order to validate a true miRNA-mRNA relation, an effect on the protein level is the ultimate proof as it gives the final phenotype of miRNA regulation. Depending on the mechanism of miRNA regulation one may anticipate different outcomes. For example, negative correlation between miRNA and mRNA may be expected if the miRNA regulation leads to degradation of the mRNA. However, if translational inhibition is the mechanism of action, such negative correlation between miRNA and mRNA may not be observed. miRNA, mRNA, and protein expression data have been integrated in order to study potentially direct and indirect effects of miRNA on protein expression [98]. In a study by Aure et al., protein expression was modeled as a function of miRNA and mRNA expression. The model considered both the effect of one miRNA at a time, and also all miRNAs combined. The resulting comprehensive "interactome" map of miRNAs in breast cancer revealed extensive coordination between miRNA and protein expression with groups of miRNAs coordinately interacting with groups of proteins, thus suggesting "block interactions" [98]. In order to suggest possible direct regulatory interactions between miRNAs and mRNAs, the use of intersected target prediction outputs aided in proposing candidates that should be further functionally assessed by biochemical experiments.

4.2 Dissecting the Functional Role of miRNAs in Breast Cancer

Altered miRNA expression in cancer has been extensively reported; however, there are still many unanswered questions regarding the role of miRNAs in cancer. miRNAs, which are differentially expressed between samples of different molecular subtypes, *TP53* mutation status, and ER status have been described in breast cancer [53]. The causes of miRNA deregulation in breast cancer have been investigated by trying to comprehensively study the effect of DNA methylation and copy number aberrations of miRNA loci and couple those to miRNA expression [84]. Identifying the various mechanisms underlying perturbation of miRNA levels will help us to understand more about the role of miRNAs in tumor development and also about miRNA biology in general.

Dissecting the functional role of miRNAs is a challenging task due to several aspects. miRNA families have likely arisen due to gene duplication events [99], and members of the same miRNA family have a high degree of similarity in sequence. In some cases, members of a miRNA family are also encoded in the same polycistronic transcript [100]. Sequence similarities suggest that they may target the same genes and thus have potentially overlapping functions. From an evolutionary perspective, the mRNA 3′-UTR where the miRNA targeting most often occurs, is not constrained by coding needs and thus has the potential to be subject to selection so that beneficial miRNA-mRNA target interactions may evolve [101]. Moreover, miRNAs originating from the same polycistronic transcript or encoded in close proximity have a high chance of being co-expressed. Hence, untangling the role of individual miRNAs is complicated.

Several target prediction algorithms have been published [80], as previously described. However, the false-positive rate of those predictions has generally been high [102], and the degree of overlap between the algorithms differs. Each miRNA is predicted to target hundreds of genes, including many with various functions. Thus, as there is a lack of high-throughput methods to validate miRNA-mRNA interactions, it can be a challenge to prioritize on which target genes to focus [96]. miRNA expression has also been shown to be both time- and context-dependent [103, 104] which can potentially reduce the transferability of validated relations from, e.g., one cell type to another. Recent studies also suggest that it is in the very nature of miRNAs to confer only subtle effects on a target's protein level (typically less than 50%) rather than conferring a total abolishment of protein expression [20, 105]. Furthermore, the effect of miRNA-mediated regulation on mRNA can generally result from two different mechanisms, either translational inhibition or mRNA degradation, which subsequently will confer different results when trying to model miRNA-mRNA interactions from high-throughput data. All these considerations should be kept in mind when studying the role of miRNAs in cancer.

As more miRNA properties emerge, the key to understanding the biological role of miRNAs may slowly be revealed as new hypotheses to be tested develop from the pieces added to the puzzle. miRNAs have many putative target genes, are expressed in a context-dependent manner, and function as cell fate switches or buffers. Integrating those properties with a view of cancer as a disease, which is driven by perturbations at the signaling network level [106], suggests that miRNAs may function as effective nodes in protein signaling networks [101, 107]. Signaling cascades, which transfer extracellular signals into cellular responses, depend on dynamic and transient action, and often involve complex feedback and feed-forward loops. Proteins are key elements in signaling networks, but miRNAs with their potential to regulate multiple targets simultaneously could play a very efficient and timely role in the tuning of signaling networks. Using mass spectrometry to investigate the effect of miRNA regulation on proteins indicated that miRNAs, for most interactions, act to make fine-scale adjustments to protein expression levels [20]. However, with a fine-tuning role in a normal cellular state, aberrant miRNA expression may represent acquired signaling capabilities [106]. Aberrant miRNA expression can have a substantial effect on pathway outcome by disturbing the tight regulation, thus contributing to a malignant phenotype of the cell. Studying both direct and indirect effects of miRNAs may unveil important core miRNAs that can further be used as markers of disease. Finally, the use of miRNA expression together with protein expression might give a robust proxy of a disease state as they together may constitute a "state-of-the-network" signature.

4.3 miRNAs as Clinical Biomarkers for Diagnostic and Predictive Purposes in Breast Cancer

The field of miRNA biology is rather new, considering that the first miRNA was discovered in 1993, and there are still new miRNAs being identified today. During the past 10 years, miRNA research has advanced rapidly, and has produced new knowledge about the molecular basis of cancer, tools for molecular classification, and new markers with diagnostic and prognostic relevance [62]. miRNAs are considered suitable biomarkers for early cancer detection because they are present and stable in human serum and plasma [108]. miRNA alterations during breast cancer progression from DCIS to invasive cancer have recently been identified within the intrinsic subtypes, Luminal A, luminal B, HER2-enriched, and basal-like [109, 110]. For immunohistochemical-based subtypes no miRNAs are differentially expressed between DCIS and the luminal subtypes. Six miRNAs were downregulated in ER−/HER2+ invasive samples compared to DCIS, of which five belong to the miR-30 family, whereas miR-139-5p was downregulated in both ER− subtypes, while miR-887-3p was downregulated in triple-negative breast cancer only [109]. This study found that subtype stratification based on molecular signatures resulted in more correct classification than stratification based on ER, PR, and HER2 alone, indicating a better representation of the intrinsic biology of the samples.

Although the focus has previously been on identifying molecular differences between cancerous and normal tissue, we often tend to forget that abnormal cell growth also occurs at benign stages. As discussed earlier in this chapter, previous studies have shown that certain types of benign tumors can increase the risk of breast cancer [22, 23, 25–28]. As the use of mammography has increased, the identification of benign breast disease has become more common. Thus, having accurate risk estimates for women who receive this diagnosis is vital. Moreover, with the distinction between benign tumors and malignant tumors, Tahiri et al. [111] identified that deregulation of known cancer-related miRNAs is evident also in fibroadenomas and fibroadenomatosis, considered as benign lesions in the breast. These cancer-related miRNAs included miR-21, members of the let-7 family and other miRNAs well known to be included in malignant transformation [111]. The level of deregulation in benign tumors was less pronounced than that observed in malignant tumors. Nevertheless, the identification of tumor-associated miRNAs in benign tumors hinted that similar processes are in place already at early stages of tumor formation. The identification of miRNAs that can be assigned to either benign or malignant groups of tumor tissue would be important for diagnostic purposes, but these results need to be further strengthened by independent confirmations.

When identifying a signature of miRNAs through expression arrays, there are different points to take into consideration. First, there is reported lack of consistency between different studies that certainly give rise to some concern of the use of miRNAs as biomarkers for the clinics [112]. Such differences might arise from sample selection or preparation, experimental design, and/or data analysis. Also, the technology used is important in the search for biomarkers. A recent study compared the expression of more than 2000 miRNAs by microarray technology and next-generation sequencing [113]. The authors observed highly significant dependency of the miRNA nucleotide composition on the expression level. Uracil-rich miRNAs showed higher expression levels when analyzed by next-generation sequencing. In contrast, guanine-rich miRNAs were detected at higher levels in microarrays. While identifying subsets of miRNAs that had high correlation with both technologies, correlation was observed only for miRNAs in the early miRBase versions (<8). Also, one of the major problems with both technologies was the elimination of low abundance miRNAs that may potentially have a great impact on overall processes. Keeping this in mind, respective bias will potentially slow down the translational process to clinical application.

Moreover, the use of different controls for data normalization can explain some of the observed variability across studies. Another possibility that must be considered is the dynamic and immediate regulation in miRNA levels in stress response and in hypoxia. As a result, time of sample collection and sample processing could further impact miRNA levels [62].

Despite those challenges, evidence reported up to date is encouraging. Even though a more comprehensive validation is still needed, the usefulness of miRNAs as biomarkers could especially be strengthened if it would be possible to identify deregulated levels of miRNAs in the circulation of patients not yet presented with the disease, or patients diagnosed with benign tumors.

4.4 Clinical Implications of miRNAs: Prospects for Therapy

As miRNAs have been reported to act as tumor-suppressor or oncogenic miRNAs, they have emerged as potential targets for therapy. miRNA expression signatures have been shown to function as classifiers for diagnosis, prognosis, and therapeutic response in cancer [56], and were associated with breast cancer subtypes and clinical subgroups. Notably, due to the tissue-specific expression of miRNAs, Rosetta Genomics has commercially launched an assay that is used to identify the tumor of origin in cancers of unknown primary origin. This assay measures the expression of 64 miRNAs which are further processed by an algorithm that can accurately identify the origin of a patient's tumor for 42 different cancer types [114]. The algorithm uses two classifiers, a binary decision tree in which the decision is made at each node by comparing the expression of a certain combination of miRNAs to a preset threshold, and a *k*-nearest-neighbor algorithm that uses a confidence measure by comparing the expression of the 64 miRNAs to the training samples.

Besides the fact that miRNAs are implicated in cancer, the above-discussed ability of miRNAs to regulate several genes in a pathway makes them interesting therapeutic agents as they may coordinate the response of an entire signaling network. Thus, by modifying miRNA activity it could be possible to restore homeostasis in cancer cells by rewiring network connections, and hence reverse a cancer phenotype [115]. Furthermore, as the absolute number of deregulated miRNAs is lower than for protein-coding genes, it might be easier to distinguish the drivers from the passengers and targeting miRNAs therapeutically may prove more successful than targeting single genes or proteins [115].

There are two main therapeutic strategies for modulation of miRNA expression. The first involves the restoration of tumor-suppressor miRNA activity, and the other is based on inhibiting the activity of oncogenic miRNAs [116]. An alternative indirect approach is to use drugs to modulate the miRNA expression by targeting steps in their biogenesis such as transcription or processing [115]. Restoration of tumor-suppressor miRNA expression can be achieved for example by introducing double-stranded miRNA mimics that are synthetic oligonucleotides with identical sequence as the selected tumor-suppressor miRNA [115]. There is, however, a long way from in vitro cell line experiments to clinical trials. The successful delivery of miRNAs to tumor cells is a major general challenge as unmodified, synthetic oligonucleotides are rapidly degraded by nucleases, and owing to their size and negative charge they may be prevented to cross the cell membrane [117]. Pharmacological blocking of oncogenic miRNAs has been achieved by using chemically modified antisense oligonucleotides [116]. These antisense strands function as competitive inhibitors of miRNAs by physically annealing to the mature miRNA and inhibiting its function. By introducing modifications to the chemical structure of the oligonucleotides such as for example locked nucleic acids (LNA), the stability, specificity, and binding affinity could be increased [115]. As an example of this antisense technology, miR-21 knockdown through LNA silencing was shown to inhibit proliferation and migration in human breast cancer cell lines and tumor growth in mice [118]. An alternative to antisense oligonucleotides are miRNA sponges that are transcribed from expression vectors and which contain multiple tandem-binding sites to a miRNA of interest [115]. They function as miRNA decoys by competing with the endogenous *bona fide* target mRNAs for miRNA binding, thus decreasing the miRNA effect.

Though their ability to regulate many genes simultaneously makes miRNA attractive as therapeutic candidates, this feature also implies that targeting miRNAs could lead to potential off-target effects. The high context dependence of miRNA action leads to potentially different functions in different tissues, which is a challenge in this regard. By designing effective systems that deliver the synthetic miRNA oligonucleotides specifically to the diseased tissue or cancer cells, these problems could be solved [115]. Another concern is the potential overloading of the endogenous miRNA processing machinery as the synthetic oligonucleotides could saturate the RISC complex and displace other endogenous miRNAs, which could potentially cause toxicity [115, 117]. Finally, due to sequence similarities between members of the same miRNA family, antisense miRNA therapy should be considered to target all family members in case of functional redundancy.

The success of miRNA-based therapy will depend on solving technical issues such as effective and specific delivery of miRNAs. Also, increased knowledge about miRNA function to identify potential therapeutic niches and to foresee the downstream effects is needed. As such, miRNA-based therapeutics could offer opportunities for a network therapy for cancer, focusing on the miRNAs rather than the protein-coding oncogenes which may be more difficult to target therapeutically [101]. miRNA signatures, e.g., from circulating miRNAs in breast cancer patients, are currently used in human clinical trials with the majority of the studies focusing on miRNA signatures as biomarkers for diagnosis, prognosis, or therapeutic response [56]. Overall, it will be exciting to follow the developments in novel efforts for therapeutically targeting miRNAs and its implications for cancer therapy and personalized medicine.

Acknowledgments

Parts of this review have been part of two doctoral theses from the University of Oslo, Norway, under the supervision of V.N.K.: one of M.R.A., fellow of the Research Council of Norway, and one of A. T., fellow of the South-Eastern Norway Regional Health Authority. Both are at present postdoctoral fellows of the South-Eastern Norway Regional Health Authority.

References

1. Crick F (1970) Central dogma of molecular biology. Nature 227(5258):561
2. Lee RC, Feinbaum RL, Ambros V (1993) The C. elegans heterochronic gene lin-4 encodes small RNAs with antisense complementarity to lin-14. Cell 75(5):843–854
3. Reinhart BJ, Slack FJ, Basson M, Pasquinelli AE, Bettinger JC, Rougvie AE, Horvitz HR, Ruvkun G (2000) The 21-nucleotide let-7 RNA regulates developmental timing in Caenorhabditis elegans. Nature 403 (6772):901–906
4. Kozomara A, Griffiths-Jones S (2011) miRBase: integrating microRNA annotation and deep-sequencing data. Nucleic Acids Res 39 (Database Issue)):D152–D157
5. Friedman RC, Farh KK-H, Burge CB, Bartel DP (2009) Most mammalian mRNAs are conserved targets of microRNAs. Genome Res 19(1):92–105
6. Lee Y, Ahn C, Han J, Choi H, Kim J, Yim J, Lee J, Provost P, Radmark O, Kim S et al (2003) The nuclear RNase III Drosha initiates microRNA processing. Nature 425 (6956):415–419
7. Ha M, Kim VN (2014) Regulation of microRNA biogenesis. Nat Rev Mol Cell Biol 15 (8):509–524

8. Kolb FA, Zhang H, Jaronczyk K, Tahbaz N, Hobman TC, Filipowicz W (2005) Human dicer: purification, properties, and interaction with PAZ PIWI domain proteins. Methods Enzymol 392:316–336
9. Forman JJ, Legesse-Miller A, Coller HA (2008) A search for conserved sequences in coding regions reveals that the let-7 micro-RNA targets Dicer within its coding sequence. Proc Natl Acad Sci U S A 105 (39):14879–14884
10. Lytle JR, Yario TA, Steitz JA (2007) Target mRNAs are repressed as efficiently by microRNA-binding sites in the 5′ UTR as in the 3′ UTR. Proc Natl Acad Sci U S A 104 (23):9667–9672
11. Dennis C (2002) The brave new world of RNA. Nature 418(6894):122–124
12. Sullivan RP, Leong JW, Fehniger TA (2013) MicroRNA regulation of natural killer cells. Front Immunol 4:44
13. Boyer LA, Lee TI, Cole MF, Johnstone SE, Levine SS, Zucker JR, Guenther MG, Kumar RM, Murray HL, Jenner RG et al (2005) Core transcriptional regulatory circuitry in human embryonic stem cells. Cell 122 (6):947–956
14. O'Donnell KA, Wentzel EA, Zeller KI, Dang CV, Mendell JT (2005) c-Myc-regulated microRNAs modulate E2F1 expression. Nature 435(7043):839–843
15. Marson A, Levine SS, Cole MF, Frampton GM, Brambrink T, Johnstone S, Guenther MG, Johnston WK, Wernig M, Newman J et al (2008) Connecting microRNA genes to the core transcriptional regulatory circuitry of embryonic stem cells. Cell 134(3):521–533
16. Mattick JS, Makunin IV (2006) Non-coding RNA. Hum Mol Genet 15:R17–R29
17. Czech B, Hannon GJ (2011) Small RNA sorting: matchmaking for Argonautes. Nat Rev Genet 12(1):19–31
18. Martin G, Schouest K, Kovvuru P, Spillane C (2007) Prediction and validation of micro-RNA targets in animal genomes. J Biosci 32 (6):1049–1052
19. Thomas M, Lieberman J, Lal A (2010) Desperately seeking microRNA targets. Nat Struct Mol Biol 17(10):1169–1174
20. Baek D, Villen J, Shin C, Camargo FD, Gygi SP, Bartel DP (2008) The impact of micro-RNAs on protein output. Nature 455 (7209):64–71
21. Jemal A, Bray F, Center MM, Ferlay J, Ward E, Forman D (2011) Global cancer statistics. CA Cancer J Clin 61(2):69–90
22. Dupont WD, Page DL (1985) Risk factors for breast cancer in women with proliferative breast disease. N Engl J Med 312(3):146–151
23. Dupont WD, Page DL, Parl FF, Vnencak-Jones CL, Plummer WD Jr, , Rados MS, Schuyler PA: Long-term risk of breast cancer in women with fibroadenoma. N Engl J Med 1994, 331(1):10–15
24. McPherson K, Steel CM, Dixon JM (2000) ABC of breast diseases. Breast cancer-epidemiology, risk factors, and genetics. BMJ 321(7261):624–628
25. Worsham MJ, Raju U, Lu M, Kapke A, Botttrell A, Cheng J, Shah V, Savera A, Wolman SR (2009) Risk factors for breast cancer from benign breast disease in a diverse population. Breast Cancer Res Treat 118(1):1–7
26. Fitzgibbons PL, Henson DE, Hutter RV (1998) Benign breast changes and the risk for subsequent breast cancer: an update of the 1985 consensus statement. Cancer Committee of the College of American Pathologists. Arch Pathol Lab Med 122 (12):1053–1055
27. McDivitt RW, Stevens JA, Lee NC, Wingo PA, Rubin GL, Gersell D (1992) Histologic types of benign breast disease and the risk for breast cancer. The Cancer and Steroid Hormone Study Group. Cancer 69 (6):1408–1414
28. Cole P, Mark Elwood J, Kaplan SD (1978) Incidence rates and risk factors of benign breast neoplasms. Am J Epidemiol 108 (2):112–120
29. Sgroi DC (2010) Preinvasive breast cancer. Annu Rev Pathol 5:193–221
30. Johnson K, Sarma D, Hwang ES (2015) Lobular breast cancer series: imaging. Breast Cancer Res 17:94
31. Perou CM, Sorlie T, Eisen MB, van de Rijn M, Jeffrey SS, Rees CA, Pollack JR, Ross DT, Johnsen H, Akslen LA et al (2000) Molecular portraits of human breast tumours. Nature 406(6797):747–752
32. Sorlie T, Perou CM, Tibshirani R, Aas T, Geisler S, Johnsen H, Hastie T, Eisen MB, van de Rijn M, Jeffrey SS et al (2001) Gene expression patterns of breast carcinomas distinguish tumor subclasses with clinical implications. Proc Natl Acad Sci U S A 98 (19):10869–10874
33. Gown AM (2008) Current issues in ER and HER2 testing by IHC in breast cancer. Mod Pathol 21(Suppl 2):S8–S15
34. de Azambuja E, Cardoso F, de Castro G, Colozza M, Mano MS, Durbecq V, Sotiriou C, Larsimont D, Piccart-Gebhart

MJ, Paesmans M (2007) Ki-67 as prognostic marker in early breast cancer: a meta-analysis of published studies involving 12 155 patients. Br J Cancer 96(10):1504–1513

35. van't Veer LJ, Dai H, van de Vijver MJ, He YD, Hart AA, Mao M, Peterse HL, van der Kooy K, Marton MJ, Witteveen AT et al (2002) Gene expression profiling predicts clinical outcome of breast cancer. Nature 415(6871):530–536
36. Enerly E, Steinfeld I, Kleivi K, Aure MR, Leivonen SK, Johnsen H, Kallioniemi O, Kristensen VN, Yakhini Z, Borresen-Dale AL (2010) Molecular characterization of breast cancer subtypes derived from joint analysis of high throughput miRNA and mRNA data. EJC Suppl 8(5):164
37. Sotiriou C, Neo SY, McShane LM, Korn EL, Long PM, Jazaeri A, Martiat P, Fox SB, Harris AL, Liu ET (2003) Breast cancer classification and prognosis based on gene expression profiles from a population-based study. Proc Natl Acad Sci U S A 100(18):10393–10398
38. Inic Z, Zegarac M, Inic M, Markovic I, Kozomara Z, Djurisic I, Inic I, Pupic G, Jancic S (2014) Difference between luminal A and luminal B subtypes according to Ki-67, tumor size, and progesterone receptor negativity providing prognostic information. Clin Med Insights Oncol 8:107–111
39. Subik K, Lee JF, Baxter L, Strzepek T, Costello D, Crowley P, Xing L, Hung MC, Bonfiglio T, Hicks DG et al (2010) The expression patterns of ER, PR, HER2, CK5/6, EGFR, Ki-67 and AR by immunohistochemical analysis in breast cancer cell lines. Breast Cancer (Auckl) 4:35–41
40. Prat A, Adamo B, Cheang MC, Anders CK, Carey LA, Perou CM (2013) Molecular characterization of basal-like and non-basal-like triple-negative breast cancer. Oncologist 18 (2):123–133
41. Atkinson AJ, Colburn WA, DeGruttola VG, DeMets DL, Downing GJ, Hoth DF, Oates JA, Peck CC, Schooley RT, Spilker BA et al (2001) Biomarkers and surrogate endpoints: preferred definitions and conceptual framework. Clin Pharmacol Therap 69(3):89–95
42. Sobin LH (2003) TNM: evolution and relation to other prognostic factors. Semin Surg Oncol 21(1):3–7
43. Karve TM, Cheema AK (2011) Small changes huge impact: the role of protein posttranslational modifications in cellular homeostasis and disease. J Amino Acids 2011:207691
44. Sharma S, Kelly TK, Jones PA (2010) Epigenetics in cancer. Carcinogenesis 31(1):27–36
45. Wu W, Zhao S (2013) Metabolic changes in cancer: beyond the Warburg effect. Acta Biochim Biophys Sin Shanghai 45(1):18–26
46. Calin GA, Dumitru CD, Shimizu M, Bichi R, Zupo S, Noch E, Aldler H, Rattan S, Keating M, Rai K (2002) Frequent deletions and down-regulation of micro-RNA genes miR15 and miR16 at 13q14 in chronic lymphocytic leukemia. Proc Natl Acad Sci U S A 99(24):15524–15529
47. Calin GA, Sevignani C, Dan Dumitru C, Hyslop T, Noch E, Yendamuri S, Shimizu M, Rattan S, Bullrich F, Negrini M et al (2004) Human microRNA genes are frequently located at fragile sites and genomic regions involved in cancers. Proc Natl Acad Sci U S A 101(9):2999–3004
48. Iorio MV, Ferracin M, Liu C-G, Veronese A, Spizzo R, Sabbioni S, Magri E, Pedriali M, Fabbri M, Campiglio M et al (2005) MicroRNA gene expression deregulation in human breast cancer. Cancer Res 65(16):7065–7070
49. Tavazoie SF, Alarcon C, Oskarsson T, Padua D, Wang Q, Bos PD, Gerald WL, Massague J (2008) Endogenous human microRNAs that suppress breast cancer metastasis. Nature 451(7175):147–152
50. Yan L-X, Huang X-F, Shao Q, Huang MAY, Deng L, Wu Q-L, Zeng Y-X, Shao J-Y (2008) MicroRNA miR-21 overexpression in human breast cancer is associated with advanced clinical stage, lymph node metastasis and patient poor prognosis. RNA 14(11):2348–2360
51. Castellano L, Giamas G, Jacob J, Coombes RC, Lucchesi W, Thiruchelvam P, Barton G, Jiao LR, Wait R, Waxman J et al (2009) The estrogen receptor-a-induced microRNA signature regulates itself and its transcriptional response. Proc Natl Acad Sci U S A 106 (37):15732–15737
52. Cittelly D, Das P, Spoelstra N, Edgerton S, Richer J, Thor A, Jones F (2010) Downregulation of miR-342 is associated with tamoxifen resistant breast tumors. Mol Cancer 9 (1):317
53. Enerly E, Steinfeld I, Kleivi K, Leivonen S-K, Aure MR, Russnes HG, Rønneberg JA, Johnsen H, Navon R, Rødland E et al (2011) miRNA-mRNA integrated analysis reveals roles for miRNAs in primary breast tumors. PLoS One 6(2):e16915
54. Deng S, Calin GA, Croce CM, Coukos G, Zhang L (2008) Mechanisms of microRNA deregulation in human cancer. Cell Cycle 7 (17):2643–2646
55. Bertoli G, Cava C, Castiglioni I (2015) MicroRNAs: new biomarkers for diagnosis,

prognosis, therapy prediction and therapeutic tools for breast cancer. Theranostics 5 (10):1122–1143

56. Nana-Sinkam SP, Croce CM (2013) Clinical applications for microRNAs in cancer. Clin Pharmacol Ther 93(1):98–104
57. Ouyang M, Li Y, Ye S, Ma J, Lu L, Lv W, Chang G, Li X, Li Q, Wang S et al (2014) MicroRNA profiling implies new markers of chemoresistance of triple-negative breast cancer. PLoS One 9(5):e96228
58. Dong Y, Wu WK, Wu CW, Sung JJ, Yu J, Ng SS (2011) MicroRNA dysregulation in colorectal cancer: a clinical perspective. Br J Cancer 104(6):893–898
59. Tumilson CA, Lea RW, Alder JE, Shaw L (2014) Circulating microRNA biomarkers for glioma and predicting response to therapy. Mol Neurobiol 50(2):545–558
60. Mazan-Mamczarz K, Gartenhaus RB (2013) Role of microRNA deregulation in the pathogenesis of diffuse large B-cell lymphoma (DLBCL). Leuk Res 37(11):1420–1428
61. Maugeri-Sacca M, Coppola V, Bonci D, De Maria R (2012) MicroRNAs and prostate cancer: from preclinical research to translational oncology. Cancer J 18(3):253–261
62. Iorio MV, Croce CM (2012) MicroRNA dysregulation in cancer: diagnostics, monitoring and therapeutics. A comprehensive review. EMBO Mol Med 4(3):143–159
63. Boeri M, Verri C, Conte D, Roz L, Modena P, Facchinetti F, Calabro E, Croce CM, Pastorino U, Sozzi G (2011) MicroRNA signatures in tissues and plasma predict development and prognosis of computed tomography detected lung cancer. Proc Natl Acad Sci U S A 108(9):3713–3718
64. Cava C, Bertoli G, Ripamonti M, Mauri G, Zoppis I, Della Rosa PA, Gilardi MC, Castiglioni I (2014) Integration of mRNA expression profile, copy number alterations, and microRNA expression levels in breast cancer to improve grade definition. PLoS One 9(5): e97681
65. Weiler J, Hunziker J, Hall J (2006) Anti-miRNA oligonucleotides (AMOs): ammunition to target miRNAs implicated in human disease? Gene Ther 13(6):496–502
66. Leivonen SK, Sahlberg KK, Makela R, Due EU, Kallioniemi O, Borresen-Dale AL, Perala M (2014) High-throughput screens identify microRNAs essential for HER2 positive breast cancer cell growth. Mol Oncol 8 (1):93–104
67. Kepp O, Galluzzi L, Lipinski M, Yuan J, Kroemer G (2011) Cell death assays for drug discovery. Nat Rev Drug Discov 10(3):221–237
68. Riss TL, Moravec RA, Niles AL, Duellman S, Benink HA, Worzella TJ, Minor L (2004) Cell viability assays. In: Sittampalam GS, Coussens NP, Brimacombe K, Grossman A, Arkin M, Auld D, Austin C, Bejcek B, Glicksman M, Inglese J et al (eds) Assay guidance manual. Eli Lilly & Company, Bethesda, MD
69. Zhu Q, Wong AK, Krishnan A, Aure MR, Tadych A, Zhang R, Corney DC, Greene CS, Bongo LA, Kristensen VN et al (2015) Targeted exploration and analysis of large cross-platform human transcriptomic compendia. Nat Methods 12(3):211–214
70. Enright A, John B, Gaul U, Tuschl T, Sander C, Marks D (2003) MicroRNA targets in Drosophila. Genome Biol 5(1):R1
71. John B, Enright AJ, Aravin A, Tuschl T, Sander C, Marks DS (2004) Human microRNA targets. PLoS Biol 2(11):e363
72. Betel D, Wilson M, Gabow A, Marks DS, Sander C (2008) The microRNA.org resource: targets and expression. Nucleic Acids Res 36(Database Issue):D149–D153
73. Lewis BP, Burge CB, Bartel DP (2005) Conserved seed pairing, often flanked by adenosines, indicates that thousands of human genes are microRNA targets. Cell 120 (1):15–20
74. Lewis BP, Shih IH, Jones-Rhoades MW, Bartel DP, Burge CB (2003) Prediction of mammalian microRNA targets. Cell 115 (7):787–798
75. Krek A, Grun D, Poy MN, Wolf R, Rosenberg L, Epstein EJ, MacMenamin P, da Piedade I, Gunsalus KC, Stoffel M et al (2005) Combinatorial microRNA target predictions. Nat Genet 37(5):495–500
76. Grun D, Wang YL, Langenberger D, Gunsalus KC, Rajewsky N (2005) microRNA target predictions across seven Drosophila species and comparison to mammalian targets. PLoS Comput Biol 1(1):e13
77. Lall S, Grun D, Krek A, Chen K, Wang YL, Dewey CN, Sood P, Colombo T, Bray N, Macmenamin P et al (2006) A genome-wide map of conserved microRNA targets in C. elegans. Curr Biol 16(5):460–471
78. Maragkakis M, Alexiou P, Papadopoulos GL, Reczko M, Dalamagas T, Giannopoulos G, Goumas G, Koukis E, Kourtis K, Simossis VA et al (2009) Accurate microRNA target

prediction correlates with protein repression levels. BMC Bioinformatics 10:295

79. Paraskevopoulou MD, Georgakilas G, Kostoulas N, Vlachos IS, Vergoulis T, Reczko M, Filippidis C, Dalamagas T, Hatzigeorgiou AG (2013) DIANA-microT web server v5.0: service integration into miRNA functional analysis workflows. Nucleic Acids Res 41(W1):W169–W173
80. Ekimler S, Sahin K (2014) Computational methods for microRNA target prediction. Genes (Basel) 5(3):671–683
81. Hanahan D, Weinberg RA (2011) Hallmarks of cancer: the next generation. Cell 144 (5):646–674
82. Hyman E, Kauraniemi P, Hautaniemi S, Wolf M, Mousses S, Rozenblum E, Ringnér M, Sauter G, Monni O, Elkahloun A et al (2002) Impact of DNA amplification on gene expression patterns in breast cancer. Cancer Res 62(21):6240–6245
83. Bergamaschi A, Kim YH, Wang P, Sørlie T, Hernandez-Boussard T, Lonning PE, Tibshirani R, Børresen-Dale A-L, Pollack JR (2006) Distinct patterns of DNA copy number alteration are associated with different clinicopathological features and gene-expression subtypes of breast cancer. Genes Chromosom Cancer 45(11):1033–1040
84. Aure MR, Leivonen SK, Fleischer T, Zhu Q, Overgaard J, Alsner J, Tramm T, Louhimo R, Alnæs GI, Perälä M, Busato F, Touleimat N, Tost J, Børresen-Dale AL, Hautaniemi S, Troyanskaya OG, Lingjærde OC, Sahlberg KK, Kristensen VN (2013) Individual and combined effects of DNA methylation and copy number alterations on miRNA expression in breast tumors. Genome Biol 14(11): R126
85. Lahti L, Schäfer M, Klein H-U, Bicciato S, Dugas M (2012) Cancer gene prioritization by integrative analysis of mRNA expression and DNA copy number data: a comparative review. Brief Bioinform 14(1):27–35
86. Louhimo R, Lepikhova T, Monni O, Hautaniemi S (2012) Comparative analysis of algorithms for integration of copy number and expression data. Nat Methods 9(4):351–355
87. Huang N, Shah PK, Li C (2011) Lessons from a decade of integrating cancer copy number alterations with gene expression profiles. Brief Bioinform 13(3):305–316
88. Zhang S, Liu C-C, Li W, Shen H, Laird PW, Zhou XJ (2012) Discovery of multi-dimensional modules by integrative analysis of cancer genomic data. Nucleic Acids Res 40(19):9379–9391
89. McDermott JE, Costa M, Janszen D, Singhal M, Tilton SC (2010) Separating the drivers from the driven: integrative network and pathway approaches aid identification of disease biomarkers from high-throughput data. Dis Markers 28(4):253–266
90. Baudot A, Real FX, Izarzugaza JMG, Valencia A (2009) From cancer genomes to cancer models: bridging the gaps. EMBO Rep 10 (4):359–366
91. Chin L, Hahn WC, Getz G, Meyerson M (2011) Making sense of cancer genomic data. Genes Dev 25(6):534–555
92. Dvinge H, Git A, Graf S, Salmon-Divon M, Curtis C, Sottoriva A, Zhao Y, Hirst M, Armisen J, Miska EA et al (2013) The shaping and functional consequences of the microRNA landscape in breast cancer. Nature 497 (7449):378–382
93. Curtis C, Shah SP, Chin SF, Turashvili G, Rueda OM, Dunning MJ, Speed D, Lynch AG, Samarajiwa S, Yuan Y et al (2012) The genomic and transcriptomic architecture of 2,000 breast tumours reveals novel subgroups. Nature 486(7403):346–352
94. Hernández Patiño CE, Jaime-Muñoz G, Resendis-Antonio O (2013) Systems biology of cancer: moving toward the integrative study of the metabolic alterations in cancer cells. Front Physiol 3:481
95. The Cancer Genome Atlas Network (2012) Comprehensive molecular portraits of human breast tumours. Nature 490(7418):61–70
96. Muniategui A, Pey J, Planes FJ, Rubio A (2012) Joint analysis of miRNA and mRNA expression data. Brief Bioinform 14 (3):263–278
97. Chi SW, Zang JB, Mele A, Darnell RB (2009) Argonaute HITS-CLIP decodes microRNA-mRNA interaction maps. Nature 460 (7254):479–486
98. Aure MR, Jernstrom S, Krohn M, Vollan H, Due E, Rodland E, Karesen R, Ram P, Lu Y, Mills G et al (2015) Integrated analysis reveals microRNA networks coordinately expressed with key proteins in breast cancer. Genome Med 7(1):21
99. Hertel J, Lindemeyer M, Missal K, Fried C, Tanzer A, Flamm C, Hofacker I, Stadler P, Students of Bioinformatics Computer Labs 2004 and 2005 (2006) The expansion of the metazoan microRNA repertoire. BMC Genomics 7:25
100. Griffiths-Jones S, Saini HK, van Dongen S, Enright AJ (2008) miRBase: tools for microRNA genomics. Nucleic Acids Res 36(Database issue):D154–D158

101. Inui M, Martello G, Piccolo S (2010) MicroRNA control of signal transduction. Nat Rev Mol Cell Biol 11(4):252–263
102. Bentwich I (2005) Prediction and validation of microRNAs and their targets. FEBS Lett 579(26):5904–5910
103. Patnaik SK, Dahlgaard J, Mazin W, Kannisto E, Jensen T, Knudsen S, Yendamuri S (2012) Expression of microRNAs in the NCI-60 cancer cell-lines. PLoS One 7(11): e49918
104. Lu J, Getz G, Miska EA, Alvarez-Saavedra E, Lamb J, Peck D, Sweet-Cordero A, Ebert BL, Mak RH, Ferrando AA et al (2005) MicroRNA expression profiles classify human cancers. Nature 435(7043):834–838
105. Selbach M, Schwanhausser B, Thierfelder N, Fang Z, Khanin R, Rajewsky N (2008) Widespread changes in protein synthesis induced by microRNAs. Nature 455(7209):58–63
106. Creixell P, Schoof EM, Erler JT, Linding R (2012) Navigating cancer network attractors for tumor-specific therapy. Nat Biotechnol 30 (9):842–848
107. Avraham R, Yarden Y (2012) Regulation of signalling by microRNAs. Biochem Soc Trans 40(1):26–30
108. Mitchell PS, Parkin RK, Kroh EM, Fritz BR, Wyman SK, Pogosova-Agadjanyan EL, Peterson A, Noteboom J, O'Briant KC, Allen A et al (2008) Circulating microRNAs as stable blood-based markers for cancer detection. Proc Natl Acad Sci U S A 105 (30):10513–10518
109. Haakensen VD, Nygaard V, Greger L, Aure MR, Fromm B, Bukholm IR, Luders T, Chin SF, Git A, Caldas C et al (2016) Subtype-specific micro-RNA expression signatures in breast cancer progression. Int J Cancer 139 (5):1117–1128
110. Lesurf R, Aure MR, Mork HH, Vitelli V, Oslo Breast Cancer Research Consortium, Lundgren S, Borresen-Dale AL, Kristensen V, Warnberg F, Hallett M et al (2016) Molecular features of subtype-specific progression from ductal carcinoma in situ to invasive breast cancer. Cell Rep 16 (4):1166–1179
111. Tahiri A, Leivonen SK, Luders T, Steinfeld I, Aure MR, Geisler J, Makela R, Nord S, Riis MLH, Yakhini Z et al (2014) Deregulation of cancer-related miRNAs is a common event in both benign and malignant human breast tumors. Carcinogenesis 35(1):76–85
112. Callari M, Dugo M, Musella V, Marchesi E, Chiorino G, Grand MM, Pierotti MA, Daidone MG, Canevari S, De Cecco L (2012) Comparison of microarray platforms for measuring differential microRNA expression in paired normal/cancer colon tissues. PLoS One 7(9):e45105
113. Backes C, Sedaghat-Hamedani F, Frese K, Hart M, Ludwig N, Meder B, Meese E, Keller A (2016) Bias in high-throughput analysis of miRNAs and implications for biomarker studies. Anal Chem 88(4):2088–2095
114. Meiri E, Mueller WC, Rosenwald S, Zepeniuk M, Klinke E, Edmonston TB, Werner M, Lass U, Barshack I, Feinmesser M et al (2012) A second-generation microRNA-based assay for diagnosing tumor tissue origin. Oncologist 17(6):801–812
115. Garzon R, Marcucci G, Croce CM (2010) Targeting microRNAs in cancer: rationale, strategies and challenges. Nat Rev Drug Discov 9(10):775–789
116. Thorsen SB, Obad S, Jensen NF, Stenvang J, Kauppinen S (2012) The therapeutic potential of microRNAs in cancer. Cancer J 18 (3):275–284
117. Aagaard L, Rossi JJ (2007) RNAi therapeutics: principles, prospects and challenges. Adv Drug Deliv Rev 59(2–3):75–86
118. Yan LX, Wu QN, Zhang Y, Li YY, Liao DZ, Hou JH, Fu J, Zeng MS, Yun JP, Wu QL et al (2011) Knockdown of miR-21 in human breast cancer cell lines inhibits proliferation, in vitro migration and in vivo tumor growth. Breast Cancer Res 13(1):R2
119. Lujambio A, Lowe SW (2012) The microcosmos of cancer. Nature 482(7385):347–355

Chapter 5

Identifying Genetic Dependencies in Cancer by Analyzing siRNA Screens in Tumor Cell Line Panels

James Campbell, Colm J. Ryan, and Christopher J. Lord

Abstract

Loss-of-function screening using RNA interference or CRISPR approaches can be used to identify genes that specific tumor cell lines depend upon for survival. By integrating the results from screens in multiple cell lines with molecular profiling data, it is possible to associate the dependence upon specific genes with particular molecular features (e.g., the mutation of a cancer driver gene, or transcriptional or proteomic signature). Here, using a panel of kinome-wide siRNA screens in osteosarcoma cell lines as an example, we describe a computational protocol for analyzing loss-of-function screens to identify genetic dependencies associated with particular molecular features. We describe the steps required to process the siRNA screen data, integrate the results with genotypic information to identify genetic dependencies, and finally the integration of protein-protein interaction data to interpret these dependencies.

Key words Cancer, siRNA screening, Synthetic lethality

1 Introduction

Recent large-scale sequencing projects and decades of small-scale studies have led to the identification of hundreds of "driver" genes in cancer—genes whose alteration through genetic or epigenetic means provides a growth or survival advantage for tumor cells [1, 2]. A key remaining challenge is to understand how these driver mutations alter cellular states to promote tumor progression and how this altered state may be exploited for the development of targeted therapeutics [3]. Identifying the set of genes that are required for growth in a given tumor cell line provides both an insight into the cellular state and suggests genes whose products may be targeted therapeutically. Toward this end, a number of laboratories have used loss-of-function screening to generate resources describing the genetic requirements of panels of tumor cell lines [4–11]. The majority of these resources use either

---

James Campbell and Colm J. Ryan contributed equally to this work.

Louise von Stechow (ed.), *Cancer Systems Biology: Methods and Protocols*, Methods in Molecular Biology, vol. 1711, https://doi.org/10.1007/978-1-4939-7493-1_5, © The Author (s) 2018 genome-scale shRNA screens carried out in a pooled format [6, 7, 10] or siRNA screens carried out in an arrayed format [4, 5, 11] to identify genetic dependencies. In the near future CRISPR-based approaches will likely be used for similar purposes, although to date the number of cell lines profiled by genome-wide CRISPR libraries remains small (e.g., five cell lines in [8]). Regardless of the experimental methodology used, the goal of loss-of-function screens is largely the same—the identification of genes required for growth in specific cancer cell lines. By integrating the results of these screens with genotypic data, it is possible to identify genes that appear specifically required for growth in the presence of a particular driver gene mutation. In some cases the driver gene mutation results in an increased dependency upon the gene itself, a phenomenon known as "oncogene addiction" [12]. Examples of this include an increased sensitivity of *ERBB2*-amplified breast cancer cell lines to siRNA reagents targeting *ERBB2* [4], and an increased sensitivity of *KRAS* mutant cell lines to shRNA reagents targeting *KRAS* [7]. More frequent are instances where the driver gene and the resulting dependency gene are different, often termed non-oncogene addictions or synthetic lethalities [12, 13]. Examples of non-oncogene addictions identified from loss-of-function screens include a dependence of *ARID1A* mutant cell lines upon the *ARID1A* paralog *ARID1B* [14], an increased sensitivity of *PTEN* mutant breast cancer cell lines to inhibition of the mitotic kinase *TTK* [4], and an increased sensitivity of *MYC* amplified breast cancer cell lines to inhibition of multiple spliceosome component coding genes [15]. Ultimately both oncogene addictions and synthetic lethalities identified in these screens may be exploited for the development of novel targeted therapeutics in cancer [13].

When these screens are analyzed, statistical approaches are used to identify significant associations between the mutation of a driver gene and an increased sensitivity to the inhibition of another gene. The interpretation of the resulting associations remains challenging—the statistical tests provide information on which genes are required in the presence of specific driver genes, but not the mechanistic explanation as to why these dependencies exist. Inspired by approaches initially developed for the interpretation of genetic interactions in yeast [16], we have recently used the integration of functional interaction networks to aid the interpretation of dependencies identified in loss-of-function screens in cancer cell lines [5]. For instance in *ERBB2*-amplified cell lines we see an increased dependency upon *ERBB2* itself and also the *ERBB2* protein-interaction partners *ERBB3* and *PIK3CA* [5]. This suggests that *ERBB2* amplified cell lines are frequently "addicted" to the functionality of *ERBB2*, the binding of *ERBB2* to its interaction partner *ERBB3*, and the function of the downstream effector *PIK3CA*.

Here, we describe a protocol for the analysis of loss-of-function screens in a panel of cancer cell lines. We use as example data a recent kinome-wide siRNA screen performed in a panel of osteosarcoma cell lines [5]. Our analysis protocol involves three main steps:

1. The conversion of siRNA screening results into gene-sensitivity scores.
2. The integration of these sensitivity scores with genotypic data to identify statistical associations between driver genes and sensitivity to the inhibition of particular genes.
3. The integration of additional data such as protein-protein interactions to interpret these associations.

Only the first step is specific to arrayed siRNA screens—we have successfully applied the latter analysis scripts to data resulting from additional screen types (e.g., pooled shRNA screens) (Fig. 1).

2 Materials

2.1 Software (See Notes 1 and 2)

1. R (available from https://www.r-project.org/).
2. R-packages:
   (a) *Gplots* (*see* **Note 3**).
   (b) cellHTS2 (*see* **Note 4**).
3. Python programming language (available from https://www.python.org/).
4. git repository containing the statistical analyses, code, and data resources discussed in the text (*see* **Note 5**) https://github.com/GeneFunctionTeam/identifying-genetic-dependencies

2.2 Input Files

1. Plate files (txt) contain the output from a loss of function screen. These each comprise three tab-separated columns of data containing the plate number (numeric), well position (e.g., B07), and the response value for the cell (e.g., luminosity readout). See the CellHTS2 documentation for further information.
2. Plate file list. This file contains three tab-separated columns with a header row listing "Filename," "Plate," and "Replicate." Filenames correspond to each plate file. The plate column defines which plate in the plate configuration file the data correspond to. The replicates column defines, which replicate a plate represents. See the CellHTS2 documentation for further information.
3. Plate configuration file. The first line defines the number of wells in each plate (e.g., "Wells: 384"). The second line defines the number of plates in the library (e.g., "Plates: 3"). The third line is a header associated with the subsequent columns (e.g.,

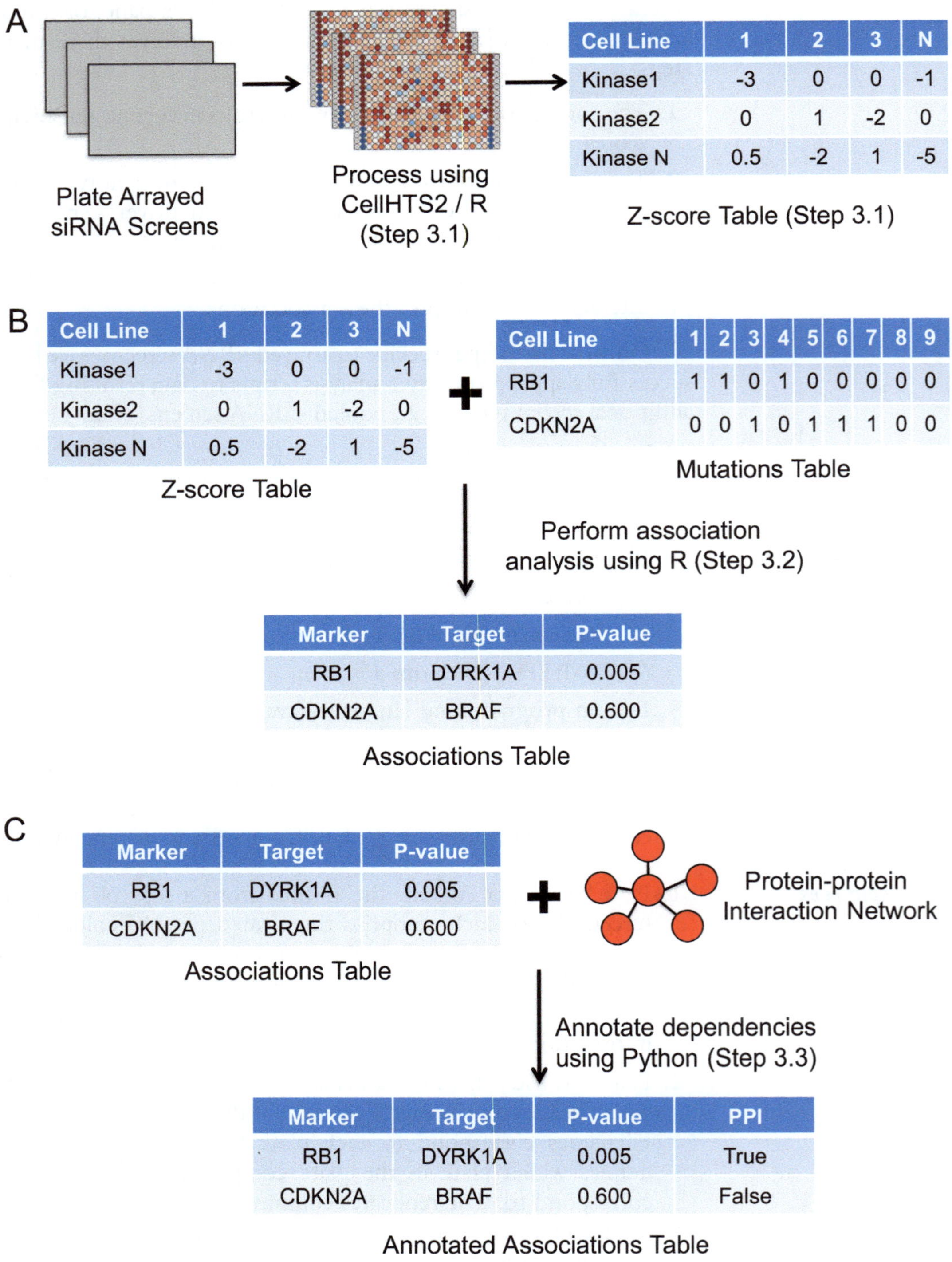

Fig. 1 Analyzing siRNA screens in Tumor Cell Line Panels. (**a**) Luminescence values derived from pooled siRNA screens are converted into *Z*-scores using CellHTS2 and custom R scripts. (**b**) *Z*-score profiles for each cell line are integrated with mutational profiles for the same set of cell lines using R. Custom R scripts are used to identify associations between the presence of particular mutations (e.g., in the RB1 gene) with increased

"Plate," "Well," "content"). The remaining lines define the wells containing samples and controls. An asterisk character (*) can be used to mean "all plates or wells." E.g., "* * sample" indicates that the all plates and all wells are "sample" unless otherwise stated. Subsequent more specific lines update the contents of other wells. E.g. "* [A-P]01 empty" indicates that on all plates (*), every row ([A-P]) of the first column (01) is marked as "empty." When defining wells as containing controls, ensure the case of the text used matches that used elsewhere. For further details on the plate configuration file, see the cellHTS2 documentation.

4. Plate annotation file. Contains at least three columns with a header. The first two columns list the plate and well IDs used in the library. The third and subsequent columns list annotations (such as the ID of the gene targeted by an siRNA). For further details on the plate configuration file, see the cellHTS2 documentation.
5. File containing functional relationships between genes (*see* **Note 6**).

3 Methods

3.1 Processing siRNA Screen Data Using CellHTS2

Typically, siRNA screens are conducted in multiwell tissue culture plates. The process of transfecting a cancer cell line with siRNAs is optimized prior to screening and once optimal conditions have been selected (described in [17]), cells are dispensed into multiwell plates containing growth media, transfection reagents, and siRNAs. The data in the example provided represent a screen of a single osteosarcoma tumor cell line using an siRNA library targeting 714 kinase and kinase-related genes. Positive and negative controls are included on each plate—typically non-targeting siRNA as a negative control and an siRNA pool targeting *PLK1* as a positive control. The full experimental protocol for this screen has been described elsewhere [4, 5]. Briefly, following siRNA transfection, the cells were cultured for 5 days, after which a luminescence assay measuring cellular ATP was used to estimate cell viability. A Victor X5 platereader was used to read luminescence values, resulting in data files in Microsoft Excel format. Prior to the analysis in R, these data files were converted to plain text *plate files.* Each plate file contains the luminescence reading from each well in one 96 or

Fig. 1 (continued) sensitivity to siRNAs targeting specific genes (e.g., DYRK1A). (**c**) The associations table is integrated with a data file describing known protein-protein interactions using Python. This results in a table of annotated dependencies—indicating whether a given association occurs between a pair of genes whose protein products are known to physically interact

384 multiwell plate. Where an siRNA library is larger than the plate format used in the screen, several plates are required for a single screen. Additionally, multiple replicate screens are typically conducted for a given cell line and siRNA library. The organization of plates into segments of an siRNA library and replicate screens is described in a *plate list file*. A plate list file contains the file names of the plate files, the replicate numbers, and plate numbers in a multi-plate screen. Annotations indicating the genes targeted by siRNAs in the library across multiple plates as well as the positions of control wells are provided in separate plain text files. The analysis protocol set out below uses the cellHTS2 [18] R package developed by Huber and Boutros to combine data from the plate files, the plate list file, the plate configuration file, and the annotation file. The luminescence data are normalized to produce *Z*-scores by first $\log_2$ transforming the values and subtracting the median log luminescence value on a plate-by-plate basis. The plate-centered data are then scaled to the median absolute deviation (MAD) value calculated across the entire siRNA library to produce *Z*-scores.

An R script named "run_cellHTS.R" in the R-scripts directory contains the following commands. The first command loads the cellHTS2 R package that provides the functions required for the analysis.

```
require(cellHTS2)
```

With cellHTS2 loaded, we then use the readPlateList() function to read the plate list file which in turn creates a cellHTS object containing the luminescence data from the plate files (*see* **Note 7**).

```
x <- readPlateList(
    filename=" platelist_p3r3.txt",
    name="CGDsExample"
    path="./"
    )
```

We next use the configure() function to add information from the plate configuration file and (optionally) the screen log and description file to the cellHTS object. The plate configuration defines the locations of samples, controls and empty wells.

```
x <- configure(
    x,
    descripFile="screen_description.txt",
    confFile="plateconf_384.txt",
    logFile="Screenlog.txt",
    path="./"
    )
```

We use the annotate() function to define the genes targeted by siRNAs in each well of the plate. This information is located in the "kinome_library.txt" file.

```
x <- annotate(
    x,
    geneIDFile="kinome_library.txt",
    path="./"
    )
```

We now process the luminescence data in the cellHTS object to normalize data values across the plates in the screen. This is done by $\log_2$ transforming the luminescence values and subtracting the median value within a plate from all the values of wells in that plate. The parameters passed to the normalizePlates() function are described in **Note 8**. The original cellHTS object "x" is passed to the normalizePlates() function and the result is saved into a new cellHTS object called "xn."

```
xn <- normalizePlates(
    x,
    scale="multiplicative",
    log=TRUE,
    method="median",
    varianceAdjust = "none",
    negControls="neg",
    posControls="pos"
    )
```

The normalized data stored in "xn" are then scaled by dividing each well's value by the median absolute deviation (MAD) calculated from the normalized values across the whole siRNA library. Control wells are excluded from the estimation of the MAD. Scaling the plate median centered normalized data by the MAD produces the robust equivalent of Studentized values or *Z*-scores (*see* **Note 9**).

```
xsc <- scoreReplicates(
    xn,
    method="zscore",
    sign="+"
    )
```

For later statistical analyses, it may be preferable to summarize the values of replicate wells targeting a specific gene as a median or some other summary statistic. This can be performed using the summarizeReplicates() function in cellHTS2.

```
xsc <- summarizeReplicates(
    xsc,
    summary="median"
    )
```

CellHTS2 also provides a function called getTopTable() that writes a plain text file containing the well annotation data as well as the luminescence data at each stage of processing. Here, we write this information to a file called “TopTable.txt.”

```
summary_info <- getTopTable(
    list(
        "raw"=x,
        "normalized"=xn,
        "scored"=xsc
        ),
    file="TopTable.txt"
    )
```

An HTML formatted report can also be generated describing the screen and the processing steps applied to it using the commands below. This HTML report provides information on the positive and negative controls included, the distribution of the resulting scores, and details of the quality of the screen (Z' scores, see below).

The contents of the HTML report can be modified using the setSettings() function. Here, we turn on the reproducibility and intensities reports (producing heatmap visualizations of well values) and set the range of heatmap colors for the screen summary scores report.

```
setSettings(
    list(
        plateList=list(
            reproducibility=list(
                include=TRUE,
                map=TRUE
                ),
            intensities=list(
                include=TRUE,
                map=TRUE)
                ),
        screenSummary=list(
            scores=list(
                range=c(-20, 10),
                map=TRUE
                )
            )
        )
    )
```

We then use the writeReport() function to generate the HTML report.

```
writeReport(
    raw=x,
    normalized=xn,
    scored=xsc,
    outdir=./report,
    force=TRUE,
    posControls="pos",
    negControls="neg",
    mainScriptFile="../R-scripts/run_cellHTS.R"
    )
```

The outputs from this cellHTS2 analysis so far have been a TopTable plain text file and a folder containing an HTML report. It is possible to extract any data in the cellHTS objects using accessor methods in order to produce customized outputs. Here, we extract information on the targeted genes, the plate numbers, well numbers, and median *Z*-scores and combine this into a data frame ("combinedz") containing four columns (compound, plate, well, and zscore).

```
genes <- geneAnno(xsc)
plates <-plate(xsc)
wells <- well(xsc)
scores <- Data(xsc)[,1,1]
combinedz <- data.frame(
    compound=compounds,
    plate=plates,
    well=wells,
    zscore=scores
    )
```

We can then write out the "combined" data frame to a text file. A use case for this is to enable joining data from multiple screens into a single file for analysis.

```
write.table(
    combinedz,
    "zscore.txt",
    sep="\t",
    quote=FALSE,
    row.names=FALSE
    )
```

This analysis needs to be performed for each screen in the experiment. Typically, multiple distinct screens would represent multiple tumor cell lines screened with a specific library of siRNAs. Quality control steps need to be applied on a screen-by-screen basis. We expect siRNA screen replicates to be strongly correlated and reject screens where no pairs of replicates have a correlation coefficient greater than 0.7 (*see* **Note 10**).

In an earlier step, we saved the output from the getTopTable() function to a data frame called "summary_info." We can extract the replicate normalized luminescence values from this data frame and calculate the Pearson correlation coefficients for each pair of replicates using the following command.

```
cor(
    summary_info[,c(
        "normalized_r1_ch1",
        "normalized_r2_ch1",
        "normalized_r3_ch1"
        )],
    use="pairwise.complete.obs"
    )
```

A further quality control step that is recommended is to examine the *Z*-prime (*Z′*) values for each screen [19]. *Z′* scores provide a measure of the separation of the positive and negative control siRNAs included in a screen and so can be considered an estimate of how much it is possible for the individual "sample" wells to vary in *Z*-scores. Larger values of *Z′* indicate better screens. Screens with *Z′* values ≥ 0.5 are considered excellent. Those with *Z′* values ≤ 0 are considered unusable and should be rejected and the experiments should be repeated. CellHTS2 calculates *Z′* scores for each replicate and these can be found in the HTML report under the "plate summaries" section.

3.2 Identification of Kinase Dependencies Associated with Driver Gene Mutation or Copy Number Alteration

We next integrate the processed results from multiple siRNA screens with data describing the genetic alterations present in each sample. For this tutorial we use the siRNA data from 18 osteosarcoma tumor cell lines and a mutations file that describes the presence or absence of genetic alterations in different members of the Retinoblastoma (RB1) pathway. In the git repository downloaded, there is a set of directories containing pre-formatted siRNA and mutation datasets as well as R scripts to process the data. Open the script R-scripts/identifying_CGDs_RB1_osteosarcoma.R and examine its contents. The first command sets the working directory to the top level of the git repository we cloned/downloaded earlier. Modification of the path given to the setwd() function is required to point to the appropriate location on your local system.

```
setwd("~/software/identifying-genetic-dependencies")
```

The next command runs R code contained in a second file in the R-scripts directory. The dot at the beginning of the path indicates that the path is relative to the current working directory. The file "identifying_CGDs_library.R" contains a set of functions that abstract the process of loading mutation and siRNA datasets as well as running a set of statistical tests. Readers familiar with R can examine the code in this file to understand the individual analysis steps in more detail.

```
source("./R-scripts/identifying_CGDs_library.R")
```

We next define the paths to the siRNA and mutation data files used in the analysis. It is a helpful to define this kind of information near the top of scripts so that in the future the files can be changed without having to find the commands where these values are used.

```
sirna_screens_file <- "./siRNA-data/Osteosarcoma_kinome_sc-
reens.txt"
rb_pathway_func_muts_file <- "./mutation-data/combined_exo-
me_cnv_func_muts_RBpathway_160418.txt"
rb_pathway_all_muts_file <- "./mutation-data/combined_exo-
me_cnv_all_muts_RBpathway_160418.txt"
```

The next command reads the siRNA and mutation datasets, identifies cell lines in common between each dataset, and returns an R list object containing analysis-ready tables. The input files comprise tab-separated data where the first row and first column represent column and row names respectively. Aside from the first row (column headings), each row contains data for a single-cell line. Each column represents a property measured across each cell line. In the "sirna_screens_file," these properties are the *Z*-scores representing the relative viability of cells treated with siRNAs targeting specific genes. In the case of the mutation datasets (rb_pathway_func_muts_file and rb_pathway_all_muts_file), these properties represent the presence or absence of a driver gene alteration. The file rb_pathway_func_muts_file contains a "1" where a cell line is considered to contain a likely functional cancer driver gene alteration (mutation or copy number alteration) and a "0" where such a change is absent. Similarly, the file rb_pathway_all_muts_file contains a "1" or "0" to indicate the presence of any driver gene alteration found in a cell line irrespective of presumed functional impact. These two files are used to identify sets of cell lines where a driver gene is considered to be functionally altered (the mutant group) or where alterations to the driver gene are entirely absent (the wild-type group) (*see* **Note 11**).

```
kinome_rb_muts <- read_rnai_mutations(
    rnai_file=sirna_screens_file,
```

```
    func_muts_file=rb_pathway_func_muts_file,
    all_muts_file=rb_pathway_all_muts_file
    )
```

With the siRNA and mutation data tables organized within kinome_rb_muts, we now run association tests between mutations or copy number alterations in RB1 pathway genes and test dependency on each gene targeted in the kinome siRNA library. The function run_univariate_tests() performs Wilcoxon Rank Sum tests between siRNA *Z*-scores of cell lines in the mutant and wild-type groups and returns a table of these test results as well as other information such as descriptive statistics (including the median *Z*-score of the mutant and wild-type group and the difference between those two values).

```
kinome_rb_mut_associations <- run_univariate_tests(
    zscores=kinome_rb_muts$rnai,
    mutations=kinome_rb_muts$func_muts,
    all_variants=kinome_rb_muts$all_muts,
    alt="less"
    )
```

We write out the results of the association tests to a text file that can be opened in a spreadsheet application or used as input for other programs such as the annotate_dependencies.py python program described in Subheading 3.3.

```
write.table(
    kinome_rb_mut_associations,
    "./results/kinome_rb_mut_associations.txt",
    sep="\t",
    col.names=TRUE,
    row.names=FALSE,
    quote=FALSE
    )
```

3.3 Annotating Molecular Dependencies According to Known Functional Relationships

In the absence of additional information, interpreting an association between the mutation of a driver gene and sensitivity to RNAi reagents targeting another gene can be difficult. One approach to aiding the interpretation of these associations is the integration of orthogonal data, including known functional relationships between genes or their protein products. We provide a simple Python script (annotate_dependencies.py) that can be used to integrate known functional relationships (e.g., protein-protein, kinase-substrate, or gene-regulatory interactions) with the associations generated by the R scripts described in Subheading 3.2. This script adds an additional column to the associations file indicating whether or not the marker-target gene pair has a known functional relationship according to a user-supplied source of interactions.

1. Create a file containing functional relationships between genes (*see* **Note 6** for potential sources of these relationships). Each line of this file should contain two gene symbols (HUGO gene names) separated by a tab. Alternatively, files in the BioGRID Tab 2.0 Format, such as those downloaded from the BioGRID database [20], can be used as input.
2. Open a command prompt/terminal and run the script as follows:

```
python annotate_dependencies.py -a <associations> -o <output>
-i <interactions> -n <column_name>
```

where <associations> is the name of the associations file created using the R scripts above, <output> is the name of the file where the annotated associations will be output to, <interactions> is the name of the file containing known functional relationships, and <column_name> is an optional name for the column where the functional annotation will be stored. If the interactions file is in the BioGRID Tab 2.0 format then add the optional "–b" argument to this command. *See* **Note 12** for additional parameters of this file.

3. View the resulting output in a text editor or spread sheet application. There should be an additional column in the file named using the <column_name> argument, with True or False values indicating whether each marker-target association involves a gene pair with a known functional relationships according to the <interactions> file
4. Additional columns can be added (e.g., to annotate the associations according to a different source of interactions) by running the script again using the output file (<output>) of the first run as input to a subsequent run. For this step it is necessary to set the <column_name> parameter to avoid overwriting previous results.

The end result of this analysis is a file containing an annotated list of associations between a particular genomic feature (indicated in the "marker" column) and increased sensitivity to siRNA reagents targeting a particular gene (indicated in the "target" column). The column titled "PPI" in this file indicates whether the marker gene and the target have a known functional relationship (e.g., protein-protein interaction) while the column "wilcox.p" gives an indication of the statistical significance of the association. These *p*-values, together with the annotation of known functional relationships, may be used to prioritize candidate genetic dependencies (synthetic lethalities) for follow-up experiments. At a minimum these follow-up experiments should involve using orthogonal means to test the observed association (e.g., alternative siRNA reagents or a small molecule targeting the protein product of the gene of interest) [21]. Ideally, the follow-up validation would test the association in additional cell lines harboring the mutation of interest. In the example provided, we found that *RB1* mutation is associated with increased sensitivity to siRNA targeting the kinase *DYRK1A*, a known *RB1* binding partner. In Campbell et al. [5] we validated this in a larger panel of osteosarcoma cell lines using four distinct siRNA reagents targeting the *DYRK1A* gene suggesting that the initial observation represents a real dependency.

4 Notes

1. All the analyses can be performed on a desktop computer. A recent version of the R statistical programming environment (available from https://www.r-project.org/) and the Python programming language (available from https://www.python.org/) are required. The Python scripts presented here have been tested with Python versions 2.7 and 3.4, while R scripts have been tested with version 3.2.5.
2. Note that we provide extensive code samples throughout this document. In these samples the tilde character (~) is used as a short cut to the user's home directory on Unix-like systems. On Microsoft Windows, the forward slash characters (/) separating the file paths will need to be substituted with back slashes (\).
3. *Gplots* is provided on the Comprehensive R Archive Network (CRAN) and can be installed by starting an R session and enter the following code:

```
install.packages(
"gplots",
dependencies=TRUE,
)
```

4. CellHTS2 [18] is an R package used to process RNAi screen data and can be installed using the following command:

```
source("https://bioconductor.org/biocLite.R")
biocLite("cellHTS2")
```

5. This repository can be downloaded as a zip file by navigating to the above URL and choosing "download ZIP." Alternatively install git (software available from https://git-scm.com), open

a console window, change directory to a suitable path that must exist (e.g., cd ~/software), and enter the following command:

```
git clone https://github.com/GeneFunctionTeam/identifying-
genetic-dependencies
```

This command should create a new directory (e.g., ~/software/identifying-genetic-dependencies) containing data and scripts. The data files include a file containing viability data from an siRNA screen of osteosarcoma cell lines [5] and driver gene mutation datasets compiled from publicly available compendia of mutations in tumor cell lines [22].

6. BioGRID is a database of experimentally determined molecular interactions [20]. The web interface to BioGRID allows users to download the entire database in Tab 2.0 format, and also the interactions associated with a specific gene. An alternative source is PathwayCommons [23], which integrates protein-protein, gene-regulatory, kinase-substrate, and other molecular relationships. More specialized data sources include PhosphoSitePlus [24] (kinase-substrate relationships) and HINT (high-confidence protein-protein interactions) [25].
7. Detailed instructions on how to use cellHTS2 can be found in an R vignette titled “End-to-end analysis of cell-based screens.” Once the cellHTS2 package is installed, the command ‘browseVignettes("cellHTS2")’ can be entered into the R console to reveal links to this and other relevant vignettes.
8. In our experience, luminescence values from multiwell siRNA screens tend to be positively skewed and show a log-normal distribution. It is thus preferable to log transform values prior to normalization. Setting the “log” argument of the “normalizePlates” function to “TRUE” and the “scale” argument to “multiplicative” instructs cellHTS2 to first log transform the luminescence values and then subtract the plate median values from each value on a plate.
9. *Z*-normalization in the classical sense refers to adjusting a set of normally distributed values such that they have a mean value of zero and a standard deviation equal to one. For idealized normally distributed *Z*-scores, 95% of the values are expected to fall between $Z = -2$ and $Z = +2$ and 99.1% of the values are expected to fall between $Z = -3$ and $Z = +3$. Log-transformed and plate-centered luminescence values from siRNA screens often have negatively skewed distributions that are not well described by statistics such as the mean and standard deviation. As an alternative to standard *Z*-score normalization we use robust *Z*-normalization where the median value is subtracted from all log-transformed plate-centered values and these values

are then divided by the median absolute deviation (MAD) of the distribution. This results in approximately 95% of the values falling between $Z = -2$ and $Z = +2$. Thus, siRNAs that produce a *Z*-score of <-2 (or more stringently, <-3) are interpreted as causing a decrease in viability.

10. At least two replicates are required for each screen in order to assess the overall reproducibility of the screen. We typically perform screens using three replicates and take the median value for each siRNA to further minimize noise.

11. Defining functional mutations in cancer driver genes can be difficult. In some cases (e.g., amplification of a gene such as *ERBB2*) the functional relevance of an alteration is well established. In many cases however, especially those involving missense mutations, the functional relevance of an alteration is uncertain. In [5] we developed a simple pipeline to classify mutations and copy number changes as either of likely functional relevance or of uncertain relevance [5]. For tumor suppressor genes we classify homozygous deletions, mutations predicted to cause a truncation (frame shift, nonsense, or splice site alteration) or missense mutations found to occur recurrently in tumors as functionally relevant. For oncogenes, we classify amplification events or recurrent missense mutations as functionally relevant. Mutations other than these are classified as of uncertain relevance and cell lines harboring these mutations are excluded from our association tests.

12. By default the "annotate_dependencies.py" script assumes that the interactions provided in the input file are undirected (i.e., the interaction (a, b) is the same as the interaction (b, a)). Using the argument "-d" changes this default behavior such that a directed network is utilized. This may be more appropriate for directed networks—e.g., for RB1 associated dependencies it may make sense to highlight associations between RB1 and genes that it regulates, but not associations involving genes that regulate RB1.

References

1. Davoli T et al (2013) Cumulative haploinsufficiency and triplosensitivity drive aneuploidy patterns and shape the cancer genome. Cell 155(4):948–962
2. Lawrence MS et al (2014) Discovery and saturation analysis of cancer genes across 21 tumour types. Nature 505(7484):495–501
3. Yaffe MB (2013) The scientific drunk and the lamppost: massive sequencing efforts in cancer discovery and treatment. Sci Signal 6(269):e13
4. Brough R et al (2011) Functional viability profiles of breast cancer. Cancer Discov 1 (3):260–273
5. Campbell J et al (2016) Large-scale profiling of kinase dependencies in cancer cell lines. Cell Rep 14(10):2490–2501
6. Cheung HW et al (2011) Systematic investigation of genetic vulnerabilities across cancer cell lines reveals lineage-specific dependencies in ovarian cancer. Proc Natl Acad Sci U S A 108 (30):12372–12377

7. Cowley GS et al (2014) Parallel genome-scale loss of function screens in 216 cancer cell lines for the identification of context-specific genetic dependencies. Sci Data 1:140035
8. Hart T et al (2015) High-resolution CRISPR screens reveal fitness genes and genotype-specific cancer liabilities. Cell 163 (6):1515–1526
9. Kim HS et al (2013) Systematic identification of molecular subtype-selective vulnerabilities in non-small-cell lung cancer. Cell 155 (3):552–566
10. Marcotte R et al (2016) Functional genomic landscape of human breast cancer drivers, vulnerabilities, and resistance. Cell 164 (1–2):293–309
11. Moser R et al (2014) Functional kinomics identifies candidate therapeutic targets in head and neck cancer. Clin Cancer Res 20 (16):4274–4288
12. Luo J, Solimini NL, Elledge SJ (2009) Principles of cancer therapy: oncogene and non-oncogene addiction. Cell 136 (5):823–837
13. Lord CJ, Tutt AN, Ashworth A (2015) Synthetic lethality and cancer therapy: lessons learned from the development of PARP inhibitors. Annu Rev Med 66:455–470
14. Helming KC et al (2014) ARID1B is a specific vulnerability in ARID1A-mutant cancers. Nat Med 20(3):251–254
15. Hsu TY et al (2015) The spliceosome is a therapeutic vulnerability in MYC-driven cancer. Nature 525(7569):384–388
16. Kelley R, Ideker T (2005) Systematic interpretation of genetic interactions using protein networks. Nat Biotechnol 23(5):561–566
17. Lord CJ et al (2008) A high-throughput RNA interference screen for DNA repair determinants of PARP inhibitor sensitivity. DNA Repair (Amst) 7(12):2010–2019
18. Boutros M, Bras LP, Huber W (2006) Analysis of cell-based RNAi screens. Genome Biol 7(7): R66
19. Zhang JH, Chung TD, Oldenburg KR (1999) A simple statistical parameter for use in evaluation and validation of high throughput screening assays. J Biomol Screen 4(2):67–73
20. Chatr-Aryamontri A et al (2015) The BioGRID interaction database: 2015 update. Nucleic Acids Res 43(Database issue): D470–D478
21. Jackson AL, Linsley PS (2010) Recognizing and avoiding siRNA off-target effects for target identification and therapeutic application. Nat Rev Drug Discov 9(1):57–67
22. Forbes SA et al (2015) COSMIC: exploring the world's knowledge of somatic mutations in human cancer. Nucleic Acids Res 43(Database issue):D805–D811
23. Cerami EG et al (2011) Pathway Commons, a web resource for biological pathway data. Nucleic Acids Res 39(Database issue): D685–D690
24. Hornbeck PV et al (2015) PhosphoSitePlus, 2014: mutations, PTMs and recalibrations. Nucleic Acids Res 43(Database issue): D512–D520
25. Das J, Yu H (2012) HINT: high-quality protein interactomes and their applications in understanding human disease. BMC Syst Biol 6:92

Open Access This chapter is licensed under the terms of the Creative Commons Attribution 4.0 International License (http://creativecommons.org/licenses/by/4.0/), which permits use, sharing, adaptation, distribution and reproduction in any medium or format, as long as you give appropriate credit to the original author(s) and the source, provide a link to the Creative Commons license and indicate if changes were made.

The images or other third party material in this chapter are included in the chapter's Creative Commons license, unless indicated otherwise in a credit line to the material. If material is not included in the chapter's Creative Commons license and your intended use is not permitted by statutory regulation or exceeds the permitted use, you will need to obtain permission directly from the copyright holder.

Part II

Systems Analyses of Signaling Networks in Cancer Cells

Chapter 6

Phosphoproteomics-Based Profiling of Kinase Activities in Cancer Cells

Jakob Wirbel, Pedro Cutillas, and Julio Saez-Rodriguez

Abstract

Cellular signaling, predominantly mediated by phosphorylation through protein kinases, is found to be deregulated in most cancers. Accordingly, protein kinases have been subject to intense investigations in cancer research, to understand their role in oncogenesis and to discover new therapeutic targets. Despite great advances, an understanding of kinase dysfunction in cancer is far from complete.

A powerful tool to investigate phosphorylation is mass-spectrometry (MS)-based phosphoproteomics, which enables the identification of thousands of phosphorylated peptides in a single experiment. Since every phosphorylation event results from the activity of a protein kinase, high-coverage phosphoproteomics data should indirectly contain comprehensive information about the activity of protein kinases.

In this chapter, we discuss the use of computational methods to predict kinase activity scores from MS-based phosphoproteomics data. We start with a short explanation of the fundamental features of the phosphoproteomics data acquisition process from the perspective of the computational analysis. Next, we briefly review the existing databases with experimentally verified kinase-substrate relationships and present a set of bioinformatic tools to discover novel kinase targets. We then introduce different methods to infer kinase activities from phosphoproteomics data and these kinase-substrate relationships. We illustrate their application with a detailed protocol of one of the methods, KSEA (Kinase Substrate Enrichment Analysis). This method is implemented in Python within the framework of the open-source Kinase Activity Toolbox (kinact), which is freely available at http://github.com/saezlab/kinact/.

Key words Phosphoproteomics, Mass-spectrometry, Kinase activity, Computational biology, Cancer systems biology, Signal transduction

1 Introduction

Protein kinases are major effectors of cellular signaling, in the context of which they form a highly complex and tightly regulated network that can sense and integrate a multitude of external stimuli or internal cues. This kinase network exerts control over cellular processes of fundamental importance, such as the decision between proliferation and apoptosis [1]. Deregulation of kinase signaling can lead to severe diseases and is observed in almost every type of cancer [2]. For instance, a single constitutively active kinase,

Louise von Stechow (ed.), *Cancer Systems Biology: Methods and Protocols*, Methods in Molecular Biology, vol. 1711, https://doi.org/10.1007/978-1-4939-7493-1_6,

© The Author(s) 2018

originating from the fusion of the *BCR* and *ABL* genes, can give rise to and sustain chronic myeloid leukemia [3]. Accordingly, the small molecule inhibitor of the BCR-ABL kinase, Imatinib, has shown unprecedented therapeutic effectiveness in affected patients [4].

Fueled by these promising clinical results, due to the essential role for kinases in the patho-mechanism of cancer, and because kinases are in general pharmacologically tractable [5], a range of new kinase inhibitors has been approved or is in development for different cancer types [6]. However, not all eligible patients respond equally well, and in addition, cancers often develop resistance to initially successful therapies. This calls for a deeper understanding of kinase signaling and opens up the possibility of exploiting this knowledge therapeutically [7].

By definition, the activity of a kinase is reflected in the occurrence of phosphorylation events catalyzed by this kinase. Thus, analysis of kinase activity was traditionally achieved by monitoring the phosphorylation status of a limited number of sites known to be targeted by the kinase of interest using immunochemical techniques [8]. This, however, requires substantial prior-knowledge and yields a comparably low throughput. Other approaches exist, e.g., protein kinase activity assays [9, 10] or attempts to measure kinase activity with chromatographic beads functionalized with ATP or small molecule inhibitors [11].

Mass spectrometry-based techniques to measure phosphorylation can identify thousands of phosphopeptides in a single sample with ever-increasing coverage, throughput, and quality, nourished by technological advances and dramatically increased performance of MS instruments in recent years [12–14]. High-coverage phosphoproteomics data should indirectly contain information about the activity of many active kinases. The high-content nature of phosphoproteomics data, however, poses challenges for computational analysis. For example, only a small subset of the described phosphorylation sites can be explicitly associated with functional impact [15].

As a means to extract functional insight, methods to infer kinase activities from phosphoproteomics data based on prior-knowledge about kinase-substrate relationships have been put forward [16–19]. The knowledge about kinase-substrate relationships, compiled in databases like PhosphoSitePlus [20] or Phospho.ELM [21], covers only a limited set of interactions. Alternatively, computational resources to predict kinase-substrate relationships based on kinase recognition motifs and contextual information have been used to enrich the collections of substrates per kinase [22, 23], but the accuracy of such kinase-substrate relationships has not been validated experimentally for most cases. The inferred kinase activities can in turn be used to reconstruct kinase network circuitry or to predict therapeutically relevant features such as sensitivity to kinase inhibitor drugs [17].

In this chapter, we start with a brief description of phosphoproteomics data acquisition, highlighting challenges for the computational analysis that may arise out of the experimental process. Subsequently, we will present different computational methods for the estimation of kinase activities based on phosphoproteomics data, preceded by the kinase-substrate resources these methods use. One of these methods, namely KSEA (Kinase-Substrate Enrichment Analysis), will be explained in more detail in the form of a guided, stepwise protocol, which is available as part of the Python open-source Toolbox kinact (for Kinase Activity Scoring) at http://www.github.com/saezlab/kinact/.

2 Phosphoproteomics Data Acquisition

For a summary of technical variations or available systems for the experimental setup of phosphoproteomics data acquisition, we would like to refer the interested reader to dedicated publications such as [24, 25]. We provide here a short overview about the experimental process to facilitate the understanding of common challenges that may arise for the data analysis that we will focus on.

Mass spectrometry-based detection of peptides with posttranslational modifications (PTM) usually requires the same steps, independent of the modification of interest: (1) cell lysis and protein extraction with special focus on PTM preservation, (2) digestion of proteins with an appropriate protease, (3) enrichment of peptides bearing the modification of interest, and (4) analysis of the peptides by LC-MS/MS [26]. After the experimental work, additional data processing steps are required to identify the position of the modification, e.g., the residue that is phosphorylated. For almost every step, different protocols are available, starting from various proteases for protein digestion to different data acquisition methods for MS [24].

2.1 Phosphopeptide Enrichment

Naturally, the enrichment of phosphopeptides is a pivotal step for phosphoproteomics. Next to the enrichment method used, the choice of the protease [27] or the MS ionization method [28] also has an impact on the part of the phosphoproteome that is sampled. For phosphopeptide enrichment, the field is dominated by immobilized metal affinity chromatography (IMAC) and metal oxide affinity chromatography (MOAC), which all exploit the affinity of the phosphorylation toward metal ions. Popular techniques include Fe^{3+}-IMAC, Ti^{4+}-IMAC [29], or TiO_2-MOAC. Alternatively, more traditional biochemical methods involving immunoaffinity purification are also in use for enrichment of phosphopeptides, although these are generally limited to studies of phosphotyrosine [30].

Of note, the different enrichment methods show little overlap in the detected phosphopeptides, although this can also be observed for replicates of runs using the identical enrichment method, as discussed below [31].

After enrichment, the phosphopeptides are separated chromatographically, usually by reversed phase liquid chromatography (RPLC), and then enter the mass spectrometer for tandem MS analysis (MS/MS), completing the workflow of LC-MS/MS. Variations in the chromatography method used as well as the multitude of mass spectrometry instrument types are reviewed in detail elsewhere [24]. Generally, the quality of the chromatographic separation will have a big impact on the number of phosphopeptides that can confidently be identified. Chromatography runs of higher quality also take more time, so that a tradeoff between resolution and throughput must be devised for each experiment.

2.2 Data Acquisition

For most phosphoproteomics studies so far, the MS instrument is operated in the data-dependent acquisition (DDA) mode. Therein, precursor ions from a first survey scan are selected—typically based on relative ion abundance—in order to generate fragmentation spectra in a second MS run [32], for which a database search yields the corresponding peptide sequences [33]. As a result, peptide detection in DDA is on the one hand biased toward high abundance species, but also considerably irreproducible due to stochastic precursor ion selection [34]. This inherent under-sampling of DDA usually leads to missing data points in LC-MS/MS datasets. However, this problem may be solved to some extent by extracting ion chromatograms of the peptides that are missing in some of the runs that are being compared [35–38], by matching across samples [39], or with the accurate mass and retention tag method [40].

In an alternative operation mode, selected reaction monitoring/multiple reaction monitoring (SRM/MRM), the presence and abundance of only a limited set of pre-specified peptides with known fragmentation spectra is surveyed [41]. This targeted approach overcomes many of the issues of shotgun methods, but is usually not feasible for large-scale investigation of the complete phosphoproteome.

Data-independent acquisition (DIA), e.g., SWATH-MS [42] tries to address the shortcoming of both established data acquisition strategies in order to combine the throughput of DDA with the reproducibility of SRM. In DIA, fragmentation spectra are generated for all precursor ions in a specific window of m/z ratios, leading to a complete map of fragmentation spectra, followed by computational extraction of quantitative information for known spectra. For phosphoproteomics, DIA-MS has already been applied to investigate insulin signaling [43] or histone modifications [44]. However, the spectra generated by DIA-MS are usually highly complex and require intricate data extraction techniques, which is even more challenging for modified peptides. Recently, a computational resource for the detection of modified peptides has been put forward [45]. Overall, the available methods for DIA have as yet to mature in order to challenge the use of DDA in large-scale studies of the phosphoproteome [24].

2.3 Quantitative Phosphoproteomics

As for regular proteomics, several experimental methods or post-acquisition tools exist to quantitate detected phosphopeptides. Those can roughly be divided into isotope labeling and label-free quantitation. In general, stable isotope labeling requires more experimental effort than label-free quantitation, but at the same time enables multiplexing of samples with different isotopes or combinations.

Stable isotope labeling by metabolic incorporation of amino acids (SILAC) is mainly used for cell cultures, in the medium of which different stable isotopes are provided that will be incorporated into the proteins of the cells. At the point of analysis, cell extracts are mixed and then jointly investigated with mass spectrometry. Mass differences between peptide pairs due to the isotopic labeling can be exploited for relative quantitation [46]. Currently, up to three conditions (light, medium, heavy) can be multiplexed. Further developments of SILAC even produced an in-vivo SILAC mouse model for the proteomic and phosphoproteomic analysis of skin cancerogenesis [47] and super-SILAC for the analysis of tissues [48], in which a metabolically labeled, tissue-specific protein mix from several cell lines, representing the complexity of the investigated proteome, is mixed with the tissue lysate as internal standard for quantification.

Chemical modification of peptides with tandem mass tags (TMT) or isobaric tags for relative and absolute quantitation (iTRAQ) are two different methods based on tags with reactive groups that bind to peptidyl amines in the peptides after protein digestion. Again, different samples are mixed before mass spectrometry analysis, whereas for TMT or iTRAQ up to eight samples can be multiplexed. In the first MS run, the peptides with different isobaric tags are indistinguishable, but upon fragmentation in the second MS run, each tag generates a unique reporter ion fragmentation spectrum, which can be used for relative quantitation of the tagged peptides [49, 50].

Label-free quantitation (LFQ), on the other hand, relies mainly on post-acquisition data analysis, so that no modification of the essential experimental workflow needs to be implemented. Comparison of an—in theory—unlimited number of different samples is therefore possible, which is associated with the downside of prolonged analysis time as multiplexing samples is not possible. While label-free approaches usually provide a deeper coverage of the proteome than label-based methods, the reproducibility and precision of quantification are inferior, so that more technical replicates are needed for confident quantification in LFQ [51]. Typically, label-free quantitation is achieved by integration of peak area measurements, i.e. the area under the curve, for individual peptides [52] or spectral counting, which reflects that the probability to sample more abundant peptides is higher [53].

For the case of phosphoproteomics, in contrast to regular proteomics, an additional challenge for quantitation arises from the fact that information from different peptides of the same protein cannot be integrated. While in regular proteomics the abundances of every peptide in the protein can be combined, the quantitation of a single phosphosite depends on direct measurements of peptides with the specific modification. Therefore, the sample sizes in phosphoproteomics quantitation are much smaller and can even consist of the measurement of only a single peptide [24].

Furthermore, different phosphosites within the same protein will in many cases not show similar pattern of phosphorylation dynamics. This may give rise to problems for subsequent analysis, if this analysis is conducted on protein rather than on phosphosite level.

2.4 Phosphosite Assignment

Phosphopeptides in large-scale phosphoproteomics experiments are identified from LC-MS/MS runs by interpreting MS/MS spectra using a suitable search engine. Several of such search engines now exist; popular ones include Mascot, Sequest, Protein Prospector, and Andromeda [54–57]. The process of determining peptide sequences from MS/MS data involves matching the mass to charge ratios of fragment ions in MS/MS spectra to the theoretical fragmentation of all protein-derived peptides in protein databases. Depending on the organism being investigated, protein databases from UniProt or NCBI are used. Each search engine has its own scoring system to reflect the confidence of peptide identification, which is a function of MS and MS/MS spectral quality. The false discovery rate (FDR) may be determined by performing parallel searches against scrambled or reversed protein databases containing the same number of sequences as the authentic protein database. The FDR is then calculated as the ratio of positive peptide identifications in the decoy database divided by those derived from the forward search. An FDR of 1% at the peptide level is normally considered adequate.

Deriving peptide sequences with these methods is a relatively straightforward process. However, site localization can be a problem when peptide sequences contain more than one amino acid residue that can be phosphorylated. To address this problem, several methods to determine precise localization of phosphorylation within a phosphopeptide have been published. Ascore uses a probabilistic approach to assess correct site assignment [58] and the algorithm has been applied alongside the Sequest search engine.

The Mascot delta score, introduced by the Kuster group, simply determines the differences in Mascot scores between the different possibilities for phosphosite localization within a phosphopeptide [59]. The larger the delta score, the greater the probability of correct site assignment. Other similar methods have been published [60] and some of them are now incorporated into search engines [61]. The output of the phosphopeptide identification step generally contains scores for both the probability of correct peptide sequence identification and phosphosite localization.

2.5 Pitfalls in the Analysis of MS-Based Phosphoproteomics Data

Although the available experimental methods for MS-based phosphoproteomics data acquisition have evolved considerably over the last years, leading to a steadily increasing number of detected phosphosites, several limitations remain for the investigation of signaling processes using phosphoproteomics data.

While it has been estimated that there are around 500,000 phosphorylation sites in the human proteome [62], the number of phosphosites that can be identified in a single MS experiment usually ranks around 10,000 to up to 40,000 [63]. Therefore, the sampled phosphoproteomic picture is incomplete. It has to be taken into account though, that, not all possible phosphorylation sites are expected to be modified at the same time point. This is caused by context-dependent regulation of phosphosites. For example, some phosphosites are controlled differentially at different cell cycle stages, while others only change under specific external stimulation such as growth factors or other effector molecules [64, 65]. The hope is therefore that a significantly larger portion of phosphosites could be mapped with improving technology and by increasing the diversity of biologically relevant conditions analyzed. So far though, in different MS runs or replicates, usually a distinct set of phosphosites is detected, as the selection of precursor ions is stochastic. This leads to incomplete datasets with a high number of missing data points, challenging computational investigation of the data such as clustering or correlation analysis. However, as discussed above, approaches in which phosphopeptide intensities are compared across MS run post-acquisition minimize this problem [38].

The functional impact of a phosphorylation event is known only in the minority of cases [15]. Indeed, it has been hypothesized that a substantial fraction of phosphorylation sites are non-functional [66], since phosphorylation sites tend to be poorly conserved throughout species [67]. Although approaches to studying the function of individual phosphorylation events have been proposed [68], it may be that a large part of the detected phosphosites serves no function at all. Thus, non-functional sites add a substantial amount of noise to phosphoproteomics data and complicate the computational analysis.

The inference of kinase activity from phosphoproteomics data that will be described in the next section aims to overcome these limitations, by the integration of the information from many phosphosites, along prior knowledge on kinases-substrate relationships, into a single measure for the kinase activity. It is important though to keep in mind that any bias in the experimental workflow will affect these scores. In particular, since highly abundant precursor ions are more likely to be selected for fragmentation and therefore detection, targets of upregulated kinases are more probably detected. Therefore, highly active kinases will be preferentially detected, although downregulated kinases may be identified when comparing different conditions.

3 Computational Methods for Inference of Kinase Activity

Traditionally, biochemical methods have been common to study kinase activities in vitro and are still broadly used [69, 70]. However, on the one hand those methods are generally limited in throughput and time-consuming. On the other hand in vitro methods might not accurately reflect the in vivo activities of kinases in a specific cellular context. MS-based methods have also been applied for assaying kinase activity [9, 10]. Here, the abundances of known target phosphosites are monitored by MS after an in vitro enzymatic reaction.

Since every phosphorylation event results—by definition—from the activity of a kinase, phosphoproteomics data should be suitable to infer the activity of many kinases from a comparably low experimental effort. This task requires computational analysis of the detected phosphorylation sites (phosphosites), since thousands of phosphosites can routinely be measured in a single experiment. Several methods have been proposed in recent years, all of which utilize prior knowledge about kinase-substrate interactions, either from curated databases or from information about kinase recognition motifs.

3.1 Resources for Kinase-Substrate Relationships

As the large-scale detection of phosphorylation events using mass spectrometry became routine, many freely available databases that collect experimentally verified phosphosites have been set up, including PhosphoSitePlus [20], Phospho.ELM [21], Signor [71], or PHOSIDA [72], to name just a few. The databases differ in size and aim; PHOSIDA for example provides a tool for the prediction of putative phosphorylation sites and recently also added acetylation and other posttranslational modification sites to its scope. Phospho.ELM computes a score for the conservation of a phosphosite. Signor is focused on interactions between proteins participating in signal transduction. PhosphoNetworks [73] is dedicated to kinase-substrate interactions, but the information is on the level of proteins, not phosphosites. The arguably most prominent database for expert-edited and curated interactions between kinases and individual phosphosites (that have not been derived from in vitro studies) is PhosphoSitePlus, currently encompassing 16,486 individual kinase-substrate relationships [04-2015]. For *Saccharomyces cerevisiae*, the database PhosphoGRID provides analogous information [74]. Specific information about targets of phosphatases can be found in DEPOD [75]. Also in the Phospho. ELM database, phosphosites have been associated with regulating kinases, although this information is available for only about 10% of the 37,145 human phosphosites in the database [04-2015].

As it has been estimated that there are between 100,000 [76] and 500,000 [62] possible phosphosites in the human proteome, the evident low coverage of the curated databases motivated the development of computational tools to predict in vivo kinase-substrate relationships. These methods identify putative new kinase-substrate relationships based on experimentally derived kinase recognition motifs, which was pioneered by Scansite [77] that uses position-specific scoring matrices (PSSMs) obtained by positional scanning of peptide libraries [78] or phage display methods [79]. Another approach, Netphorest [80] tries to classify phosphorylation sites according to the relevant kinase family instead of predicting individual kinase-substrate links. However, the in vitro specificity of kinases differs significantly from the kinase activity inside of the cell, biasing the experimentally identified kinase recognition motifs [81]. The integration of contextual information, for example co-expression, protein-protein interactions, or subcellular colocalization, markedly improves the accuracy of the predictions [69]. The software packages NetworKIN [82] (recently extended in the context of the resource KinomeXplorer [22], correcting for biases caused by over-studied proteins) and iGPS [23] are examples for methods that combine information about kinase recognition motifs, in vivo phosphorylation sites, and contextual information, e.g., from the STRING database [83]. Recently, Wagih et al. presented a method to predict kinase specificity for kinases without any known phosphorylation sites [84]. Based on the assumption that functional interaction partners of kinases (derived from the STRING database) are more likely to be phosphorylated by the respective kinase, they should therefore contain an amino acid motif conferring kinase specificity. This can then be uncovered by motif enrichment.

The described methods provide predictions that are very valuable but not free from error, for example due to the described differences in in vitro and in vivo kinase specificity or the influence of subcellular localization. Thus, the predicted kinase-substrate interactions should be considered hypotheses to be tested experimentally [85].

We hereafter present four computational methods to infer kinase activities from phosphoproteomics data, which use either curated or computationally predicted kinase-substrate interactions.

3.2 GSEA

Methodologically, inference of kinase activity from phosphoproteomics data is related to the inference of transcription factor activity based on gene expression data. A plethora of different methods has been developed for the prediction of transcription factor activity, e.g., the classical gene set enrichment analysis [86] or elaborated machine learning methods [87].

For example, Drake et al. [88] analyzed the kinase signaling network in castration-resistant prostate cancer with GSEA. They predicted the kinases responsible for each phosphosite with kinase-substrate interactions from PhosphoSitePlus, kinase recognition motifs from PHOSIDA, and predictions from NetworKIN. Subsequently, they computed the enrichment of each kinase' targets with the gene set enrichment algorithm after Subramanian et al. [86], which corresponds to a Kolmogorov–Smirnov-like statistic. The significance of the enrichment score is determined based on permutation tests, whereas the *p*-value depends on the number of permutations.

Alternatively, the gene set enrichment web-tool Enrichr [89, 90] can also be used for enrichment of kinases [91]. The authors compiled kinases-substrate interactions from different databases and extracted additional interactions manually from the literature in order to generate kinase-targets sets. Furthermore, they added protein-protein interactions involving kinases from the Human Protein Reference Database (HPRD) [92], based on the assumption that those are highly enriched in kinase-substrate interactions. Using this prior knowledge, the enrichment of the targets of a kinase is then computed with Fisher's exact test as described in [89].

3.3 KAA

Another approach to link phosphoproteomics data with the activity of kinases was presented in a publication from Qi et al. [16], which they termed kinase activity analysis (KAA).

In this study, the authors collected phosphoproteomics data from adult mouse testis in order to investigate the process of mammalian spermatogenesis. With the software package iGPS [23] they predicted putative kinase-substrate relationships for the detected phosphosites. The authors hypothesized that the number of links for a given kinase in the predicted kinase-substrate network can serve as proxy for the activity of this kinase in the specific cell type. This activity was then compared to the kinase activity background which was calculated by computing the number of links in the background kinase-substrate network based on the mouse phosphorylation atlas by Huttlin et al. [93]. Qi and colleagues predicted high activity of PLK kinases in adult mouse testis and could validate this prediction experimentally.

However, there are several limitations of KAA. For once, it is mainly based on computational predictions of kinase substrate relationships, which are known to be susceptible to errors

[69, 85]. Additionally, in their method the activity of a kinase is only dependent on the number of detected, putative targets. The abundance of the individual phosphosites or the fold change between conditions is not taken into account.

De Graaf et al. [94] chose a comparable approach in a study of the phosphoproteome of Jurkat T cells after stimulation with prostaglandin E_2. However, they did not explicitly calculate kinase activities. Instead, they grouped phosphosites into different clusters with distinct temporal profiles and used the NetworKIN algorithm [82] to calculate the enrichment of putative targets of a given kinase in a specific cluster. As a result, they associated kinases with temporal activity profiles based on the enrichment in one of the detected clusters.

3.4 CLUE

A method designed specifically for time-course phosphoproteomics data is the knowledge-based CLUster Evaluation approach, in short CLUE [18]. This method is based on the assumption that phosphosites targeted by the same kinase will show similar temporal profiles, which is utilized to guide a clustering algorithm and infer kinases associated with these clusters. As in the study by de Graaf et al. [94], kinases are not associated with distinct values for activities but rather with temporal activity profiles. The notable distinction of CLUE is that the clustering is found based on the prior knowledge about kinase-substrate relationships, as outlined below.

Methodologically, CLUE uses the *k*-means clustering algorithm to group the phosphoproteomics data into clusters in which the phosphosites show similar temporal kinetics. The performance of *k*-means clustering is particularly sensitive to the parameter *k*, i.e., the number of clusters. CLUE therefore tests a range of different values for *k* and evaluates them based on the enrichment of kinase-substrate relationships in the identified clusters. The method utilizes the data from the PhosphoSitePlus database in order to derive prior knowledge about kinase-substrate relationships. With Fisher's exact test the enrichment of the targets of a given kinase in a specific cluster is tested for significance. The implemented scoring system penalizes distribution of the targets of a single kinase throughout several clusters, as well as the combination of unrelated phosphosites in the same cluster.

CLUE is freely available as R package in the Comprehensive R Archive Network CRAN under https://cran.r-project.org/web/packages/ClueR/index.html.

A limitation of CLUE is represented by the fact that possible 'noise' in the prior knowledge, i.e., incorrect annotations, potentially derived from cell type-specific kinase-substrate relationships, can affect the performance of the clustering, although simulations showed reasonable robustness. CLUE is tailored toward time-course phosphoproteomics data and associates kinases with temporal activity profiles. Since the method does not provide singular activity scores for each kinase, it may be only partly applicable to experiments in which the individual responses of kinases to different treatments or conditions are of interest.

3.5 KSEA

Casado et al. [17] presented a method for kinase activity estimation based on kinase-substrate sets. Using kinase-substrate relationships derived from the databases PhosphoSitePlus and Phospho.ELM, all phosphosites that are targeted by a given kinase can be grouped together into a substrate set (*see* Fig. 1 for an outline of the workflow). In theory, these phosphosites should show similar values, since they are targeted by the same kinase. However, due to the transient and therefore inherently noisy nature of phosphorylation, Casado and colleagues proposed integrating the information from all phosphosites in the substrate set in order to enhance the signal-to-noise ratio by signal averaging [95].

For KSEA, log2-transformed fold change data is needed, i.e., the change of the abundance of a phosphosite between the initial and treated states, initial and later time points, or between two different cell types. Therefore, KSEA activity scores describe the activity of a kinase in one condition relative to another.

The authors suggested three possible metrics (mean score, alternative mean score, and delta score) that can be extracted out of the substrate set and serve as proxy for kinase activity: (1) The

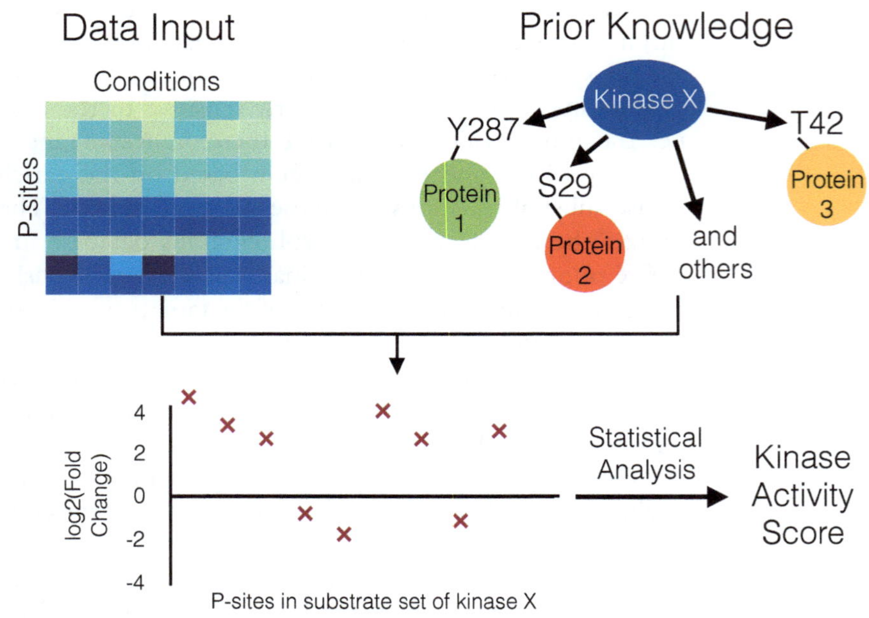

Fig. 1 Work-flow of methods to obtain Kinase activity scores such as KSEA. As prior knowledge, the targets of a given kinase are extracted out of curated databases like PhosphoSitePlus. Together with the data of the detected phosphosites, substrate sets are constructed for each kinase, from which an activity score can be calculated

main activity score, also used in following publications [96], is defined as the mean of the log2 fold changes of the phosphosites in the substrate set; (2) alternatively, only phosphosites with significant fold changes can be considered for the calculation of the mean; and (3) for the last approach, termed "delta count," the occurrence of significantly upregulated phosphosites in the substrate set is counted, from which the number of significantly downregulated sites is subtracted. For each method, the significance of the kinase activity score is tested with an appropriate statistical test. In the publication of Casado et al., all three measures were in good agreement, even if spanning different numerical ranges (*see* Fig. 2). The implementation of these three methods is discussed in detail in the following section.

Like the other methods described in this section, KSEA strongly depends on the prior knowledge kinase-substrate relationships available in the freely accessible databases. These are far from complete and therefore limit the analytical depth of the kinase activity analysis. Additionally, databases are generally biased toward well-studied kinases or pathways [22], so that the sizes of the different substrate sets differ considerably. Casado et al. tried to address these limitations by integrating information about kinase recognition motifs and obtained comparable results.

A detailed protocol on how to use KSEA is provided in Subheading 4.

3.6 IKAP

Recently, Mischnik and colleagues introduced a machine-learning method to estimate kinase activities and to predict putative kinase-substrate relationships from phosphoproteomics data [19].

In their model for kinase activity, the effect e of a given kinase j on a single phosphosite i is modeled with

$$e_{ji} = k_j \times p_{ji}$$

as a product of the kinase activity k and the affinity p of kinase j for phosphosite i. The abundance P of the phosphosite i is expressed as mean of all effects acting on it, since several kinases can regulate the same phosphosite:

$$P_i = \sum_{j=1}^{m} e_{ji} / \sum_{j=1}^{m} p_{ji}$$

The information about the kinase-substrate relationships is also derived from the PhosphoSitePlus database. Using a nonlinear optimization routine, IKAP estimates the described parameters while minimizing a least square cost function between predicted and measured phosphosite abundance throughout time points or conditions. For this optimization, the affinity parameters are estimated globally, while the kinase activities are fitted separately for each time point.

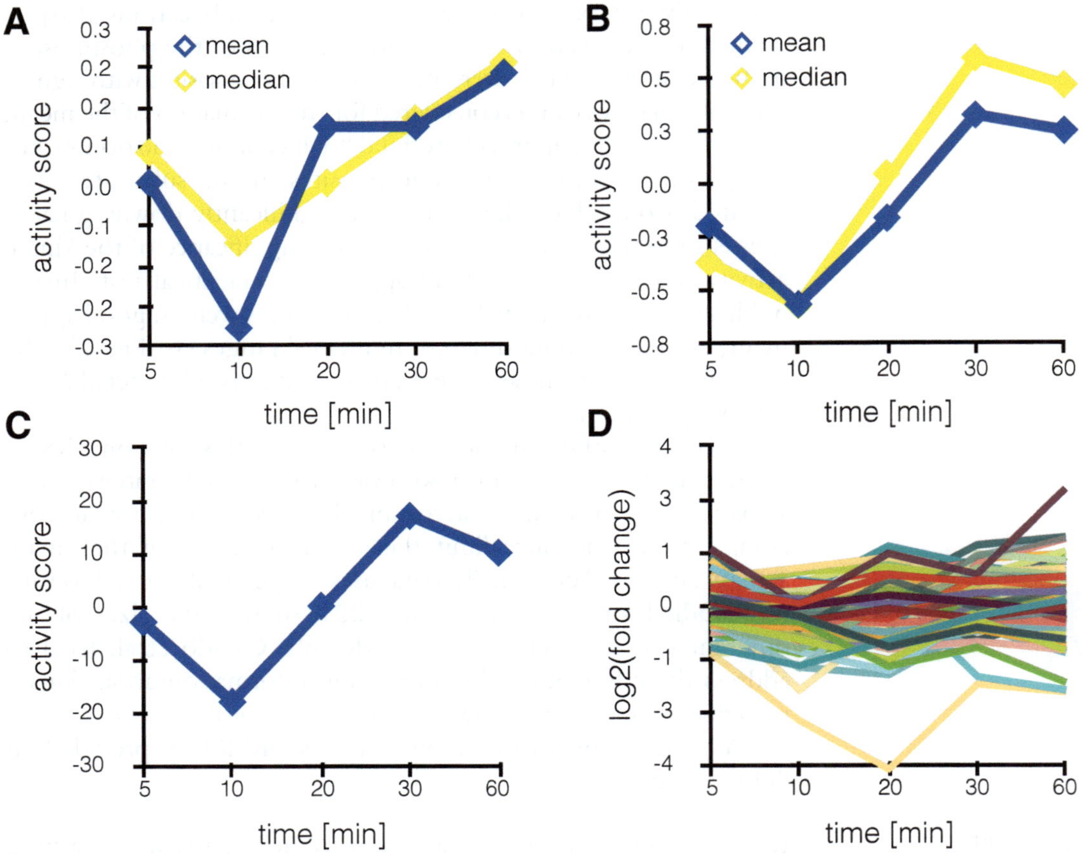

Fig. 2 KSEA activity scores for Casein kinase II subunit alpha. (**a**) Activity scores for Casein kinase II subunit alpha over all time points of the de Graaf dataset [94], calculated as the mean of all phosphosites in the substrate set. In *yellow*, the median has been used. (**b**) Activity scores for Casein kinase II subunit alpha over all time points of the de Graaf dataset, calculated as the mean of all significantly regulated phosphosites in the substrate set. The median is again shown in *yellow*. (**c**) Delta score for Casein kinase II subunit alpha over all time points of the de Graaf dataset, calculated as number of significantly upregulated phosphosites minus the number of significantly downregulated phosphosites in the substrate set. (**d**) The log2 fold changes for all time points for all phosphosites in the substrate set of the Casein kinase II subunit alpha

In a second step, putative new kinase-substrate relationships are predicted based on the correlation of a phosphosite with the estimated activity of a kinase throughout time points or conditions. These predictions are then tested by database searches and by comparison to kinase recognition motifs from NetworKIN.

In contrast to KSEA, which computes the kinase activity based on the fold changes of the phosphosites in the respective substrate set, IKAP is built on a heuristic machine learning algorithm and tries to fit globally the described model of kinase activity and affinity to the phosphoproteomics data. Therefore, the output of IKAP is not only a score for the activity of a kinase, but also a value representing the strength of a specific kinase-substrate interaction in the investigated cell type. On the other hand, the amount of parameters that have to be estimated is rather large, so that a fair number of experimental conditions or time points are needed for unique solutions. Mischnik et al. included a function to perform an identifiability analysis of the obtained kinase activities and could show in the case of the two investigated example datasets that the found solutions are indeed unique on the basis of the phosphoproteomics measurements.

The MATLAB code for IKAP can be found online under www.github.com/marcel-mischnik/IKAP/, accompanied by an extensive step-by-step documentation, which we recommend as additional reading to the interested reader.

4 Protocol for KSEA

In this section, we present a stepwise, guided protocol for the KSEA approach to infer kinase activities from phosphoproteomics data. This protocol (part of the Kinase Activity Toolbox under https://github.com/saezlab/kinact) is accompanied by a freely available script, written in the Python programming language (Python version 2.7.x) that should enable the use of KSEA for any phosphoproteomics dataset. We plan to expand Kinact to other methods in the future. We are going to explain the performed computations in detail in the following protocol to facilitate understanding and to enable a potential re-implementation into other programming languages.

As an example application, we will use KSEA on the phosphoproteomics dataset from de Graaf et al. [94], which was derived from Jurkat T cells stimulated with prostaglandin E_2 and is available as supplemental information to the article online at http://www.mcponline.org/content/13/9/2426/suppl/DC1

4.1 Quick Start

As a quick start for practiced Python users, we can use the utility functions from kinact to load the example dataset. The data should be organized as Pandas DataFrame containing the log2-transformed fold changes, while the columns represent different conditions or time points and the row individual phosphosites. The *p*-value of the fold change is optional, but should be organized in the same way as the data.

```
import kinact
data_fc, data_p_value = kinact.get_example_data()
print data_fc.head()
>>>                 5min      10min     20min     30min     60min
>>> ID
>>> A0AVK6_S71   -0.319306 -0.484960 -0.798082 -0.856103
-0.928753
```

```
>>> A0FGR8_S743  -0.856661  -0.981951  -1.500412  -1.441868
-0.861470
>>> A0FGR8_S758  -1.445386  -2.397915  -2.692994  -2.794762
-1.553398
>>> A0FGR8_S691   0.271458   0.264596   0.501685   0.461984
0.655501
>>> A0JLT2_S226  -0.080786   1.069710   0.519780   0.520883
-0.296040
```

The kinase-substrate relationships have to be loaded as well with the function get_kinase_targets(). In this function call, we can specify with the 'sources'-parameter, from which databases we want to integrate the information about kinase-substrate relationships, e.g., PhosphoSitePlus, Phospho.ELM, or Signor. The function uses an interface to the pypath python package, which integrates several resources for curated signaling pathways [97] (*see* also **Note 1**).

```
kin_sub_interactions = kinact.get_kinase_targets(sources=
['all'])
```

An important requirement for the following analysis is that the structure of the indices of the rows of the data and the prior knowledge need to be the same (see below for more detail). As an example, KSEA can be performed for the condition of 5 min after stimulation in the de Graaf dataset using:

```
activities, p_values = kinact.ksea.ksea_mean(data_fc['5min'],
kin_sub_interactions, mP=data_fc.values.mean(),
delta=data_fc.values.std())
print activities.head()
>>>    AKT1          0.243170
>>>    AKT2          0.325643
>>>    ATM          -0.127511
>>>    ATR          -0.141812
>>>    AURKA         1.783135
>>>    dtype: float64
```

Besides the data (data_fc['5min']) and kinase-substrate interactions (kin_sub_interactions), the variables 'mP' and 'delta' are needed to determine the *z*-score of the enrichment. The *z*-score builds the basis for the *p*-value calculation. The *p*-values for all kinases are corrected for multiple testing with the Benjamini-Hochberg procedure [98].

In Fig. 2, the different activity scores for the Casein kinase II alpha, which de Graaf et al. had associated with increased activity after prolonged stimulation with prostaglandin E_2, are shown together with the log2 fold change values of all phosphosites that are known to be targeted by this kinase. For methods, which use the mean, the median as more robust measure can be calculated alternatively. The qualitative changes of the kinase activities (Fig. 2a–c) are quite similar regardless of the method, and would not be apparent from looking at any specific substrate phosphosite alone (Fig. 2d).

4.2 Loading the Data

In the following, we walk the reader step by step through the procedure for KSEA. First, we need to organize the data so that the KSEA functions can interpret it.

In Python, the library Pandas [99] provides useful data structures and powerful tools for data analysis. Since the provided script depends on many utilities from this library, we would strongly advice the reader to have a look at the Pandas documentation, although it will not be crucial in order to understand the presented protocol. The library, together with the NumPy [100] package, can be loaded with:

```
import pandas as pd
import numpy as np
```

The data accompanying the article is provided as Excel spreadsheet and can be imported to python using the pandas 'read_excel' function or first be saved as csv-file, using the 'Save As' function in Excel in order to use it as described below. For convenience, in the referenced Github repository, the data is already stored as csv-file, so that this step is not necessary. The data can be loaded with the function 'read_csv', which will return a Pandas DataFrame containing the data organized in rows and columns.

```
data_raw = pd.read_csv('FILEPATH', sep=',')
```

In the DataFrame object 'data_raw', the columns represent the different experimental conditions or additional information and the row's unique phosphosites. A good way to gain an overview about the data stored in a DataFrame and to keep track of changes are the following functions:

print data_raw.head() to show the first five rows of the DataFrame or print data_raw.shape in order to show the dimensions of the DataFrame.

Phosphosites that can be matched to different proteins or several positions within the same protein are excluded from the analysis. In this example, ambiguous matching is indicated by the presence of a semicolon that separates multiple possible identifiers, and can be removed like this:

```
data_reduced = data_raw[~data_raw['Proteins'].str.contains
(';')]
```

For more convenient data handling, we will index each phosphosite with an unambiguous identifier comprising the UniProt accession number, the type of the modified residue, and the position within the protein. For the example of a phosphorylation of the serine 59 in the Tyrosine-protein kinase Lck, the identifier would be P06239_S59. The identifier can be constructed by concatenating the information that should be provided in the dataset. In the example of de Graaf et al., the UniProt accession number can be found in the column 'Proteins', the modified residue in 'Amino acid', and the position in 'Positions within proteins'.

The index is used to access the rows in a DataFrame and will later be needed to construct the kinase-substrate sets. After the creation of the identifier, the DataFrame is indexed by calling the function 'set_index'.

```
data_reduced['ID'] = data_reduced['Proteins'] + '_' +
data_reduced['Amino acid'] +
data_reduced['Positions within proteins']
data_indexed = data_reduced.set_index(data_reduced['ID'])
```

Mass spectrometry data is usually accompanied by several columns containing additional information about the phosphosite (e.g., the sequence window) or statistics of the database search (for example the posterior error probability), which are not necessarily needed for KSEA. We therefore extract only the columns of interest containing the processed data. In the example dataset, the names of the crucial columns start with 'Average', enabling selection by a simple 'if' statement. Generally, more complex selection of column names can be achieved by regular expressions with the python module 're'.

```
data_intensity = data_indexed[[x for x in data_indexed
    if x.startswith('Average')]] # (see Note 2)
```

Now, we can compute the fold change compared to the control, which is the condition of 0 min after stimulation. With $\log(a/b) = \log(a) - \log(b)$, we obtain the fold changes by subtracting the column with the control values from the rest using the 'sub' function of Pandas (*see* **Note 3**).

```
data_fc = data_intensity.sub(data_intensity['Average Log2 In-
tensity 0min'], axis=0)
```

Further data cleaning (re-naming columns and removal of the columns for the control time point) results in the final dataset:

```
data_fc.columns = [x.split()[-1] for x in data_fc] # Rename
columns
```

```
data_fc.drop('0min', axis=1, inplace=True) # Delete control
column
print data_fc.head()
>>>                  5min      10min      20min      30min      60min
>>> ID
>>> A0AVK6_S71   -0.319306  -0.484960  -0.798082  -0.856103
-0.928753
>>> A0FGR8_S743  -0.856661  -0.981951  -1.500412  -1.441868
-0.861470
>>> A0FGR8_S758  -1.445386  -2.397915  -2.692994  -2.794762
-1.553398
>>> A0FGR8_S691   0.271458   0.264596   0.501685   0.461984
0.655501
>>> A0JLT2_S226  -0.080786   1.069710   0.519780   0.520883
-0.296040
```

If the experiments have been performed with several replicates, statistical analysis enables estimation of the significance of the fold change compared to a control expressed by a *p*-value. The *p*-value will be needed to perform KSEA using the 'Delta count' approach but may be dispensable for the mean methods. The example dataset contains a *p*-value (transformed as negative logarithm with base 10) in selected columns and can be extracted using:

```
data_p_value = data_indexed[[x for x in data_indexed
if x.startswith('p value')]]
data_p_value = data_p_value.astype('float') # (see Note 4)
```

4.3 Loading the Kinase-Substrate Relationships

Now, we load the prior knowledge about kinase-substrate relationships. In this example, we use the information provided in the PhosphoSitePlus database (*see* **Note 5**), which can be downloaded from the website www.phosphosite.org. The organization of the data from comparable databases, e.g., Phospho.ELM, does not differ drastically from the one from PhosphoSitePlus and therefore requires only minor modifications. Using 'read_csv' again, we load the downloaded file with:

```
ks_rel = pd.read_csv('FILEPATH', sep='\t') # (see Note 6)
```

In this file, every row corresponds to an interaction between a kinase and a unique phosphosite. However, it must first be restricted to the organism of interest, e.g., 'human' or 'mouse', since the interactions of different organisms are reported together in PhosphoSitePlus.

```
ks_rel_human = ks_rel.loc[(ks_rel['KIN_ORGANISM'] == 'human') &
        (ks_rel['SUB_ORGANISM'] == 'human')]
```

Next, we again construct unique identifiers for each phosphosite using the information provided in the dataset. The modified residue and its position are already combined in the provided data.

```
ks_rel_human['psite'] = ks_rel_human['SUB_ACC_ID'] +
                   '_' + ks_rel_human['SUB_MOD_RSD']
```

Now, we construct an adjacency matrix for the phosphosites and the kinases. In this matrix, an interaction between a kinase and a phosphosite is denoted with a *1*, all other fields are filled with a *0*. For this, the Pandas function 'pivot_table' can be used:

```
ks_rel_human['value'] = 1 # (see Note 7)
adj_matrix = pd.pivot_table(ks_rel_human, values='value',
             index='psite', columns='GENE', fill_value=0)
```

The result is an adjacency matrix of the form $m \times n$ with m being the number of phosphosites and n the number of kinases. If a kinase is known to phosphorylate a given phosphosite, the corresponding entry in this matrix will be a *1*, otherwise a *0*. A *0* does not mean that there cannot be an interaction between the kinase and the respective phosphosite, but rather that this specific interaction has not been reported in the literature. As sanity check, we can print the number of known kinase-substrate interactions for each kinase saved in the adjacency matrix:

```
print adj_matrix.sum(axis=0).sort_values(ascending=False).
head()
>>> GENE
>>> CDK2        541
>>> CDK1        458
>>> PRKACA      440
>>> CSNK2A1     437
>>> SRC         391
>>> dtype: int64
```

4.4 KSEA

In the accompanying toolbox, we provide for each method of KSEA a custom python function that automates the analysis for all kinases in a given condition. Here, however, we demonstrate the principle of KSEA by computing the different activity scores for a single kinase and a single condition. As an example, the Cyclin-dependent kinase 1 (CDK1, *see* **Note 8**) and the condition of 60 min after prostaglandin stimulation shall be used.

```
data_condition = data_fc['60min'].copy()
p_values = data_p_value['p value_60vs0min']
kinase = 'CDK1'
```

First, we determine the overlap between the known targets of the kinase and the detected phosphosites in this condition, because we need it for every method of KSEA. Now, we benefit from having the same format for the index of the dataset and the adjacency matrix. We can use the Python function 'intersection' to determine the overlap between two sets.

```
substrate_set = adj_matrix[kinase].replace(
0, np.nan).dropna().index # (see Note 9)
detected_p_sites = data_condition.index
intersect=list(set(substrate_set).intersection(detected_p_-
sites))
print len(intersect)
>>> 114
```

4.4.1 KSEA Using the "Mean" Method

For the "mean" method, the KSEA score is equal to the mean of the fold changes in the substrate set *mS*.

The significance of the score is tested with a *z*-statistic using

$$z = \frac{mS - mP\sqrt{m}}{\delta}$$

with *mP* as mean of the complete dataset, *m* being the size of the substrate set, and δ the standard deviation of the complete dataset, adapted from the PAGE method for gene set enrichment [101]. The "mean" method has established itself as the preferred method in the Cutillas lab that developed the KSEA approach.

```
mS = data_condition.ix[intersect].mean()
mP = data_fc.values.mean()
m = len(intersect)
delta = data_fc.values.std()
z_score = (mS - mP) * np.sqrt(m) * 1/delta
```

The *z*-score can be converted into a *p*-value with a function from the SciPy [102] library:

```
from scipy.stats import norm
p_value_mean = norm.sf(abs(z_score))
print mS, p_value_mean
>>> -0.441268760191 9.26894825183e-07
```

4.4.2 KSEA Using the Alternative 'Mean' Method

Alternatively, only the phosphosites in the substrate set that change significantly between conditions can be considered when computing the mean of the fold changes in the substrate set. Therefore, we need a cutoff, determining a significant increase or decrease, respectively, which can be a user-supplied parameter. Here, we use a standard level to define a significant change with a cutoff of 0.05. The significance of the KSEA score is tested as before with the *z*-statistic.

```
cut_off = -np.log10(0.05)
set_alt = data_condition.ix[intersect].where(
p_values.ix[intersect] > cut_off).dropna()
mS_alt = set_alt.mean()
z_score_alt = (mS_alt - mP) * np.sqrt(len(set_alt)) * 1/delta
p_value_mean_alt = norm.sf(abs(z_score_alt))
print mS_alt, p_value_mean_alt
>>> -0.680835732551 1.26298232031e-13
```

4.4.3 KSEA Using the "Delta Count" Method

In the "Delta count" method, we count the number of phosphosites in the substrate set that are significantly increased in the condition versus the control and subtract the number of phosphosites that are significantly decreased.

```
cut_off = -np.log10(0.05)
score_delta = len(data_condition.ix[intersect].where(
(data_condition.ix[intersect] > 0) &
(p_values.ix[intersect] > cut_off)).dropna()) -
len(data_condition.ix[intersect].where(
(data_condition.ix[intersect] < 0) &
(p_values.ix[intersect] > cut_off)).dropna()) # (see Note 10)
```

The *p*-value of the score is calculated with a hypergeometric test, since the number of significantly regulated phosphosites is a discrete variable. To initialize the hypergeometric distribution, we need as variables $M =$ the total number of detected phosphosites, $n =$ the size of the substrate set, and $N =$ the total number of phosphosites that are in an arbitrary substrate set and significantly regulated.

```
from scipy.stats import hypergeom
M = len(data_condition)
n = len(intersect)
N = len(np.where(
p_values.ix[adj_matrix.index.tolist()] > cut_off)[0])
hypergeom_dist = hypergeom(M, n, N)
p_value_delta = hypergeom_dist.pmf(len(
p_values.ix[intersect].where(
p_values.ix[intersect] > cut_off).dropna()))
print score_delta, p_value_delta
>>> -58 8.42823410966e-119
```

5 Closing Remarks

In summary, the methods described in this review use different approaches to calculate kinase activities or to relate kinases to activity profiles from phosphoproteomics datasets. All of them utilize prior knowledge about kinase-substrate relationships, either from curated databases or from computational prediction tools. Using these methods, the noisy and complex information from the vast amount of detected phosphorylation sites can be condensed into a much smaller set of kinase activities that is easier to interpret. Modeling of signaling pathways or prediction of drug responses can be performed in a straightforward way with these kinase activities as shown in the study by Casado et al. [17].

The power of the described methods strongly depends on the available prior knowledge about kinase-substrate relationships. As our knowledge increases due to experimental methods like in vitro kinase selectivity studies [103] or the CEASAR (Connecting Enzymes And Substrates at Amino acid Resolution) approach [104], the utility and applicability of methods for inference of kinase activities will grow as well. Additionally, the computational approaches for the prediction of possible kinase-substrate relationships are under on-going development [84, 105], increasing the reliability of the in silico predictions.

Phosphoproteomic data is not only valuable for the analysis of kinase activities: for example, PTMfunc is a computational resource that predicts the functional impact of posttranslational modifications based on structural and domain information [15], and PHONEMeS [96, 106] combines phosphoproteomics data with prior knowledge kinase-substrate relationships, in a similar fashion as kinase-activity methods. However, instead of scoring kinases, PHONEMeS derives logic models for signaling pathways at the phosphosite level.

For the analysis of deregulated signaling in cancer, mutations in key signaling molecules can be of crucial importance. Recently, Creixell and colleagues presented a systematic classification of genomic variants that can perturb signaling, either by rewiring of the signaling network or by the destruction of phosphorylation sites [107]. Another approach was introduced in the last update of the PhosphoSitePlus database, in which the authors reported with PTMVar [20] the addition of a dataset that can map missense mutation onto the posttranslational modifications. With these tools, the challenging task of creating an intersection between genomic variations and signaling processes may be addressed.

It remains to be seen how the different scoring metrics for kinase activity relate to each other, as they utilize different approaches to extract a kinase activity score out of the data. IKAP is based on a nonlinear optimization for the model of kinase-dependent phosphorylation, KSEA on statistical analysis of the values in the substrate set of a kinase, and CLUE on the *k*-means clustering algorithm together with Fisher's exact test for enrichment. In a recent publication by Hernandez-Armenta et al. [108], the authors compiled a benchmark dataset from the literature, consisting of phosphoproteomic experiments under perturbation. For each experiment, specific kinases are expected to be regulated, e.g., EGFR receptor tyrosine kinase after stimulation with EGF. Using this "gold standard," the authors assessed how well different methods for the inference of kinase activities could recapitulate the expected kinase regulation in the different conditions. All of the assessed methods performed comparably strongly, but the authors observed a strong dependency on the prior knowledge about kinase-substrate relationships. This is a first effort to assess the applicability, performance, and drawbacks of the different methods, thereby guiding the use of phosphoproteomics data to infer kinase activities, from which to derive insights into molecular cancer biology and many other processes controlled by signal transduction.

6 Notes

1. To the sources parameter in the function get_kinase_targets, either a list of kinase-substrate interaction sources that are available in pypath or 'all' in order to include all sources can be passed. If no source is specified, only the interactions from PhosphoSitePlus and Signor will be used. The available sources in pypath are "ARN" (Autophagy Regulatory Network) [109], "CA1" (Human Hippocampal CA1 Region Neurons Signaling Network) [110], "dbPTM" [111], "DEPOD" [75], "HPRD" (Human Protein Reference Database) [92], "MIMP" (Mutation IMpact on Phosphorylation) [112], "Macrophage" (Macrophage pathways) [113], "NRF2ome" [114], "phosphoELM" [21], "PhosphoSite" [20], "SPIKE" (Signaling Pathway Integrated Knowledge Engine) [115], "SignaLink3" [116], "Signor" [71], and "TRIP" (Mammalian Transient Receptor Potential Channel-Interacting Protein Database) [117].
2. The provided code is equivalent to:

```
intensity_columns = []
for x in data_indexed:
...if x.starstwith('Average'):
... ...intensity_columns.append(x)
data_intensity = data_indexed[intensity_columns]
```

3. In our example, it is not necessary to transform the data to log2 intensities, since the data is already provided after log2-transformation. But for raw intensity values, the following function from the NumPy module can be used:

```
data_log2 = np.log2(data_intensity)
```

4. Due to a compatibility problem with the output of Excel, Python recognizes the *p*-values as string variables, not as floating point numbers. Therefore, this line is needed to convert the type of the *p*-values.
5. The adjacency matrix can also be constructed based on kinase recognition motifs or kinase prediction scores and the amino acid sequence surrounding the phosphosite. To use NetworKIN scores for the creation of the adjacency matrix, kinact will provide dedicated functions. In the presented example, however, we focus on the curated kinase-substrate relationships from PhosphoSitePlus.
6. The file from PhosphoSitePlus is provided as text file in which a tab ('\t') delimits the individual fields, not a comma. The file contains a disclaimer at the top, which has to be removed first. Alternatively, the option 'skiprows' in the function 'read_csv' can be set in order to ignore the disclaimer.
7. This column is needed, so that in the matrix resulting from pd. pivot_table the value from this column will be entered.
8. If necessary, mapping between protein names, gene names, and UniProt-Accession numbers can easily be performed with the Python module 'bioservices', to the documentation of which we want the refer the reader [118].
9. In this statement, we first select the relevant columns of the kinase from the connectivity matrix (adj_matrix[kinase]). In this column, we replace all *0* values with NAs (replace(0, np. nan)), which are then deleted with dropna(). Therefore, only those interactions remain, for which a *1* had been entered in the matrix. Of these interactions, we extract the index, which is a list of the phosphosites known to be targeted by the kinase of interest.
10. The where method will return a copy of the DataFrame, in which for cases where the condition is not true, NA is returned. dropna will therefore delete all those occurrences, so that len will count how often the condition is true.

Acknowledgments

Thanks to Emanuel Gonçalves, Aurélien Dugourd, and Claudia Hernández-Armenta for comments on the manuscript. For help with the code, thanks to Emanuel Gonçalves.

References

1. Jørgensen C, Linding R (2010) Simplistic pathways or complex networks? Curr Opin Genet Dev 20:15–22
2. Hanahan D, Weinberg RA (2011) Hallmarks of cancer: the next generation. Cell 144:646–674
3. Sawyers CL (1999) Chronic myeloid leukemia. N Engl J Med 340:1330–1340
4. Sawyers CL, Hochhaus A, Feldman E et al (2002) Imatinib induces hematologic and cytogenetic responses in patients with chronic myelogenous leukemia in myeloid blast crisis: results of a phase II study. Blood 99:3530–3539
5. Zhang J, Yang PL, Gray NS (2009) Targeting cancer with small molecule kinase inhibitors. Nat Rev Cancer 9:28–39
6. Gonzalez de Castro D, Clarke PA, Al-Lazikani B et al (2012) Personalized cancer medicine: molecular diagnostics, predictive biomarkers and drug resistance. Clin Pharmacol Ther 93:252–259
7. Cutillas PR (2015) Role of phosphoproteomics in the development of personalized cancer therapies. Proteomics Clin Appl 9:383–395
8. Bertacchini J, Guida M, Accordi B et al (2014) Feedbacks and adaptive capabilities of the PI3K/Akt/mTOR axis in acute myeloid leukemia revealed by pathway selective inhibition and phosphoproteome analysis. Leukemia 28:2197–2205
9. Cutillas PR, Khwaja A, Graupera M et al (2006) Ultrasensitive and absolute quantification of the phosphoinositide 3-kinase/Akt signal transduction pathway by mass spectrometry. Proc Natl Acad Sci U S A 103:8959–8964
10. Yu Y, Anjum R, Kubota K et al (2009) A site-specific, multiplexed kinase activity assay using stable-isotope dilution and high-resolution mass spectrometry. Proc Natl Acad Sci U S A 106:11606–11611
11. McAllister FE, Niepel M, Haas W et al (2013) Mass spectrometry based method to increase throughput for kinome analyses using ATP probes. Anal Chem 85:4666–4674
12. Doll S, Burlingame AL (2015) Mass spectrometry-based detection and assignment of protein posttranslational modifications. ACS Chem Biol 10:63–71
13. Choudhary C, Mann M (2010) Decoding signalling networks by mass spectrometry-based proteomics. Nat Rev Mol Cell Biol 11:427–439
14. Sabidó E, Selevsek N, Aebersold R (2012) Mass spectrometry-based proteomics for systems biology. Curr Opin Biotechnol 23:591–597
15. Beltrao P, Albanèse V, Kenner LR et al (2012) Systematic functional prioritization of protein posttranslational modifications. Cell 150:413–425
16. Qi L, Liu Z, Wang J et al (2014) Systematic analysis of the phosphoproteome and kinase-substrate networks in the mouse testis. Mol Cell Proteomics 13:3626–3638
17. Casado P, Rodriguez-Prados J-C, Cosulich SC et al (2013) Kinase-substrate enrichment analysis provides insights into the heterogeneity of signaling pathway activation in leukemia cells. Sci Signal 6:rs6
18. Yang P, Zheng X, Jayaswal V et al (2015) Knowledge-based analysis for detecting key signaling events from time-series Phosphoproteomics data. PLoS Comput Biol 11: e1004403
19. Mischnik M, Sacco F, Cox J et al (2015) IKAP: a heuristic framework for inference of kinase activities from Phosphoproteomics data. Bioinformatics 32(3):424–431
20. Hornbeck PV, Zhang B, Murray B et al (2015) PhosphoSitePlus, 2014: mutations, PTMs and recalibrations. Nucleic Acids Res 43:D512–D520
21. Dinkel H, Chica C, Via A et al (2011) Phospho.ELM: a database of phosphorylation sites—update 2011. Nucleic Acids Res 39: D261–D267
22. Horn H, Schoof EM, Kim J et al (2014) KinomeXplorer: an integrated platform for kinome biology studies. Nat Methods 11:603–604

23. Song C, Ye M, Liu Z et al (2012) Systematic analysis of protein phosphorylation networks from phosphoproteomic data. Mol Cell Proteomics 11:1070–1083
24. Riley NM, Coon JJ (2016) Phosphoproteomics in the age of rapid and deep proteome profiling. Anal Chem 88:74–94
25. Nilsson CL (2012) Advances in quantitative phosphoproteomics. Anal Chem 84:735–746
26. Hennrich ML, Gavin A-C (2015) Quantitative mass spectrometry of posttranslational modifications: keys to confidence. Sci Signal 8:re5
27. Giansanti P, Aye TT, van den Toorn H et al (2015) An augmented multiple-protease-based human phosphopeptide atlas. Cell Rep 11:1834–1843
28. Ruprecht B, Roesli C, Lemeer S et al (2016) MALDI-TOF and nESI Orbitrap MS/MS identify orthogonal parts of the phosphoproteome. Proteomics 16(10):1447–1456
29. Zhou H, Ye M, Dong J et al (2013) Robust phosphoproteome enrichment using monodisperse microsphere-based immobilized titanium (IV) ion affinity chromatography. Nat Protoc 8:461–480
30. Rush J, Moritz A, Lee KA et al (2005) Immunoaffinity profiling of tyrosine phosphorylation in cancer cells. Nat Biotechnol 23:94–101
31. Ruprecht B, Koch H, Medard G et al (2015) Comprehensive and reproducible phosphopeptide enrichment using iron immobilized metal ion affinity chromatography (Fe-IMAC) columns. Mol Cell Proteomics 14:205–215
32. Domon B, Aebersold R (2006) Mass spectrometry and protein analysis. Science (New York, NY) 312:212–217
33. Nesvizhskii AI (2007) Protein identification by tandem mass spectrometry and sequence database searching. Methods Mol Biol (Clifton, NJ) 367:87–119
34. Liu H, Sadygov RG, Yates JR (2004) A model for random sampling and estimation of relative protein abundance in shotgun proteomics. Anal Chem 76:4193–4201
35. Cutillas PR, Vanhaesebroeck B (2007) Quantitative profile of five murine core proteomes using label-free functional proteomics. Mol Cell Proteomics 6:1560–1573
36. Cutillas PR, Geering B, Waterfield MD et al (2005) Quantification of gel-separated proteins and their phosphorylation sites by LC-MS using unlabeled internal standards: analysis of phosphoprotein dynamics in a B cell lymphoma cell line. Mol Cell Proteomics 4:1038–1051
37. Bateman NW, Goulding SP, Shulman NJ et al (2014) Maximizing peptide identification events in proteomic workflows using data-dependent acquisition (DDA). Mol Cell Proteomics 13:329–338
38. Alcolea MP, Casado P, Rodríguez-Prados J-C et al (2012) Phosphoproteomic analysis of leukemia cells under basal and drug-treated conditions identifies markers of kinase pathway activation and mechanisms of resistance. Mol Cell Proteomics 11:453–466
39. Cox J, Hein MY, Luber CA et al (2014) Accurate proteome-wide label-free quantification by delayed normalization and maximal peptide ratio extraction, termed MaxLFQ. Mol Cell Proteomics 13:2513–2526
40. Strittmatter EF, Ferguson PL, Tang K et al (2003) Proteome analyses using accurate mass and elution time peptide tags with capillary LC time-of-flight mass spectrometry. J Am Soc Mass Spectrom 14:980–991
41. Lange V, Picotti P, Domon B et al (2008) Selected reaction monitoring for quantitative proteomics: a tutorial. Mol Syst Biol 4:222
42. Gillet LC, Navarro P, Tate S et al (2012) Targeted data extraction of the MS/MS spectra generated by data-independent acquisition: a new concept for consistent and accurate proteome analysis. Mol Cell Proteomics 11:O111.016717
43. Parker BL, Yang G, Humphrey SJ et al (2015) Targeted phosphoproteomics of insulin signaling using data-independent acquisition mass spectrometry. Sci Signal 8:rs6
44. Sidoli S, Fujiwara R, Kulej K et al (2016) Differential quantification of isobaric phosphopeptides using data-independent acquisition mass spectrometry. Mol BioSyst 12 (8):2385–2388
45. Keller A, Bader SL, Kusebauch U et al (2016) Opening a SWATH window on posttranslational modifications: automated pursuit of modified peptides. Mol Cell Proteomics 15:1151–1163
46. Ong S-E, Blagoev B, Kratchmarova I et al (2002) Stable isotope labeling by amino acids in cell culture, SILAC, as a simple and accurate approach to expression proteomics. Mol Cell Proteomics 1:376–386
47. Zanivan S, Meves A, Behrendt K et al (2013) In vivo SILAC-based proteomics reveals phosphoproteome changes during mouse skin carcinogenesis. Cell Rep 3:552–566

48. Shenoy A, Geiger T (2015) Super-SILAC: current trends and future perspectives. Expert Rev Proteomics 12:13–19
49. Thompson A, Schäfer J, Kuhn K et al (2003) Tandem mass tags: a novel quantification strategy for comparative analysis of complex protein mixtures by MS/MS. Anal Chem 75:1895–1904
50. Ross PL, Huang YN, Marchese JN et al (2004) Multiplexed protein quantitation in Saccharomyces cerevisiae using amine-reactive isobaric tagging reagents. Mol Cell Proteomics 3:1154–1169
51. Li Z, Adams RM, Chourey K et al (2012) Systematic comparison of label-free, metabolic labeling, and isobaric chemical labeling for quantitative proteomics on LTQ Orbitrap Velos. J Proteome Res 11:1582–1590
52. Chelius D, Bondarenko PV (2002) Quantitative profiling of proteins in complex mixtures using liquid chromatography and mass spectrometry. J Proteome Res 1:317–323
53. Neilson KA, Ali NA, Muralidharan S et al (2011) Less label, more free: approaches in label-free quantitative mass spectrometry. Proteomics 11:535–553
54. Perkins DN, Pappin DJ, Creasy DM et al (1999) Probability-based protein identification by searching sequence databases using mass spectrometry data. Electrophoresis 20:3551–3567
55. Clauser KR, Baker P, Burlingame AL (1999) Role of accurate mass measurement (+/−10 ppm) in protein identification strategies employing MS or MS/MS and database searching. Anal Chem 71:2871–2882
56. MacCoss MJ, Wu CC, Yates JR (2002) Probability-based validation of protein identifications using a modified SEQUEST algorithm. Anal Chem 74:5593–5599
57. Cox J, Neuhauser N, Michalski A et al (2011) Andromeda: a peptide search engine integrated into the MaxQuant environment. J Proteome Res 10:1794–1805
58. Beausoleil SA, Villén J, Gerber SA et al (2006) A probability-based approach for high-throughput protein phosphorylation analysis and site localization. Nat Biotechnol 24:1285–1292
59. Savitski MM, Lemeer S, Boesche M et al (2011) Confident phosphorylation site localization using the Mascot Delta Score. Mol Cell Proteomics 10:M110.003830
60. Chalkley RJ, Clauser KR (2012) Modification site localization scoring: strategies and performance. Mol Cell Proteomics 11:3–14
61. Baker PR, Trinidad JC, Chalkley RJ (2011) Modification site localization scoring integrated into a search engine. Mol Cell Proteomics 10:M111.008078
62. Lemeer S, Heck AJR (2009) The phosphoproteomics data explosion. Curr Opin Chem Biol 13:414–420
63. Sharma K, D'Souza RCJ, Tyanova S et al (2014) Ultradeep human phosphoproteome reveals a distinct regulatory nature of Tyr and Ser/Thr-based signaling. Cell Rep 8:1583–1594
64. Olsen JV, Blagoev B, Gnad F et al (2006) Global, in vivo, and site-specific phosphorylation dynamics in signaling networks. Cell 127:635–648
65. Olsen JV, Vermeulen M, Santamaria A et al (2010) Quantitative phosphoproteomics reveals widespread full phosphorylation site occupancy during mitosis. Sci Signal 3:ra3
66. Landry CR, Levy ED, Michnick SW (2009) Weak functional constraints on phosphoproteomes. Trends Genet 25:193–197
67. Beltrao P, Trinidad JC, Fiedler D et al (2009) Evolution of phosphoregulation: comparison of phosphorylation patterns across yeast species. PLoS Biol 7:e1000134
68. Beltrao P, Bork P, Krogan NJ et al (2013) Evolution and functional cross-talk of protein post-translational modifications. Mol Syst Biol 9:714
69. Newman RH, Zhang J, Zhu H (2014) Toward a systems-level view of dynamic phosphorylation networks. Front Genet 5:263
70. Glickman JF (2012) Assay development for protein kinase enzymes. Eli Lilly & Company and the National Center for Advancing Translational Sciences, Bethesda, MD. http://www.ncbi.nlm.nih.gov/books/NBK91991/
71. Perfetto L, Briganti L, Calderone A et al (2016) SIGNOR: a database of causal relationships between biological entities. Nucleic Acids Res 44:D548–D554
72. Gnad F, Gunawardena J, Mann M (2011) PHOSIDA 2011: the posttranslational modification database. Nucleic Acids Res 39: D253–D260
73. Hu J, Rho H-S, Newman RH et al (2014) PhosphoNetworks: a database for human phosphorylation networks. Bioinformatics (Oxford, England) 30:141–142
74. Sadowski I, Breitkreutz B-J, Stark C et al (2013) The PhosphoGRID Saccharomyces cerevisiae protein phosphorylation site database: version 2.0 update. Database 2013: bat026

75. Duan G, Li X, Köhn M (2015) The human DEPhOsphorylation database DEPOD: a 2015 update. Nucleic Acids Res 43: D531–D535
76. Zhang H, Zha X, Tan Y et al (2002) Phosphoprotein analysis using antibodies broadly reactive against phosphorylated motifs. J Biol Chem 277:39379–39387
77. Obenauer JC, Cantley LC, Yaffe MB (2003) Scansite 2.0: proteome-wide prediction of cell signaling interactions using short sequence motifs. Nucleic Acids Res 31:3635–3641
78. C. Chen and B.E. Turk (2010) Analysis of serine-threonine kinase specificity using arrayed positional scanning peptide libraries., Curr Protoc Mol Biol Chapter 18:Unit 18.14
79. Sidhu SS, Koide S (2007) Phage display for engineering and analyzing protein interaction interfaces. Curr Opin Struct Biol 17:481–487
80. Miller ML, Jensen LJ, Diella F et al (2008) Linear motif atlas for phosphorylation-dependent signaling. Sci Signal 1:ra2
81. Hjerrild M, Stensballe A, Rasmussen TE et al (2004) Identification of phosphorylation sites in protein kinase A substrates using artificial neural networks and mass spectrometry. J Proteome Res 3:426–433
82. Linding R, Jensen LJ, Pasculescu A et al (2008) NetworKIN: a resource for exploring cellular phosphorylation networks. Nucleic Acids Res 36:D695–D699
83. Szklarczyk D, Franceschini A, Wyder S et al (2015) STRING v10: protein-protein interaction networks, integrated over the tree of life. Nucleic Acids Res 43:D447–D452
84. Wagih O, Sugiyama N, Ishihama Y et al (2016) Uncovering phosphorylation-based specificities through functional interaction networks. Mol Cell Proteomics 15:236–245
85. Linding R, Jensen LJ, Ostheimer GJ et al (2007) Systematic discovery of in vivo phosphorylation networks. Cell 129:1415–1426
86. Subramanian A, Tamayo P, Mootha VK et al (2005) Gene set enrichment analysis: a knowledge-based approach for interpreting genome-wide expression profiles. Proc Natl Acad Sci U S A 102:15545–15550
87. Schacht T, Oswald M, Eils R et al (2014) Estimating the activity of transcription factors by the effect on their target genes. Bioinformatics (Oxford, England) 30:i401–i407
88. Drake JM, Graham NA, Stoyanova T et al (2012) Oncogene-specific activation of tyrosine kinase networks during prostate cancer progression. Proc Natl Acad Sci 109:1643–1648
89. Chen EY, Tan CM, Kou Y et al (2013) Enrichr: interactive and collaborative HTML5 gene list enrichment analysis tool. BMC Bioinformatics 14:128
90. Kuleshov MV, Jones MR, Rouillard AD et al (2016) Enrichr: a comprehensive gene set enrichment analysis web server 2016 update. Nucleic Acids Res 44(W1):W90–W97
91. Lachmann A, Ma'ayan A (2009) KEA: kinase enrichment analysis. Bioinformatics (Oxford, England) 25:684–686
92. Keshava Prasad TS, Goel R, Kandasamy K et al (2009) Human Protein Reference Database—2009 update. Nucleic Acids Res 37: D767–D772
93. Huttlin EL, Jedrychowski MP, Elias JE et al (2010) A tissue-specific atlas of mouse protein phosphorylation and expression. Cell 143:1174–1189
94. de Graaf EL, Giansanti P, Altelaar AFM et al (2014) Single-step enrichment by Ti4+-IMAC and label-free quantitation enables in-depth monitoring of phosphorylation dynamics with high reproducibility and temporal resolution. Mol Cell Proteomics 13:2426–2434
95. Wilm M, Mann M (1996) Analytical properties of the nanoelectrospray ion source. Anal Chem 68:1–8
96. Wilkes EH, Terfve C, Gribben JG et al (2015) Empirical inference of circuitry and plasticity in a kinase signaling network. Proc Natl Acad Sci U S A 112:7719–7724
97. Türei D, Korcsmáros T, Saez-Rodriguez J (2016) OmniPath: guidelines and gateway for literature-curated signaling pathway resources. Nat Methods 13:966–967
98. Benjamini Y, Hochberg Y (2000) On the adaptive control of the false discovery rate in multiple testing with independent statistics. J Educ Behav Stat 25:60–83
99. Mckinney W (2010) Data structures for statistical computing in python. Proceedings of the 9th python in science conference
100. Van Der Walt S, Colbert SC, Varoquaux G (2011) The NumPy Array: A Structure for Efficient Numerical Computation, Comput Sci Eng 13:22–30. https://doi.org/10.1109/MCSE.2011.37
101. Kim S-Y, Volsky DJ (2005) PAGE: parametric analysis of gene set enrichment. BMC Bioinformatics 6:144
102. Jones E, Oliphant TE, Peterson P (2007) Python for scientific computing. Comput Sci Eng 9:10–20
103. Imamura H, Sugiyama N, Wakabayashi M et al (2014) Large-scale identification of

phosphorylation sites for profiling protein kinase selectivity. J Proteome Res 13:3410–3419

104. Newman RH, Hu J, Rho H-S et al (2013) Construction of human activity-based phosphorylation networks. Mol Syst Biol 9:655

105. Creixell P, Palmeri A, Miller CJ et al (2015) Unmasking determinants of specificity in the human kinome. Cell 163:187–201

106. Terfve CDA, Wilkes EH, Casado P et al (2015) Large-scale models of signal propagation in human cells derived from discovery phosphoproteomic data. Nat Commun 6:8033

107. Creixell P, Schoof EM, Simpson CD et al (2015) Kinome-wide decoding of network-attacking mutations rewiring cancer signaling. Cell 163:202–217

108. Hernandez-Armenta C, Ochoa D, Goncalves E et al (2016) Benchmarking substrate-based kinase activity inference using phosphoproteomic data. Bioinformatics 33 (12):1845–1851

109. Türei D, Földvári-Nagy L, Fazekas D et al (2015) Autophagy Regulatory Network - a systems-level bioinformatics resource for studying the mechanism and regulation of autophagy. Autophagy 11:155–165

110. Ma'ayan A, Jenkins SL, Neves S et al (2005) Formation of regulatory patterns during signal propagation in a Mammalian cellular network. Science (New York, NY) 309:1078–1083

111. Huang K-Y, Su M-G, Kao H-J et al (2016) dbPTM 2016: 10-year anniversary of a resource for post-translational modification of proteins. Nucleic Acids Res 44: D435–D446

112. Wagih O, Reimand J, Bader GD (2015) MIMP: predicting the impact of mutations on kinase-substrate phosphorylation. Nat Methods 12:531–533

113. Raza S, McDerment N, Lacaze PA et al (2010) Construction of a large scale integrated map of macrophage pathogen recognition and effector systems. BMC Syst Biol 4:63

114. Türei D, Papp D, Fazekas D et al (2013) NRF2-ome: an integrated web resource to discover protein interaction and regulatory networks of NRF2. Oxidative Med Cell Longev 2013:737591

115. Paz A, Brownstein Z, Ber Y et al (2011) SPIKE: a database of highly curated human signaling pathways. Nucleic Acids Res 39: D793–D799

116. Fazekas D, Koltai M, Türei D et al (2013) SignaLink 2 - a signaling pathway resource with multi-layered regulatory networks. BMC Syst Biol 7:7

117. Chun JN, Lim JM, Kang Y et al (2014) A network perspective on unraveling the role of TRP channels in biology and disease. Pflugers Arch 466:173–182

118. Cokelaer T, Pultz D, Harder LM et al (2013) BioServices: a common Python package to access biological Web Services programmatically. Bioinformatics 29:3241–3242

Open Access This chapter is licensed under the terms of the Creative Commons Attribution 4.0 International License (http://creativecommons.org/licenses/by/4.0/), which permits use, sharing, adaptation, distribution and reproduction in any medium or format, as long as you give appropriate credit to the original author(s) and the source, provide a link to the Creative Commons license and indicate if changes were made.

The images or other third party material in this chapter are included in the chapter's Creative Commons license, unless indicated otherwise in a credit line to the material. If material is not included in the chapter's Creative Commons license and your intended use is not permitted by statutory regulation or exceeds the permitted use, you will need to obtain permission directly from the copyright holder.

Chapter 7

Perseus: A Bioinformatics Platform for Integrative Analysis of Proteomics Data in Cancer Research

Stefka Tyanova and Juergen Cox

Abstract

Mass spectrometry-based proteomics is a continuously growing field marked by technological and methodological improvements. Cancer proteomics is aimed at pursuing goals such as accurate diagnosis, patient stratification, and biomarker discovery, relying on the richness of information of quantitative proteome profiles. Translating these high-dimensional data into biological findings of clinical importance necessitates the use of robust and powerful computational tools and methods. In this chapter, we provide a detailed description of standard analysis steps for a clinical proteomics dataset performed in Perseus, a software for functional analysis of large-scale quantitative omics data.

Key words Perseus, Software, Omics data analysis, Translational bioinformatics, Cancer proteomics

1 Introduction

High-resolution mass spectrometry-based proteomics, aided by computational sciences, is continuously pushing the boundaries of systems biology. Obtaining highly accurate quantitative proteomes on a genome-wide scale is becoming feasible within realistic measurement times [1]. Similar to the clinical goals of genomics and transcriptomics to provide a deeper understanding of a certain disease that goes beyond the standard clinical parameters of cancer diagnosis, proteomics offers a comprehensive view of the molecular players in a cell at a particular moment and in a specific state [1]. The maturation of the technology together with the development of suitable methods for quantification of human tissue proteomes [2–4] has opened new doors for employing proteomics in medical applications and is shaping the growing field of clinical proteomics [5, 6]. Following these advances, proteomic approaches have been used to address multiple clinical questions in the context of various cancer types. The major area of application is the

Louise von Stechow (ed.), *Cancer Systems Biology: Methods and Protocols*, Methods in Molecular Biology, vol. 1711, https://doi.org/10.1007/978-1-4939-7493-1_7, © The Author(s) 2018

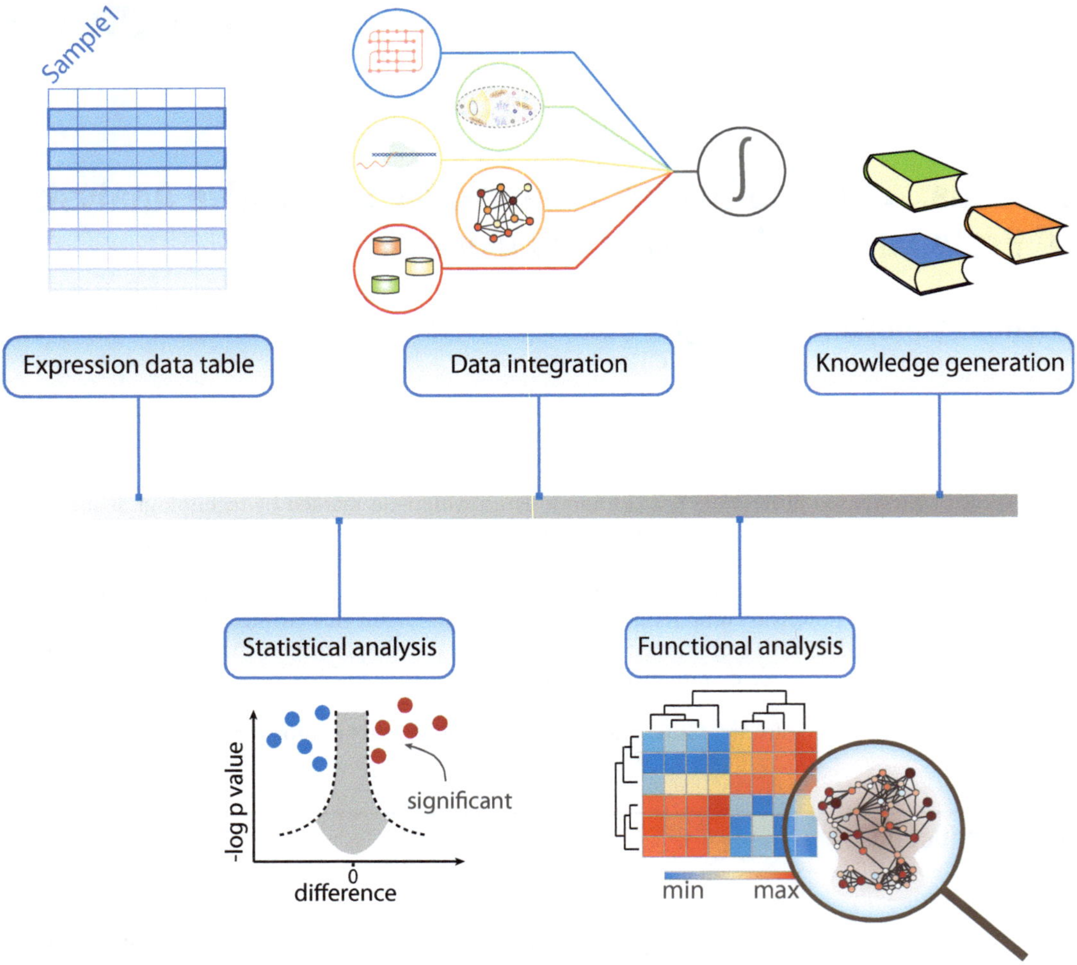

Fig. 1 Outline of a typical analysis workflow in Perseus. The workflow shows the process of converting data into information and knowledge. Statistical analysis can be used to guide the identification of biologically relevant hits and drive hypotheses generation. Various external databases, annotation sources, and multiple omics types can be loaded and matched within the software and together with powerful enrichment techniques allow for smooth data integration

profiling of cancer-relevant tissues—including the proteomes of colorectal cancer [7, 8] and prostate cancer [9], as well as the subtyping of lymphoma [10] and breast cancer [11, 12] patients. Although proteomics has become an extremely powerful approach for studying biomedical questions, offering unique advantages compared to other omics techniques, the functional interpretation of the vast amounts of data of a typical proteomics experiment often poses analytical challenges to the biological domain experts.

The aim of data analysis is to translate large amounts of proteomic data that cover numerous samples, conditions and time points into structured, domain-specific knowledge that can guide clinical decisions (Fig. 1). Prior to any statistical analysis, data cleansing is usually performed which includes normalization, to ensure that different samples are comparable, and missing value handling to enable the use of methods that require all data points to be present. A plethora of imputation methods developed for microarray data [13] can be applied to proteomics as well [14]. Among these, methods with the underlying assumption that missing values result from protein expression that lies under the detection limit of modern mass spectrometers are frequently used. A typical task of clinical proteomics studies is to identify proteins that show differential expression between healthy and diseased states or between different subtypes of a disease. Although commonly established statistical methods, which achieve this task exist, distinguishing between expression differences due to technical variability, genetic heterogeneity, or even intra-sample variability and true disease-related changes require deep knowledge of statistical tools and good understanding of the underlying problems in the analysis of omics data.

For instance, testing thousands of proteins for differential expression is hampered by the multiple hypothesis-testing problem, which results in an increased probability of calling a protein a significant hit when there is no actual difference in expression. This necessitates the use of correction methods to increase the confidence of the identified hits. The choice of the appropriate correction method depends on the balance between wrongly accepted hits (error type I) and wrongly rejected hits (error type II) that an experimentalist is willing to accept. For instance, permutation-based FDR [15] has a reduced error type II rate compared to the Benjamini-Hochberg correction [16]. Once the initial list of quantified proteins is narrowed down to only the significantly changing hits the question of their functional relevance arises. Enrichment analysis of protein annotations is the preferred method for deriving functional implications of sets of proteins and is applicable to both categorical (Fisher's exact test [17]) and expression/numerical data (1D enrichment test [18]). The outcome of such an analysis often offers a comprehensive view of the biological roles of the selected proteins through highlighting key pathways and cellular processes in which they are involved.

In this chapter, we provide a step-by-step workflow of bioinformatic analysis of proteomics data of luminal-type breast cancer progression. Commonly used analytical practices are described including data cleansing and preprocessing, exploratory analysis, statistical methods and guidelines, as well as functional enrichment techniques. All the steps are implemented as processes in Perseus [19], a comprehensive software for functional analysis of omics data.

2 Materials

2.1 Software Download and Installation

Written in C#, Perseus achieves optimal performance when run on Windows operating systems. The latest versions require 64 bit system and .NET Framework 4.5 to be installed on the same computer. To use the software on MacOS set up BootCamp and optionally in addition Parallels. Registration and acceptance of the Software License Agreement are required prior to downloading Perseus from the official website: http://www.coxdocs.org/doku.php?id=perseus:start . Once the download has finished, decompress the folder, locate the Perseus.exe file, and double-click it to start the program.

2.2 Data Files

In the subsequent analysis, we used a subset of the data measured by Pozniak et al. [20]. The authors provide a genome-wide proteomic analysis of progression of breast cancer in patients by studying major differences at the proteome level between healthy, primary tumor, and metastatic tissues. The data were measured as ratios between an optimized heavy-labeled mix of cell lines representing different breast cancer stages and the patient proteome [2]. This constitutes an accurate relative quantification approach used especially in the analysis of clinical samples. Peptide and protein identification and quantification was performed using the MaxQuant suite for the analysis of raw mass spectrometry data [21] at peptide spectrum match and protein false discovery rate of 1%. The subset used in this protocol contains proteome profiles of 22 healthy, 21 lymph node negative, and 25 lymph node metastatic tissue samples and spans over 10,000 protein groups and can be found in the proteinGroups.txt file provided as supplementary material to the Pozniak et al. study (*see* **Note 1**).

3 Methods

The Methods section contains several modules covering the most frequently performed steps in the analysis of proteomics data. Often, a proteomics study benefits from a global overview of the data, which usually includes the total number of identified and quantified proteins, dynamic range, coverage of specific pathways, and groups of proteins. A good practice in data analysis is to start with exploratory statistics in order to check for biases in the data, undesirable outliers, and experiments with poor quality data and to make sure that all requirements for performing the subsequent statistical tests are met. Once the data are filtered and normalized appropriately, statistical and bioinformatic analyses are performed in order to identify proteins that are likely to be functionally-important. When the list of such proteins is small enough and direct links to the question of interest can be inferred using prior knowledge, follow-up experiments can be performed after this step to confirm the results of the statistical analysis. However, one of the advantages of mass spectrometry-based proteomics is the ability to unravel new discoveries in an unbiased way, for instance, through functional analysis. This analysis is often based on enrichment tests, which can highlight guiding biological processes and mechanisms.

3.1 Loading the Data

1. Go to the "Load" section in Perseus and click the "Generic matrix upload" button.
2. In the pop-up window, navigate to the file to be loaded (*see* **Note 2**).
3. Select all the expression columns and transfer them to the *Main* columns window (*see* **Note 3**). Select all additional numerical data that may be needed in the analysis and transfer them to the *Numerical* columns window. Make sure that the columns containing identifiers (e.g., protein IDs) are selected as *Text* columns. Click *ok*.

3.2 Summary Statistics

Get familiar with the Software and its five main sections: Load, Processing, Analysis, Multi-processing, and Export (*see* Fig. 2).

1. In the workflow panel, change the name of the data matrix from *matrix 1* to *InitialData* by right-clicking the node and changing the *Alternative name* box. Close the pop-up window. Explore the right-most panel of Perseus, which contains useful information such as number of main columns and number of rows.
2. Go to "Processing → Filter rows → Filter rows based on categorical column" to exclude proteins identified by site, matching to the reverse database or contaminants (*see* **Note 4**).
3. Transform the data to a logarithmic scale by going to "Processing → Basic → Transform" and specifying the transformation function (e.g., log2(x)).
4. In the "Processing" section, select the "Basic" menu and click on the "Summary statistics (columns)" button. Select all expression columns by transferring them to the right-hand side. Click *ok* and explore the new matrix.

3.3 Filtering

1. Use the workflow window to select the *InitialData* matrix data by clicking on it (*see* **Note 5**).
2. In the "Processing" section, go to the "Filter rows" menu and select "Filter rows based on valid values." Change the *Min. valids* parameter to *Percentage* and keep the default value of *70%* for the *Min. percentage of values* parameter. Click *ok*. Check how many protein groups were retained after the filtering (*see* **Note 6**).

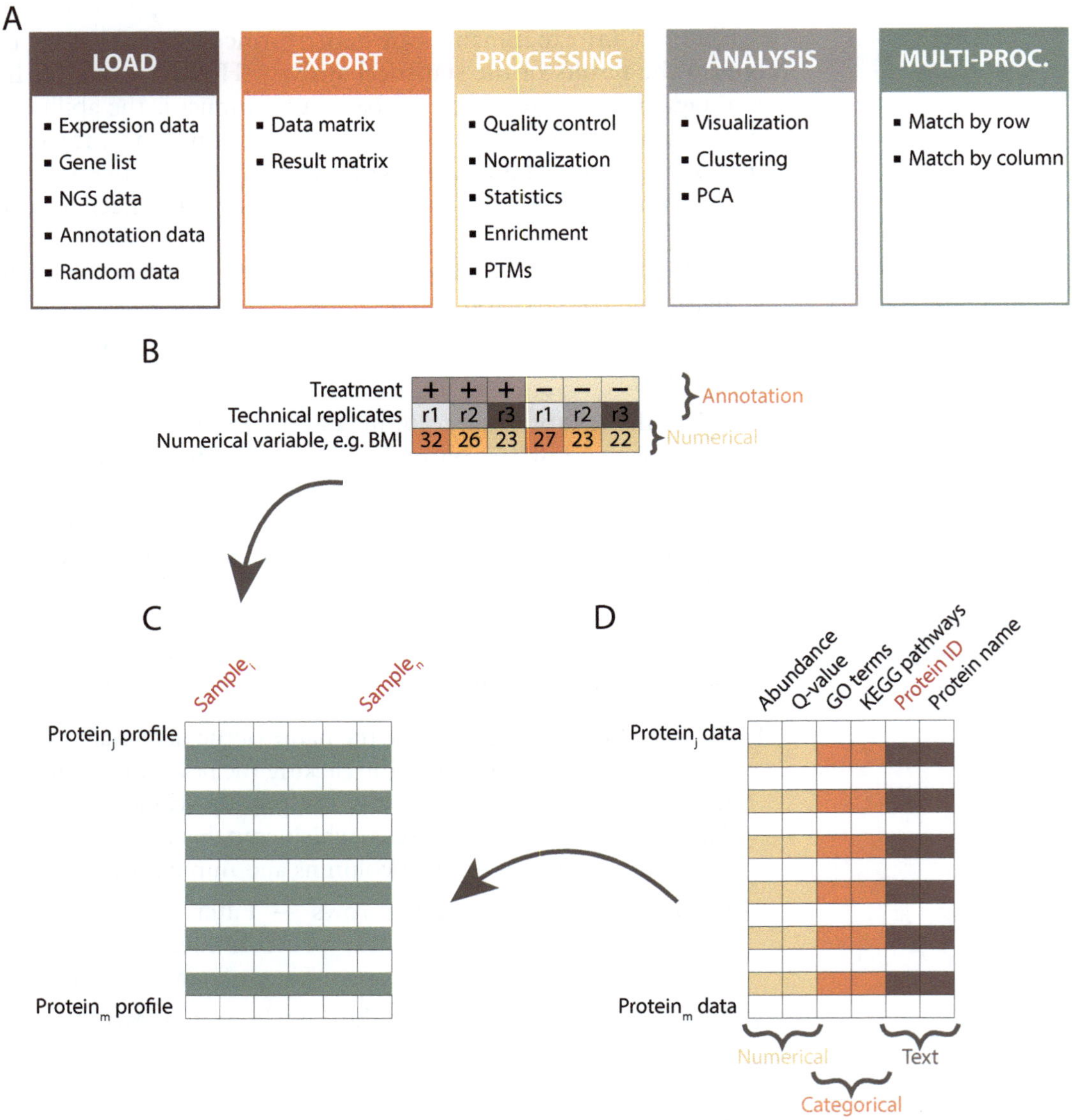

Fig. 2 Interfaces of Perseus and the augmented data matrix format. (**a**) Perseus extends over five interfaces, each of which includes various analysis and transformation functionalities and visualization possibilities. (**b**) Experimental design is specified as annotation (e.g., treatment vs. control groups) or numerical rows (e.g., variable concentration). Multiple annotation rows can be specified that allow biological and technical replicates to be analyzed together. (**c**) The data is organized in a matrix format where typically all samples are displayed as columns and all proteins as rows. (**d**) Additional protein information can be added in the form of Numerical, Categorical, or Text annotations

3.4 Exploratory Analysis

1. To visually inspect the data, go to "Analysis → Visualization → Histograms." Select all the samples of interest by transferring them to the right-hand side. Click *ok*.
2. Explore the visualization options in the Histogram panel by testing the functionality of each of the buttons (e.g., *Properties*, *Fit width*, *Fit height*).

3. Click on the *pdf* button to export the plot (*see* **Note 7**).
4. Switch the view to the "Data" tab.
5. Go to "Analysis → Visualization → Multi scatter plot." Select the desired samples by transferring them to the right-hand side. Click *ok* (*see* Fig. 3).
6. Adjust the plot using the *Fit width* and *Fit height* options and resizing the plot window.
7. In the drop-down menu "Display in plots" in the plot window, select *Pearson correlation*.
8. Select a scatter plot by clicking on it. The selected plot will be shown in an enlarged view.
9. Select a number of proteins from the "Point" table on the right of the multi scatter plot and examine their position in all pairwise sample comparisons.
10. Switch back to the "Data" tab to continue with the analysis.
11. "Go to Processing → Basic → Column correlation." Make sure that the *Type* is set to *Pearson correlation*. The output table contains all pairwise correlations between the selected columns.
12. To visualize the sample correlations, go to "Analysis → Clustering/PCA → Hierarchical clustering." Use the *Change color gradient* to set a continuous gradient similar to the one in Fig. 3a.
13. Export the plot by clicking on the *pdf* button.
14. Navigate back to the previous data matrix by clicking on it in the workflow panel.
15. Principal component analysis requires all values to be valid. To remove all protein groups with missing values, repeat Subheading 3.3, **step 2** setting the percentage parameter to *100* (*see* **Note 8**).
16. Go to "Analysis → Clustering/PCA → Principal component analysis" and click *ok*. Explore the sample separation (dot plot in the upper panel) and the corresponding loadings (dot plot in the lower panel).
17. In the table on the right of the PCA plot, select a set of samples (e.g., all samples that belong to one experimental condition) and change their color by clicking on the *Symbol color* button and selecting the desired color.
18. Check the contribution of other components by substituting Component 1 and 2 with other components from the drop-down menu. Find the components that show sample separation according to the experimental conditions (*see* Fig. 3c).

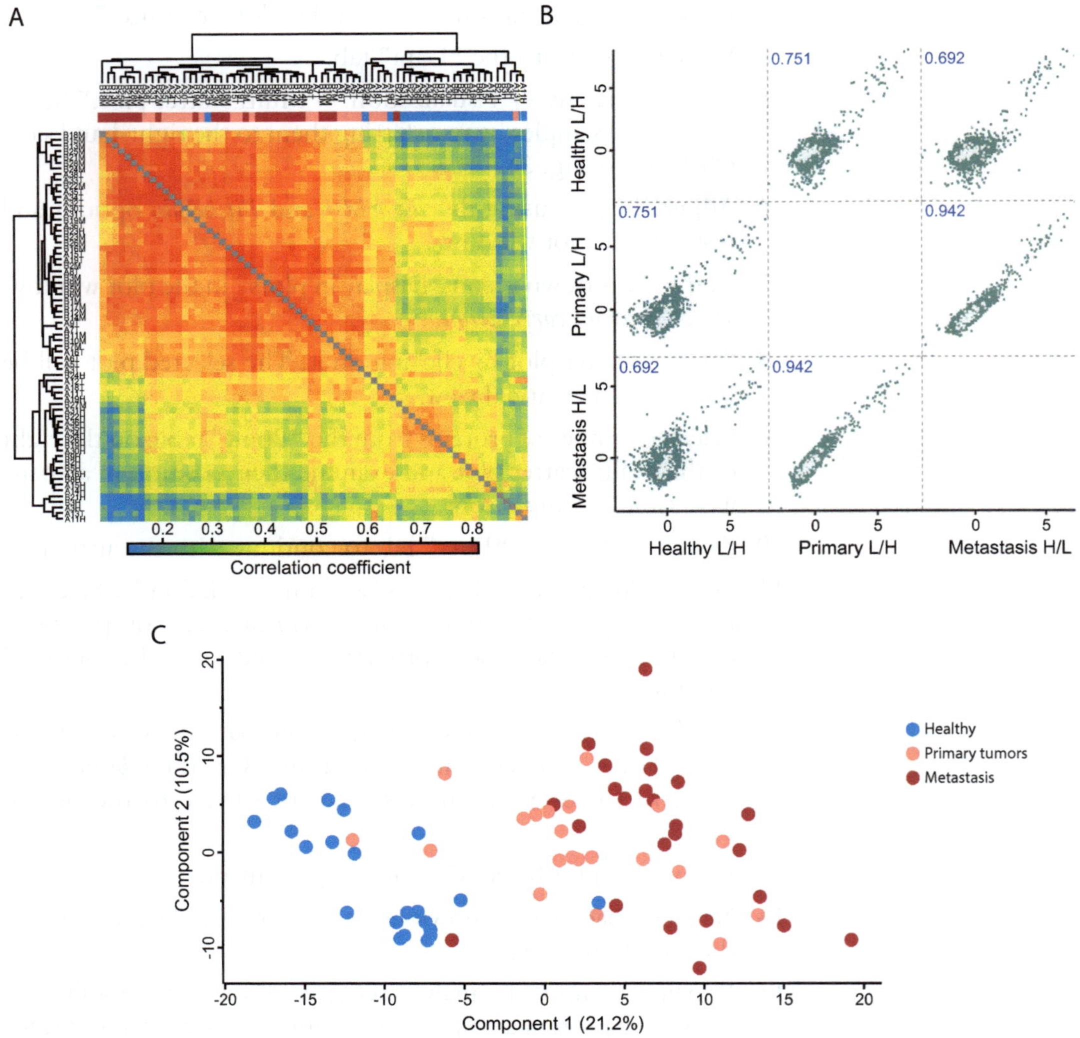

Fig. 3 Exploratory analysis outputs in Perseus. (**a**) Hierarchical clustering of all the samples based on the correlation coefficients between them reveals higher similarity between primary and metastatic tumors versus healthy tissue samples. (**b**) Multi-scatter plot of averaged profiles among the three main groups clearly represents the disease progression by highlighting strong similarities between subsequent stages, e.g., healthy tissue samples are more similar to primary tumors than to metastasis (correlation coefficient 0.76 vs 0.69), whereas primary tumors are most similar to metastasis ($R = 0.94$). The category Cell division is highlighted in bright green in all pairwise comparison plots. (**c**) Principal component analysis (PCA) attributes the largest variance to the difference between healthy (blue dots) and cancer tissues (pink and red dots) (Component 1, 21.1%) and shows that primary and metastatic tumors (pink and red dots respectively) are difficult to distinguish

19. Explore the proteins driving this separation. In the loadings plot beneath the PCA, change the selection *Mode* to *rectangular selection*. Hold the left mouse key down and draw a rectangle around the dots in the upper right corner and then release the mouse. The selected proteins are highlighted in the table to the right and their labels are displayed in the plot.

3.5 Normalization

1. Navigate back to the data matrix before filtering for 100% valid values (Subheading 3.3, **step 2**).
2. Go to "Processing → Normalization → Z-score." Change the *Matrix access* parameter to *Columns* and select the *Use median* option. In the new data table, plot histograms for the same subset of samples as in Subheading 3.4, **step 1** (*see* **Note 9**).

3.6 Experimental Design

1. Go to "Processing → Annot. rows → Categorical annotation rows." Use the *Create action* option to manually specify the experimental condition to which a sample belongs (i.e., indicate control versus stimulus, or different stages of a disease). All the samples belonging to one condition should have the same annotation. A new row will be added under the column names in the newly generated data matrix (*see* **Note 10**).

3.7 Loading Annotations

1. Go to the drop-down menu indicated with a white arrow at the top left corner of Perseus and select "Annotation download."
2. Click on the link in the pop-up window. Select the appropriate annotation file (e.g., "PerseusAnnotaion → FrequentlyUsed → mainAnnot.homo_sapiens.txt.gz," if the organism of interest is *homo sapiens*).
3. Download the file to the *Perseus/conf/annotations* folder.
4. Go to "Processing → Annot. columns → Add annotation." Select the file from the previous step as a *Source*.
5. Set the *UniProt column* parameter to the column that contains UniProt identifiers. These identifiers will be used for overlaying the annotation data with the expression matrix (e.g., *Protein IDs)*.
6. Select several categories of interest to be overlaid with the main matrix and move them to the right-hand side. Click *ok*.

3.8 Differential Expression Analysis

1. Go to "Processing → Tests." From the menu select the appropriate test. For the data set used in this chapter, the *Multiple-sample tests* option should be chosen, as there are more than two conditions that are compared. The default parameters do not have to be changed (*see* **Note 11**).
2. Specify the categorical row that contains information about the experimental conditions of the samples that will be used in the differential analysis in the *Grouping* parameter.
3. Keep the default value of *0* for the *S0* parameter, to use the standard t-test statistic. Change the parameter to use the modified test statistic approach described by Tusher et al. [15].
4. Select the multiple hypothesis testing correction method to be used by specifying the *Use for truncation* parameter (*see* **Note 12**, Fig. 4a).

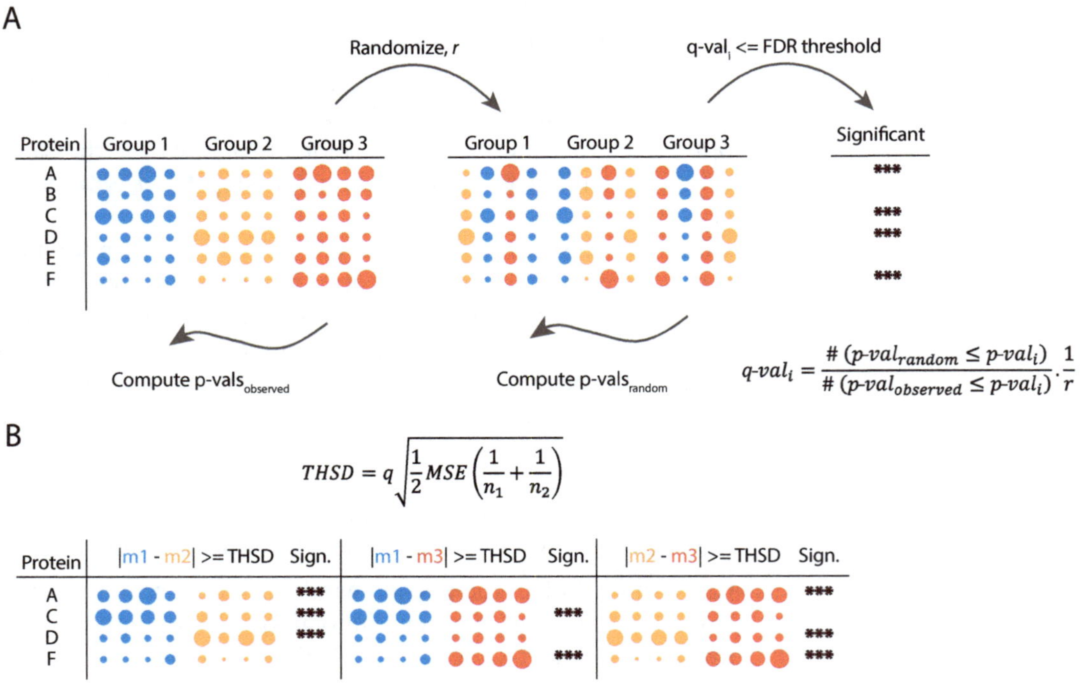

Fig. 4 Differential expression and multiple hypothesis testing. (**a**) Multiple hypothesis testing correction using a permutation-based false discovery rate approach is shown. Labels are randomly swapped between the three groups (blue, yellow, and red). The Randomization is repeated *r* times. ANOVA *p*-values are computed both on the measured and the permutated data and local FDR values (*q*-values) are computed as the fraction of accepted hits from the permuted data over accepted hits from the measured data normalized by the total number of randomizations *r*. All hits with a *q*-value smaller than a threshold are considered significant. (**b**) To determine the exact pairwise differences of protein expression Tukey's Honest Significant Difference (THSD) test is used on the ANOVA significant hits. If the mean difference between two groups is greater than or equal to the corresponding THSD, the difference is considered significant between the compared groups. q: constant depending on the number of treatments and the degrees of freedom that can be found in a Studentized range q table; MSE: mean squared error; n1, n2, number of data points in each group

5. Specify if a suffix should be added to the output columns produced by Perseus. This option is relevant when multiple tests are conducted, e.g., with different parameter settings, as it helps to distinguish between them in the output table.
6. Inspect the output table. It contains three new columns: *ANOVA significant*, *−Log ANOVA p-value*, and *ANOVA q-value* (*see* **Note 13**).
7. Go to "Processing → Filter rows → Filter rows based on categorical column." Set the *Column* parameter to *ANOVA Significant* and the *Mode* parameter to *Keep matching rows* to retain all differentially expressed proteins.
8. Go to "Processing → Tests → Post-hoc tests." Set the *Grouping* parameter to the same grouping that was used for the ANOVA test (*see* Subheading 3.6, **step 1**) and the FDR to

the desired threshold. Tukey's honestly significant difference (THSD) is computed for all proteins and all pairwise comparisons and the significant hits within the corresponding pairs are marked (*see* **Note 14**, Fig. 4b).

3.9 Clustering and Profile Plots

1. Go to "Analysis → Clustering/PCA → Hierarchical clustering." Keep the default parameters and click *ok*.
2. Inspect the resulting heatmap and the relationship between the groups and the proteins.
3. Click on the *Change color gradient* button in the button ribbon above the heatmap to examine the color scale usage (red means high and green low expression) and to modify them.
4. Click on several node junctions in the protein tree that represent potentially interesting clusters of proteins (i.e., upregulation in a certain experimental condition). The selected clusters are highlighted and appear in the "Row clusters" table displayed to the right of the heatmap (*see* **Note 15**).
5. Inspect the different profile plots as you navigate through the different clusters in the table. Change the color by modifying the *Color scale* and export the profile plots by clicking on the *Export image* button (*see* Fig. 5).
6. From the ribbon menu in the heat map view, click on the *Export row clustering* button to add the cluster information to a new data matrix.

3.10 Functional Analysis

1. Go to "Multi-proc. → Matching rows by name." Both *Base* and *Other* matrices point to the last matrix.
2. Click on *Base matrix* and then in the workflow window select the data matrix that was generated before filtering for ANOVA significant hits (Subheading 3.9, **step 6**).
3. In the pop-up window set *Matching column in matrix 1* and *2* to a common identifier (e.g., *Protein IDs)*.
4. In the categorical columns section, transfer the category *Cluster* to the right hand-side. Click *ok* (*see* **Note 16**).
5. Go to "Processing → Annot. columns → Fisher exact test." Change the *Column* parameter to *Cluster* and click *ok*. The resulting table contains information about all annotation categories that were found to be significantly enriched or depleted using a Fisher's exact test and multiple hypotheses correction (*see* **Note 17**).

In summary, this chapter provides a complete protocol for fundamental analysis of proteomic data, starting from data upload and transformation and ending with identification of proteins, characteristic of the specific disease progression stage, and the underlying processes in which they are involved. The

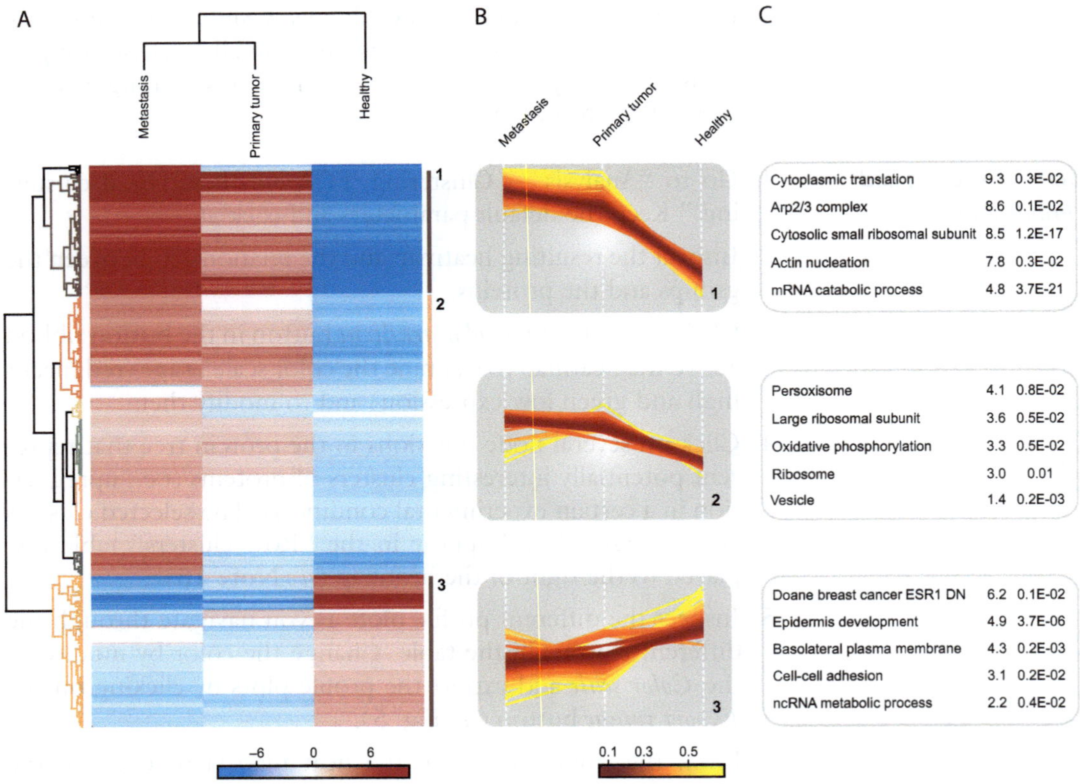

Fig. 5 Enrichment analysis highlighting important pathways and processes. (**a**) Hierarchical clustering of proteins found to have differential expression between pairs of disease states. High and low expression are shown in red and blue respectively. Various clusters of protein groups are highlighted in the dendrogram. (**b**) Profile plots of three selected clusters showing distinct behavior with respect to the three disease states are shown: *1* strongly increased expression in tumor tissues; *2* moderate increase in tumor tissue; and *3* decreased expression in tumor samples. (**c**) Functional analysis of protein annotation terms resulted in multiple categories that were enriched in the three selected clusters. The enriched terms and the corresponding enrichment factor and *p*-value are shown

described analytical methods and visualization tools are integrated in Perseus, a freely available platform for analysis of omics data, which provides a comprehensive portfolio of analysis tools with a user-friendly interface [19]. A special emphasis is placed on employing statistically sound methods in the analysis of large data, avoiding wrong interpretation and extracting maximum information. More advanced computational techniques such as supervised learning are also supported and are often instrumental for the analysis of complex data where genetic and intra-tumor variability may pose challenges. Moreover, Perseus is being continuously developed to integrate analysis of various data types, including posttranslation modifications, sequence information, as well as to allow deeper functional interpretation through network and pathways analysis.

4 Notes

1. Proteins with shared peptides that cannot be distinguished based on the peptides identified in a bottom-up proteomics approach are often reported together as a protein group [21].
2. The input file format of Perseus is a tab-delimited txt file that contains a header row with the names of all columns. The type of data is specified during file loading. Make sure that the "Regional and Language Options" are set to English to avoid errors while reading numerical data such as decimal separators being wrongly interpreted.
3. Different expression and meta data can be imported in Perseus and used for subsequent analysis. Common expression data are in one of the following formats: normalized intensities (e.g., LFQ intensity as described in [4], iBAQ as described in [22]) or ratios between heavy standard and light/non-labeled sample. Other data types that can be analyzed with Perseus are shown in Fig. 2.
4. Reverse, identified by site and contaminant proteins have to be marked in a categorical column before these filters can be applied. These are automatically set when MaxQuant output tables are used for analysis in Perseus. Additional filtering options can be used to remove proteins based on a quantitative measure such as a minimum number of quantified peptides or a maximum q-value.
5. Different activities have different output results including a data matrix with the same expression values and additional columns containing the results of the analysis or a new data matrix containing only the output of an analysis activity. An activity is always performed on the data matrix and specific tab for that matrix that is active at the moment.
6. Depending on the nature of missing values, different filtering strategies may be employed and are supported in Perseus. For example, if large differences between groups are expected with proteins having very low expression level in one of the groups, filtering based on a minimum number of valid values in at least one group would be a more suitable approach than filtering for a minimum number of valid values in the complete matrix.
7. All the plots can be exported in figure-ready formats such as pdf, tiff, or png.
8. Very stringent filtering is usually not recommended, as a large amount of the data will be lost. Instead milder filtering combined with imputation may be more appropriate.
9. Data normalization is not always necessary. Different types of normalization can be applied on the data to correct for systematic shifts or skewness and to make samples comparable.

10. Regular expressions can be used to derive the experimental design from the sample names ("Action → Create from experiment name"). Additionally, a template txt file can be written out, edited in an external editor program, and read in to indicate the experimental design.
11. Analysis of differentially expressed proteins depends on the number of compared conditions, the underlying distribution properties, and the availability of biological replicates. For example, data sets with one condition with replicates should be analyzed with *One-sample tests*, with two conditions—with *Two-sample tests*, and with more than two conditions—*Multiple-sample tests.* Paired samples test and tests abolishing the requirement for equal variance are also available.
12. The method with largest power *Permutation-based FDR* is recommended and at least 250 repetitions are suggested. In case of technical replicates, these have to be specified as a separate grouping (*see* Subheading 3.7, **step 1**) and selected with the "Preserve grouping in randomizations" option. Failure to specify technical replicates will result in wrong FDR calculation. The more conservative Benjamini-Hochberg correction can also be used when a lower number of Type I errors at the cost of lower sensitivity are desired.
13. The "Significant" column contains a "+" if a protein met the selected significance threshold (usually *q*-value). Additionally, *p*-values (probability of type I error) and the corresponding *q*-values (corrected *p*-value) are provided in the output table.
14. Tukey's honestly significant difference (THSD) is a post-hoc test that when performed on ANOVA significant hits determines in exactly which pairwise group comparisons a given protein is differentially expressed.
15. Clusters can be defined by clicking on the respective nodes in the protein tree or based on the precise distance measure used to compute the tree. To use the latter option, click on the "Define row clusters" button and specify the desired number of clusters, which will then be automatically defined.
16. The matching step is necessary in order to define the correct background against which enrichment will be computed. Too small (only significant hits) or too large (the complete proteome, even if not detected with MS analysis) introduces bias in the enrichment results.
17. The enrichment output table contains information about the values used to compute the contingency table for the Fisher's exact test (e.g., category and intersection size), the enrichment factor, and the statistical significance of the hit indicated by a *p*-value and the associated false discovery rate.

References

1. Mann M, Kulak NA, Nagaraj N, Cox J (2013) The coming age of complete, accurate, and ubiquitous proteomes. Mol Cell 49 (4):583–590. https://doi.org/10.1016/j.molcel.2013.01.029
2. Geiger T, Cox J, Ostasiewicz P, Wisniewski JR, Mann M (2010) Super-SILAC mix for quantitative proteomics of human tumor tissue. Nat Methods 7(5):383–385. https://doi.org/10.1038/nmeth.1446
3. Shenoy A, Geiger T (2015) Super-SILAC: current trends and future perspectives. Expert Rev Proteomics 12(1):13–19. https://doi.org/10.1586/14789450.2015.982538
4. Cox J, Hein MY, Luber CA, Paron I, Nagaraj N, Mann M (2014) Accurate proteome-wide label-free quantification by delayed normalization and maximal peptide ratio extraction, termed MaxLFQ. Mol Cell Proteomics 13(9):2513–2526. https://doi.org/10.1074/mcp.M113.031591
5. Ellis MJ, Gillette M, Carr SA, Paulovich AG, Smith RD, Rodland KK, Townsend RR, Kinsinger C, Mesri M, Rodriguez H, Liebler DC, Clinical Proteomic Tumor Analysis C (2013) Connecting genomic alterations to cancer biology with proteomics: the NCI Clinical Proteomic Tumor Analysis Consortium. Cancer Discov 3(10):1108–1112. https://doi.org/10.1158/2159-8290.CD-13-0219
6. Hanash S, Taguchi A (2010) The grand challenge to decipher the cancer proteome. Nat Rev Cancer 10(9):652–660. https://doi.org/10.1038/nrc2918
7. Wisniewski JR, Dus-Szachniewicz K, Ostasiewicz P, Ziolkowski P, Rakus D, Mann M (2015) Absolute proteome analysis of colorectal mucosa, adenoma, and cancer reveals drastic changes in fatty acid metabolism and plasma membrane transporters. J Proteome Res 14(9):4005–4018. https://doi.org/10.1021/acs.jproteome.5b00523
8. Zhang B, Wang J, Wang X, Zhu J, Liu Q, Shi Z, Chambers MC, Zimmerman LJ, Shaddox KF, Kim S, Davies SR, Wang S, Wang P, Kinsinger CR, Rivers RC, Rodriguez H, Townsend RR, Ellis MJ, Carr SA, Tabb DL, Coffey RJ, Slebos RJ, Liebler DC, Nci C (2014) Proteogenomic characterization of human colon and rectal cancer. Nature 513 (7518):382–387. https://doi.org/10.1038/nature13438
9. Iglesias-Gato D, Wikstrom P, Tyanova S, Lavallee C, Thysell E, Carlsson J, Hagglof C, Cox J, Andren O, Stattin P, Egevad L, Widmark A, Bjartell A, Collins CC, Bergh A, Geiger T, Mann M, Flores-Morales A (2016) The proteome of primary prostate cancer. Eur Urol 69(5):942–952. https://doi.org/10.1016/j.eururo.2015.10.053
10. Deeb SJ, Tyanova S, Hummel M, Schmidt-Supprian M, Cox J, Mann M (2015) Machine learning based classification of diffuse large B-cell lymphoma patients by their protein expression profiles. Mol Cell Proteomics 14 (11):2947–2960. https://doi.org/10.1074/mcp.M115.050245
11. Tyanova S, Albrechtsen R, Kronqvist P, Cox J, Mann M, Geiger T (2016) Proteomic maps of breast cancer subtypes. Nat Commun 7:10259. https://doi.org/10.1038/ncomms10259
12. Mertins P, Mani DR, Ruggles KV, Gillette MA, Clauser KR, Wang P, Wang X, Qiao JW, Cao S, Petralia F, Kawaler E, Mundt F, Krug K, Tu Z, Lei JT, Gatza ML, Wilkerson M, Perou CM, Yellapantula V, Huang KL, Lin C, McLellan MD, Yan P, Davies SR, Townsend RR, Skates SJ, Wang J, Zhang B, Kinsinger CR, Mesri M, Rodriguez H, Ding L, Paulovich AG, Fenyo D, Ellis MJ, Carr SA, Nci C (2016) Proteogenomics connects somatic mutations to signalling in breast cancer. Nature 534(7605):55–62. https://doi.org/10.1038/nature18003
13. Troyanskaya O, Cantor M, Sherlock G, Brown P, Hastie T, Tibshirani R, Botstein D, Altman RB (2001) Missing value estimation methods for DNA microarrays. Bioinformatics 17(6):520–525
14. Lazar C, Gatto L, Ferro M, Bruley C, Burger T (2016) Accounting for the multiple natures of missing values in label-free quantitative proteomics data sets to compare imputation strategies. J Proteome Res 15(4):1116–1125. https://doi.org/10.1021/acs.jproteome.5b00981
15. Tusher VG, Tibshirani R, Chu G (2001) Significance analysis of microarrays applied to the ionizing radiation response. Proc Natl Acad Sci U S A 98(9):5116–5121. https://doi.org/10.1073/pnas.091062498
16. Benjamini Y, Hochberg Y (1995) Controlling the false discovery rate: a practical and powerful approach to multiple testing. J R Stat Soc Series B 57:289–300
17. Fisher RA (1922) On the interpretation of x (2) from contingency tables, and the calculation of P. J R Stat Soc 85:87–94. https://doi.org/10.2307/2340521
18. Cox J, Mann M (2012) 1D and 2D annotation enrichment: a statistical method integrating

quantitative proteomics with complementary high-throughput data. BMC Bioinformatics 13(Suppl 16):S12. https://doi.org/10.1186/1471-2105-13-S16-S12

19. Tyanova S, Temu T, Sinitcyn P, Carlson A, Hein MY, Geiger T, Mann M, Cox J (2016) The Perseus computational platform for comprehensive analysis of (prote)omics data. Nat Methods 13(9):731–740. https://doi.org/10.1038/nmeth.3901

20. Pozniak Y, Balint-Lahat N, Rudolph JD, Lindskog C, Katzir R, Avivi C, Ponten F, Ruppin E, Barshack I, Geiger T (2016) System-wide clinical proteomics of breast cancer reveals global remodeling of tissue homeostasis. Cell Syst 2(3):172–184. https://doi.org/10.1016/j.cels.2016.02.001

21. Cox J, Mann M (2008) MaxQuant enables high peptide identification rates, individualized p.p.b.-range mass accuracies and proteome-wide protein quantification. Nat Biotechnol 26(12):1367–1372. https://doi.org/10.1038/nbt.1511

22. Schwanhäusser B, Busse D, Li N, Dittmar G, Schuchhardt J, Wolf J, Chen W, Selbach M (2011) Global quantification of mammalian gene expression control. Nature 473(7347):337–342. https://doi.org/10.1038/nature10098

Open Access This chapter is licensed under the terms of the Creative Commons Attribution 4.0 International License (http://creativecommons.org/licenses/by/4.0/), which permits use, sharing, adaptation, distribution and reproduction in any medium or format, as long as you give appropriate credit to the original author(s) and the source, provide a link to the Creative Commons license and indicate if changes were made.

The images or other third party material in this chapter are included in the chapter's Creative Commons license, unless indicated otherwise in a credit line to the material. If material is not included in the chapter's Creative Commons license and your intended use is not permitted by statutory regulation or exceeds the permitted use, you will need to obtain permission directly from the copyright holder.

Chapter 8

Quantitative Analysis of Tyrosine Kinase Signaling Across Differentially Embedded Human Glioblastoma Tumors

Hannah Johnson and Forest M. White

Abstract

Glioblastoma is the most aggressive primary brain tumor with a poor mean survival even with the current standard of care. Kinase signaling analyses of clinical glioblastoma samples provide a physiologically relevant view of oncogenic signaling networks. Here, we describe the methods that enable the quantification of protein expression profiles and phosphotyrosine signaling across flash frozen and optimal cutting temperature (OCT) compound embedded tumor specimens. The data derived from these experiments can be used to identify the intra- and inter-patient heterogeneity present in these tumors. Correlation and functional analyses on the quantitative protein expression and phosphotyrosine signaling data obtained from clinical samples can be used to identify tyrosine kinase signaling networks present in these tumors and reveal the differential expression of functionally related proteins. This chapter provides the quantitative mass spectrometry methods required for the identification of in vivo oncogenic signaling networks from human tumor specimens.

Key words Glioblastoma, iTRAQ labeling, Heterogeneity, Phosphorylation, Mass spectrometry

1 Introduction

Glioblastoma (GBM) is the most common primary brain tumor with the current standard of care consisting of surgical removal, radiotherapy, and chemotherapy [1]. Despite these invasive interventions the median survival time remains at approximately 15 months following diagnosis [2]. Molecular classification of GBM tumors has led to the identification of four sub-classes of GBM, based largely on differences in transcriptional profiles: classical, mesenchymal, neural, and proneural. While each subtype is associated with the mutation/dysregulation of selected molecular drivers, the intra-tumoral heterogeneity is such that different tumors within each sub-type may still have different oncogenic drivers [3–6]. Intriguingly, activation of receptor tyrosine kinase (RTK) signaling, through overexpression or mutation, is a common feature in >80% of all glioblastomas, thereby implicating

Louise von Stechow (ed.), *Cancer Systems Biology: Methods and Protocols*, Methods in Molecular Biology, vol. 1711, https://doi.org/10.1007/978-1-4939-7493-1_8, © Springer Science+Business Media, LLC 2018

kinase signaling in the pathogenesis of glioblastoma [6]. Moreover, most RTKs are attractive targets for therapeutic intervention, as their activation potentiates survival through MAPK and PI3K/AKT signaling [7]. Unfortunately, it is often difficult to determine which RTK(s) are activated in a given tumor from genomic profiling alone, as RTK activation is typically regulated at the protein posttranslational modification level rather than at the transcriptional level. Therefore, in order to directly identify RTK activity and thereby select potential RTK-targeted therapeutic strategies for a given patient tumor, we have recently developed an approach to quantifying protein tyrosine phosphorylation and protein expression profiles in human tumor tissue specimens [8].

Using this approach, it is possible to quantify phosphorylation events across patient samples with high sensitivity and throughput. Profiling tyrosine phosphorylation by mass spectrometry has been demonstrated to identify activated tyrosine kinase signaling pathways across a number of tumor types [9, 10]. Accurate characterization of dynamic phosphorylation signaling can prove challenging due to the limited availability of clinical samples for proteomic analysis. Differential embedding of human tumors before storage often compounds the limited availability of clinical samples [8]. Embedding tissues using formalin-fixed paraffin (FFPE) and optimal cutting temperature (OCT) compound is routine in pathology labs to allow sectioning histopathologic analysis [11]. Evaluation of the ability to quantify activated signaling networks and protein expression profiles across these differentially embedded tumors will allow the utlization of available tissue samples [8, 12]. Furthermore, these analyses allow the identification of (i) signaling changes that can occur in the tumor between resection and freezing [13], and (ii) the presence of intra-tumoral heterogeneity [14–16]. We have previously quantified phosphotyrosine signaling across a panel of glioblastoma patient-derived xenograft (PDX) tumors with differing expression of the epidermal growth factor (EGFR) variant vIII [17], a panel of differentially embedded human glioblastoma tumors [8], and across a panel of colorectal and ovarian tumors [13]. Throughout these analyses we have identified inter and intra-tumoral heterogeneity at the tyrosine kinase signaling level.

In this chapter, we describe the quantification of tyrosine kinase signaling using iTRAQ labeling of human glioblastoma tumors. The methodology described here can be readily applied to other tumor types. To investigate the effect of alternate storage methods on protein stability or protein posttranslational modifications, we have quantified activated phosphotyrosine networks and profiled global protein expression across pairs of human glioblastoma tumor sections that have been either embedded in OCT compound followed by flash freezing in liquid nitrogen (LN_2) or immediately flash frozen in LN_2. Samples were labeled with iTRAQ8plex

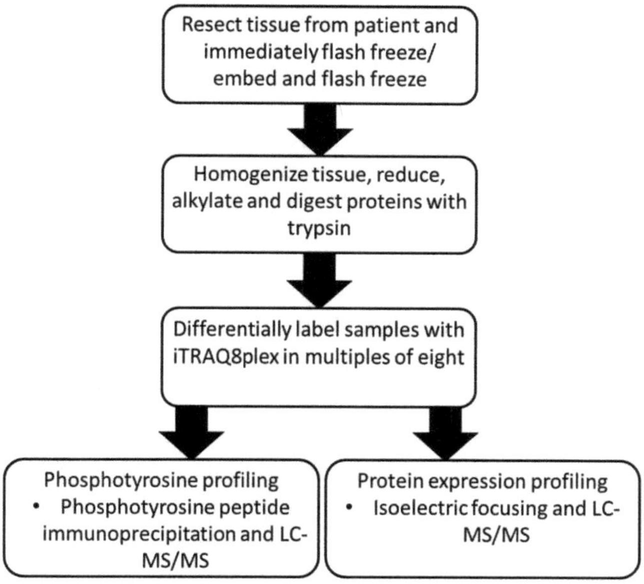

Fig. 1 Quantification of tyrosine phosphorylation signaling and protein expression profiles across human glioblastoma tumors. Experimental mass spectrometry workflow. Human glioblastoma tumor sections are homogenized, reduced, alkylated, and digested with trypsin and peptides labeled with iTRAQ8plex. Phosphotyrosine peptide enrichment was carried out by immunoprecipitation using anti-phosphotyrosine antibodies and analyzed by LC-MS/MS. For protein expression profiling, peptides are fractionated by isoelectric focusing and analyzed by LC-MS/MS

followed by phosphotyrosine peptide enrichment and protein expression profiling across the panel of human glioblastoma tumors (Fig. 1) [8]. A quantitative proteomic analysis of these clinical samples allows the identification of the effects of sample storage on tyrosine kinase signaling and protein expression profiles and enables the identification of oncogenic kinase signaling. To identify activated phosphotyrosine signaling related to glioblastoma biology, we describe correlation analysis and functional analysis of the quantitative proteomic data to highlight groups of related proteins that are co-expressed in glioblastoma tumors. Ultimately, the methods described here allow the identification of activated tyrosine kinases and downstream signaling in vivo in the context of inter- and intra-tumoral heterogeneity.

2 Materials

Prepare all the solutions using HPLC grade solvents unless indicated otherwise. Prepare and store all the reagents at room temperature unless indicated otherwise. Follow waste disposal regulations when disposing of chemicals.

2.1 Tumor Homogenization

1. Polytron hand held homogenizer.
2. Timer.
3. Phosphate-buffered saline (PBS).
4. Protein phosphotyrosyl phosphatase inhibitor: 200 mM sodium orthovanadate (Na_3VO_4) stock, make 100 μL aliquots and store at −20 °C until ready to use.
5. Complete protease and phosphatase inhibitors.
6. Mass spectrometry lysis buffer: 8 M urea. Add urea to MilliQ water and vortex to dissolve. Supplement with 1 mM sodium orthovanadate, 0.1% NP-40, and protease and phosphatase inhibitor cocktail tablets.
7. Immunoblotting lysis buffer: Radioimmunoprecipitation assay (RIPA) buffer supplemented with 1 mM sodium orthovanadate, 0.1% NP-40, and protease and phosphatase inhibitor cocktail tablets.
8. Bicinchoninic acid (BCA) assay.
9. Methanol.
10. Liquid nitrogen.
11. Scales.
12. Dry ice.

2.2 Immunoblotting

1. 7.5% polyacrylamide gels.
2. Nitrocellulose.
3. Quantitative Tyrosine Kinase Signaling in GlioblastomaBlocking buffer: 5% BSA in Tris-buffered saline with tween (TBS-T) (150 mM NaCl, 0.1% Tween 20, 50 mM Tris–HCl, pH 8.0).
4. Primary antibodies: anti-phosphotyrosine (4G10, Millipore), anti-EGFR (BD Biosciences), anti-Her3/ErbB3 (CST), anti-PDGFRα (CST), anti-PDGFRβ (CST), anti-Met (CST), anti-AKT (CST), anti-AKT pS473 (CST), anti-p53 (CST), and anti-β-tubulin (CST).
5. Secondary antibodies: goat anti-rabbit or goat anti-mouse conjugated to horseradish peroxidase.
6. Enhanced chemiluminescence (ECL) detection kit.

2.3 Reduction, Alkylation, and Tryptic Digestion

1. 100 mM ammonium acetate in water, pH 8.9.
2. Reducing solution: 1 M dithiothreitol (DTT) in 100 mM ammonium acetate pH 8.9.Store at −20 °C in aliquots.
3. Alkylation solution: 1 M iodoacetamide (IAA) in 100 mM ammonium acetate pH 8.9.
4. Sequencing grade trypsin.
5. Formic acid.

6. C18 cartridges.
7. Acetonitrile.

2.4 iTRAQ 8plex Labeling

1. iTRAQ reagents.
2. Dissolution buffer: 500 mM triethylammonium bicarbonate (TEAB), pH 8.5.
3. Isopropanol.
4. 0.1% acetic acid.
5. Vacuum centrifuge.

2.5 Phosphotyrosine Peptide Immunoprecipitation

1. Protein G agarose.
2. Immunoprecipitation (IP) buffer: 100 mM Tris–HCl, 1% NP-40, pH 7.4.
3. Tris buffer: 500 mM Tris–HCl, pH 8.5.
4. Phosphotyrosine antibodies: 4G10 (Millipore), PY100 (CST), and PT66 (Sigma).
5. Rinse buffer: 100 mM Tris–HCl, pH 7.4.
6. Elution buffer: 100 mM glycine, pH 2–2.5.
7. pH paper.

2.6 Phosphopeptide Enrichment by IMAC

1. Self-packed IMAC columns (15 cm in length): Pack columns with Poros 20MC beads. Capillary type: OD 360 μm: ID 200 μm.
2. Easy-nLC 1000 Nano HPLC.
3. 100 mM iron(III) chloride.
4. MilliQ water.
5. 100 mM EDTA pH 8.0.
6. 0.1% acetic acid.
7. 250 mM sodium phosphate pH 8.0.

2.7 Peptide Isoelectric Focusing and Protein Expression Profiling

1. MilliQ water.
2. Formic acid.
3. Acetonitrile.
4. ZOOM isoelectric focusing (IEF) fractionator.
5. Six ZOOM disks: pH 3.0, pH 4.6, pH 5.4, pH 6.2, pH 7.0, and pH 10.0.
6. Anode buffer: 7 M urea, 2 M thiourea, Novex IEF anode buffer, pH 3.0.
7. Cathode buffer: 7 M urea, 2 M thiourea, Novex IEF cathode buffer, pH 10.4.
8. ZOOM carrier ampholytes.

9. C18 cartridges.
10. Acetic acid.

2.8 Mass Spectrometric Analyses

1. Self-packed pre-columns (15 cm in length): Pack pre-columns with 10 μm YMC gel, ODS-A, 12 nm beads. Capillary type: OD 360 μm: ID 50 μm.
2. Self-packed analytical columns (15 cm in length): Pack analytical columns with 5 μm beads (YMC gel, ODS-AQ, 12 nm, S-5 μm, AQ12S05). (Capillary type: OD 360 μm: ID 100 μm).
3. Easy-nLC 1000 Nano HPLC.
4. Mass Spectrometer, e.g., Orbitrap QExactive Plus mass spectrometer (Thermo Fisher Scientific).
5. Buffer A: 200 mM acetic acid.
6. Buffer B: 70% Acetonitrile, 200 mM acetic acid.

2.9 Protein Expression Data Analysis and Functional Data Analysis

1. Proteome Discoverer can be obtained from Thermo scientific.
2. MASCOT search engine software can be obtained from Matrix Science; http://www.matrixscience.com/
3. Human protein sequence database, downloadable from NCBI.
4. CAMV. CAMV is open source software that can be downloaded from http://white-lab.mit.edu/software/camv
5. GENE-E. GENE-E is open source software that can be downloaded from http://www.broadinstitute.org/cancer/software/GENE-E/
6. PANTHER. is an online gene ontology tool that can be accessed here: http://www.pantherdb.org/
7. STRING protein-protein interaction functional database can be accessed here; http://string-db.org/
8. Phosphosite online database for phosphorylation sites can be accessed here; http://www.phosphosite.org/

3 Methods

3.1 Tumor Homogenization and Removal of Optimal Cutting Temperature

It is essential that tumors are flash frozen immediately following resection as cold ischemia can lead to significant changes in the tyrosine kinase signaling network [13]. Perform **steps 3–8** in the chemical hood on ice.

1. Immediately flash freeze tumor samples in liquid nitrogen following resection or embed in OCT compound and flash freeze in liquid nitrogen as soon as possible (ideally within 5 min).
2. Take tissue out of the tube using tweezers and deposit it in a weighting tray that has been previously tared. Record the weight of the tumor, the size, and the shape (*see* **Note 1**).

3. Rinse OCT compound embedded tumors in ice-cold PBS to remove the OCT compound surrounding the tissue prior to homogenization. Work on ice and minimize the time taken to carry out this step. Place the samples on dry ice once they have been thoroughly rinsed.
4. Homogenize tumors in ice-cold MS lysis buffer or RIPA buffer for immunoblotting, on ice, with 6 × 10 s pulses (full speed), separated by 10 s intervals. Homogenize tumors weighing approximately 50–200 mg in 3 mL of lysis buffer. Carefully assess the tube to identify any visible tissue left in the lysis buffer at the end of this procedure. Apply additional 10 s pulses if necessary.
5. Centrifuge tissue homogenate at 1070 × *g* for 5 min at 4 °C.
6. Take a 50 μL aliquot for BCA assay and place the rest of the lysate immediately on dry ice and store at −80 °C.
7. Quantify protein concentrations using BCA.
8. Rinse the polytron homogenizer thoroughly with PBS and methanol between samples.

3.2 Immunoblotting

The RTK status of the tumors can be used to help understand the tyrosine phosphorylation results (e.g., to help define the relative stoichiometry of phosphorylation between samples). RTK expression can be assessed using standard immunoblotting (*see* **Note 2**).

1. Separate tissue homogenates on 7.5% polyacrylamide gels and electrophoretically transfer to nitrocellulose.
2. Block nitrocellulose with blocking buffer for 1 h.
3. Dilute primary antibodies in blocking buffer and incubate with nitrocellulose overnight at 4 °C.
4. Dilute secondary antibodies (either goat anti-rabbit or goat anti-mouse conjugated to horseradish peroxidase) in TBS-T at a 1:10,000 ratio and incubate at room temperature for 1 h.
5. Wash nitrocellulose 3× 10 min with TBS-T.
6. Detect antibody binding with ECL, film, and a standard developer.

3.3 Reduction, Alkylation, and Tryptic Digestion

1. Dilute tissue homogenates 1:10 with 100 mM ammonium acetate pH 8.9 to reduce the urea concentration to less than 800 mM.
2. Reduce protein disulfide bridges by adding 10 mM DTT to each tumor homogenate to be analyzed and incubate at 56 °C for 45 min.
3. Alkylate reduced cysteines with 50 mM IAA at room temperature in the dark for 1 h.

4. Digest proteins with sequencing grade trypsin at an enzyme/substrate ratio of 1:100, on rotator at room temperature overnight.
5. Quench trypsin activity by adding formic acid to a final concentration of 5%.
6. Remove urea from the samples by reverse phase desalting using a C18 cartridge. Elute the peptides from the C18 cartridge into 80% acetonitrile, 0.1% formic acid.
7. Lyophilize the peptides in 400 μg aliquots and store at −80 °C.

3.4 iTRAQ 8plex Labeling

iTRAQ labeling currently allows the multiplexed quantification across up to eight different samples. Multiple iTRAQ8plex experiments can be combined to quantify across multiples of eight tumors. This multiplexing strategy requires the presence of a common sample in each experiment in order to compare across different experiments. This multiplex labeling strategy can also be performed with TMT reagents, available in 6-plex or 10-plex.

1. Label 400 μg peptide (quantified by BCA before C18 desalting) from each of the tumors with one tube of iTRAQ 8plex reagent (*see* **Note 3**).
2. Dissolve 400 μg lyophilized peptides in 30 μL dissolution buffer. Vortex each sample for 1 min and spin at 12,000 × *g* for 1 min.
3. Dissolve each tube of iTRAQ reagent in 70 μL of isopropanol. Vortex each tube for 1 min and spin at 12,000 × *g* for 1 min.
4. Add the isopropanol and iTRAQ 8plex reagent to the 400 μg peptide in dissolution buffer and vortex. Incubate at room temperature for 2 h.
5. Concentrate the eight tubes of peptide/iTRAQ mix to 40 μL using a vacuum centrifuge (speed-vac).
6. Combine the eight differentially labeled samples into a single tube.
7. Sequentially rinse out all the tubes with 3× 60 μL 0.1% acetic acid and add to the sample.
8. Concentrate the combined iTRAQ sample using a vacuum centrifuge (spin to dryness) and store at −80 °C. At this point the sample is stable for long-term storage (*see* **Note 4**).

3.5 Phosphotyrosine Profiling

Due to the low abundance of phosphotyrosine in the cell, it is necessary to carry out a series of steps to selectively enrich peptides that contain a phosphotyrosine residue. All the steps should be carried out at 4 °C unless otherwise stated.

1. Wash 60 μL protein G agarose with 300 μL IP buffer. Centrifuge for 5 min at 2850 × *g* and remove the supernatant to waste.
2. To the washed protein G agarose add 12 μg 4G10, 12 μg PY100, and 12 μg PT66.
3. Allow the antibody to conjugate to the protein G agarose at 4 °C for 6–8 h with rotation.
4. Spin down the mixture at 2850 × *g* for 5 min. Remove the supernatant and discard.
5. Wash the beads with 400 μL IP buffer, for 5 min on the rotor. Spin the beads down in the centrifuge at 2850 × *g* for 5 min.
6. Resuspend the iTRAQ 8plex labeled peptides in 400 μL IP buffer, vortex until the sample is completely dissolved and adjust the pH to 7.4 using Tris buffer (500 mM, pH 8.5) (*see* **Note 5**).
7. Remove the supernatant from the beads and replace it with the sample.
8. Incubate the sample with the beads on the rotor at 4 °C overnight.
9. The next day, centrifuge in the cold room at 2850 × *g* for 5 min.
10. Save the supernatant in a new tube and store it at −80 °C until carrying out protein expression profiling (*see* **Note 6**).
11. Wash the beads with 1× 400 μL IP buffer and then 3× 400 μL rinse buffer. Place the tube on the rotator for 5 min, spin down at 2850 × *g* for 5 min, and remove the supernatant in between each wash step. Discard the supernatant.
12. Add 70 μL of elution buffer and incubate at room temperature on the rotor for 30 min.
13. Load the eluate onto an immobilized metal affinity chromatography (IMAC) column.

3.6 Phosphopeptide Enrichment by IMAC

IMAC columns are packed and used according to the previously described protocol [18]. The steps required to enrich for phosphopeptides are briefly highlighted here.

1. Rinse the IMAC column with 100 mM EDTA pH 8.0 for 10 min at 10 μL/min.
2. Rinse the IMAC column with MilliQ water for 10 min at 10 μL/min.
3. Load 100 mM iron(III) chloride onto the column for 30 min at 10 μL/min (*see* **Note 7**).
4. Rinse the IMAC column with 0.1% acetic acid for 10 min at 10 μL/min.

5. Load iTRAQ 8plex labeled phosphotyrosine immunoprecipitation eluate to the IMAC column at a rate of 1–2 μL/min.
6. Rinse with 0.1% acetic acid at 10–12 μL/min for 10 min.
7. Elute peptides retained on the IMAC column onto a C18 reverse-phase pre-column with 250 mM sodium phosphate pH 8.0.
8. Rinse pre-column with 0.1% acetic acid to remove excess phosphate buffer and analyze by MS.

3.7 Analysis of Tyrosine Phosphorylation by MS

1. After rinsing with 0.1% acetic acid, attach the pre-column to a C18 reverse-phase analytical column with integrated electrospray emitter tip.
2. Chromatographically separate peptides by reverse phase HPLC over a 140 min gradient, with the eluent ionized by nanoelectrospray into an Orbitrap QExactive Plus instrument.
3. Operate the instrument in positive ion mode. Record full scans in the Orbitrap mass analyzer (resolution- FWHM 60,000) at a mass/charge (m/z) range of 350–2000 in profile mode. Select the top 15 most intense ions per scan for higher-energy C-trap dissociation (HCD)-based MS/MS analysis for peptide fragmentation and for iTRAQ reporter ion quantification, recording MS/MS scans in the Orbitrap mass analyzer (resolution-FWHM 60,000) at a mass/charge (m/z) range of 100–2000 in profile mode.

3.8 Peptide Isoelectric Focusing and Protein Expression Profiling

Understanding the protein expression profile within heterogeneous tumors can provide additional biological insight to accompany phosphorylation changes. Additionally, protein expression profiling can often provide a molecular basis for the observed phosphorylation changes, due to the dramatically different genetic backgrounds of each tumor.

1. Fractionate iTRAQ labeled peptides into five fractions using the ZOOM IEF fractionator with a set of 6 ZOOM disks (pH 3.0, pH 4.6, pH 5.4, pH 6.2, pH 7.0, and pH 10).
2. Use anode buffers and cathode buffers as per the manufacturer's instructions.
3. Add MilliQ water to the iTRAQ 8plex labeled peptide sample to a final volume of 3.35 mL.
4. Add ZOOM carrier ampholytes to each diluted sample at 1:100 and DTT to a final concentration of 20 mM.
5. Perform fractionation using the following parameters: 2 mA current limit, 2 W power limit, with 100 V for 20 min, 200 V for 80 min, and 600 V for 80 min.

6. Following fractionation, rinse each chamber with 500 μL water, and add the wash to the appropriate fraction.
7. Acidify each fraction with formic acid to 0.2% and desalt using C18 Cartridges. Elute peptides into 90% acetronitrile in 0.1% acetic acid.
8. Concentrate fractions to near dryness in a vacuum centrifuge.
9. Resuspend each fraction in 200 μL 0.1% acetic acid.
10. Dilute each resuspended fraction 1:100 with 0.1% acetic acid and load 20 μL (approximately 600 ng protein) onto an acidified pre-column.

3.9 Proteome Analysis by MS

1. Attach the peptide loaded pre-column to a C18 reverse-phase analytical column with integrated electrospray emitter tip.
2. Chromatographically separate each fraction by reverse phase HPLC over a 240 min gradient with the eluent ionized by nanoelectrospray into an Orbitrap QExactive Plus mass spectrometer.
3. Operate the instrument in positive ion mode. Record full scans in the Orbitrap mass analyzer (resolution- FWHM 60,000) at a mass/charge (m/z) range of 350–2000 in profile mode. Select the top 15 most intense ions per scan for HCD-based MS/MS analysis for peptide fragmentation and for iTRAQ reporter ion quantification, recording MS/MS scans in the Orbitrap mass analyzer (resolution-FWHM 60,000) at a mass/charge (m/z) range of 100–2000 in profile mode.

3.10 Protein Expression Data Analysis

1. Relative quantification and protein identification can be performed using Proteome Discoverer with MASCOT as the search engine.
2. Search MS/MS spectra against the human protein sequence database, downloadable from NCBI.
3. Search parameters should be set to "carbamidomethylation of cysteines" by IAA as a static modification, with "oxidation of methionine" and "iTRAQ 8plex labeling" as additional dynamic modifications.
4. Relative quantitation of protein expression can be performed by selecting for proteins containing a minimum of at least two peptides with MASCOT score above 20. Only peptides unique for a given protein should be considered for relative quantitation, those common to other isoforms or proteins of the same family should be excluded. Identified peptides should be excluded from quantitative analyses if (1) the peaks corresponding to the iTRAQ labels are not detected, (2) the same peptide sequence is shared by multiple proteins, or (3) the peptide sequence is discordant.

5. A decoy database search strategy can be used to estimate the false discovery rate (FDR), defined as the percentage of decoy proteins identified against the total protein identification. In this case, the MASCOT score threshold for peptide or protein identification can be established by setting the FDR to 1% following a search of the spectra against the NCBI non redundant *Homo sapiens* decoy database.

3.11 Phosphotyrosine Data Analysis

Understanding the tyrosine kinase signaling within heterogeneous tumors is critical to define the activated networks responsible for tumor cell proliferation, migration, and invasion.

1. Relative quantification and phosphotyrosine peptide identification can be performed using Proteome Discoverer with MASCOT software as the search engine.
2. Search MS/MS spectra against human protein sequence database, downloadable from NCBI.
3. Search parameters should be set to "carbamidomethylation of cysteines by IAA" as a static modification, with additional dynamic modifications; "oxidation of methionine," "iTRAQ 8plex labeling" and "phosphorylation of serine, threonine, and tyrosine" (*see* **Note 8**).
4. Peptides identified by MASCOT with an ion score >25 should be considered for further manual validation and quantification using CAMV (*see* **Note 9**) [19].

3.12 Functional Data Analysis

1. To identify groups of similarly expressed proteins and phosphorylation sites, perform unsupervised hierarchical clustering.
   (a) Clustering of the mean normalized and $\log_2$ transformed phosphotyrosine and protein expression quantitative iTRAQ data (using one minus Pearson correlation as a distance metric) can be performed using GENE-E (*see* **Note 10**).
2. To visualize quantitative phosphotyrosine and protein expression profiles across tumors, generate heat maps of mean normalized and $\log_2$ transformed phosphotyrosine and protein expression quantitative iTRAQ data (*see* **Note 11**).
   (a) When using GENE-E, upload an excel file with mean normalized and $\log_2$ transformed phosphotyrosine and protein expression quantitative iTRAQ data with the quantitative information specified in a data matrix, where phosphorylation sites are row metadata and corresponding iTRAQ labels are column.
   (b) Heat maps can be aesthetically modified under "preferences."

3. To identify differences between differentially embedded tumors (inter-tumoral heterogeneity) and between different patients tumors (intra-tumoral heterogeneity) carry out Pearson's correlation analysis of the quantitative phosphotyrosine or protein expression profiles using Excel. *P*-values can be calculated using *t* approximation (*see* **Note 12**).
   (a) Pearson's correlation (r) can be calculated in Excel using the PEARSON function.
   (b) To calculate the p-value for any value of r, calculate the associated t-value (t) using the following formula:

   $$t = \frac{r \times \sqrt{n-2}}{\sqrt{1-r^2}}$$

   (c) Once the t-value is calculated, use the TDIST function in Excel to find the associated p-value. This function requires the value of t, the degrees of freedom (i.e., $n-2$) and the number of tails.
4. Gene ontology (GO) annotations can be identified by uploading gene lists to the Protein Analysis Through Evolutionary Relationships (PANTHER) online classification system (*see* **Note 13**).
   (a) Protein names acquired from MS data output can be converted to Gene names and symbols using the online phosphorylation site database Phosphosite.
5. The Search Tool for the Retrieval of Interacting Genes/Proteins (STRING) database can be queried to identify known and predicted protein-protein interactions within clusters of co-expressed proteins and phosphotyrosine sites.

4 Notes

1. Recording the weight and morphology of each tumor before sectioning allows the assessment of intratumoral heterogeneity at the cellular level. Further sectioning of the tumor prior to the analysis of proteins and phosphorylation sites enables a complete understanding of the tumor origin (i.e., was the tumor section derived from the center of the tumor? the edge of the tumor?). Assessing the tumor content of each tumor section enables accurate assessment of the protein and phosphotyrosine level across multiple samples of the same tumor.
2. Due to the low abundance of tyrosine kinases in tissue samples it is common to not detect these proteins in whole proteome mass spectrometry profiling analyses. Immunoblotting presents a simple and effective method of identifying the relative levels of tyrosine kinases present within tissues.

3. Depending on the level of tyrosine phosphorylation in the sample and the sensitivity of the LC-MS/MS system, it may be necessary to label 800 μg peptide with two tubes of iTRAQ8plex reagent.
4. Due to the stability of peptides in a dry state peptides can be stored long term (> a year) at −80 °C. Repeated freeze-thaw cycles should be avoided, if peptide samples need to be periodically taken from the stock, make a series of aliquots from the stock prior to drying and store at −80 °C.
5. Add low volumes of the Tris buffer pH 8.5 (i.e., 1–5 μL increments) and measure pH using pH paper.
6. Peptide solutions are prone to degradation. To minimize this, aliquot peptide solutions and store at −80 °C. Avoid repeated freeze-thaw cycles, as this can lead to peptide degradation.
7. Iron(III) chloride should be kept dry due to the exothermic reaction that takes place when iron(III) chloride undergoes hydrolysis.
8. Phosphorylation of tyrosine needs to be a dynamic modification, as many of the tyrosines in the sample are not phosphorylated.
9. Given the importance of site identification, manual validation is critical to properly localize the phosphorylation site within the peptide.
10. Hierarchical clustering can be used to identify groups of phosphotyrosine sites and proteins that are similarly and differentially expressed across the different tumor samples. Alternatively, the quantitative data can be clustered using affinity propagation clustering [17].
11. Heat maps can be generated using software packages such as GENE-E, MATLAB (www.mathworks.com/), or R (https://www.r-project.org/).
12. Pearson's correlation analysis can alternatively be carried out using software packages such as MATLAB (www.mathworks.com/) or R (https://www.r-project.org/).
13. Functional grouping of proteins and/or phosphorylation sites can also be assessed using DAVID functional annotation (https://david.ncifcrf.gov) or Cytoscape (www.cytoscape.org/).

Acknowledgments

This work was supported in part by a generous gift from the James S. McDonnell Foundation and by NIH grants P30 CA014051 and R01 CA184320. The authors would like to thank Ms. Marcela White at the brain tumor bank (www.Braintumourbank.com) for access to patient materials.

References

1. Stupp R, Mason WP, van den Bent MJ, Weller M, Fisher B, Taphoorn MJ, Belanger K, Brandes AA, Marosi C, Bogdahn U, Curschmann J, Janzer RC, Ludwin SK, Gorlia T, Allgeier A, Lacombe D, Cairncross JG, Eisenhauer E, Mirimanoff RO (2005) Radiotherapy plus concomitant and adjuvant temozolomide for glioblastoma. N Engl J Med 352(10):987–996. https://doi.org/10.1056/NEJMoa043330
2. Stupp R, Hegi ME, Mason WP, van den Bent MJ, Taphoorn MJ, Janzer RC, Ludwin SK, Allgeier A, Fisher B, Belanger K, Hau P, Brandes AA, Gijtenbeek J, Marosi C, Vecht CJ, Mokhtari K, Wesseling P, Villa S, Eisenhauer E, Gorlia T, Weller M, Lacombe D, Cairncross JG, Mirimanoff RO (2009) Effects of radiotherapy with concomitant and adjuvant temozolomide versus radiotherapy alone on survival in glioblastoma in a randomised phase III study: 5-year analysis of the EORTC-NCIC trial. Lancet Oncol 10 (5):459–466. https://doi.org/10.1016/S1470-2045(09)70025-7
3. Verhaak RG, Hoadley KA, Purdom E, Wang V, Qi Y, Wilkerson MD, Miller CR, Ding L, Golub T, Mesirov JP, Alexe G, Lawrence M, O'Kelly M, Tamayo P, Weir BA, Gabriel S, Winckler W, Gupta S, Jakkula L, Feiler HS, Hodgson JG, James CD, Sarkaria JN, Brennan C, Kahn A, Spellman PT, Wilson RK, Speed TP, Gray JW, Meyerson M, Getz G, Perou CM, Hayes DN (2006) Integrated genomic analysis identifies clinically relevant subtypes of glioblastoma characterized by abnormalities in PDGFRA, IDH1, EGFR, and NF1. Cancer Cell 17(1):98–110. https://doi.org/10.1016/j.ccr.2009.12.020
4. Brennan C, Momota H, Hambardzumyan D, Ozawa T, Tandon A, Pedraza A, Holland E (2009) Glioblastoma subclasses can be defined by activity among signal transduction pathways and associated genomic alterations. PLoS One 4(11):e7752. https://doi.org/10.1371/journal.pone.0007752
5. Phillips HS, Kharbanda S, Chen R, Forrest WF, Soriano RH, TD W, Misra A, Nigro JM, Colman H, Soroceanu L, Williams PM, Modrusan Z, Feuerstein BG, Aldape K (2006) Molecular subclasses of high-grade glioma predict prognosis, delineate a pattern of disease progression, and resemble stages in neurogenesis. Cancer Cell 9(3):157–173. https://doi.org/10.1016/j.ccr.2006.02.019
6. Network TCGAR (2008) Comprehensive genomic characterization defines human glioblastoma genes and core pathways. Nature 455(7216):1061–1068
7. Krakstad C, Chekenya M (2010) Survival signalling and apoptosis resistance in glioblastomas: opportunities for targeted therapeutics. Mol Cancer 9:135. https://doi.org/10.1186/1476-4598-9-135
8. Johnson H, White FM (2014) Quantitative analysis of signaling networks across differentially embedded tumors highlights interpatient heterogeneity in human glioblastoma. J Proteome Res 13(11):4581–4593. https://doi.org/10.1021/pr500418w
9. Rikova K, Guo A, Zeng Q, Possemato A, Yu J, Haack H, Nardone J, Lee K, Reeves C, Li Y, Hu Y, Tan Z, Stokes M, Sullivan L, Mitchell J, Wetzel R, Macneill J, Ren JM, Yuan J, Bakalarski CE, Villen J, Kornhauser JM, Smith B, Li D, Zhou X, Gygi SP, TL G, Polakiewicz RD, Rush J, Comb MJ (2007) Global survey of phosphotyrosine signaling identifies oncogenic kinases in lung cancer. Cell 131 (6):1190–1203. https://doi.org/10.1016/j.cell.2007.11.025
10. Drake JM, Graham NA, Lee JK, Stoyanova T, Faltermeier CM, Sud S, Titz B, Huang J, Pienta KJ, Graeber TG, Witte ON (2013) Metastatic castration-resistant prostate cancer reveals intrapatient similarity and interpatient heterogeneity of therapeutic kinase targets. Proc Natl Acad Sci U S A 110(49): E4762–E4769. https://doi.org/10.1073/pnas.1319948110
11. Steu S, Baucamp M, von Dach G, Bawohl M, Dettwiler S, Storz M, Moch H, Schraml P (2008) A procedure for tissue freezing and processing applicable to both intra-operative frozen section diagnosis and tissue banking in surgical pathology. Virchows Arch 452 (3):305–312. https://doi.org/10.1007/s00428-008-0584-y
12. Loken SD, Demetrick DJ (2005) A novel method for freezing and storing research tissue bank specimens. Hum Pathol 36(9):977–980. https://doi.org/10.1016/j.humpath.2005.06.016
13. Gajadhar AS, Johnson H, Slebos RJ, Shaddox K, Wiles K, Washington MK, Herline AJ, Levine DA, Liebler DC, White FM (2015) Phosphotyrosine signaling analysis in human tumors is confounded by systemic ischemia-driven artifacts and intra-specimen heterogeneity. Cancer Res 75(7):1495–1503. https://doi.org/10.1158/0008-5472.CAN-14-2309
14. Snuderl M, Fazlollahi L, Le LP, Nitta M, Zhelyazkova BH, Davidson CJ, Akhavanfard S,

Cahill DP, Aldape KD, Betensky RA, Louis DN, Iafrate AJ (2011) Mosaic amplification of multiple receptor tyrosine kinase genes in glioblastoma. Cancer Cell 20(6):810–817. https://doi.org/10.1016/j.ccr.2011.11.005

15. Szerlip NJ, Pedraza A, Chakravarty D, Azim M, McGuire J, Fang Y, Ozawa T, Holland EC, Huse JT, Jhanwar S, Leversha MA, Mikkelsen T, Brennan CW (2012) Intratumoral heterogeneity of receptor tyrosine kinases EGFR and PDGFRA amplification in glioblastoma defines subpopulations with distinct growth factor response. Proc Natl Acad Sci U S A 109(8):3041–3046. https://doi.org/10.1073/pnas.1114033109

16. Sottoriva A, Spiteri I, Piccirillo SG, Touloumis A, Collins VP, Marioni JC, Curtis C, Watts C, Tavare S (2013) Intratumor heterogeneity in human glioblastoma reflects cancer evolutionary dynamics. Proc Natl Acad Sci U S A 110(10):4009–4014. https://doi.org/10.1073/pnas.1219747110

17. Johnson H, Del Rosario AM, Bryson BD, Schroeder MA, Sarkaria JN, White FM (2012) Molecular characterization of EGFR and EGFRvIII signaling networks in human glioblastoma tumor xenografts. Mol Cell Proteomics 11(12):1724–1740. https://doi.org/10.1074/mcp.M112.019984

18. Zhang Y, Wolf-Yadlin A, Ross PL, Pappin DJ, Rush J, Lauffenburger DA, White FM (2005) Time-resolved mass spectrometry of tyrosine phosphorylation sites in the epidermal growth factor receptor signaling network reveals dynamic modules. Mol Cell Proteomics 4 (9):1240–1250. https://doi.org/10.1074/mcp.M500089-MCP200

19. Curran TG, Bryson BD, Reigelhaupt M, Johnson H, White FM (2013) Computer aided manual validation of mass spectrometry-based proteomic data. Methods 61 (3):219–226. https://doi.org/10.1016/j.ymeth.2013.03.004

Part III

Systems Analysis of Cancer Cell Metabolism

Chapter 9

Prediction of Clinical Endpoints in Breast Cancer Using NMR Metabolic Profiles

Leslie R. Euceda, Tonje H. Haukaas, Tone F. Bathen, and Guro F. Giskeødegård

Abstract

Metabolic profiles reflect biological conditions as a result of biochemical changes within a living system. It is therefore possible to associate metabolic signatures with clinical endpoints of diseases, such as breast cancer. Nuclear magnetic resonance (NMR) spectroscopy is one of the most common techniques used for metabolic profiling, and produces high dimensional datasets from which meaningful biological information can be extracted. Here, we present an overview of data analysis techniques used to achieve this, describing key steps in the procedure. Moreover, examples of clinical endpoints of interest are provided. Although these are specific for breast cancer, the procedures for the analysis of NMR spectra as described here are applicable to any type of cancer and to other diseases.

Key words Breast cancer, Cross validation, Hierarchical clustering, Hypothesis testing, Metabolites, Model diagnostic statistics, Multivariate analysis, NMR spectroscopy, Partial least squares, Principle component analysis, Permutation testing

1 Introduction

Metabolic profiling refers to the large-scale measurement of low molecular weight metabolites and their intermediates generated within a living system at a given moment. Those metabolites reflect a biological condition as a result of biochemical changes caused by genetic modification and physiological or pathophysiological stimuli [1]. Metabolites include substances such as carbohydrates, amino acids, nucleotides, fatty acids, and vitamins. The complete set of metabolites in a living system is termed the metabolome, comparable to the terms genome, transcriptome, and proteome. These four molecular levels interact with each other, generating an information exchange flow known as the omics cascade

Leslie R. Euceda and Tonje H. Haukaas contributed equally to this work.

Louise von Stechow (ed.), *Cancer Systems Biology: Methods and Protocols*, Methods in Molecular Biology, vol. 1711, https://doi.org/10.1007/978-1-4939-7493-1_9, © Springer Science+Business Media, LLC 2018

[2]. Metabolites are downstream products affected by the omics signaling cascade (DNA, RNA, and proteins), but can also affect upstream signaling processes, such as gene expression. They are representative for the functional phenotype observed and provide information of the active biological pathways.

Nuclear magnetic resonance (NMR) spectroscopy is one of the most commonly used techniques for metabolic profiling, and can be applied to both biofluids and intact tissue samples. For details on the NMR theory, we refer to [3]. Briefly, NMR spectroscopy is based on the intrinsic property of spin possessed by certain atomic nuclei, such as protons (^{1}H), which gives rise to a small magnetic field. This magnetic field is called the magnetic moment of a nucleus, and can be thought of as a vector with direction and magnitude. When a sample is subjected to an external magnetic field, the magnetic moments will align either in the direction of that field or opposite to it, bringing the nuclei to a low or high energy spin state, respectively. Energies from opposite spin states counterbalance each other. However, the low energy spin state is slightly more energetically favorable, and thus a higher population of nuclei in a system will exist in this state. This generates a residual magnetization component parallel to the external magnetic field. In addition to nuclear spin, nuclei precess, i.e., rotate, about the external magnetic field axis. The rate of precession, also known as the resonance frequency or Larmor frequency, corresponds to the energy difference between the energy spin states. By applying a radio frequency (RF) pulse that matches the Larmor frequency of the nuclei of interest, the nuclei will absorb the energy and transition to a higher energy state. When the RF pulse is switched off, the nuclei recover to their original energy state, and the released energy can be measured by receiver coils. The observed signal can be mathematically converted to an NMR spectrum, which is a plot of the intensity of emission as a function of the resonance frequency (*see* Fig. 1).

Atomic nuclei of the same isotope experience small variations or shifts in their resonance frequencies depending on the chemical environment of each nucleus. This will cause the signals to appear at different positions in the NMR spectrum; this position is termed the chemical shift. In addition, the signals can be split in different ways as an effect of the chemical bonds of the nuclei. Therefore, an NMR spectrum can provide detailed information about molecular structure of the metabolites to which the observed nuclei belong. Moreover, the observed signal intensity is proportional to the number of nuclei giving rise to that signal [4], and can be exploited for quantification purposes.

Because the resonance frequency of a nucleus is proportional to the external magnetic field, it can be calibrated relative to the resonance frequency of a signal from a reference at the applied magnetic field. This makes NMR spectra comparable independent

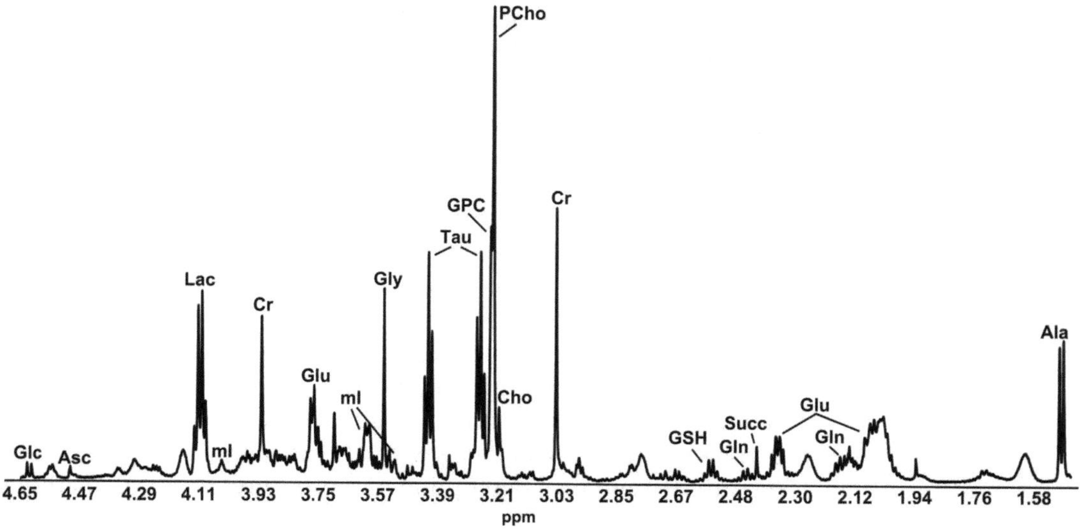

Fig. 1 Example ^{1}H NMR spectrum of breast tumor tissue. Observable metabolite peaks include glucose (Glc), ascorbate (Asc), lactate (Lac), myo-inositol (mI), creatine (Cr), glutamate (Glu), glycine (Gly), taurine (Tau), glycerophosphocholine (GPC), phosphocholine (PCho), choline (Cho), glutathione (GSH), glutamine (Gln), succinate (Succ), and alanine (Ala)

of the strength of the magnet employed. Since differences in resonance frequencies are very small, the chemical shift scale is expressed in parts per million (ppm). Trimethylsilyl propionic acid (TSP) is a common reference compound that is typically calibrated to 0 ppm [5].

Sample preparation of both liquid and solid state NMR is simple and non-destructive. High resolution (HR) magnetic resonance spectroscopy (MRS) of biofluids, cell extracts, and culture media is suited for high-throughput analysis and can typically detect 20–70 metabolites. For a description of procedures for recording metabolic profiles of liquid solutions using ^{1}H NMR, we refer to [6].

In NMR, anisotropic interactions between nuclei are those that are dependent on the direction of molecules with respect to the external magnetic field. In solution NMR, these interactions are averaged out due to high molecular mobility. In solids and semi-solids, such as tissue samples, molecular motion is restricted and so anisotropic interactions can give rise to peak broadening that may ultimately lead to signal overlap. It is possible to overcome this by rapidly spinning the sample at an angle of 54.7°, known as the magic angle, which imitates molecular motion in liquid solution. This method, called high resolution (HR) magic angle spinning (MAS) MRS, yields highly resolved spectra of tissue comparable with those obtained for biofluids using conventional MRS. For a detailed protocol describing HR MAS MRS of intact tissue, we refer to [7].

NMR acquisition of metabolic profiles results in complex, high dimensional datasets. Prior to statistical analysis, biologically irrelevant differences caused for example by instrumental or experimental artifacts must be removed from the raw data. This is known as data preprocessing and involves different computational procedures to convert the acquired data into a format that is usable to extract meaningful information. Because high intensity lipid peaks arising from normal breast adipose cells are often present in spectra from breast tumor tissue, spectral preprocessing may be challenging and is seldom straightforward [8]. Careful inspection of the preprocessed spectra is therefore essential to evaluate each individual preprocessing step. Modifications to the original preprocessing strategy are often required. For an overview of frequently used preprocessing tools and a general metabolomics workflow, we refer to [9], and for a description of specific preprocessing methods we refer to [10].

In this chapter, we provide an overview of data analysis strategies used to associate metabolic signatures with clinical endpoints of diseases, with a focus on breast cancer. Additionally, key steps in those data analysis strategies are described. We furthermore provide guidance for the interpretation of results.

2 Materials

2.1 Data Input

Choosing the optimal approach for statistical analysis is dependent on the type and structure of the data input as well as the hypothesis of interest. The NMR spectral data should, prior to statistical analysis, have been through proper preprocessing procedures. For multivariate analysis, the preprocessed data forms the X-matrix, where each row represents one sample and each column represents one variable or point in an NMR spectrum. Alternatively, metabolites can be quantified to make the data applicable for both multivariate and univariate analysis. In such cases, quantified metabolites from the same sample can be combined so that each variable of the X-matrix used for multivariate analysis represents one individual metabolite.

The Y-matrix or vector, used in supervised analysis, contains information about the clinical endpoint that should be predicted. The clinical endpoint is defined as the relevant patient information of interest to test for correlation with metabolic signature. Examples of clinical outcome variables of interest in breast cancer are patient 5-year survival, tumor size, tumor cell percentage, lymph node status, metastatic status, pathological response to treatment, and hormone (estrogen or progesterone) receptor status. These can either be categorical (e.g., lymph node status) or continuous variables (e.g., tumor cell percentage) (*see* **Note 1**).

In summary, the different data input structures are:
X-matrix:

- Preprocessed NMR spectra.
- Relative or absolute metabolite concentrations.

Y-matrix/vector:

- Prediction analysis: Categorical clinical endpoint.
- Regression analysis: Continuous clinical endpoint.

2.2 Software

There are several different softwares available for univariate or multivariate analysis of metabolomics data, differing in their flexibility and user-friendliness (*see* Table 1). Software such as Matlab and R can be used for all types of data analysis, but require that the user has knowledge on programming.

3 Methods

3.1 Multivariate Analysis

3.1.1 Unsupervised Methods

Unsupervised methods are exploratory and useful tools for getting to know your dataset in terms of possible groupings, patterns, and outliers, without taking a response variable into account. Examples of common methods are principal component analysis (PCA) and hierarchical cluster analysis (HCA).

Principal Component Analysis (PCA)

PCA is a method that through linear combinations of the original independent variables $\boldsymbol{X}$, constructs a new lower dimension coordinate system made up by latent variables (LVs), which in PCA are called principal components (PCs) [11]. These variables explain variance within the dataset with the aim of capturing the main trends in the data. The position of each sample in the new coordinate system is reflected by the scores matrix ($\boldsymbol{T}$), while the influence of the original variables on the PCs is defined by the loadings matrix ($\boldsymbol{P}$) such that:

$$\boldsymbol{X} = \boldsymbol{T}\boldsymbol{P}^{\mathrm{T}} + \boldsymbol{E} \tag{1}$$

where $\boldsymbol{E}$ is the residual matrix of variance not explained by the model, and ^{T} indicates the transpose of a matrix. The results can thus be visualized in scores and loadings plots (*see* Fig. 2).

Protocol for PCA

1. Additional preprocessing of variables. Although the spectral data was preprocessed prior to data analysis, PCA is sensitive to the scaling of the variables.

   Spectral data: Mean center the data by subtracting the variable mean from each variable value to make the mean zero. Mean centering of spectral data removes the offset from each variable

Table 1
Examples of available software and interfaces to perform multivariate and/or univariate metabolomics analyses described here

Software/ Interface	Reference/URL	Methods Implemented
Amix	https://www.bruker.com/products/mr/nmr/nmr-software/software/amix/overview.html	Multivariate
Knime[a]	https://www.knime.org/knime	Univariate and multivariate
Matlab	http://www.mathworks.com/products/matlab/	Univariate and multivariate
MetaboAnalyst[a]	http://www.metaboanalyst.ca/faces/docs/Format.xhtml	Univariate and multivariate
PLS toolbox[b]	http://www.eigenvector.com/software/pls_toolbox.htm	Multivariate
R	https://www.r-project.org/	Univariate and multivariate
SIMCA	http://umetrics.com/products/simca	Multivariate
SIRIUS	http://www.prs.no/Sirius/Sirius.html	Multivariate
SPSS	http://www-01.ibm.com/software/analytics/spss/products/statistics/	Univariate
STATA	http://www.stata.com/	Univariate and multivariate
The Unscrambler	http://www.camo.com/rt/Products/Unscrambler/unscrambler.html	Univariate and multivariate

[a]Uses R packages
[b]Requires Matlab

so that PC1 will not capture the mean of the data but the direction of maximum variance.

Quantified metabolites: Autoscale the metabolite concentrations by normalizing each value to the variable standard deviation after mean centering. Autoscaling allows each variable to have the same influence on the model, and the resulting variables have mean zero and standard deviation of one (*see* **Note 2**).

2. Perform PCA using the software of choice (*see* Table 1).
3. Select number of components to include in the model. There are two alternative approaches:

   Cumulative variance plot: Evaluate the cumulative variance explained by the model with increasing number of PCs. Choose the number of PCs that explain a certain predetermined amount of variance (*see* **Note 3**).

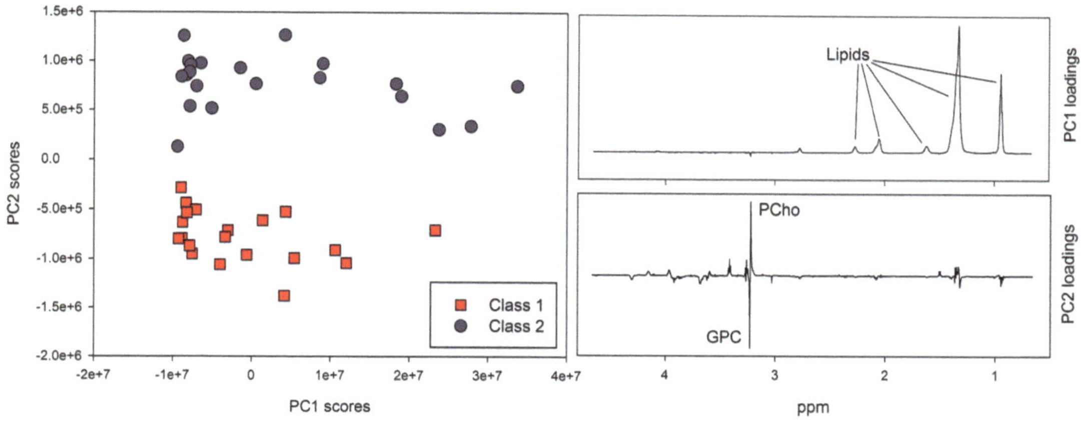

Fig. 2 Result from principal component analysis of breast cancer tissue from two different groups. The two classes are perfectly separated in the second principal component (PC2). Samples from class 2 have high PC2 scores compared to class 1, thus they have higher levels of the metabolite phosphocholine (PCho) and lower levels of glycerophosphocholine (GPC) compared to the class 1 samples. The first principal component (PC1) shows that the largest variation in the dataset is due to differences in lipid concentrations between the samples, as the samples to the right in the scores plot, having high scores on PC1, have high lipid levels compared to the remaining samples

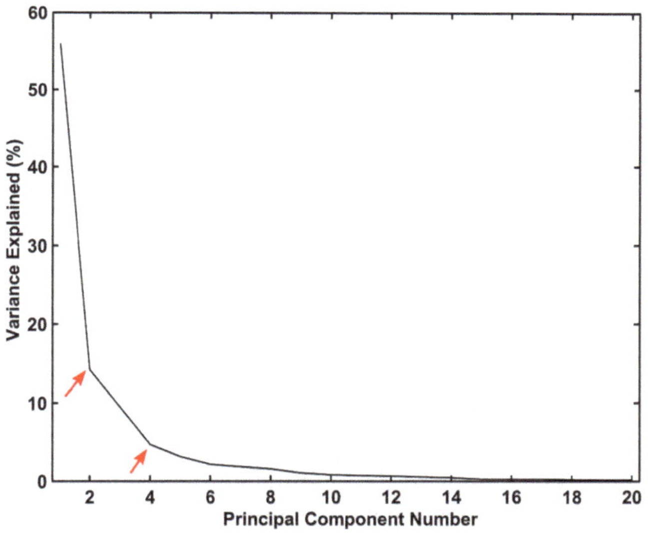

Fig. 3 Scree plot example. Two "knees," marked by red arrows, are observed, suggesting two or four principal components (PCs) to be the optimal number. In this case, the cumulative variance plot can aid in the determination of the best "knee," selecting the one that represents the number of PCs that explains a certain predetermined amount of variance

Scree plot: Plot the variance explained by each PC (*see* **Note 4**). The variance will decrease for each PC. Choose the PC representing the "knee" in the curve (*see* Fig. 3).

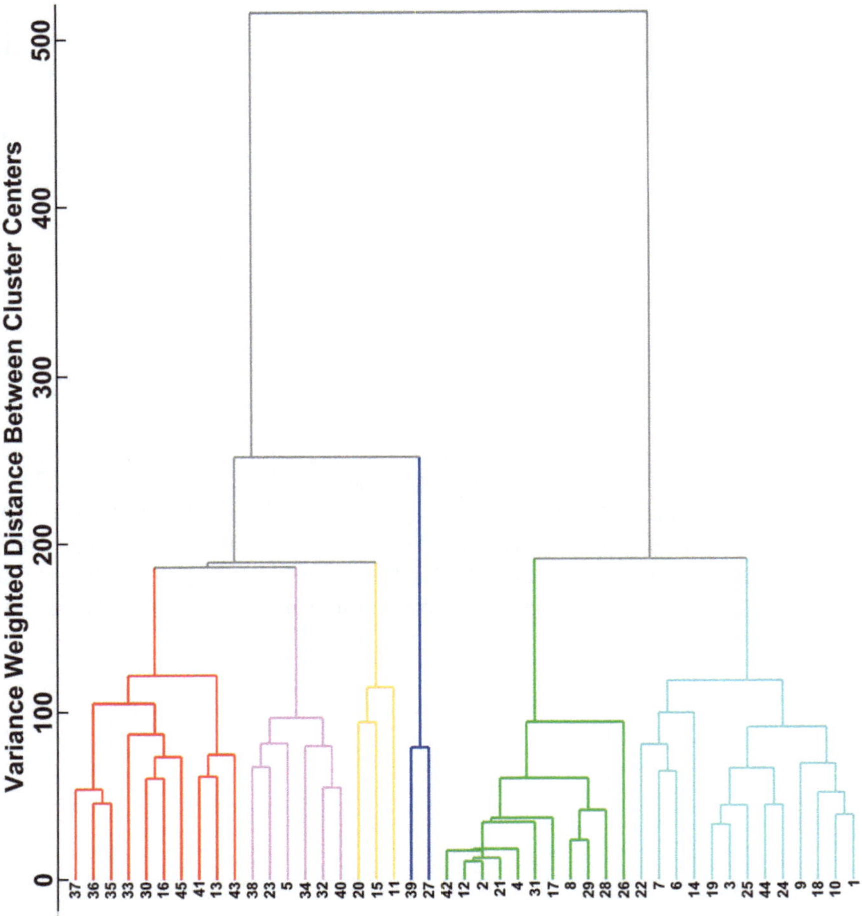

Fig. 4 Dendogram example. Samples, whose ID numbers are specified in the *x*-axis, are divided into six clusters, shown in different colors, by manually setting a cutoff at height 150

4. Review the scores plot to look for possible groupings, e.g., of clinical outcome, or outliers in your dataset.

Hierarchical Cluster Analysis (HCA)

HCA aims to find natural clusters among samples using a hierarchical approach where samples are grouped according to calculated similarities and dissimilarities. The result is visualized as a dendrogram (*see* Fig. 4). At the bottom of the dendrogram, all objects represent individual clusters. For each level, the two closest objects are joined into one cluster. This continues until all clusters are joined by one branch. There are different measures for determining the distance between individual samples or between clusters of samples. Common measures for individual samples include Euclidean distance, Manhattan distance, and sample correlations. Common measures for distance between clusters include single linkage, average linkage, complete linkage, and Ward's method. The procedure is done automatically by most software. The steps for the HCA algorithm are listed below.

Protocol for HCA

1. Calculate the distance between all possible pairs of clusters using the chosen distance measure for individual samples.
2. Merge the two clusters with the smallest distance.
3. Calculate the new clusters' distance to other clusters using a chosen distance measure for clusters.
4. Repeat **steps 2–3** until all samples are merged into one cluster.

Alternatively, a top down approach can be used where all objects are considered one cluster initially and subsequently divided into smaller clusters depending on their dissimilarities.

To decide which matrices are optimal for distance measurements and assessing how well the dendrogram reflects your data, the cophenetic correlation coefficient [12] can be used. This coefficient calculates the correlation of the original pairwise distance between two objects and the level/height at which the two objects were joined in one cluster.

The resulting dendrogram can be used to divide the samples into clusters. The number of resulting clusters can be defined beforehand or a cutoff can be set at a decided level of the dendogram, either manually or using for instance Gap statistics [13]. All the samples joined by branches below the cutoff are considered one cluster. The resulting clusters can be evaluated in terms of clinical endpoints of interest. Prediction of cluster labels for new samples can be achieved based on the shortest distance to each of the cluster centroids or using validated supervised models (e.g., PLS-DA, *see* Subheading 3.1.2).

3.1.2 Supervised Methods

Supervised multivariate methods are used to identify correlations and build models that can predict characteristics of new data. These methods model the relationship between the independent variables or X-matrix (e.g., spectral data or quantified metabolites), and a response variable(s) or Y-matrix/vector (e.g., clinical endpoints) by identifying patterns in the input data and making decision rules that can be applied to new data. Several supervised analysis methods exist, some of which build models by linear combinations of the input variables (e.g., partial least squares methods), while others model more complex, nonlinear relationships (e.g., neural networks and support vector machines). An advantage of linear models is that biological interpretation is more easily achievable.

1. In case of continuous response variables such as blood pressure or tumor size, a supervised regression method that predicts a response value should be used.

2. For categorical response variables (e.g., treatment and control groups), a classification method that can assign samples to two or more groups should be used.

A common feature of supervised analysis methods is that these methods are prone to overfitting to the input data [14]. This would result in a model that makes good predictions for the data used, but predicts badly for new data. To avoid overfitted models and over-optimistic results, validation is a crucial step for supervised analysis methods. Validation procedures are described in Subheading 3.1.4.

Partial Least Squares (PLS) Methods

Partial least squares (PLS) is commonly used both for regression and for classification problems. PLS defines underlying structures that maximize the covariance between the independent variables and the response variable [15]. Instead of only modeling the independent variables ***X***, as is the case for PCA, the dependent response variables ***Y*** are also modeled:

$$\boldsymbol{X} = \boldsymbol{TP}^{\mathrm{T}} + \boldsymbol{E} \tag{2}$$

$$\boldsymbol{Y} = \boldsymbol{UQ}^{\mathrm{T}} + \boldsymbol{F} \tag{3}$$

where ***T*** and ***U*** are the score matrices, ***P*** and ***Q*** are the loading matrices, and ***E*** and ***F*** are the residuals for ***X*** and ***Y***, respectively. ^{T} indicates the transpose of a matrix. The X-scores, ***T***, are predictors of ***Y*** and will also model ***X***, thus both ***X*** and ***Y*** are assumed to be modeled by the same latent variables. Hence, ***Y*** can be written as

$$\boldsymbol{Y} = \boldsymbol{TGQ}^{\mathrm{T}} + \boldsymbol{F} \tag{4}$$

where ***G*** is the diagonal matrix resulting from ***U*** = ***TG***.

There are several algorithms that can be used to estimate these parameters, all of which provide more or less similar results [16].

The covariance between ***X*** and ***Y*** is optimized by defining PLS LVs, which are linear combinations of the original ***X*** variables, and the dimensionality of the resulting PLS model is equal to the number of LVs used in the model. The optimal number of LVs to use is chosen based on different model diagnostic terms used to evaluate the overall quality of the model for different numbers of LVs. For PLSR, the Q^2 statistic (Eq. 5) and the root mean square error (RMSE) (Eq. 6) are typically used. These statistics reflect the differences between the predicted value ($\breve{y}$) and the known y:

$$Q^2 = 1 - \frac{\sum_i \left(y_i - \breve{y}_i\right)^2}{\sum_i \left(y_i - \bar{y}\right)^2} \tag{5}$$

$$\mathrm{RMSE} = \sqrt{\frac{\sum_i \left(y_i - \breve{y}_i\right)^2}{m}} \quad (6)$$

where $i = 1,2,\ldots,m$ for m included samples and $\bar{y}$ is the mean of all y values. If RMSE is to be compared between datasets, the value should be normalized (NRMSE) to make it independent of scale (Eq. 7):

$$\mathrm{NRMSE} = \frac{\mathrm{RMSE}}{y_{\max} - y_{\min}} \quad (7)$$

A Q^2 of 1 corresponds to perfect prediction, while a very low or negative Q^2 indicates a poor prediction. A NRMSE value of 0 corresponds to perfect prediction, while a value close to 1 indicates poor prediction.

For PLS-DA, commonly used diagnostic statistics include the classification error, sensitivity, and specificity. The number of correctly classified samples, i.e., true positives (TP) and true negatives (TN), and the number of incorrectly classified samples, i.e., false negatives (FN) and false positives (FP), is subsequently recorded. The prediction error (*see* **Note 5**) relates the number of incorrectly classified samples with the total number of samples (Eq. 8). The model accuracy equals one minus the classification error.

$$\begin{aligned}\mathrm{Error} &= \frac{\text{Number of incorrectly classifed samples}}{\text{Total number of samples}} \\ &= \frac{\mathrm{FN} + \mathrm{FP}}{\mathrm{TP} + \mathrm{TN} + \mathrm{FN} + \mathrm{FP}}\end{aligned} \quad (8)$$

Sensitivity measures the ability to correctly predict the case class, or true positive samples (Eq. 9). A highly sensitive model is one that generates few false negatives.

$$\begin{aligned}\mathrm{Sensitivity} &= \frac{\text{Number of correctly classified cases}}{\text{Total number of cases}} \\ &= \frac{\mathrm{TP}}{\mathrm{TP} + \mathrm{FN}}\end{aligned} \quad (9)$$

Specificity measures the ability to correctly classify the control class, or true negative samples (Eq. 10). A highly specific model is one that generates few false positives.

$$\begin{aligned}\mathrm{Specificity} &= \frac{\text{Number of correctly classified controls}}{\text{Total number of controls}} \\ &= \frac{\mathrm{TN}}{\mathrm{TN} + \mathrm{FP}}\end{aligned} \quad (10)$$

The scores and loadings are calculated from the LVs, and are used for biological interpretation of the data in similar manners as for PCA.

Partial Least Squares Regression (PLSR)

PLSR is used for regression problems where the intention is to model correlations and/or make prediction on a continuous response variable.

Protocol for PLSR

1. Perform additional scaling of variables as described in the protocol for PCA, **step 1**.
2. Perform PLSR using the spectral data or quantified metabolites as independent variables X and the sample characteristic to be modeled as a continuous response variable Y. Depending on the number of samples in your dataset, choose the type of cross-validation that suits your dataset (*see* Subheading 3.1.4). Make PLSR models for a restricted number of LVs (*see* **Note 6**).
3. Examine the cross-validated RMSE or Q^2 for the different numbers of LVs.
4. Choose the number of LVs giving the first minimum in cross-validated RMSE or the first maximum in Q^2 (*see* **Note** 7).
5. If you have an independent validation set, make predictions of the independent test set using the model obtained in **step 4**.

Partial Least Squares Discriminant Analysis (PLS-DA)

PLS-DA is used for classification problems where the intention is to model correlations and/or make prediction between two or more groups of samples.

Protocol for PLS-DA

1. Perform additional scaling of variables as described in the **Protocol for PCA, step 1**.
2. Perform PLS-DA using the spectral data or quantified metabolites as independent variables X and the sample characteristic to be modeled as a categorical response variable Y, with discrete numbers representing each class (e.g., 1 for treatment group and 2 for control group). Depending on the number of samples in your dataset, choose the type of cross-validation that suits your dataset (*see* Subheading 3.1.4). Make PLS-DA models for a restricted number of LVs (*see* **Note 6**).
3. Examine the cross-validated classification error for the different numbers of LVs.
4. Choose the number of LVs giving the first minimum in cross-validated classification error (*see* **Note** 7).
5. If you have an independent validation set, make predictions of the independent test set using the model obtained in **step 4**.

Orthogonal PLS (OPLS)

In orthogonal PLSR (OPLSR) and OPLS-DA, the response orthogonal variations in X are separated out before the model is built. Orthogonalizing the model gives identical model performance to that of original PLS as the OPLS components are rotations of the original ones. However, interpretations are easier as all relevant information will be present in the first LV, hence only the first score and loading vector will be used for interpretations.

Multilevel PLS-DA

For longitudinal or cross-over studies, where each individual serves as its own control, multilevel PLS-DA can be used to separate the within-patient variation from the between-patient variation [17]. By focusing on the within-patient variation, representing metabolic changes due to the intervention (e.g., samples before and after treatment), metabolic changes that would otherwise be masked by the often much larger between-patient variation can be revealed. In the example of samples of the same individuals before and after intervention, the within-patient variation would be separated according to:

$$\text{Control} = A - B \tag{11}$$

$$\text{Intervention} = B - A \tag{12}$$

where $\boldsymbol{A}$ is the metabolic data before intervention and $\boldsymbol{B}$ is the metabolic data after intervention. These new matrices from Eqs. 11 and 12 are concatenated and used as independent variables in PLS-DA with a categorical variable representing control and intervention as the $\boldsymbol{Y}$ vector.

Other Multivariate Analysis Methods

In addition to PLS-based methods, several other multivariate analysis methods are suitable for the analysis of breast cancer metabolomics data. Neural networks (NNs) can model complex, nonlinear relationships between the input variables and the problem to be solved, and can be used for both regression and classification problems [18]. NNs consist of three or more layers: an input layer, one or several hidden layers, and an output layer. The nodes of each layer are connected through weights, and the weights and hidden layer(s) will be adapted to the input data through learning. Another suitable method is support vector machines (SVMs) [19]. SVMs do not learn like NNs, but instead aim to find boundaries that separate different groups. The boundary determined by SVMs will be a line in 2D, a plane in 3D, or a hyperplane in n dimensions. By choosing different kernel functions, SVMs can be applied to nonlinear problems by transformation of the input space into a higher dimension where the classes are linearly separable. Although these methods can be powerful for making predictions, a main drawback of NNs and SVMs is the difficulty in interpreting the resulting models.

3.1.3 Variable Selection

Metabolic datasets are made up of several variables, or columns, each one representing a point in a spectrum or an individual metabolite. Variables that are biologically irrelevant add noise to the model and can impair model performance. A variable selection procedure is therefore often performed when analyzing metabolomics data. Although variable selection can be carried out using several methods that evaluate individual variables or subsets of them at a time, all of them use information obtained from statistical modeling (e.g., variable importance in projection [20], prediction error [21], selectivity ratio [22]) as a variable importance measure. It is therefore important to perform suitable validation of the selected variables (*see* Subheading 3.1.4).

Protocol for Variable Selection

1. Define the variable selection method to use. For an overview of available methods, *see* Ref. 23.
2. Perform variable selection using the algorithm or script defined.
3. Build models using only the selected variables.

It is worth mentioning that if many variables provide the same information, only one (or very few) of these will be selected to minimize redundancy. Highly correlated metabolites involved in the same pathway could thus be discarded while still being biologically important.

3.1.4 Validation

Supervised multivariate models (*see* Subheading 3.1.2) tend to overfit the data, which means they may capture even the noise of the dataset used for building, or training, the model. If this occurs, the model will perform well only for the training samples, and not be applicable to new, similar samples [14]. Proper validation is therefore essential to assess model quality and robustness. A second purpose for validation is to optimize model parameters, such as optimal dimensionality, e.g., the number of LVs in PLS or the number of hidden layers in neural networks.

Cross Validation

Multivariate model validation is typically performed using training and test data. The test set is a group of samples with known independent and dependent variable values not used for model building, while the remaining data constitutes the training set used for modeling. The training and test set should be representative of each other, i.e., sampled from a similar population and handled identically. The resampling procedure known as cross validation [24] is commonly used for defining training and test data, building models, and if applicable, optimizing model parameters. Using the test sets generated with cross validation (CV) to simultaneously assess the quality of the model and to determine the optimal dimensionality (e.g., number of LVs for PLS) is prone to produce an over-optimistic error as the validation for both purposes should be independent of each other. It is therefore preferable to use a completely independent validation set comprised of a large group of new samples for final validation. However, due to, e.g., budget, technical, or time constraints, the number of available samples is often not sufficient to do this, and the model performance will be based on the cross-validation results.

Protocol for Cross Validation

1. Divide your data into *k* subsets.
2. Exclude the first subset as a test set, and make a multivariate model of the remaining data (the training data).
3. Test the performance of the resulting model on the test set.
4. Repeat **steps 2–3** for all *k* subsets.
5. Assess the mean model diagnostic statistic (e.g., RMSE or prediction error) for all *k* test sets. If applicable determine the correct dimensionality or other parameters of the model as described in the **Protocols for PLSR** and **PLS-DA, steps 2–4**.
6. If an independent validation set is present, build a model on all data from **step 1**, using the dimensionality or parameters determined in **step 5**, and test the model using the independent validation set.

The above procedure is designated k-fold CV, according to the number of subsets, or folds (*k*) defined. Defining *k* to be the total number of samples results in leave-one-out (LOO) CV. This procedure is particularly useful for datasets with low sample number (*see* **Note 8**).

Alternative to defining a number of folds to divide the data into, a fixed number or percentage of samples to be left out from model building for each *n* validation can be defined. By using random sampling, a different total error will be obtained if the CV procedure is repeated, since the test sets will vary at random with each repetition. This provides a less biased validation result.

Alternative to using an independent validation set, a double CV procedure should be employed when there are a sufficient number of samples. The procedure includes an inner loop nested in an outer loop for separate model parameter optimization and model quality assessment (*see* Fig. 5) and is carried out as follows:

Protocol for Double CV

1. Divide your data into *k* subsets.
2. Exclude the first subset as a validation set, and use the remaining data as input for the inner CV loop.
3. Perform the inner CV loop as **steps 1** through **5** described for a typical single-layered k-fold CV above (*see* **Note 9**).
4. Build a model on all data inputted in the inner CV loop, using the dimensionality or parameters determined in **step 3**, and test the model using the excluded validation set in **step 2**.
5. Repeat **steps 2–4** for all *k* subsets.
6. Assess the mean model diagnostic statistic (e.g., RMSE or prediction error) for all *k* validation sets.

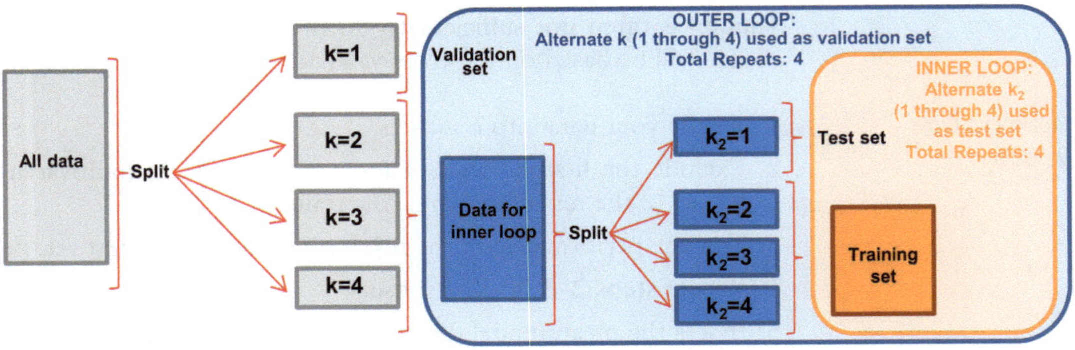

Fig. 5 Illustration of data splitting for a four-fold double cross validation procedure through which the number of latent variables (LVs) is optimized in the inner loop and model quality is assessed in the outer loop. Samples are divided into four different outer loop groups or folds ($k = 1$–4). At each outer loop repetition, three folds comprise the data input for the inner loop, while one is left out as a validation set. The inner loop is then partitioned into four inner loop folds ($k_2 = 1$–4), which at each inner loop repetition alternate the role of test set while the remaining folds comprise the training set. The samples comprising the validation set in the outer loop are therefore unseen to the latent variable optimization procedure, reducing the risk of over optimistic results when using them to assess the model built with the inner loop data. A classical, single-layered CV procedure consists only of the outer loop, with the inner loop data being the training set

Model quality assessment employing the described double CV procedure is thus performed using validation samples completely unseen during LV optimization, reducing the risk of overfitting. Double CV is usually not available for direct implementation in data analysis software without the creation of in-house scripts being required. In most multivariate analysis software, however, the typical single-layered k-fold CV previously described can be performed using ready-made scripts or graphical user interphases (GUIs) which provide the option to use an independent validation set to assess the quality of the cross-validated model. It is important to emphasize that when optimizing model parameters using single-layered CV, unless model quality is assessed using new samples, diagnostic statistics obtained cannot be considered reliable.

Permutation Testing

To ensure that obtained model diagnostic statistics are significantly better than those that would be obtained by chance, permutation testing can be performed. By rearranging the y response variable in a random order, the y continuous values or classes are no longer associated with their true corresponding metabolic information ($\boldsymbol{X}$); thus, any relationships between $\boldsymbol{X}$ and y are lost [24]. The procedure can be performed to evaluate a double CV procedure as follows.

Protocol for Permutation Testing

1. Permute or rearrange the values in the original y variable in a random order to obtain $y_{permuted}$. Replace the original y variable with $y_{permuted}$.

2. Perform **steps 1** through **5** described for a typical single-layered k-fold CV above using the optimized parameter value determined for the model being tested.
3. Repeat **steps 1–2** a defined number of times (n_{perm}) (*see* **Note 10**). A total of n_{perm} errors from different permuted models will be obtained.
4. Count the number of permuted models achieving an equal or better diagnostic statistic than the unpermuted model being tested and define it as a (e.g., $\text{error}_{\text{nonperm}} \geq \text{error}_{\text{perm}}$).
5. Calculate a p-value as: a/n_{perm} (*see* **Note 11**).

3.2 Univariate Analysis

Univariate analysis can be performed to search for statistically significant differences in individual metabolites between groups.

3.2.1 Selection Criteria for Univariate Tests

Prior to univariate analysis, it should be decided whether the data is prone to parametric or nonparametric tests. This is decided based on at least three check points: normality (*see* **Note 12**), homogeneity of variances, i.e., homoscedasticity (in case of heteroscedasticity, *see* **Note 13**), and independency of samples (for dependent samples, e.g., repeated measurements, samples from the same hospital, etc., linear mixed-effects models can be used (*see* Subheading 3.2.2)). Figure 6 shows a simplified overview of tests to select according to data distribution and number of groups to test. For more extensive details regarding selection criteria for univariate tests, refer to [25].

3.2.2 Linear Mixed-Effects Model

Linear mixed-effects models (LMM) are an extension of general linear models taking into account both fixed and random effects, where fixed effects often are those of primary interest, e.g., effect of treatment type, while random effects are results of random selection, e.g., age, hospital, or individual. The modeling of random effects enables inclusion of repeated measurements. An additional advantage is that LMM can handle missing values, thus improving the power in multilevel analysis where some observations are missing.

In longitudinal studies, where samples have been collected from individuals over time, LMM can be used to evaluate which metabolites are significantly different with respect to one or more outcomes of interest. In such cases, metabolite levels are set as individual response variables, clinical outcome as a fixed effect, and patient number as a random effect.

To perform LMM, first define the fixed and random effects. Categorical fixed effects are set as factors. To decide whether or not to model interactions between the fixed effects, a likelihood ratio test comparing the reduced model (without interactions) to the full

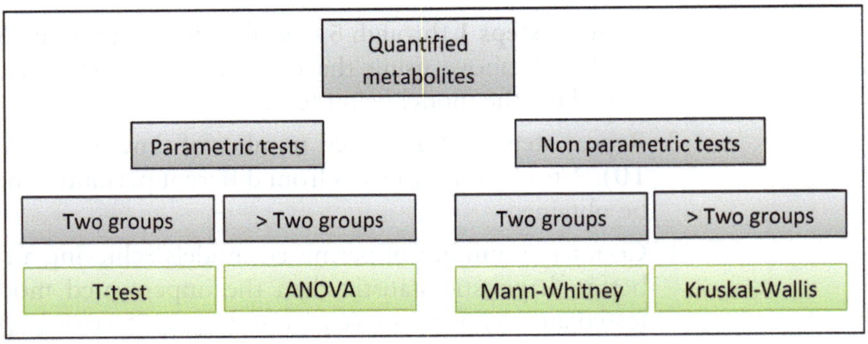

Fig. 6 Examples of univariate tests that can be used for evaluating group differences in the level of quantified metabolites

model (with interactions) can be performed. The resulting *p*-value will reflect the significance of the interaction.

1. Perform LMM using the software of choice.
2. Check that residuals are normally distributed. If not: try transforming the data to achieve normally distributed residuals.

The resulting *p*-values from LMM indicate the significance of the fixed effects after the correction of the random effect(s). Furthermore, LMM estimates show metabolite increasing or decreasing trends for continuous fixed effects or whether metabolites are higher or lower in one level or group compared to another for categorical fixed effects.

3.2.3 Multiple Testing Correction

When performing tests to associate a *p*-value to each individual metabolite separately, such as those described in Subheadings 3.2.1 and 3.2.2, the same test is repeated for all metabolites. The likelihood of significant *p*-values being achieved by chance will increase with the number of tests performed. Hence, the number of false positives (i.e., type I errors) should be controlled for. Here lies the purpose of multiple testing corrections, which can be achieved via different approaches. A widely used approach is the Bonferroni adjustment [26].

Protocol for Bonferroni Adjustment

1. Generate *p*-values for all *n* metabolites using a suitable statistical test.
2. Multiply each *p*-value by *n*.

The Bonferroni method controls for the family-wise error rate (FWER), which is the probability of producing at least one false positive. Although simple, the Bonferroni adjustment is generally unnecessarily strict for the purposes of metabolic analyses. Alternatively, one can implement less stringent correction methods that control for the false discovery rate (FDR), which is the expected proportion of false positives to be generated. One such method is the Benjamini-Hochberg adjustment [27, 28], whose procedure is as follows.

Protocol for Benjamini-Hochberg Adjustment:

1. Input p-values and record their order, referred to as the "original order."
2. Rank and sort the inputted p-values in an ascending order, such that the rank i of the smallest p-value is 1, the second smallest has an $i = 2$, etc.
3. Calculate an intermediate q-value (q_{int}) for each sorted p-value (p_{val}): $q_{int} = (p_{val}/i) \times n$, where n is the number of inputted p-values.
4. Sort the q_{int} in an ascending order, recording their corresponding p-value ranks.
5. The sorted q_{int} values will now be adjusted according to their p-value rank. The first q_{int} value (q_{int0}) remains the same. If the rank of the second sorted q_{int} value (q_{int1}) is lower than that of the previous value (q_{int0}), overwrite q_{int1} with q_{int0}. Next, look to the rank of q_{int2}. If its rank is lower than q_{int1} then replace q_{int2} with the new value of q_{int1}, if the rank is higher, then q_{int2} remains unchanged. Next, compare the rank of q_{int2} with q_{int3} and repeat the previous steps until all q_{int} values have been adjusted. The result is a list of the final q-values (*see* **Note 14**).
6. Reorder the final q-values so that they correspond to the original order of the inputted p-values recorded in **step 1**.

The adjusted p-value (q-value) represents the smallest FDR at which the corresponding test will be significant. So for a q-value of 0.02, the test would be considered significant (null hypothesis rejected) when allowing a maximum of 2% of all significant tests to be false positives (i.e., FDR threshold is 2%). As for all statistical tests, the desired FDR threshold value to base significance on should be defined prior to testing.

3.3 Multivariate Versus Univariate Analysis

An overview of the key steps to analyze metabolic profiles in breast cancer using both multivariate and univariate methods has been provided. To conclude, a comparison of these methods regarding their advantages and disadvantages is presented in Table 2.

4 Notes

1. Refer to the following for example studies where metabolic signature was related to specific clinical endpoints in breast cancer: hormone receptor status and axillary lymph node status [29] and treatment response and 5-year survival [30] studying

Table 2
Advantages and disadvantages of univariate and multivariate methods

	Advantages	Disadvantages
Univariate methods	• Widely used/known in all scientific fields. • Usually simple and straightforward to perform and interpret. • Useful for targeted approaches when one or a few metabolites have been defined to be tested. • Variables (i.e., metabolites) do not affect the outcome of each other's tests (with the exception of multiple testing procedures).	• Accurate measure of absolute or relative concentrations is essential. • Untargeted approaches present challenges, particularly the risk of false discoveries increasing with increasing number of univariate paralleled tests performed. Although this can be addressed by applying multiple testing corrections, these in turn may be too strict, thereby risking to miss a true discovery. • Does not account for variable correlation.
Multivariate methods	• Useful for exploratory purposes, such as outlier detection. • Applicable for untargeted approaches as they can handle large numbers of variables and evaluate their importance, i.e., relevance to the scientific problem at hand. • Takes proper account of the correlation between spectral points/metabolites. • No need to correct for multiple testing, as all variables are analyzed simultaneously. • Evidence of individual metabolites can accumulate to reveal findings that would not be detected separately with univariate methods. • Quantification not necessary	• Not widely known in clinical fields • Computationally intensive, time-consuming algorithms • Interpretation might not be straightforward • Unimportant variables that are mainly noise can obscure information from important variables that would be detected using univariate tests. • When using the metabolic profile as input, scaling will increase the influence of the noise and might not be optimal. Thus, differences in metabolites of lower abundance may be obscured by those of higher abundance.

tissue; early recurrence [31, 32] and weight change [33] studying serum; risk of disease development [34] studying plasma.

2. Autoscaling should not be performed on spectral data, as this will scale up the noise regions between metabolite signals.
3. Choosing a number of PCs explaining 80–90% of the data variation will usually be sufficient to get a good overview of the data.
4. Alternatively, scree plots can be plotted as the eigenvalue versus the corresponding principal component (PC). An eigenvalue describes the amount of variance accounted for by its associated PC [35].
5. For unbalanced datasets, i.e., those with very few samples of one class and many of the other, the prediction error may be misleading. For example, if 90% of samples in a dataset are of class A, a model that predicts every sample as class A will achieve

an error of 0.1. In these cases, the sensitivity and specificity provide a more correct assessment.

6. A maximum of 20 LVs is sufficient for most datasets.
7. In certain cases, outlying samples might give a small increase in RMSE/classification error or a decrease in Q^2 before the RMSE/classification error continues to decrease or the Q^2 continues to increase. If a sample appears to be an outlier in a score plot, try to remove this sample and see if that changes the results.
8. Typically $n < 20$ samples is too few to perform a cross validation other than LOOCV.
9. Data inputted in the inner loop is partitioned into a training set to build models and a test set to assess the models built at each iteration.
10. n_{perm} is usually at least 1000 so that the obtained p-values are in the thousandth order of magnitude (10^{-3}).
11. If $a = 0$, i.e., no permuted model performed better than the unpermuted model, report the p-value as lower than $1/n_{perm}$. For example, for $n_{perm} = 1000$, $p < 1/1000$, $p < 0.001$.
12. Statistical tests or graphical visualization can be used to evaluate data distribution. Examples of statistical tests used to check normality are Shapiro-Wilk or Kolmogorov-Smirnov [36]. Graphical visualization can be performed by plotting histograms (symmetrical and bell-shaped indicates normal distribution) or normal probability plots (q-q plot) (a line at $y = x$ indicates normal distribution). For non-normally distributed data, *see* **Note 13**.
13. Transformation of the data (e.g., log transformation) can be applied to allow for parametric testing. Alternatively, nonparametric tests can be chosen.
14. Due to the overwriting of q-values based on rank, it is typical for one or more q-values to be repeated when adjusting using this method.

References

1. Clarke CJ, Haselden JN (2008) Metabolic profiling as a tool for understanding mechanisms of toxicity. Toxicol Pathol 36 (1):140–147. https://doi.org/10.1177/0192623307310947
2. Bujak R, Struck-Lewicka W, Markuszewski MJ, Kaliszan R (2015) Metabolomics for laboratory diagnostics. J Pharm Biomed Anal 113:108–120. https://doi.org/10.1016/j.jpba.2014.12.017.
3. Keeler J, Understanding NMR (2010) Spectroscopy, 2nd edn. Wiley, Chichester, UK
4. Hu K, Westler WM, Markley JL (2011) Simultaneous quantification and identification of individual Chemicals in Metabolite Mixtures by two-dimensional extrapolated time-zero (1)H–(13)C HSQC (HSQC(0)). J Am Chem Soc 133(6):1662–1665. https://doi.org/10.1021/ja1095304
5. Nicholson JK (1989) High resolution nuclear magnetic resonance spectroscopy in clinical

chemistry and disease diagnosis. In: den Boer NC, van der Heiden C, Leijnse B, Souverijn JHM (eds) Clinical chemistry, an overview. Plenum Press, New York, NY
6. Le Gall G (2015) NMR spectroscopy of biofluids and extracts. In: Bjerrum TJ (ed) Metabonomics: methods and protocols. Springer, New York, NY, pp 29–36
7. Giskeødegård GF, Cao MD, Bathen TF (2015) High-resolution magic-angle-spinning NMR spectroscopy of intact tissue. In: Bjerrum TJ (ed) Metabonomics: methods and protocols. Springer, New York, NY, pp 37–50
8. Bathen TF, Geurts B, Sitter B, Fjøsne HE, Lundgren S, Buydens LM et al (2013) Feasibility of MR metabolomics for immediate analysis of resection margins during breast cancer surgery. PLoS One 8(4):e61578. https://doi.org/10.1371/journal.pone.0061578
9. Vettukattil R (2015) Preprocessing of raw metabonomic data. In: Bjerrum TJ (ed) Metabonomics: methods and protocols. Springer, New York, NY, pp 123–136
10. Euceda LR, Giskeødegard GF, Bathen TF (2015) Preprocessing of NMR metabolomics data. Scand J Clin Lab Invest 75(3):193–203. https://doi.org/10.3109/00365513.2014.1003593
11. Wold S, Esbensen K, Geladi P (1987) Principal component analysis. Chemometr Intell Lab Syst 2(1):37–52. https://doi.org/10.1016/0169-7439(87)80084-9.
12. The Mathworks Inc. Cophenetic correlation coefficient. http://www.mathworks.com/help/stats/cophenet.html#zmw57dd0e176726. Accessed 13 Apr 2016
13. Tibshirani R, Walther G, Hastie T (2001) Estimating the number of clusters in a data set via the gap statistic. J R Stat Soc Series B Stat Methodol 63(2):411–423. https://doi.org/10.1111/1467-9868.00293
14. Hawkins DM (2004) The problem of Overfitting. J Chem Inf Comput Sci 44(1):1–12. https://doi.org/10.1021/ci0342472
15. Wold S, Sjöström M, Eriksson L (2001) PLS-regression: a basic tool of chemometrics. Chemometr Intell Lab Syst 58(2):109–130. https://doi.org/10.1016/S0169-7439(01)00155-1.
16. Andersson M (2009) A comparison of nine PLS1 algorithms. J Chemometr 23 (10):518–529. https://doi.org/10.1002/cem.1248
17. van Velzen EJJ, Westerhuis JA, van Duynhoven JPM, van Dorsten FA, Hoefsloot HCJ, Jacobs DM et al (2008) Multilevel data analysis of a crossover designed human nutritional intervention study. J Proteome Res 7 (10):4483–4491. https://doi.org/10.1021/pr800145j
18. Brougham DF, Ivanova G, Gottschalk M, Collins DM, Eustace AJ et al (2011) Artificial neural networks for classification in metabolomic studies of whole cells using 1H nuclear magnetic resonance. J Biomed Biotechnol 2011:8. https://doi.org/10.1155/2011/158094.
19. Gromski PS, Muhamadali H, Ellis DI, Xu Y, Correa E, Turner ML et al (2015) A tutorial review: metabolomics and partial least squares-discriminant analysis – a marriage of convenience or a shotgun wedding. Anal Chim Acta 879:10–23. https://doi.org/10.1016/j.aca.2015.02.012.
20. Wold S, Johansson E, Cocchi M (1993) PLS: partial least-squares projections to latent structures. In: Kubinyi H (ed) 3D QSAR in Drug Design. ESCOM, Leiden, The Netherlands, pp 523–550
21. Li H-D, Zeng M-M, Tan B-B, Liang Y-Z, Q-S X, Cao D-S (2010) Recipe for revealing informative metabolites based on model population analysis. Metabolomics 6(3):353–361. https://doi.org/10.1007/s11306-010-0213-z
22. Rajalahti T, Arneberg R, Kroksveen AC, Berle M, Myhr K-M, Kvalheim OM (2009) Discriminating variable test and selectivity ratio plot: quantitative tools for interpretation and variable (biomarker) selection in complex spectral or chromatographic profiles. Anal Chem 81(7):2581–2590. https://doi.org/10.1021/ac802514y
23. Mehmood T, Liland KH, Snipen L, Sæbø S (2012) A review of variable selection methods in partial least squares regression. Chemometr Intell Lab Syst. 118:62–69. https://doi.org/10.1016/j.chemolab.2012.07.010.
24. Westerhuis JA, Hoefsloot HCJ, Smit S, Vis DJ, Smilde AK, Velzen EJJ et al (2008) Assessment of PLSDA cross validation. Metabolomics 4 (1):81–89. https://doi.org/10.1007/s11306-007-0099-6
25. Riffenburgh RH (2006) Statistics in medicine, 2nd edn. Elsevier Academic Press, Burlington, MA
26. Bonferroni CE (1936) Teoria statistica delle classi e calcolo delle probabilità, Pubblicazioni del R Istituto Superiore di Scienze Economiche e Commerciali di Firenze, vol 8. Seeber, Firenze, pp 3–62. doi:citeulike-article-id:1778138
27. Benjamini Y, Hochberg Y (1995) Controlling the false discovery rate: a practical and powerful approach to multiple testing. J R Stat Soc Series B Stat Methodol 57(1):289–300

28. Benjamini Y, Yekutieli D (2001) The control of the false discovery rate in multiple testing under dependency. Ann Stat 29(4):1165–1188
29. Giskeødegård GF, Grinde MT, Sitter B, Axelson DE, Lundgren S, Fjøsne HE et al (2010) Multivariate modeling and prediction of breast cancer prognostic factors using MR metabolomics. J Proteome Res 9(2):972–979. https://doi.org/10.1021/pr9008783
30. Cao MD, Giskeødegård GF, Bathen TF, Sitter B, Bofin A, Lønning PE et al (2012) Prognostic value of metabolic response in breast cancer patients receiving neoadjuvant chemotherapy. BMC Cancer 12(1):1–11. https://doi.org/10.1186/1471-2407-12-39.
31. Asiago VM, Alvarado LZ, Shanaiah N, Gowda GAN, Owusu-Sarfo K, Ballas RA et al (2010) Early detection of recurrent breast cancer using metabolite profiling. Cancer Res 70 (21):8309–8318. https://doi.org/10.1158/0008-5472.can-10-1319
32. Tenori L, Oakman C, Morris PG, Gralka E, Turner N, Cappadona S et al (2015) Serum metabolomic profiles evaluated after surgery may identify patients with oestrogen receptor negative early breast cancer at increased risk of disease recurrence. Results from a retrospective study. Mol Oncol 9(1):128–139. https://doi.org/10.1016/j.molonc.2014.07.012.
33. Keun HC, Sidhu J, Pchejetski D, Lewis JS, Marconell H, Patterson M et al (2009) Serum molecular signatures of weight change during early breast cancer chemotherapy. Clin Cancer Res 15(21):6716–6723. https://doi.org/10.1158/1078-0432.ccr-09-1452
34. Bro R, Kamstrup-Nielsen MH, Engelsen SB, Savorani F, Rasmussen MA, Hansen L et al (2015) Forecasting individual breast cancer risk using plasma metabolomics and biocontours. Metabolomics 11(5):1376–1380. https://doi.org/10.1007/s11306-015-0793-8
35. Bernstein IH, Garvin CP, Teng GK (1988) Applied multivariate analysis. Springer, New York, NY
36. Ghasemi A, Zahediasl S (2012) Normality tests for statistical analysis: a guide for non-statisticians. Int J Endocrinol Metab 10 (2):486–489. https://doi.org/10.5812/ijem.3505

Part IV

Systems Biology of Metastasis and Tumor/Microenvironment Interactions

Chapter 10

Stochastic and Deterministic Models for the Metastatic Emission Process: Formalisms and Crosslinks

Christophe Gomez and Niklas Hartung

Abstract

Although the detection of metastases radically changes prognosis of and treatment decisions for a cancer patient, clinically undetectable micrometastases hamper a consistent classification into localized or metastatic disease. This chapter discusses mathematical modeling efforts that could help to estimate the metastatic risk in such a situation. We focus on two approaches: (1) a stochastic framework describing metastatic emission events at random times, formalized via Poisson processes, and (2) a deterministic framework describing the micrometastatic state through a size-structured density function in a partial differential equation model. Three aspects are addressed in this chapter. First, a motivation for the Poisson process framework is presented and modeling hypotheses and mechanisms are introduced. Second, we extend the Poisson model to account for secondary metastatic emission. Third, we highlight an inherent crosslink between the stochastic and deterministic frameworks and discuss its implications. For increased accessibility the chapter is split into an informal presentation of the results using a minimum of mathematical formalism and a rigorous mathematical treatment for more theoretically interested readers.

Key words Poisson process, Structured population equation, Metastasis, Mathematical modeling

1 Introduction

Metastasis is the spread of cancer cells to distant tissues broadly divided into two steps, physical dissemination and tissue-specific colonization [1]. While the first part is facilitated by a reversible phenotypic change of cancer cells [2], successful colonization involves complex tumor–microenvironment interactions and is still not well understood [3, 4].

Being responsible for most cancer-related deaths, metastasis is a pivotal point in disease history [5]. However, since metastases smaller than approximately 10^7 cells remain undetectable by medical imaging or other diagnostic tools, the clinical appearance of nonmetastatic disease may not reflect the true metastatic state of a

Both authors equally contributed to this work.

Louise von Stechow (ed.), *Cancer Systems Biology: Methods and Protocols*, Methods in Molecular Biology, vol. 1711, https://doi.org/10.1007/978-1-4939-7493-1_10, © Springer Science+Business Media, LLC 2018

patient. Therefore, estimating the metastatic risk in cancer patients without visible metastases is of major clinical importance [6]. In this respect, mathematical and statistical techniques have the potential to derive risk scores from clinical data.

Today, there is a large body of mathematical oncology literature; a recent review specifically focuses on the metastatic process [7]. Here, we will briefly summarize modeling efforts focusing on analyzing metastatic risk.

The emergence of a metastatic phenotype is governed by a number of key mutations [8, 9]. In [10] and [11] this assumption is translated into mathematical models, thereby deriving a metastatic risk score from evolutionary principles. In an opinion paper, approaches explaining emergent behavior through lower-level mechanisms were qualified as "the essence of integrated mathematical oncology" [12]. While acknowledging the importance of such approaches for improving our understanding of cancer biology, we will focus here on more data-driven models featuring simpler principles and thereby better matching the limited amount of information available in a clinical situation.

In an early work, a link between primary tumor size at surgery and risk of recurrence was established from a large cohort of breast cancer patients [13]. Later, these and other data were explained heuristically [14] (see also Subheading 2.1). In addition to such phenotypic characteristics, specific genetic signatures of the primary tumor have been found to be associated with increased metastatic risk [15]. These risk prediction models are static; they do not aim at representing the time evolution of the disease.

In contrast, dynamic models allow to predict the modeled system at different times, and more easily integrate data obtained at different observation times. To represent the dependency of the metastatic process on the primary tumor, a dynamic description of primary tumor growth is also integrated into metastatic models. The simplest dynamic model for tumor growth is the exponential model, which adequately describes growth under no restrictions (e.g., in vitro). In many cases of interest however, especially in vivo, sigmoidal (s-shaped) models with an initial exponential phase and subsequent deceleration are better suited to describe growth dynamics. The Gompertz model is a classical example that has been commonly used [16–18]. Power growth models have also been used for the description of clinical [19] and preclinical tumor growth data [20]. Although much more complicated models have been developed, e.g. describing the spacial evolution of a tumor. Here we restrict our discussion of primary tumor growth to the simple models introduced above.

A stochastic dynamic model for metastasis was proposed in [21]. Their approach described the emission times of metastases as random events, formalized through a so-called non-homogeneous Poisson process with an emission rate increasing with primary tumor size (*see* Subheading 2.2). A variant of this model successfully described data on bone lesions from a metastatic breast cancer case [22]. The Poisson distribution allows for an interpretation of the emission process as being "memoryless" and many of its properties can be analyzed mathematically.

Dynamic models for metastasis can also be used to infer parts of the process that cannot be observed experimentally or clinically. In this respect, a partial differential equation (PDE) model was proposed to describe the size distribution of metastatic colonies [23] (*see* Subheading 3.1). Thereby, they characterized the micrometastatic state of a hepatocellular carcinoma patient from clinical information on visible metastatic colonies. Later, this size-structured model was also successfully incorporated into mixed-effects models to predict the size evolution of metastases in animal models without [24] and with [25] surgical removal of the primary tumor. A unique feature of this approach is that it allows to integrate secondary metastatic emission into the model. Indirect evidence on the capacity of metastases to spread further was given both through cancer network models [26–28] and through the discovery of self-seeding mechanisms [29, 30].

In this chapter, we focus on the two dynamical frameworks for metastasis described above, the Poisson process and the size-structured model. Three aspects are covered:

- an accessible introduction to the Poisson process framework with an example motivating the approach,
- an extension of the Poisson model to account for secondary metastatic emission,
- the inherent link between the (extended) Poisson model and the size-structured model.

Finally, we exploit the crosslink between the two frameworks in order to evaluate the adequacy of the modeling assumptions in the deterministic model and to realize simulations using both frameworks together.

We restrict our discussion to models describing the natural history of metastatic progression. While surgery of the primary tumor can be represented in these models, to incorporate the effect of systemic treatments a more general formalism would be required. For the ease of presentation, we will not include this layer of complexity here, although we point out that both frameworks have been extended to cover much more general cases, including systemic treatment [31, 32] and more complex interactions between primary tumor and metastases [33].

For increased accessibility for readers with a non-mathematical background, the present work is split into two parts. First, the concepts behind these approaches are presented informally in Subheadings 2 and 3 breaking down the mathematical formalism as much as possible. Subheading 5 contains rigorous definitions of all mathematical objects and the precise statement of the mathematical results; detailed proofs of these results are provided in Appendix.

2 A Probabilistic Framework for Metastatic Emission

2.1 Metastatic Risk

Predicting the probability of metastatic disease at diagnosis of the primary tumor is of major clinical importance since it is strongly linked to survival expectancy. One possibility to build such a prediction model is by using large databases to correlate information on the presence of metastases to primary tumor characteristics at diagnosis or surgery. As an example, we show a relationship established in [14] between primary tumor size at surgery and probability of metastasis based on clinical data on breast cancer:

$$\mathbb{P}(\text{no metastases}) = \exp(-c\, d^z), \tag{1}$$

where d is the largest diameter of the primary tumor at surgery, and c, $z > 0$ are parameters which were determined from the cohort data. At first view, these parameters are merely empirical and cannot be associated to any mechanism. However, a mechanistic interpretation as the combined effect of growth and emission dynamics is possible, and is based on the following premises:

Power growth. The growth of the largest primary tumor diameter d follows a power law, i.e. it is the solution of the ordinary differential equation $d'(t) = a_{\mathrm{PG}}\, d(t)^\alpha$. In this equation, a_{PG} determines the growth speed and α allows to describe different growth shapes: exponential growth ($\alpha = 1$), linear growth ($\alpha = 0$), and a spectrum of sigmoidal growth patterns in between ($0 < \alpha < 1$).

Power law of emission. During each (infinitesimal) time interval, there is a chance that the primary tumor emits a metastasis. The emission intensity λ depends on the current size of the primary tumor through a power law: $\lambda(t) = b\, d(t)^\beta$. In this context, the parameter b can be interpreted as the metastatic aggressiveness of the emitting tumor. In [23], β was linked to the mode of vascularization of the primary tumor: a uniform vascularization would correspond to $\beta = 3$ (dimension of space) and a surficial vascularization to $\beta = 2$ (dimension of a surface) (*see* **Note 1**).

Memorylessness. The probability of emission of a metastasis is independent of the previous emission history.

A typical model allowing a mathematical formalization of the above premises is the **Poisson process**. In principle, randomness of the metastatic emission process could also be represented through different probability laws, thereby dropping the memorylessness property as done, e.g., by Bethge et al. [34]. However, the Poisson model has several advantages. It does not require any additional statistical parameters, it has a high degree of analytical tractability (i.e., many of its properties can be investigated through mathematical analysis and not only by simulations), and there are efficient numerical routines such as thinning to simulate the process (see, e.g., [35]). The detailed derivation of the empirical relationship is presented in Subheading 5.2.

2.2 Poisson Processes

A Poisson process (PP) is a model for counting a series of events occurring at random times. The precise definition of this process is given in Subheading 5, but its basic properties are the two following ones (*see* Fig. 1 for an illustration):

1. The number of events in disjoint time intervals is independent. This translates the memorylessness property since given some time t, the number of future events (those happening at any time $t_{\text{future}} > t$) does not depend on the past events (those happening at any time $t_{\text{past}} < t$), but only depend on the present state of the system at time t. For example, in Fig. 1 the time elapsed between t and $T^{(4)}$ is independent of when exactly $T^{(3)}$ occurred. In other words, the system forgot what happened up to time t.
2. The number of events N_t that occurred by time t has a *Poisson distribution* with parameter

$$\Lambda(t) = \int_0^t \lambda(s) ds,$$

i.e., the integral over each emission intensity $\lambda(s)$ for s in the time

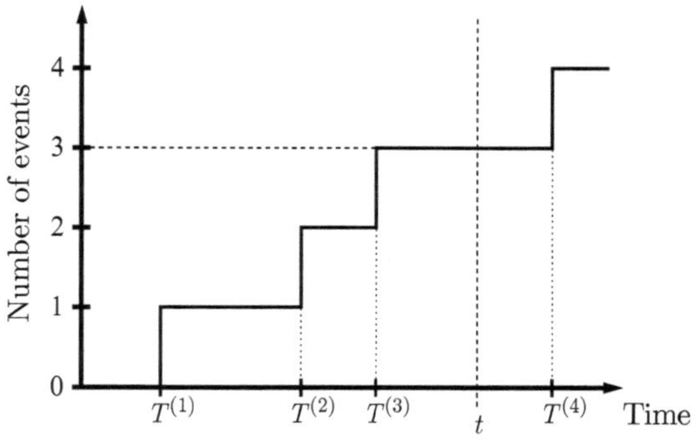

Fig. 1 Schematic trajectory of a Poisson process. Here, $T^{(1)}, \ldots, T^{(4)}$ are the times at which events occur, and by time t we have 3 events, i.e. $N_t = 3$

interval $[0, t]$. This means that the probability of having observed exactly k events until time t is given by

$$\mathbb{P}(N_t = k) = \frac{\Lambda(t)^k}{k!} e^{-\Lambda(t)}.$$

These two properties characterize the PP, and can even serve as a definition in addition to $N_0 = 0$. From these two properties, one can show that the probability that the next event time lies between times t and $t + \Delta t$ is approximately $\lambda(t)\Delta t$. Hence, λ determines the event frequency, and this is the reason why it is called the *intensity function*.

In the setting of this chapter, we are interested in describing the inception times of new metastatic lesions via PPs. This means that N_t is the number of metastases emitted until time t in our context. Following [21, 22], we will first suppose that only the primary tumor has the capacity of seeding metastases.

A constant emission intensity λ (called a homogeneous PP) would mean that a tumor consisting of a few cells is equally likely to shed a metastasis as a large tumor of several grams. Since such a model is not realistic, we need to consider time-varying intensities λ (called non-homogeneous PPs). We will consider an emission intensity λ that depends on some measure of primary tumor size $X_p(t)$ (diameter, volume, number of cells, etc.). The relationship between primary tumor size and emission intensity is given by a size-dependent emission law γ, i.e. $\lambda(t) = \gamma(X_p(t))$.

Before going into more detail, let us introduce a set of clinical parameters (summarized in Table 1), which will be used throughout this chapter to further illustrate those concepts. These parameters were estimated in [23] from clinical data on a hepatocellular carcinoma with multiple liver metastases. Although derived within the deterministic framework of the size-structured model

Table 1
Growth and emission laws derived in [23] from clinical data of a hepatocellular carcinoma case with multiple liver metastases

Model	Parameter	Symbol	Value	Unit
Growth (Gompertz law)	Initial size	x_0	1	Cells
$g(x) = a_{\mathrm{Gomp}} x \log(x_p^{\infty}/x)$	Growth rate	a_{Gomp}	0.00286	Days^{-1}
	Maximum size	x_p^{∞}	7.3×10^{10}	Cells
Emission (power law)	Rate constant	b	5.3×10^{-8}	Days^{-1} cells^{-1}
$\gamma(x) = bx^{\beta}$	Emission power	β	0.663	–

Primary tumor size is given by $X_p' = g(X_p)$, $X_p(0) = x_0$, and primary tumor emission rate is given by $\lambda(t) = \gamma(X_p(t))$. This set of parameters is used throughout the chapter; when used in the PP framework the emission rate λ is taken as the emission intensity

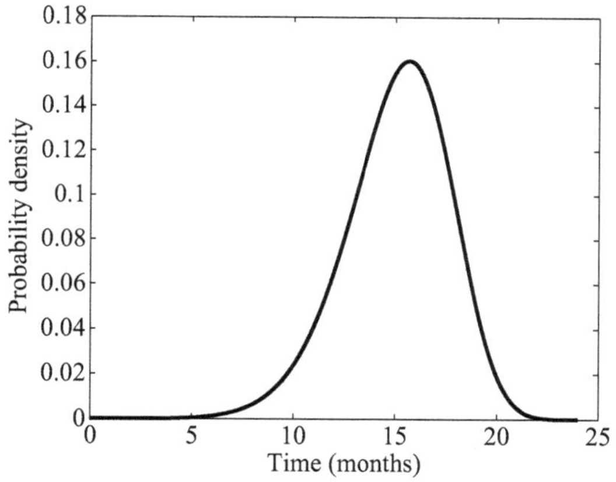

Fig. 2 Probability density function (pdf) for the emission time of the first metastasis $T^{(1)}$, using the clinical parameters in Table 1. The analytical formula of the pdf is $f_{T^{(1)}}(t) = \lambda(t)e^{-\Lambda(t)}$

(*see* Subheading 3.1 for more details), the inherent link with the PP framework described in Subheading 3.2 ensures that these parameters are also relevant in the PP model; we will therefore use the same set of parameters in both frameworks. Also, we will make use of a slight modification of this clinical setting to predict the risk of distant metastasis after surgery. To represent the impact of a surgery at time t_{surgery}, the emission intensity will be set to zero for all times larger than t_{surgery}. Randomness of emission means that each emission time can be represented via its probability density function; this is illustrated for the emission time of the first metastasis $T^{(1)}$ in Fig. 2.

The number of metastases N_t is itself random in this model, but relevant deterministic quantities can be derived from N_t, such as the expected number of metastases $\mathbb{E}[N_t]$ or the probability of metastatic disease $\mathbb{P}(N_t > 0)$. Exploiting the memorylessness property of PPs, these quantities can be computed without any need to simulate the process (all the following formulas are proven in Appendix:

$$\mathbb{E}[N_t] = \int_0^t \lambda(s)ds \tag{2}$$

and

$$\mathbb{P}(N_t > 0) = 1 - \exp(-\int_0^t \lambda(s)ds).$$

Also, a formula for the variance of N_t is obtained readily:

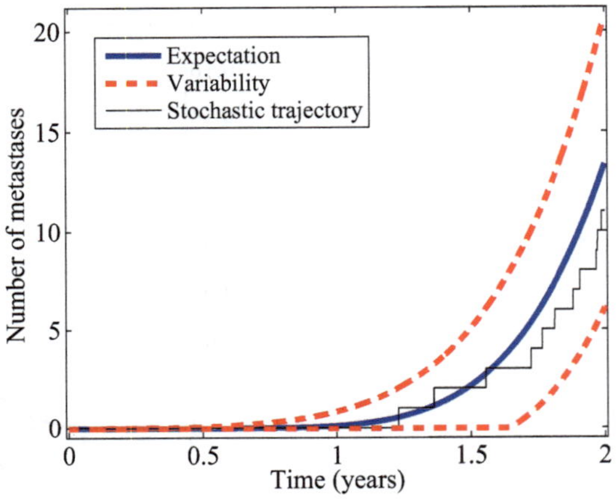

Fig. 3 Illustration of the non-homogeneous Poisson process N_t, representing the number of metastases emitted by the primary tumor (clinical parameters, Table 1). To simulate the process, a set of random times is simulated, which then yields a random trajectory. Repeated simulations would lead to different trajectories, and for a large number of random trajectories the "average trajectory" is approximately given by the expectation of the process $\mathbb{E}[N_t]$, which can be directly computed via Eq. 2. Variability around $\mathbb{E}[N_t]$ can be computed as $\mathbb{E}[N_t]p\backslash pm2 \cdot \sqrt{\text{var}[N_t]}$, where var[$N_t$] is the variance of N_t, computed via Eq. 3

$$\text{var}[N_t] = \int_0^t \lambda(s)ds. \tag{3}$$

The concepts N_t, $\mathbb{E}[N_t]$ and var[N_t] are illustrated in Fig. 3.

If a metastatic growth law is added to the model, the total metastatic mass (or total cell count, sum of lesion volumes) M_t—again a random quantity—can be represented via the emission times of the PP. M_t can be compared to quantitative measures of total metastatic biomass, obtainable, e.g., via bioluminescence imaging [24]. We will assume that all metastases follow the same deterministic growth law X_m, but which can be different from the primary tumor growth law. Therefore, the size difference among metastases is entirely explained by differences in metastatic inception times, and M_t can be written as

$$M_t = \sum_{k=1}^{N_t} X_m(t - T^{(k)}),$$

where $X_m(0) = {x_m}^0$ is the initial size of a metastasis. Expectation and variance of the metastatic burden can also be calculated analytically:

$$\mathbb{E}[M_t] = \int_0^t \lambda(s) X_m(t-s) ds. \tag{4}$$

$$\mathrm{var}[M_t] = \int_0^t \lambda(s) (X_m(t-s))^2 ds. \tag{5}$$

The assumption of equal growth law for the metastases greatly simplifies the model, which is both an advantage (for identifiability from clinical data) and drawback (for correct representation of cancer biology). Beyond the scope of this chapter, it could be replaced by a less restrictive assumption, e.g. by supposing that individual growth parameters are drawn randomly from a given probability distribution. However, even if easily integrated into numerical algorithms, such a feature would be prohibitive for any characterization of the model through mathematical analysis.

2.3 Secondary Emission

In the model described above, metastases do not have the capacity to emit metastases themselves. However, it is easy to think of a case in which such a property would make a difference in the model. For example, suppose that only a single metastasis is emitted prior to surgery of the primary tumor (*see* **Note 2**). If this metastasis cannot emit further metastases, its successful removal cures the patient but the second surgery may fail if the metastasis is able to seed as well. Of course, there are other mechanisms potentially leading to treatment failure (e.g., local recurrence, surgery impossible, etc., see, e.g., [36, 37]), but for simplicity these are not considered here.

In this section, we extend the previously shown PP model to account for secondary metastatic emission using PPs as building blocks. Many of the advantages and limitations of the PP model carry over to the extended model, and we do not claim that a comprehensive framework for cancer metastasis is built in that way. The model is, however, simple enough to have a chance to be parametrized reasonably from clinical data.

Conceptually, the extension is straightforward: as before, the primary tumor grows according to X_p and metastatic emission by the primary tumor is represented by a PP with intensity λ_p. In addition, any emitted metastasis has the same capacities as the primary tumor, but possibly with different growth and emission rates (X_m instead of X_p, and λ_m instead of λ_p). If we consider a metastasis emitted at time s, this means that at a later time t it reaches the size $X_m(t-s)$ and emits metastases with intensity $\lambda_m(t-s)$. Every newly emitted metastasis starts a new PP. Also, each metastasis has a precursor (either the primary tumor or another metastasis). The whole model then consists of the metastatic emission times from all of these PPs. Since each PP can start other PPs, we call this model a *PP cascade*. We then need to make a hypothesis on how the different emission processes play together.

Independence of Emissions: We assume that each metastasis emits independently of any other metastasis, and also independently of the primary tumor. In other words, all the PPs involved in the dynamics are independent.

Such an assumption is very important to be able to characterize the properties of the PP cascade with mathematical techniques. Note that simply ordering all emission events including secondary emissions by increasing emission time would not allow to use the above made independence assumption since the inception time of each metastasis depends on its level in the generational hierarchy (primary tumor, metastases emitted from the primary tumor, metastases emitted from the metastases emitted from the primary tumor, etc.). Therefore, in our model we have to account for the filiation of each metastasis. For example, the emission time of the first metastasis emitted by the primary tumor is denoted by $T^{(1)}$, and the emission time of the first metastasis emitted by the first metastasis emitted by the primary tumor is denoted by $T^{(1,1)}$, which depends on $T^{(1)}$. More precisely, $T^{(1,1)} = T^{(1)} + \tilde{T}^{(1,1)}$ where $\tilde{T}^{(1,1)}$ is the first emission time for the PP generated at time $T^{(1)}$. Filiation in the cascaded model is further illustrated in Fig. 4, and a rigorous definition is provided in Subheading 5.4.

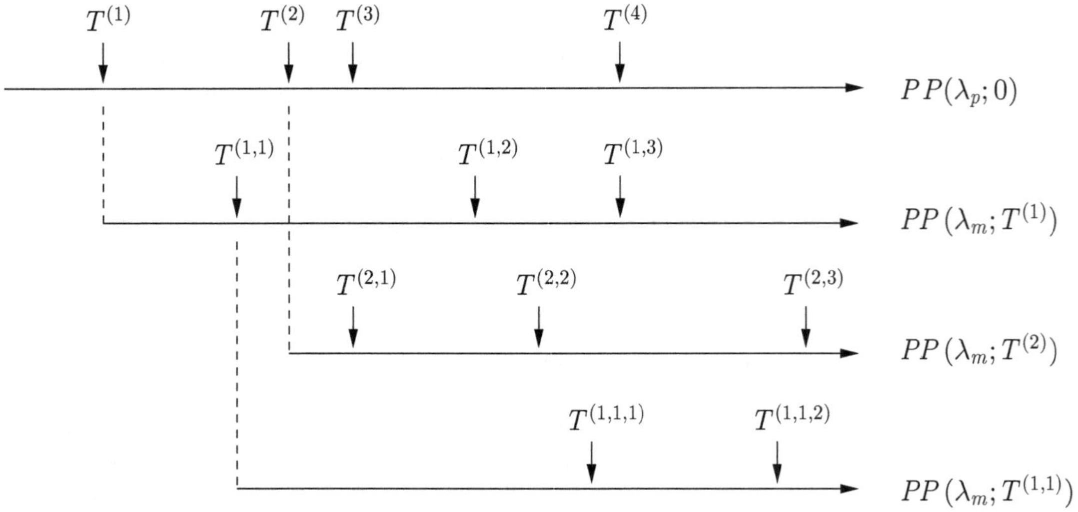

Fig. 4 Illustration of the first three generations for a Poisson process (PP) cascade. Each long horizontal arrow represents a PP (from top to bottom: primary tumor, first metastasis of first generation, second metastasis of first generation, first metastasis of second generation emitted by first metastasis of first generation). Each short vertical arrow represents an emission by the PP it points towards. This starts a new PP, connected by a dashed line. In the notation PP(λ; T), λ is the intensity of the PP and T is its starting time for the new PP (emission times are counted from the start of the respective PP and not from zero)

3 Crosslink to a Structured Population Model

3.1 Size-Structured Model

Let us consider a different framework for the description of metastasis, which also represents the metastatic process purely as growth and emission dynamics. To describe the micrometastatic state of cancer patients a *size-structured model* was developed [23]. The model describes the time evolution of a density function $\rho(x, t)$ representing the size distribution of metastatic colonies: the integral $\int_{x_1}^{x_2} \rho(x, t)\,dx$ represents the number of metastases at time t with size between x_1 and x_2. Therefore, ρ is like a smoothed histogram of the number of metastases within different size ranges.

To better understand why a size density is considered, it is instructive to draw an analogy to Lagrangian and Eulerian description of a fluid flow (see, e.g., [38] for a comprehensive discussion). In a Lagrangian description, the observer follows individual particles through the flow field. In contrast, for the Eulerian point of view the observer considers the flow density through fixed reference points. These two frames of reference are illustrated in Fig. 5. In this picture, metastatic growth becomes "flow through size space." In the PP model this is represented in a Lagrangian fashion: a growth function is associated to each individual metastasis. In the size-structured model an Eulerian frame of reference is used: the entire population of metastatic tumors is described through a density function moving through size space at a "speed" $g(x)$ (i.e., the growth rate), in other words a size-structured density.

Formalizing metastatic growth from an Eulerian perspective leads to a PDE model. Metastatic emission is the boundary

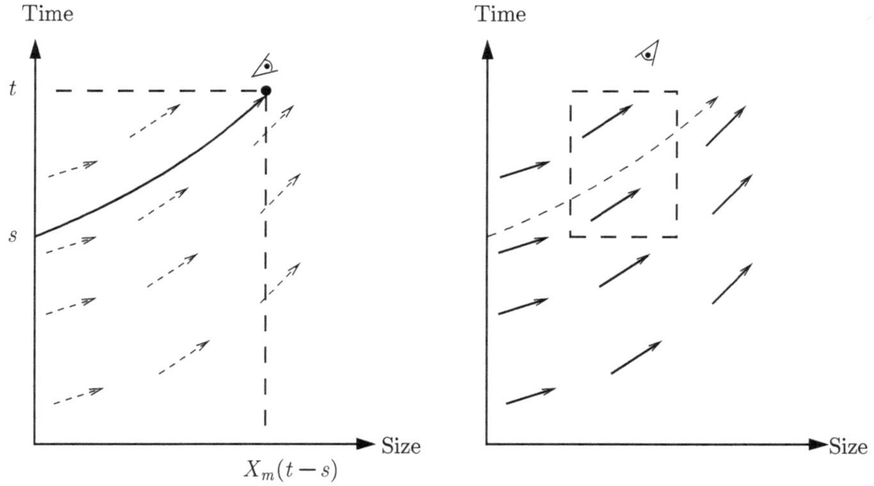

Fig. 5 Representation of the Lagrangian (left) and Eulerian (right) frames of reference for describing a population of growing metastases. Left: the observer (the eye symbol) follows the growth curves of individual metastases; time and size coordinates determine the observer's position. Right: a static observer looks from the outside at the growth speed g in fixed time-size areas. The relationship $X_m'(t) = g(X_m(t))$ holds

condition of the PDE, which means that it describes the "arrival of new particles into size space" (*see* Eq. 9). In contrast to the PP cascade model, where metastatic emission was a stochastic process, in the size-structured model (the PDE model) emission is deterministic. The emission dynamics consists of a primary tumor contribution and a contribution of the metastases themselves, both depending on the size of the emitting tumor. The size-dependency of the metastatic emission rate entwines metastatic growth with metastatic emission dynamics, which requires special attention during mathematical analysis of the model [39] as well as for designing an efficient numerical resolution scheme [40].

To illustrate the model dynamics, the clinical parameters of Table 1 were used to simulate the metastatic density function at different times (*see* Fig. 6). The model equations, together with relevant properties of the size-structured model, are presented in Subheading 5.3.

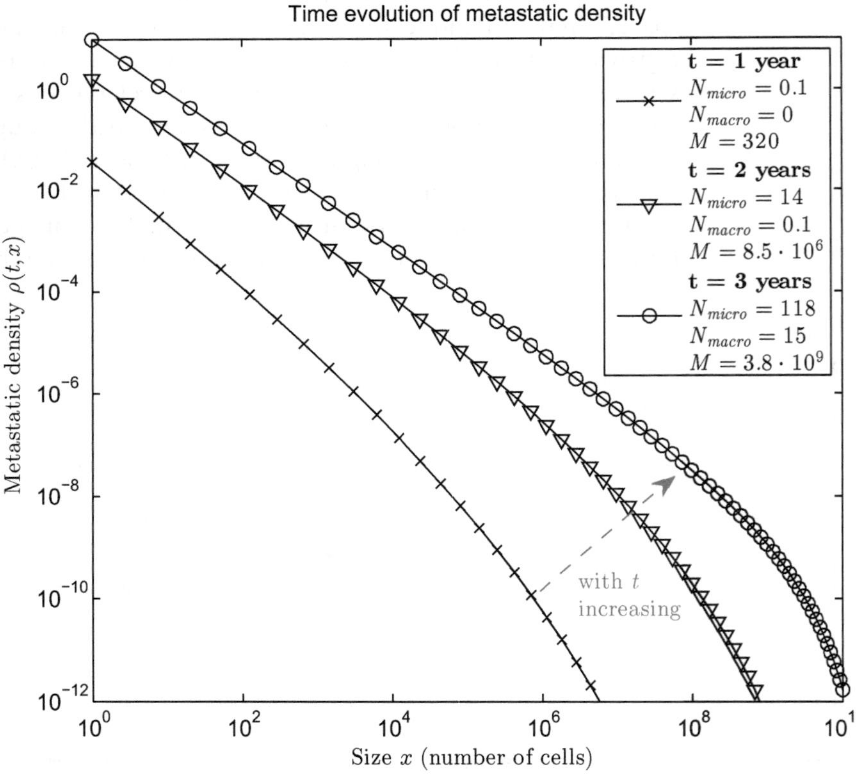

Fig. 6 Time evolution of the metastatic density in the size-structured model for metastasis. Each solid line represents a snapshot of the metastatic density ρ at a particular point in time (1/2/3 years after inception of the primary tumor). Due to the growth dynamics, the density is transported to the right. Several quantities computable from the density are represented in the legend: N_{micro}, number of metastases smaller than 10^8 cells; N_{macro}, number of metastases larger than 10^8 cells; M, total metastatic mass (number of cells of all metastases together)

3.2 Bridging the Gap: Model Observables

We now describe how the size-structured model and the PP cascade model are related. At first view, the two frameworks describe quite different objects. While the PP cascade is concerned with a collection of emission times with a generational hierarchy, the size-structured model features a density function. Nevertheless, as we will see, the latter can be seen as the expectation of the PP cascade model. To describe precisely the relationship between the models, we need to introduce *model observables* as a common theme. In fact, we have already introduced some model observables without naming them so. The model observables include the number of metastases, the number of micro-/macro-metastases, and the total metastatic mass.

Let us start with the size-structured model. For each function f, a *model observable* (MO) is defined by

$$\mathrm{MO}_f(t) := \int_{x_m^0}^{x_m^\infty} f(x)\rho(x,t)dx, \tag{6}$$

where ${x_m}^0$ is the size of a newly emitted metastasis and ${x_m}^\infty$ denotes the theoretical upper boundary, i.e. it is integrated over all possible sizes of metastases. Different choices for f are possible, and each of them corresponds to one observable (this dependency is made explicit through the subscript$_f$ in MO_f). The definition includes the above-mentioned quantities:

- The number of metastases is obtained for $f = \mathbf{1}$, i.e. the function that equals 1 for all x: $\mathrm{MO}_{\mathbf{1}}(t) = \int_{x_m^0}^{x_m^\infty} 1 \cdot \rho(x,t)dx = N(t)$.
- Similarly, the number of macrometastases is obtained with

$$f_{\text{macro}}(x) = \begin{cases} 1 & \text{if} \quad x \geq c \\ 0 & \text{if} \quad x < c, \end{cases}$$

  and the number of micrometastases with

$$f_{\text{micro}}(x) = \begin{cases} 0 & \text{if} \quad x \geq c \\ 1 & \text{if} \quad x < c. \end{cases}$$

  Here c stands for the detectability threshold, which depends on the imaging modality.
- The total metastatic mass M is obtained with the identity function $f_{\text{Id}}(x) = x$ for all x:

$$\mathrm{MO}_{f_{\text{Id}}}(t) = \int_{x_m^0}^{x_m^\infty} x\rho(x,t)dx = M(t).$$

Apart from allowing us to consider all these model-derived quantities at once, it is important for the mathematical proofs in Subheading 5 to consider such a general notion of observable.

Writing down the model observables in the PP cascade model is slightly more complicated and it will be easier to illustrate it with an observable in the PP model without secondary emission. There, a *stochastic model observable* (SMO) is defined by

$$\mathrm{SMO}_f(t) := \sum_{k=1}^{N_t} f(X_m(t - T^{(k)})). \tag{7}$$

The observables are defined in such a way that their interpretation is the same in both frameworks. For example, $f = \mathbf{1}$ yields the number of metastases N_t

$$\mathrm{SMO}_{\mathbf{1}}(t) = \sum_{k=1}^{N_t} 1 = N_t,$$

and f_{Id} yields the metastatic mass M_t

$$\mathrm{SMO}_{f_{\mathrm{Id}}}(t) = \sum_{k=1}^{N_t} X_m(t - T^{(k)}) = M_t. \tag{8}$$

If we ordered all emission events including secondary emissions by increasing emission time (and still called these times $T^{(1)}$, $T^{(2)}$, etc.), this could also be used as the definition of a stochastic model observable in the PP cascade. However, in order to carry out the calculations required for bridging the gap between the two frameworks, we need to account for the filiation of a metastasis, i.e. its level in the generational hierarchy. An explicit definition of the SMO using filiation is provided by Eq. 13 in Subheading 5.4.

Similarly to Eq. 2, where the expected number of metastases $\mathbb{E}[N_t]$ was computed in the PP model without secondary emission, an expression for the expectation and variance of each SMO can be derived in the PP cascade model. These computations are more complicated and are presented in detail in Appendix. It is then shown that the expected value of each SMO is equal to the corresponding MO in the size-structured model; in this sense, the size-structured model describes the mean behavior of the PP cascade model:

$$\mathrm{MO}_f(t) = \mathbb{E}[\mathrm{SMO}_f(t)].$$

A rigorous mathematical statement of these results is given in Subheading 5.4.

The relationship between model observables in the two frameworks is a consequence of a relationship between more fundamental mathematical objects (a random measure in the PP cascade and an absolutely continuous measure in the size-structured model). For the sake of simplicity, we do not present this additional layer here and refer to Subheading 5.4 for more details.

3.3 Implications

In physics, a density is usually derived on the hypothesis of a large number of constituting particles. In their derivation of the size-structured model, these principles were applied to metastasis [23]. However, this density notion is challengeable during the early phase of metastasis where the number of metastases is low: what is one single metastasis spread over the whole size range? The alternative interpretation as the expected value of a cascade of PPs provides a more flexible framework. For any model observable (e.g., the number of metastases), the adequacy of the size-structured model can be evaluated by quantifying the variance of the corresponding PP cascade.

Let us illustrate this approach by an example. When parametrizing the size-structured model from clinical data on the size distribution of metastatic colonies, the model authors did not represent randomness inherent in the emission process [23]. To account for this neglected source of variability, we use the crosslink between size-structured and PP cascade models. By simulating the PP cascade model with the same parameters (Table 1), standard deviation as well as typical trajectories of the stochastic model can be taken as a measure of variability around the prediction by the size-structured model. We choose the observables used in [23] to parametrize the model, i.e. the number of metastases exceeding certain size thresholds c (i.e., f_{macro} with different thresholds). Simulation results are shown in Fig. 7.

The average deviation of the data from the size-structured model prediction is much smaller than the stochastic fluctuation of the PP cascade model, and we can interpret this from two different perspectives. On the one hand, since these deviations are consistent with typical trajectories, the data are in principle explainable by stochasticity of emission. On the other hand, since the range of plausible trajectories is relatively wide using the estimated model parameters, different sets of model parameters would also be compatible with the same observed trajectory. To put it differently, the precision of the parameters of the size-structured model is probably overestimated since the variability by randomness of emission is not taken into account.

Parameter estimation is much easier in deterministic than in stochastic models. In special cases, such as in a PP model without secondary emission [22], it is possible to estimate model parameters in a stochastic model. However, if the statistical model becomes more complicated, e.g. a mixed-effects model to deal with population data [41], the computational and even methodological feasibility limit is quickly reached with a stochastic structural model [42]. In this case, the crosslink described in Subheading 3.2

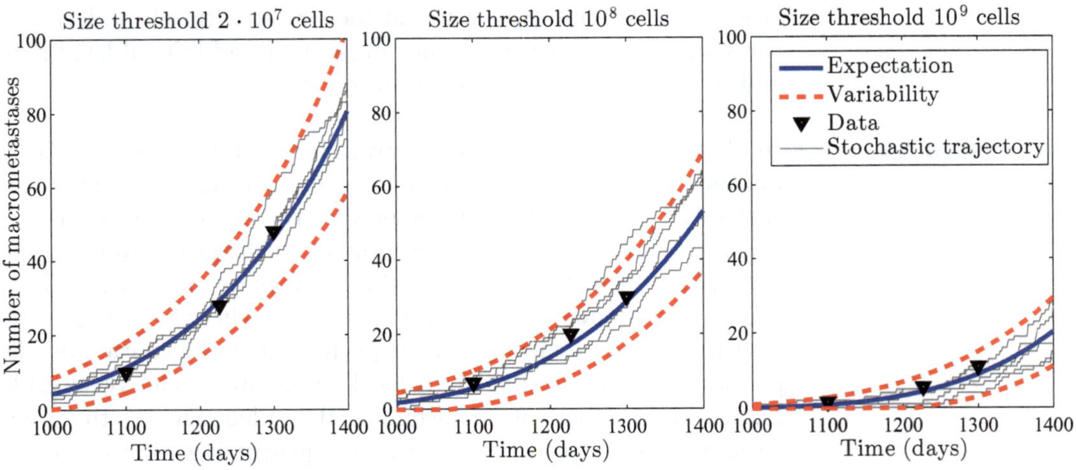

Fig. 7 Comparison of residual variability from the size-structured model fit and stochastic variability of the PP cascade model. Expectation (bold solid line) is the size-structured model prediction $N_{\text{macro}}(t)$, which was used to fit the clinical data from [23] (computed via Eq. 10). Variability of the corresponding PP cascade model is displayed in two ways: through stochastic trajectories (thin lines) and $\mathbb{E}[N_{\text{macro},t}]p\backslash pm2 \cdot \sqrt{\text{var}[N_{\text{macro},t}]}$, with var[$N_{\text{macro},t}$] computed via Eq. 15 (Variability, bold dashed line). As in [23], we count time from inception of the first primary tumor cell, which was back-calculated from primary tumor data assuming Gompertzian growth (hence the first CT scan with metastatic disease is approximately 3 years post-inception)

can be exploited to derive reasonable parameters for simulations from the PP cascade model by estimating the parameters of its mean, i.e. the size-structured model.

We applied this reasoning already in Fig. 7, in which the variability due to stochasticity of emission was discussed. To give an example using the stochastic nature of the PP cascade model more explicitly, we now use the stochastic framework to assess the impact of secondary metastatic emission following surgery of the primary tumor. Using the clinical parameters stated above, we assume that the primary tumor is surgically removed 500 days after its inception, where it has reached a tumor mass of 180 g, and assess the number of metastases another 500 days later (Fig. 8). Since every secondary emission is preceded by at least one primary emission, the probability of metastatic disease is the same in both models (with and without secondary emission). However, on average a much larger number of metastases is predicted from the model with secondary emission ($\mathbb{E}[N_t] = 4.7$ with secondary emission vs. $\mathbb{E}[N_t] = 1.2$ without).

4 Summary and Outlook

This chapter focuses on Poisson processes as possible frameworks describing metastatic emission, usable to predict metastatic risk or micrometastatic dynamics. Although representing randomness in a

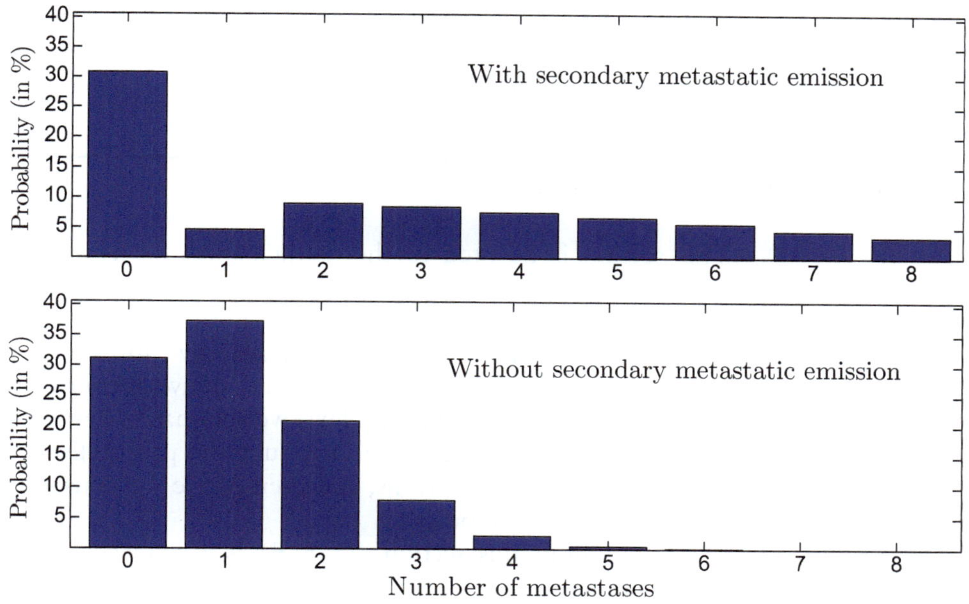

Fig. 8 Probability of metastatic disease after surgery with (top panel) or without (bottom panel) secondary metastatic emission (each based on 10.000 simulations). In addition to the clinical parameters derived in [23], it is assumed that the primary tumor is surgically removed 500 days after its inception, and that the number of metastases is evaluated another 500 days later

relatively simple way, PPs have appealing properties that have been illustrated here. They can be easily included as building blocks in larger models, which has been shown with the PP cascade model, but which applies in a much more general way. Also, they allow for a high degree of analytical tractability, which was exploited here to characterize the mean behavior of the PP cascade model.

Without doubt, further improvements of these techniques are required. In particular, to make individualized risk predictions with the model we have to match patient characteristics to model parameters. In this respect, circulating biomarkers, such as circulating tumor cells or circulating tumor DNA can be a useful source of information, especially since quantification methods are rapidly getting more reliable [43, 44, 2]. Both frameworks presented here allow for such an extension. Once validated, a mathematical model can serve as a powerful tool for informed treatment decisions for cancer patients by integrating case-specific information into a consistent quantitative framework.

While this chapter has focused on the natural metastatic emission kinetics, it is possible to extend the formalism to cover systemic treatments such as chemotherapy (represented as a size function $X_m(t; t_{\text{incept}})$ depending on inception time in the stochastic context of Subheading 2.2, or a time-varying growth rate $g(x, t)$ in the deterministic context of Subheading 3.1). However, although both the deterministic and stochastic settings can be extended in this way, a direct link between these two extended frameworks (as in Subheading 3.2) has not been established yet.

5 Mathematical Formalism and Results

This section is devoted to the mathematical formalism and the derivations of the results of Subheadings 2 and 3. We provide here precise definitions of the mathematical objets and rigorous computations and results. We start by introducing the non-homogeneous Poisson process and derive formula (Eq. 1) from the Poisson assumption. Next, we summarize key results for the size-structured model, we introduce the probabilistic framework for secondary emissions, and then derive rigorously the link between these two models.

Throughout this section $(\Omega, \mathcal{F}, \mathbb{P})$ is a probability space on which all the random variables we consider are defined.

5.1 Definition of a Poisson Process and Basic Properties

The Poisson distribution is a standard way to count the occurrences of some events.

Definition 5.1 (Poisson Distribution).: *Let* $\mu \geq 0$. *A random variable* Y *with values in* $\mathbb{N}$ *is said to have a Poisson distribution with parameter* μ, *that we denote by* $\Upsilon \sim \mathcal{P}(\mu)$, *if for all* $k \in \mathbb{N}$

$$\mathbb{P}(\Upsilon = k) = e^{-\mu} \frac{\mu^k}{k!}$$

for $\mu > 0$, *and if* $\mathbb{P}(\Upsilon = 0) = 1$ *in the case* $\mu = 0$.

The parameter $\mu \in \mathbb{R}^+$ can be interpreted as the expected number of occurrences since

$$\mathbb{E}[\Upsilon] = \mu \qquad \text{with} \qquad \Upsilon \sim \mathcal{P}(\mu).$$

In our context, it counts the number of metastases. However, at this level, we have no information on the event times we are counting, nor how this number evolves with respect to time. To handle the random nature of these times, let us introduce the Poisson processes.

Definition 5.2 (Non-homogeneous Poisson Process).: *Let* $\lambda : \mathbb{R}^+ \to \mathbb{R}^+$ *be a continuous function. We say that* $(N_t)_{t \geq 0}$ *is a non-homogeneous Poisson process with intensity* λ *if:*

1. $N_0 = 0$;
2. *the number of occurrences in disjoint time intervals is independent, i.e. for* $t_0 < \ldots < t_n$, *the random variables* $N_{t_k} - N_{t_{k-1}}$, $k = 1, \ldots, n$ *are independent;*
3. *For all* $t > 0$, N_t *has a Poisson distribution with parameter* $\Lambda(t)$, *given by*

$$\Lambda(t) = \int_0^t \lambda(u)du.$$

The terminology *non-homogeneous* results from the fact that the intensity function λ can vary in time, as opposed to a homogeneous Poisson process for which λ is constant. Also, there are several equivalent definitions for a non-homogenous Poisson process. For instance, the third item above can be replaced by the following properties:

$$\mathbb{P}(N_{t+\Delta t} - N_t = 1) = \lambda(t)\Delta t + o(\Delta t)$$
$$\text{and} \qquad \mathbb{P}(N_{t+\Delta t} - N_t \geq 2) = o(\Delta t),$$

where $o(\Delta t)$ stands for a function satisfying $o(\Delta t)/\Delta t \to 0$ as $\Delta t \to 0$. In the previous definition, we chose λ as a continuous function since we would not expect any discontinuities in rate of metastatic emission. Nevertheless, mathematically this assumption can be relaxed to more general nonnegative functions.

Finally, from a Poisson process $(N_t)_{t \geq 0}$, one can define the event times recursively as

$$T^{(k)} = \inf(t > T^{(k-1)}, \quad N_t = N_{t-} + 1) \qquad \text{for} k = 1, 2, \ldots$$

with $T^{(0)} = 0$. We refer to Fig. 1 for an illustration of the relation between $(N_t)_{t \geq 0}$ and the event times. From these times, one can consider the following (random) measure on $\mathbb{R}^+$:

$$P(du) := \sum_{k=1}^{+\infty} \delta_{T^{(k)}},$$

where δ_x stands for the Dirac distribution at point x. This measure is called the Poisson random measure associated to $(N_t)_{t \geq 0}$. From this definition of P, it is direct to see that for any $t \geq 0$,

$$N_t = P([0, t]) = \sum_{k=1}^{+\infty} \mathbf{1}_{(T^{(k)} \leq t)}.$$

Here, $\mathbf{1}_A$ is the indicator function of A, that is it takes the value 1 if A is true and 0 otherwise. In the same way, the total metastatic biomass (Eq. 8), for instance, can be rewritten as

$$M_t = \sum_{k=1}^{+\infty} \mathbf{1}_{(T^{(k)} \leq t)} X_m(t - T^{(k)}) = \int_0^t X_m(t - u) P(du).$$

Expressions involving an integral with respect to the Poisson measure allow convenient manipulations as we will see in Appendix using Proposition A.1.

5.2 Derivation of Empirical Formula from Poisson Assumptions

Let us assume that the primary tumor diameter $d(t)$ follows a power law:

$$d'(t) = a\, d(t)^{\alpha}, \quad d(0) = d_0, \quad 0 < \alpha < 1.$$

Power growth of volume with a power between 2/3 and 1 has been described in the literature, leading to the above model if we assume a spherical shape of the tumor.

Furthermore, let us assume that metastatic emission is governed by a Poisson process with intensity $\lambda(t) = b\, d(t)^{\beta}$. We will require $\beta > 0$, since the emission rate should increase with primary tumor size. Then, the number of metastases N_t by time t is Poisson distributed with parameter $\Lambda(t) = \int_0^t \lambda(s) ds$ and the probability of metastasis-free disease at time t is given by

$$\mathbb{P}(\text{no metastases}) = \mathbb{P}(N_t = 0) = \exp\Big(-b \int_0^t d(s)^{\beta} ds\Big).$$

Here, we have

$$b \int_0^t d^{\beta}(s) ds = \frac{b}{a} \int_0^t d'(s) d(s)^{\beta-\alpha} ds = \frac{b}{a} \int_0^t \frac{d}{dt}\Big(\frac{d(s)^{\beta-\alpha+1}}{\beta-\alpha+1}\Big) ds$$

$$= \frac{b}{a(\beta-\alpha+1)}\big(d(t)^{\beta-\alpha+1} - d_0^{\beta-\alpha+1}\big),$$

and assuming that the tumor is initiated with a negligible size ($d_0 \ll d(t)$), one obtains

$$b \int_0^t d(s)^{\beta} ds \approx \frac{b}{a(\beta-\alpha+1)} d(t)^{\beta-\alpha+1}.$$

This then yields the empirical formula

$$\mathbb{P}(\text{no metastases}) = \exp\big(-c\, d(t)^{z}\big),$$

with $c = \frac{b}{a(\beta-\alpha+1)}$, and $z = \beta - \alpha + 1$. Since $\beta > 0$ and $\alpha \leq 1$, we have $z > 0$, and the above manipulation is justified.

It should be noted that although c and z can be determined unambiguously from information on metastatic status and primary tumor size at surgery if the patient cohort is large enough, this does not apply for the growth and emission parameters of the underlying Poisson process. To distinguish the growth and emission processes additional information is required, such as repeated tumor size measurements over time.

5.3 Summary of Key Results on the Size-Structured Model

The framework proposed in [23] focuses on the evolution of a size-structured metastatic density ρ. Originally, it was assumed that primary and secondary tumors have the same growth and emission patterns. Here, we present an extended version, described, e.g., in [40], where primary and secondary growth and emission dynamics can be different.

As before, X_p denotes the size of the primary tumor and γ_p the primary tumor emission law. The size of a metastasis is given by X_m, which is the solution of an autonomous ordinary differential equation

$$X'_m(t) = g(X_m(t)), \qquad X_m(0) = x_m^0.$$

The emission law of the metastases is γ_m. Then, the metastatic density function ρ solves the following equation:

$$\begin{cases} \partial_t \rho(x,t) + \partial_x[g(x)\rho(x,t)] = 0, & (x,t) \in (x_m^0, x_m^\infty) \times (0, +\infty), \\ g(x_m^0)\rho(x_m^0, t) = \gamma_p(X_p(t)) + \int_{x_m^0}^{x_m^\infty} \gamma_m(x)\rho(x,t)\,dx, & t \in (0, +\infty), \\ \rho(x, 0) = 0, & x \in [x_m^0, x_m^\infty]. \end{cases} \tag{9}$$

We also introduce the emission rates of the primary tumor and the metastases, respectively:

$$\lambda_p(t) := \gamma_p(X_p(t)) \qquad \text{and} \qquad \lambda_m(t) := \gamma_m(X_m(t)).$$

Existence of a unique weak solution ρ to this model has been shown under general conditions in [39]. For the purpose of this chapter, it is sufficient to assume that λ_p, γ_m, and g are continuously differentiable nonnegative functions, and that $\lim_{t \to +\infty}\lambda_p(t)$ < $+\infty$ and $g({x_m}^\infty) = 0$.

Model observables for the size-structured model have been introduced in Eq. 6. As shown in [40], they can be characterized as the solutions of a Volterra convolution equation:

Theorem 5.1.: For any $f \in L^\infty([x_m^0, x_m^\infty])$, $MO_f$is the unique solution of the following renewal equation:

$$MO_f(t) = \int_0^t \lambda_p(s) f(X_m(t-s))\,ds + \int_0^t \lambda_m(s) MO_f(t-s)\,ds. \tag{10}$$

This alternative formulation will be important to bridge the gap between the stochastic and deterministic frameworks. Of note, it is also the basis of an efficient numerical resolution algorithm [40].

5.4 Probabilistic Framework for Secondary Emission

Let us first remind the reader that we assume all emissions of the primary tumor and of the metastases to be independent. To exploit this property in calculations, we have to take care of the filiation of each metastasis, i.e. the generational hierarchy (the primary tumor, the metastases emitted from the primary tumor, the metastases emitted from the metastases emitted from the primary tumor, etc.). We will therefore introduce a cascade of independent PPs, and define recursively the emission times with respect to the generational hierarchy.

- The emission times for the first generation of metastases, that is, the ones emitted by the primary tumor, are the event times of a PP $(N_t^{(1)})_{t \geq 0}$ with intensity λ_p; we will write $\Pi^{(1)} := (T^{(j)})_{j \geq 1}$ for the set of random emission times.

  The emission times for the next generations of metastases are defined recursively.

- Let $k \geq 2$ and $n_1, \ldots, n_{k-1} \geq 1$. The jth emission time for the kth generation of metastasis with filiation $n_1, \ldots, n_{k-1}$ is defined by

$$T^{(n_1,\ldots,n_{k-1},j)} := T^{(n_1,\ldots,n_{k-1})} + \tilde{T}^{(n_1,\ldots,n_{k-1},j)} \tag{11}$$

  This is the time it takes for the offspring with filiation $n_1, \ldots, n_{k-1}$ to give birth to its jth offspring. Here, the family $\left(\tilde{T}^{(n_1,\ldots,n_{k-1},j)}\right)_{j\geq 1}$ is formed by the event times of a PP $\left(N^{(n_1,\ldots,n_{k-1})}\right)_{t\geq 0}$ with intensity λ_m.

We refer to Fig. 4 for an illustration of these emission times, but for instance, $T^{(2,3,4)}$ is the inception time of the fourth offspring produced by the third offspring of the second offspring of the primary tumor. Using biologically relevant parameters, the expected emission times for all but the first few generations are larger than any reasonable observation timeframe. However, even if the contribution of late generations is very small, we need to consider the whole cascade of emission times to bridge the gap to the size-structured model.

Finally, assuming that

$$\left\{\left(N_t^{(n_1,\ldots,n_k)}\right)_{t\geq 0}, \quad k \geq 1, \quad n_1, \ldots, n_k \geq 1\right\}$$

is a family of independent PPs implies the biological assumption we made, which is that the primary tumors and all the metastases emit independently from each other.

In the PP model without secondary emission, model observables were defined in Eq. 7. With the PP cascade defined above, we are now able to extend this concept to secondary emission constructively. For $f \in L^\infty([x_m{}^0, x_m{}^\infty])$, the SMO for the kth generation can be expressed by

$$\mathrm{SMO}_f^{(k)}(t) := \sum_{n_1,\ldots,n_k \geq 1} \mathbf{1}_{\left(T^{(n_1,\ldots,n_k)} \leq t\right)} f(X_m(t - T^{(n_1,\ldots,n_k)})), \quad (12)$$

and we have the following definition.

Definition 5.3 (Model Observables with Secondary Emission).: *The SMOs with secondary emission are given by*

$$SMO_f(t) := \sum_{k=1}^{+\infty} SMO_f^{(k)}(t). \quad (13)$$

In this definition, $\mathrm{SMO}_f^{(k)}$ describes the contribution of the *k*th generation to the SMO.

The following proposition links the MOs from the stochastic and deterministic frameworks.

Proposition 5.1 (Link to the Model Observables).: Let $f \in L^\infty([x_m^0, x_m^\infty])$. The SMO (Eq. 13) is well defined in the sense that

$$\mathbb{P}(\forall t \geq 0, \quad 0 \leq SMO_f(t) < +\infty. = 1.$$

Moreover, the expected value

$$e_f(t) := \mathbb{E}[SMO_f(t)]$$

is finite for all t $\geq$ *0* and satisfies (Eq. 10), so that

$$e_f(t) = MO_f(t).$$

Let us remark that the SMO (Eq. 13) may also be seen as integrals w.r.t. a random measure

$$\mathrm{SMO}_f(t) = \int f(x) \mathcal{M}_t(dx),$$

for any $t \geq 0$, with

$$\mathcal{M}_t := \sum_{k \geq 1} \sum_{n_1,\ldots,n_k \geq 1} \delta_{X_m\left(t - T^{(n_1,\ldots,n_k)}\right)}. \quad (14)$$

This description is the key point to bridge the gap to the description of metastasis via a structured population equation.

Theorem 5.2 (Link to the Structured Population Model).: For all t $\geq$ *0*, the measure

$$\mu_t := \mathbb{E}[\mathcal{M}_t]$$

is σ-finite, absolutely continuous with respect to the Lebesgue measure, and its Radon–Nikodýn density is given by ρ*(· , t),*

$$\frac{d\mu_t}{dx} = \rho(\cdot, t),$$

where ρ is the solution of the structured population equation (Eq. 9).

The last result we present here concerns the variability of the SMO (Eq. 13) with respect to its mean MO_f.

Proposition 5.2 (Variance of Observables).: Let $f \in L^\infty([x_m^0, x_m^\infty])$. The variance of the SMO

$$v_f(t) := var[SMO_f(t)]$$

is finite for any t $\geq$ *0*, and satisfies a renewal equation:

$$v_f(t) = \int_0^t \lambda_p(s)(f(X_m(t-s)) + e_{m,f}(t-s))^2 + \int_0^t \lambda_m(s) v_f(t-s). \quad (15)$$

Here,

$$e_{m,f}(t) := \mathbb{E}[SMO_{m,f}(t)],$$

and where $SMO_{m,f}$ is defined as (Eq. 13), but for a different cascade of PPs, which has only λ_m for intensity (both for the first and subsequent generations).

This result is of great interest to design confidence intervals as illustrated in Subheading 3.3. In fact, the renewal equation (Eq. 15) allows the use of an efficient numerical resolution algorithm [40].

6 Notes

1. The interpretation of β depends on the unit of the primary tumor measure. As an example, a surficial vascularization would correspond to $\beta = 2$ if size is measured in diameter, but to $\beta = 2/3$ (the fractal dimension of a surface in space) if size is measured in volume.
2. We remind the reader that a surgery at time $t_{surgery}$ is represented by setting the emission intensity λ to zero for all times larger than $t_{surgery}$.

Acknowledgements

The authors thank Florence Hubert, Charlotte Kloft, and Andrea Henrich for suggestions and critical reading of the manuscript. NH gratefully acknowledges financial support by the Agence Nationale de la Recherche under grant ANR-09-BLAN-0217-01.

A Appendix: Proofs of Results of Section 5.4

The proofs provided in this section are based on the following classical result on Poisson random measures. We refer to [45, Chap. 6, pp. 251] for further details. Also, note that this result directly yields Eqs. 2 through 5.

Proposition A.1.: *Let* $(N_t)_{t \geq 0}$ be a PP with intensity λ and P the corresponding Poisson random measure. We have for $\phi, \psi \in L^1(\mathbb{R}^+, \lambda(u)du) \cap L^2(\mathbb{R}^+, \lambda(u)du)$

$$\mathbb{E}\left[\int \phi(u)P(du)\right] = \int \phi(u)\lambda(u)du,$$

and

$$\mathbb{E}\left[\int \phi(u_1)P(du_1) \int \phi(u_2)P(du_2)\right] = \int \phi(u_1)\lambda(u_1)du_1 \int \psi(u_2)\lambda(u_2)du_2 + \int \phi(u)\psi(u)\lambda(u)du.$$

In other words, we can write the first order moment of the Poisson random measure P in a more compact form

$$\mathbb{E}[P(du)] = \lambda(u)du,$$

as well as its second order moment

$$\mathbb{E}[P(du_1)P(du_2)] = \lambda(u_1)\lambda(u_2)du_1du_2 + \delta(u_1 - u_2)\lambda(u_1)du_1du_2.$$

Moreover, to simplify notations in the forthcoming computations, we introduce the following convolution-like notation: for functions ϕ, ψ

$$\phi*\psi(t) := \int_0^t \phi(t-u)\psi(u)du. \tag{16}$$

A.1 Proof of Proposition 5.1

We first need to establish the following lemma, which is proven further below.

Lemma A.1.: We have

$$e_f = \lambda_p*(f(X_m) + e_{m,f}), \tag{17}$$

where $e_{m,f}$ has been introduced in Proposition 5.2.

This is not exactly the renewal equation we want. To derive the desired equation (Eq. 10) we just have to make the following remark. Taking $\lambda_p = \lambda_m$, Lemma A.1 gives that $e_{m,f}$ satisfies

$$e_{m,f} = \lambda_m*(f(X_m) + e_{m,f}).$$

Hence, from (Eq. 17), we have

$$\begin{aligned} e_f &= \lambda_p * f(X_m) + \lambda_p * (\lambda_m * f(X_m) + \lambda_m * e_{m,f}) \\ &= \lambda_p * f(X_m) + \lambda_m * (\lambda_p * f(X_m) + \lambda_p * e_{m,f}) \\ &= \lambda_p * f(X_m) + \lambda_m * e_f. \end{aligned} \tag{18}$$

Now, let $T > 0$, we have from the last line of (Eq. 18) that for all $t \in [0, T]$,

$$e_f(t) \leq f_{L^\infty([x_m^0, x_m^\infty])} \lambda_{pL^\infty([0,T])} T + \lambda_{mL^\infty([0,T])} \int_0^t e_f(u) du,$$

which gives using Gronwall's inequality

$$\sup_{t \in [0, T]} e_f(t) \leq C_{T,f,\lambda_p,\lambda_m} f_{L^\infty([x_m^0, x_m^\infty])} < +\infty. \tag{19}$$

As a result, $e_f(t)$ < $+\infty$ for all $t \geq 0$ since T is arbitrary, and also

$$\mathbb{P}(\mathrm{SMO}_f(T) < +\infty. = 1.$$

Finally, using that $t \mapsto \mathrm{SMO}_f(t)$ is an increasing non-negative function, we have

$$\mathbb{P}(\forall t \in [0, T], \quad \mathrm{SMO}_f(t) < +\infty. = 1,$$

and then

$$\mathbb{P}(\forall t \geq 0, \quad \mathrm{SMO}_f(t) < +\infty. = \lim_{n \to +\infty} \mathbb{P}(\forall t \in [0, n], \quad \mathrm{SMO}_f(t) < +\infty. = 1.$$

Proof (of Lemma A.1).: *Let us start with the following remark. According to the recursive definition (Eq. 11) of our PP cascade, one has*

$$T^{(n_1, \ldots, n_k)} = T^{(n_1)} + \overline{T}^{(n_1, \ldots, n_k)}, \tag{20}$$

where all the times

$$\{\overline{T}^{(n_1, \ldots, n_k)}, \quad k \geq 2, \quad n_1, \ldots, n_k \geq 1\}$$

are independent of $\Pi^{(1)} := (T^{(n_1)})_{n_1 \geq 1}$.

Now, from this consideration, by taking apart the first generation of metastasis, we can rewrite SMO_f *as follows:*

$$\begin{aligned} \mathrm{SMO}_f(t) &= \sum_{n_1 \geq 1} \mathbf{1}_{(T^{(n_1)} \leq t)} f(X_m(t - T^{(n_1)})) \\ &\quad + \sum_{n_1 \geq 1} \mathbf{1}_{(T^{(n_1)} \leq t)} \mathrm{SMO}_{n_1,f}(t - T^{(n_1)}) \\ &:= I + J, \end{aligned} \tag{21}$$

with

$$\mathrm{SMO}_{n_1,f}(t) := \sum_{k\geq 2}\sum_{n_2,\ldots,n_k\geq 1} \mathbf{1}_{\left(\overline{T}^{(n_1,\ldots,n_k)}\leq t\right)} f(X_m(t-\overline{T}^{(n_1,\ldots,n_k)})),$$

which are independent of $\Pi^{(1)}$. Note that all the times

$$\overline{\Pi}_{n_1} := \{\overline{T}^{(n_1,\ldots,n_k)},\quad k\geq 2,\quad n_2,\ldots,n_k\geq 1\},$$

can be defined following (Eq. 11), but with λ_m *as intensity for all the PPs since we consider all the times from the second generation. Therefore,* $(\mathrm{SMO}_{n_1,f})_{n_1\geq 1}$ *are all independent. Moreover, the shape of all the* $\mathrm{SMO}_{n_1,f}$ *is similar to* SMO_f *except that the PPs in the cascade have all* λ_m *for intensity. Hence, all the* $\mathrm{SMO}_{n_1,f}$ *have the same law as* $SMO_{m,\,f}$.

Using Proposition A.1 with the Poisson random measure $P^{(1)}(du)$ *associated to* $(N_t^{(1)})_{t\,\geq\,0}$ *(with intensity* λ_p*), it is direct to see that*

$$\mathbb{E}[I] = \mathbb{E}\left[\int_0^t f(X_m(t-u))P^{(1)}(du)\right] = \int_0^t \lambda_p(u)f(X_m(t-u))du$$
$$= \lambda_p * f(X_m)(t).$$

For the second term, using standard properties of the conditional expectation (especially $\mathbb{E}[X]=\mathbb{E}[\mathbb{E}[X|Y]]$), one has

$$\mathbb{E}[II] = \mathbb{E}\Big[\sum_{n_1\geq 1}\mathbf{1}_{(T^{(n_1)}\leq t)}\mathbb{E}\big[\mathrm{SMO}_{n_1,f}\Big(t-T^{(n_1)}\Big)|\Pi^{(1)}\big]\Big]$$

with

$$\begin{aligned}\mathbb{E}\big[\mathrm{SMO}_{f,n_1}\big(t-T^{(n_1)}\big)|\Pi^{(1)}. &= \mathbb{E}[\mathrm{SMO}_{n_1,f}(t-u)]_{|u=T^{(1)}_{n_1}}\\ &= \mathbb{E}[\mathrm{SMO}_{m,f}(t-u)]_{|u=T^{(n_1)}}\\ &= e_{m,f}\Big(t-T^{(n_1)}\Big),\end{aligned}\qquad (22)$$

and then

$$\sum_{n_1\geq 1}\mathbf{1}_{(T^{(n_1)}\leq t)}\mathbb{E}\big[\mathrm{SMO}_{n_1,f}\Big(t-T^{(n_1)}\Big)|\Pi^{(1)}\big] = \int_0^t e_{m,f}(t-u)P^{(1)}(du).$$

This, together with Proposition A.1, yields

$$\mathbb{E}[II] = \mathbb{E}\Big[\int_0^t e_{m,f}(t-u)P^{(1)}(du). = \int_0^t \lambda_p(u)e_{m,f}(t-u)du$$
$$= \lambda_p * e_{m,f}(t),$$

which concludes the proof of (Eq. 17). □

A.2 Proof of Proposition 5.2

Using the same strategy as for (Eq. 18), the proof of (Eq. 15) consists only in proving the following relation:

$$v_f = \lambda_p * (f(X_m) + e_{m,f}.^2 + \lambda_p * v_{m,f}, \tag{23}$$

with $v_{m,f}(t) := \text{var}[\text{SMO}_{m,f}(t)]$. Knowing Proposition 5.1 and the formula of the variance, one can focus on the term $e_f^{(2)}(t) := \mathbb{E}[\text{SMO}_f^2(t)]$. Using (Eq. 21), we have to compute three terms

$$e_f^{(2)}(t) = \mathbb{E}[I^2] + 2\mathbb{E}[IJ] + \mathbb{E}[J^2].$$

The Term $\mathbb{E}[I^2]$: *Reminding that* $I = \int_0^t f(X_s(t-u).P^{(1)}(du)$, *and using Proposition A.1, it is direct that*

$$\mathbb{E}[I^2] = \Big(\int_0^t \lambda_p(u) f(X_m(t-u))du\Big)^2 + \int_0^t \lambda_p(u) f^2(X_m(t-u))du.$$

The Term $\mathbb{E}[IJ]$: *Using standard properties of the conditional expectation, and that for all* $n_1 \geq 1$

$$e_{m,f}(t) = \mathbb{E}[\text{SMO}_{n_1,f}(t)],$$

we have using (Eq. 22)

$$\begin{aligned}\mathbb{E}[IJ] &= \mathbb{E}\left[\sum_{n_1^1, n_1^2 \geq 1} \mathbf{1}_{(T^{(n_1^1)} \leq t)} \mathbf{1}_{(T^{(n_1^2)} \leq t)} f(X_m(t - T^{(n_1^1)}))\mathbb{E}\left[\text{SMO}_{n_1^2,f}(t - T^{(n_1^2)})\Big|\pi^{(1)}\right]\right] \\ &= \mathbb{E}\left[\sum_{n_1^1, n_1^2 \geq 1} \mathbf{1}_{(T^{(n_1^1)} \leq t)} \mathbf{1}_{(T^{(n_1^2)} \leq t)} f(X_m(t - T^{(n_1^1)}))e_{m,f}(t - T^{(n_1^2)})\right].\end{aligned}$$

As result, according to Proposition A.1, we have

$$\begin{aligned}\mathbb{E}[IJ] &= \mathbb{E}\left[\int_0^t f(X_m(t-u).P^{(1)}(du) \int_0^t e_{m,f}(t-u)\, P^{(1)}(du)\right] \\ &= \int_0^t \lambda_p(u_1) f(X_m(t-u_1).du_1 \int_0^t \lambda_p(u_2) e_{m,f}(t-u_2)du_2 \\ &+ \int_0^t \lambda_p(u) f(X_m(t-u).e_{m,f}(t-u)du.\end{aligned}$$

The Term $\mathbb{E}[J^2]$: *To compute this term we have to consider two cases*

$$J^2 = \sum_{n_1 \geq 1} \mathbf{1}_{(T^{(n_1)} \leq t)} \mathrm{SMO}^2_{n_1,f}\left(t - T^{(n_1)}\right)$$

$$+ \sum_{\substack{n_1^1, n_1^2 \geq 1 \\ n_1^1 \neq n_1^2}} \mathbf{1}_{\left(T^{\left(n_1^1\right)} \leq t\right)} \mathbf{1}_{\left(T^{\left(n_1^2\right)} \leq t\right)} \mathrm{SMO}_{n_1^1,f}\left(t - T^{\left(n_1^1\right)}\right) \mathrm{SMO}_{n_1^2,f}\left(t - T^{\left(n_1^2\right)}\right)$$

$$:= J_1 + J_2.$$

Following (Eq. 22), but with $\mathrm{SMO}_{n,f}^2$ instead, we have

$$\mathbb{E}[J_1] = \mathbb{E}\left[\sum_{n_1 \geq 1} \mathbf{1}_{\left(T^{\left(n_1^1\right)} \leq t\right)} \mathbb{E}\left[\mathrm{SMO}^2_{n_1,f}\left(t - T^{(n_1)}\right) \middle| \Pi^{(1)}\right]\right]$$

$$= \mathbb{E}\left[\int_0^t e^{(2)}_{m,f}(t-u) P^{(1)}(du)\right]$$

$$= \int_0^t \lambda_p(u) e^{(2)}_{m,f}(t-u)\,du,$$

where $e^{(2)}_{m,f}(t) := \mathbb{E}[\mathrm{SMO}^2_{m,f}(t)]$. Now using that $\mathrm{SMO}_{n_1^1,f}$ and $\mathrm{SMO}_{n_1^2,f}$ are independent for $n_1^1 \neq n_1^2$, *we have, using (Eq. 22) and Proposition A.1 one more time,*

$$\begin{aligned}
&\mathbb{E}[J_2] \\
&= \mathbb{E}\left[\sum_{n_1^1 \neq n_1^2} \mathbf{1}_{\left(T^{\left(n_1^1\right)} \leq t\right)} \mathbf{1}_{\left(T^{\left(n_1^2\right)} \leq t\right)} \mathbb{E}\left[\mathrm{SMO}_{n_1^1,f}\left(t - T^{\left(n_1^1\right)}\right) \middle| \Pi^{(1)}\right] \mathbb{E}\left[\mathrm{SMO}_{n_1^2,f}\left(t - T^{\left(n_1^2\right)}\right) \middle| \Pi^{(1)}\right]\right] \\
&= \mathbb{E}\left[\left(\sum_{n_1 \geq 1} \mathbf{1}_{(T^{(n_1)} \leq t)} e_{m,f}\left(t - T^{(n_1)}\right)\right)^2\right] - \mathbb{E}\left[\sum_{n_1 \geq 1} \mathbf{1}_{(T^{(n_1)} \leq t)} e^2_{m,f}\left(t - T^{(n_1)}\right)\right] \\
&= \mathbb{E}\left[\left(\int_0^t e_{m,f}(t-u) P^{(1)}(du)\right)^2\right] - \mathbb{E}\left[\int_0^t e^2_{m,f}(t-u) P^{(1)}(du)\right] \\
&= \left(\int_0^t \lambda_p(u) e_{m,f}(t-u)\,du\right)^2.
\end{aligned}$$

Combining the three previous computations, we obtain

$$e_f^{(2)} = \left(\lambda_p * (f(X_m) + e_{m,f})\right)^2 + \lambda_p * f^2(X_m) + 2\lambda_p * (f(X_m) e_{m,f}) + \lambda_p * e^{(2)}_{m,f}. \quad (24)$$

Considering this equation for $\lambda_p = \lambda_m$, *we obtain as for the expectation a renewal equation for* $e_{m,f}^{(2)}$, *which yields for all* $t > 0$

$$e^{(2)}_{m,f}(t) \leq C_1 + C_2 \int_0^t e^{(2)}_{m,f}(u)\,du,$$

and then for all $T > 0$,

$$\sup_{t\in[0,T]} e^{(2)}_{m,f}(t) \le C_{1,T} + C_{2,T} \sup_{t\in[0,T]} e^2_{m,f} < +\infty,$$

using Gronwall's inequality and Proposition 5.1. This proves that $\mathbb{E}[\mathrm{SMO}^2_{m,f}(t)] < +\infty$ for all t $\geq$ *0, and then* $\mathbb{E}[\mathrm{SMO}^2_f(t)] < +\infty$ *by going back to (Eq. 24). Now, rewriting (Eq. 24), we obtain*

$$e^{(2)}_f = e^2_f + \lambda_p * (f(X_m) + e_{m,f})^2 + \lambda_p * v_{m,f},$$

which is (Eq. 23).

A.3 Proof of Theorem 5.2

Using that $X_m(s) \in [x_m{}^0, x_m{}^\infty]$ for all $s \in \mathbb{R}^+$, the σ-finiteness and absolute continuity of μ_t (for any $t \geq 0$) are direct consequences of (Eq. 19). Denoting by $\tilde{\rho}_t$ its Radon–Nikodým density, Proposition 5.1 and Theorem 5.1 then yield

$$\int_{x^0_m}^{x^\infty_m} f(x)\mu_t(dx) = \int_{x^0_m}^{x^\infty_m} f(x)\tilde{\rho}_t(x)dx = \int_{x^0_m}^{x^\infty_m} f(x)\rho(t,x)dx,$$

for all $f \in C([x^0_m, x^\infty_m]) \cap L^\infty([x^0_m, x^\infty_m])$, which concludes the proof.

References

1. Çınlar E (2011) Probability and stochastics. Graduate texts in mathematics, vol 261. Springer, New York
2. Yu M, Bardia A, Wittner BS, Stott SL, Smas ME, Ting DT, Isakoff SJ, Ciciliano JC, Wells MN, Shah AM, Concannon KF, Donaldson MC, Sequist LV, Brachtel E, Sgroi D, Baselga J, Ramaswamy S, Toner M (2013) Circulating breast tumor cells exhibit dynamic changes in epithelial and mesenchymal composition. Science 339(6119):580–584
3. Nguyen DX, Bos PD, Massagué J (2009) Metastasis: from dissemination to organ-specific colonization. Nat Rev Cancer 9 (4):274–284
4. Sahai E (2007) Illuminating the metastatic process. Nat Rev Cancer 7(10):737–749
5. WHO (2015) Cancer fact sheet. http://www.who.int/mediacentre/factsheets/fs297/en/ . Accessed 14 Jan 2016
6. Pantel K, Cote RJ, Fodstad O (1999) Detection and clinical importance of micrometastatic disease. J Natl Cancer Inst 91(13):1113–1124
7. Scott JG, Gerlee P, Basanta D, Fletcher AG, Maini PK, Anderson ARA (2013) Mathematical modeling of the metastatic process. In: Malek A (ed) Experimental metastasis: modeling and analysis. Springer, Dordrecht, pp 189–208
8. Gupta GP, Massagué J (2006) Cancer metastasis: building a framework. Cell 127 (4):679–695
9. Hanahan D, Weinberg RA (2011) Hallmarks of cancer: the next generation. Cell 144 (5):646–674
10. Michor F, Nowak MA, Iwasa Y (2006) Stochastic dynamics of metastasis formation. J Theor Biol 240(4):521–530
11. Haeno H, Michor F (2010) The evolution of tumor metastases during clonal expansion. J Theor Biol 263(1):30–44
12. Anderson AR, Quaranta V (2008) Integrative mathematical oncology. Nat Rev Cancer 8 (3):227–234
13. Koscielny S, Tubiana M, Lê MG, Valleron J, Mouriesse H, Contesso G, Sarrazin D (1984) Breast cancer: relationship between the size of the primary tumour and the probability of metastatic dissemination. Br J Cancer 49 (6):709–715
14. Michaelson JS, Silverstein M, Wyatt J, Weber G, Moore R, Halpern E, Kopans DB, Hughes K (2002) Predicting the survival of patients with breast carcinoma using tumor size. Cancer 95(4):713–723
15. van de Vijver MJ, He YD, van't Veer LJ, Dai H, Hart AAM, Voskuil DW, Schreiber GJ, Peterse JL, Roberts C, Marton MJ, Parrish M, Atsma D, Witteveen A, Glas A, Delahaye L,

van der Velde T, Bartelink H, Rodenhuis S, Rutgers ET, Friend SH, Bernards R (2002) A gene-expression signature as a predictor of survival in breast cancer. N Engl J Med 347 (25):1999–2009

16. Hahnfeldt P, Panigrahy D, Folkman J, Hlatky L (1999) Tumor development under angiogenic signaling: a dynamical theory of tumor growth, response and postvascular dormancy. Cancer Res 59:4770–5
17. Norton L (1988) A Gompertzian model of human breast cancer growth. Cancer Res 48:7067–7071
18. Verga F (2010) Modélisation mathématique de processus métastatiques. Ph.D. thesis, Aix-Marseille Université
19. Hart D, Shochat E, Agur Z (1998) The growth law of primary breast cancer as inferred from mammography screening trials data. Br J Cancer 78:382–387
20. Benzekry S, Lamont C, Beheshti A, Tracz A, Ebos JML, Hlatky L, Hahnfeldt P (2014) Classical mathematical models for description and prediction of experimental tumor growth. PLoS Comput Biol 10(8):e1003800
21. Bartoszyński R, Edler L, Hanin L, Kopp-Schneider A, Pavlova L, Tsodikov A, Zorin A, Yakovlev A (2001) Modeling cancer detection: tumor size as a source of information on unobservable stages of carcinogenesis. Math Biosci 171:113–142
22. Hanin L, Rose J, Zaider M (2006) A stochastic model for the sizes of detectable metastases. J Theor Biol 243:407–417
23. Iwata K, Kawasaki K, Shigesada N (2000) A dynamical model for the growth and size distribution of multiple metastatic tumors. J Theor Biol 203:177–186
24. Hartung N, Mollard S, Barbolosi D, Benabdallah A, Chapuisat G, Henry G, Giacometti S, Iliadis A, Ciccolini J, Faivre C, Hubert F (2014) Mathematical modeling of tumor growth and metastatic spreading: validation in tumor-bearing mice. Cancer Res 74:6397–6407
25. Benzekry S, Tracz A, Mastri M, Corbelli R, Barbolosi D, Ebos JML (2016) Modeling spontaneous metastasis following surgery: an in vivo-in silico approach. Cancer Res 76 (3):535–547
26. Chaffer CL, Weinberg RA (2011) A perspective on cancer cell metastasis. Science 331 (6024):1559–1564
27. Newton PK, Mason J, Bethel K, Bazhenova LA, Nieva J, Kuhn P (2012) A stochastic Markov chain model to describe lung cancer growth and metastasis. PLoS One 7(4):e34637
28. Newton PK, Mason J, Bethel K, Bazhenova L, Nieva J, Norton L, Kuhn P (2013) Spreaders and sponges define metastasis in lung cancer: a Markov chain Monte Carlo mathematical model. Cancer Res 73(9):2760–2769
29. Comen E, Norton L, Massague J (2011) Clinical implications of cancer self-seeding. Nat Rev Clin Oncol 8(6):369–377
30. Scott JG, Basanta D, Anderson AR, Gerlee P (2013) A mathematical model of tumour self-seeding reveals secondary metastatic deposits as drivers of primary tumour growth. J R Soc Interface 10(82):20130011
31. Hanin L, Zaider M (2011) Effects of surgery and chemotherapy on metastatic progression of prostate cancer: evidence from the natural history of the disease reconstructed through mathematical modeling. Cancers 3 (3):3632–3660
32. Wheldon TE (1988) Mathematical models in cancer research. Medical science series. Adam Hilger, Bristol/Philadelphia
33. Benzekry S, Gandolfi A, Hahnfeldt P (2014) Global dormancy of metastases due to systemic inhibition of angiogenesis. PLoS One 9(1): e84249
34. Bethge A, Schumacher U, Wedemann G (2015) Simulation of metastatic progression using a computer model including chemotherapy and radiation therapy. J Biomed Inform 57:74–87
35. Lewis PAW, Shedler GS (1979) Simulation of nonhomogeneous poisson processes by thinning. Nav Res Log Q 26(3):403
36. Sadahiro S, Suzuki T, Ishikawa K, Nakamura T, Tanaka Y, Masuda T, Mukoyama S, Yasuda S, Tajima T, Makuuchi H, Murayama C (2003) Recurrence patterns after curative resection of colorectal cancer in patients followed for a minimum of ten years. Hepatogastroenterology 50 (53):1362–1366
37. Siegel R, DeSantis C, Virgo K, Stein K, Mariotto A, Smith T, Cooper D, Gansler T, Lerro C, Fedewa S, Lin C, Leach C, Cannady RS, Cho H, Scoppa S, Hachey M, Kirch R, Jemal A, Ward E (2012) Cancer treatment and survivorship statistics, 2012. CA Cancer J Clin 62(4):220–241
38. Batchelor GK (1967) An introduction to fluid dynamics. Cambridge University Press, Cambridge
39. Barbolosi D, Benabdallah B, Hubert F, Verga F (2009) Mathematical and numerical analysis for a model of growing metastatic tumors. Math Biosci 218:1–14
40. Hartung N (2015) Efficient resolution of metastatic tumour growth models by

reformulation into integral equations. Discrete Contin Dyn Syst B 20:445–467

41. Lavielle M (2014) Mixed effects models for the population approach. models, tasks, methods and tools. Chapman & Hall/CRC biostatistics series. Chapman & Hall/CRC, Boca Raton
42. Tornøe CW, Overgaard RV, Agersø H, Nielsen HA, Madsen H, Jonsson EN (2005) Stochastic differential equations in NONMEM: implementation, application, and comparison with ordinary differential equations. Pharm Res 22 (8):1247–1258
43. Bulfoni M, Gerratana L, Del Ben F, Marzinotto S, Sorrentino M, Turetta M, Scoles G, Toffoletto B, Isola M, Beltrami CA, Di Loreto C, Beltrami AP, Puglisi F, Cesselli D (2016) In patients with metastatic breast cancer the identification of circulating tumor cells in epithelial-to-mesenchymal transition is associated with a poor prognosis. Breast Cancer Res 18(1):30
44. Paoletti C, Hayes DF (2016) Circulating tumor cells. Adv Exp Med Biol 882:235–258
45. Chen LL, Blumm N, Christakis NA, Barabasi AL, Deisboeck TA (2009) Cancer metastasis networks and the prediction of progression patterns. Br J Cancer 101(5):749–758

Chapter 11

Mechanically Coupled Reaction-Diffusion Model to Predict Glioma Growth: Methodological Details

David A. Hormuth II, Stephanie L. Eldridge, Jared A. Weis, Michael I. Miga, and Thomas E. Yankeelov

Abstract

Biophysical models designed to predict the growth and response of tumors to treatment have the potential to become a valuable tool for clinicians in care of cancer patients. Specifically, individualized tumor forecasts could be used to predict response or resistance early in the course of treatment, thereby providing an opportunity for treatment selection or adaption. This chapter discusses an experimental and modeling framework in which noninvasive imaging data is used to initialize and parameterize a subject-specific model of tumor growth. This modeling approach is applied to an analysis of murine models of glioma growth.

Key words Cancer, Biophysical stress, Diffusion, Invasion, MRI, Finite difference method

1 Introduction

Biophysical models of tumor growth and treatment response have the potential to fundamentally change the clinical care for cancer patients by providing clinicians with accurate and precise patient-specific predictive models. Through the use of noninvasive imaging data, these biophysical models can be parameterized by the unique characteristics of an individual's tumor to provide a "forecast" of future tumor growth and treatment response [1]. We [2–6] and others [7–11] have begun investigating the development of patient-specific mathematical models of cancer. In this work, we provide a detailed guide to the implementation of a mechanically coupled reaction-diffusion model [4, 6, 12] applied to glioma growth in rats.

The standard reaction-diffusion equation, Eq. 1, is commonly used to model glioma growth [5, 7] and describes the spatial-temporal evolution of tumor cell number due to the random movement of tumor cells (diffusion, first term on the right-hand side)

Louise von Stechow (ed.), *Cancer Systems Biology: Methods and Protocols*, Methods in Molecular Biology, vol. 1711, https://doi.org/10.1007/978-1-4939-7493-1_11, © Springer Science+Business Media, LLC 2018

and the proliferation of cells (reaction, second term on the right-hand side):

$$\frac{\partial N(x,y,z,t)}{\partial t} = \nabla \cdot \left(D(x,y,z)\nabla N(x,y,z,t)\right) + k(x,y,z)N(x,y,z,t)\left(1 - \frac{N(x,y,z,t)}{\theta}\right), \quad (1)$$

where $N(x, y, z, t)$ is the number of tumor cells at the three-dimensional position (x, y, z) and time t, $D(x, y, z)$ is the tumor cell diffusion coefficient, $k(x, y, z)$ is the net tumor cell proliferation, and θ is the tumor cell carrying capacity. One important limitation of the standard reaction-diffusion equation is that tumor growth is only restricted by the boundaries of the simulation domain (i.e., the skull for gliomas). In reality, as the tumor expands it interacts with the healthy brain tissue causing increased mechanical stress and the displacement of surrounding tissue, a phenomena termed the "mass effect" [13] and observed in several types of brain tumors [14]. The increased stress experienced by the tumor can impede further growth as demonstrated in the seminal work by Helmlinger et al. [15]. In Helmlinger et al.'s [15] contribution multi-cellular spheroids were grown in agar gel concentrations ranging from 0% to 1%. Increasing the agar concentration resulted in inhibited expansion of the spheroid as the substrate stiffness increased. More specifically, similar spheroid interactions with the surrounding environment would require increased force at elevated levels of stiffness. This phenomenon can also result in the preferential growth of tumors in areas of increasing mechanical compliance. To incorporate this effect, we first describe the mechanical equilibrium, Eq. 2:

$$\nabla \cdot \sigma - \lambda_f \cdot \nabla N = 0, \quad (2)$$

where σ is the stress tensor and λ_f is tumor cell-force coupling constant. For implementation, Eq. 2 is rewritten in terms of the tissue displacement ($\vec{u}$) under a linear elastic isotropic material assumption in Eq. 3:

$$\nabla \cdot G\nabla\vec{u} + \nabla\frac{G}{1-2\nu}\cdot\left(\nabla\cdot\vec{u}\right) - \lambda_f \nabla N = 0, \quad (3)$$

where G is the shear modulus (a material property that represents the constant of proportionality between shear stress to shear strain) and ν is Poisson's ratio (a material property that is a ratio relating lateral to longitudinal strain). The first two terms on the left-hand side in Eq. 3 represent the linear-elastic description of tissue displacement, while the third term represents a local body force generated by the invading tumor. $\vec{u}$ is then used to calculate the local normal (ε_{xx}, ε_{yy}, ε_{zz}) and shear strains (ε_{xy}, ε_{xz}, ε_{yz}). For small deformations, strain $\varepsilon_{i,j}$ is defined as the total deformation in the direction i divided by the original length in direction j and is calculated using Eq. 4:

$$\begin{bmatrix} \varepsilon_{xx} \\ \varepsilon_{yy} \\ \varepsilon_{zz} \\ \varepsilon_{xy} \\ \varepsilon_{xz} \\ \varepsilon_{yz} \end{bmatrix} = \begin{bmatrix} \partial u/\partial x \\ \partial v/\partial y \\ \partial w/\partial z \\ \partial u/\partial y \\ \partial u/\partial z \\ \partial v/\partial z \end{bmatrix}, \tag{4}$$

where u, v, and w represent the deformation in the x-, y-, and z-directions, respectively. The normal and shear strains are then used to calculate the normal and shear stresses using Hooke's law, Eq. 5:

$$\begin{bmatrix} \sigma_{xx} \\ \sigma_{yy} \\ \sigma_{zz} \\ \sigma_{xy} \\ \sigma_{xz} \\ \sigma_{yz} \end{bmatrix} = \frac{2G}{1-2\nu} \begin{bmatrix} 1-\nu & \nu & \nu & 0 & 0 & 0 \\ \nu & 1-\nu & \nu & 0 & 0 & 0 \\ \nu & \nu & 1-\nu & 0 & 0 & 0 \\ 0 & 0 & 0 & (1-2\nu) & 0 & 0 \\ 0 & 0 & 0 & 0 & (1-2\nu) & 0 \\ 0 & 0 & 0 & 0 & 0 & (1-2\nu) \end{bmatrix} \begin{bmatrix} \varepsilon_{xx} \\ \varepsilon_{yy} \\ \varepsilon_{zz} \\ \varepsilon_{xy} \\ \varepsilon_{xz} \\ \varepsilon_{yz} \end{bmatrix}. \tag{5}$$

The normal and shear stresses for a given voxel are then incorporated into a single term called the Von Mises stress, $\sigma_{\mathrm{vm}}(x, y, z, t)$, in Eq. 6:

$$\sigma_{\mathrm{vm}}(x,y,z,t) = \left[\frac{1}{2}\left(\begin{array}{c} (\sigma_{xx}(x,y,z,t) - \sigma_{yy}(x,y,z,t))^2 + (\sigma_{xx}(x,y,z,t) - \sigma_{zz}(x,y,z,t))^2 \\ +(\sigma_{zz}(x,y,z,t) - \sigma_{yy}(x,y,z,t))^2 \\ +6\left(\sigma_{xy}(x,y,z,t)^2 + \sigma_{xz}(x,y,z,t)^2 + \sigma_{yz}(x,y,z,t)^2\right) \end{array}\right)\right]^{1/2}. \tag{6}$$

The Von Mises stress is a term that reflects the total experienced stress for a given section of tissue, and is often used within failure criterion strategies in materials. We use the Von Mises stress to reflect the interaction between the growing tumor and its environment, that is, in our approach we use the Von Mises stress to spatially and temporally restrict tumor cell diffusion [4, 6, 12] using Eq. 7:

$$D(x, y, z, t) = D_0 e^{-\lambda_{\mathrm{D}} \cdot \sigma_{\mathrm{vm}}(x,y,z,t)}, \tag{7}$$

where D_0 represents the diffusion coefficient of tumor cells in the absence of mechanical restrictions and λ_{D} is a stress-tumor cell diffusion coupling constant.

In this chapter, we will discuss how to implement this model system using the finite difference method as well as how to individualize this model using an individual patient's imaging data. Noninvasive imaging measurements from diffusion-weighted magnetic resonance imaging (DW-MRI [16]) and contrast enhanced MRI (CE-MRI, [17]) are used to estimate the spatial distribution of tumor cell number in a murine model of glioma at several experimental time points. The in vivo estimated cell number then provides the initial tumor cell distribution and is also used to solve an inverse problem to return estimates of the model parameters. The estimated model parameters can then be used to simulate future tumor growth.

2 Materials

2.1 Dataset

The numerical methods presented in this chapter use an in vivo dataset acquired in rats with intracranially inoculated glioma cells [5, 18, 19]. Alternatively, an in silico dataset can also be used [5]. For both approaches the dataset should contain:

1. Three-dimensional estimates of the distribution of tumor cells at several time points.
2. Three-dimensional map of k (or initial guess).
3. Single value for D_0 (or initial guess).
4. Values for G, ν, λ_D, λ_f, and θ (based on literature, calculation, or assignment, *see* **Note 1**).

For use in Matlab this dataset should be saved as a ".mat" file consisting of a 4D array of tissue cellularity, a 3D array of k values, and one-element arrays of D_0, G, ν, λ_D, λ_f, and θ all with double precision.

2.2 Software/Hardware Requirements

The forward evaluation and parameter optimization of the mechanically coupled model was ran on a Dell PowerEdge R820 server consisting of four Intel Xenon E5–4610 2.3 GHz processors with a total of 256 GB of memory using Matlab 2015b. The forward evaluation is relatively less computationally intensive and takes less than 16 s for a 10 day simulation on a laptop with 8 GB of memory and an Intel i5-2550 M 2.5 GHz processor. The parameter optimization computation time, however, depends on both the number of parameters being estimated and the number of iterations of the optimization algorithm until stopping criteria are met. Parallelization of the parameter perturbation code can reduce computation time by a factor approximately equal to the number of parallel threads. (For example parameter perturbation for 100 parameters takes 13.1 min with 1 thread, 3.1 min with 4 threads, 1.7 min with 8 threads, 0.9 min with 16 threads, and 0.7 min with 32 threads.)

3 Methods

3.1 Animal Experiments

While details are presented in [5], we here discuss the salient features of the experimental procedure (*see* Fig. 1). The in vivo

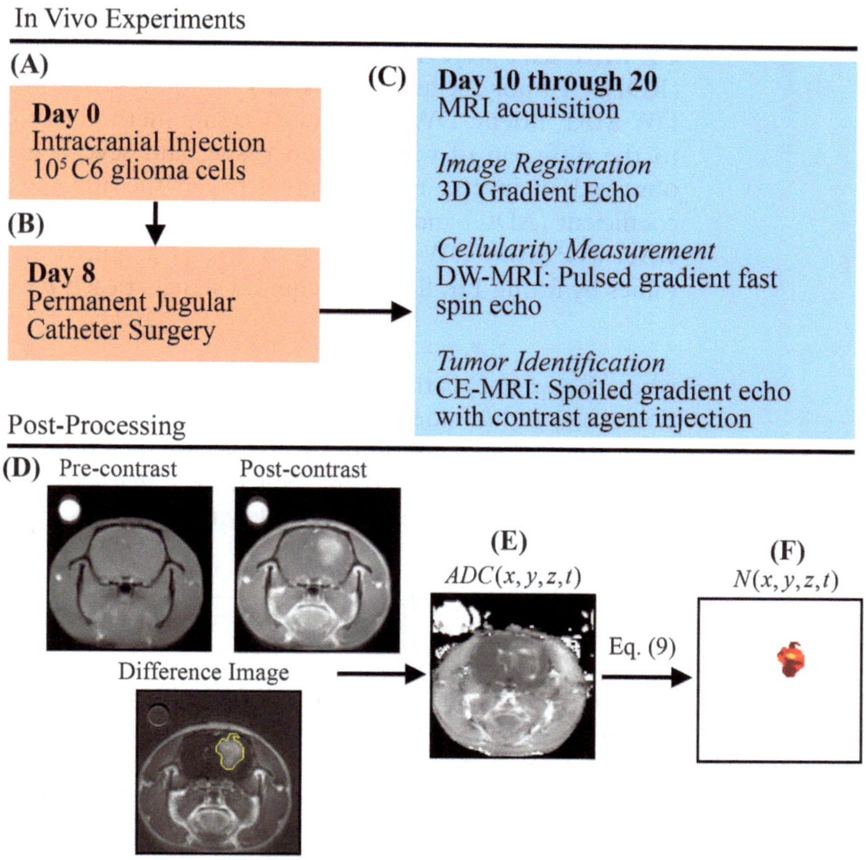

Fig. 1 Experimental timeline and estimation of in vivo cell number from DW-MRI data. (**a**) On day 0, rats are injected intracranially with 10^5 C6 glioma cells. (**b**) Jugular catheters are then inserted on day 8. (**c**) On days 10 through 20, rats are imaged with MRI with 3D gradient echo, DW-MRI, and CE-MRI. (**d**) CE-MRI is used to identify tumor tissue by subtracting pre-contrast image from the post-contrast image. (**e**) ADC(x, y, z, t) is then estimated from DW-MRI data. Finally, $N(x, y, z, t)$ is estimated (**f**) within the tumor tissue using Eq. 9 and ADC(x, y, z, t)

dataset described in this section was acquired in female Wistar rats inoculated intracranially with C6 Glioma cells (1×10^5) via stereotaxic injection on day 0 (Fig. 1a). On day 8, permanent jugular catheters were placed in each rat (Fig. 1b). Beginning on day 10, rats are imaged (Fig. 1c), with a 3D gradient echo, DW-MRI and CE-MRI (*see* **Note 2** for remarks on the experiment timeline and measurement frequency). The 3D gradient echo data was collected with a larger field of view (45 mm × 45 mm × 45 mm) and larger sampling matrix (256 × 256 × 128) for image registration purposes. The DW-MRI and CE-MRI data was acquired with a 32 mm × 32 mm × 16 mm field of view and a 128 × 128 × 16 sampling matrix. During the CE-MRI acquisition, a 200 μL bolus (0.05 mmol/kg) of gadolinium-diethylenetriamine pentaacetic acid, an MRI contrast agent, is injected to identify tumor regions

(Fig. 1d). Areas of signal enhancement in the post-contrast CE-MRI data were used to identify tumor regions of interest (ROI). Tumor cellularity ($N(x, y, z, t)$) was estimated from DW-MRI. Briefly, DW-MRI is an imaging method that is sensitive to the diffusion of water within tissue, and several groups have observed an inverse relationship between the apparent diffusion coefficient (ADC) and cellularity [20–24]. The ADC is estimated voxel-wise from DW-MRI (Fig. 1e) data acquired at several b-values by fitting Eq. 8 to the acquired signal at each b-value:

$$S(x, y, z, b) = S_0(x, y, z) \cdot e^{-b \cdot \mathrm{ADC}(x,y,z)}, \tag{8}$$

where $S(x, y, z, b)$ is the acquired signal at three-dimensional position (x,y,z) and b-value b, $S_0(x, y, z)$ is the intrinsic signal, and $\mathrm{ADC}(x, y, z)$ is the apparent diffusion coefficient. The tumor ROI identified from CE-MRI is then applied as a mask to $\mathrm{ADC}(x, y, z)$ (Fig. 1f), to estimate cellularity only within the tumor using Eq. 9:

$$N(x, y, z) = \theta\left(\frac{\mathrm{ADC_w} - \mathrm{ADC}(x, y, z)}{\mathrm{ADC_w} - \mathrm{ADC_{min}}}\right), \tag{9}$$

where θ is the maximum tumor cell carrying capacity, $\mathrm{ADC_w}$ is the ADC of water at 37 °C ($2.5 \times 10^{-3}\ \mathrm{mm^2/s}$) [25], $\mathrm{ADC}(x, y, z)$ is the ADC value at position (x,y,z), and ADC_{min} is the minimum ADC value which corresponds to the voxel with the largest number of cells [2]. θ can be calculated using the imaging voxel dimensions (0.25 mm × 0.25 mm × 1.00 mm), and assuming spherical tumor cells with a packing density of 0.7405 [26] and an average cell volume of 908 μm^3 [27] (*see* **Note 3** for further remarks on packing density and cell volume).

A voxel-wise k and a global D_0 are estimated from serial measurements of $N(x, y, z, t)$ in a parameter optimization procedure [5]. G is assigned from literature values to anatomical regions identified in imaging data (e.g., cortex, corpus callosum, hippocampus, thalamus, putamen) [28, 29], while ν is set to 0.45 (as we assume that tissue is nearly incompressible). λ_D can be assigned or a range of values can be evaluated to apply different degrees of mechanical coupling, while λ_f is set to 1.

3.2 Modeling

We now discuss the details of the finite difference simulation for Eqs. 1 and 2, the forward evaluation of the model system, and the parameter optimization and the tumor growth prediction approach. Figure 2 shows an overview of the data collection, parameter optimization, and prediction approach. Briefly, data is acquired from t_i to t_f. A subset of the total data (days t_i to t_n, where t_n is less than t_f) are first used to determine the optimal model parameters. Once the stopping criteria are met for the parameter optimization approach, the optimized model parameters are then

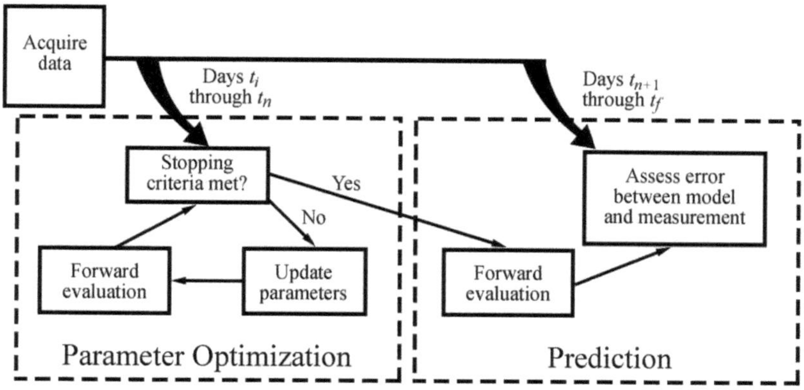

Fig. 2 Tumor growth modeling and prediction flow chart. DW-MRI and CE-MRI data is first acquired in rats at days t_i to t_f. A subset of the total data (t_i to t_n) is used to first estimate model parameters using an iterative optimization algorithm. The optimized model parameters are then used in a forward evaluation of the model system to predict tumor growth at the remaining data points (t_{n+1} to t_f). The error is then assessed between the model and measured values of $N(x, y, z, t)$

used in a forward evaluation of the model to simulate future tumor growth. The measured data is then compared to the model predicted growth on days t_{n+1} to t_f. With respect to the clinical context, t_n would represent the time point at which early-course of therapy data could be collected, and calibrated to the patient. Once complete, assessments on efficacy of therapy would be forecasted in silico for future time point t_f and perhaps lead to changes to therapy regimen or alternate therapies.

3.2.1 Finite Difference Simulation Setup

As an illustrative example for clarity, we show the derivation of the finite difference model for a 1D implementation, followed by extending the model to the full 3D implementation. A Taylor series expansion is used to derive the finite difference approximation of the tumor cell model (Eq. 1) as shown for the 1D implementation in Eq. 10:

$$\begin{aligned}\frac{N(x,t+h_t)-N(x,t)}{h_t} &= \left(\frac{\delta N(x,t)}{2h_x}\cdot\frac{\delta D(x)}{2h_x}\right)+D(x)\\ &\quad\cdot\left(\frac{\delta^2 N(x,t)}{h_x^2}\right)+k(x)\cdot N(x,t)\\ &\quad\cdot\left(1-\frac{N(x,t)}{\theta}\right),\end{aligned} \tag{10}$$

where h_t is the time step, and h_x is the grid spacing in the x-direction, and δ represents the central difference operator, defined below in Eqs. 11 and 12. Finite difference approximations are derived using a full grid approach to take advantage of the natural, voxelized gridding from the experimental imaging data measurements. The central difference approximation of the first derivative in (for example) the x-direction is shown in Eq. 11:

$$\frac{\partial N(x,t)}{\partial x} \approx \frac{\delta N(x,t)}{2h_x} = \frac{N(x+h_x,t) - N(x-h_x,t)}{2h_x}. \quad (11)$$

Similarly, the central difference approximation of the second derivative in (for example) the x-direction is shown in Eq. 12:

$$\frac{\partial^2 N(x,t)}{\partial x^2} \approx \frac{\delta^2 N(x,t)}{h_x^2} = \frac{N(x+h_x,t) - 2\cdot N(x,t) + N(x-h_x,t)}{h_x^2}. \quad (12)$$

In the case of a mesh boundary, where the node at either $(x+1)$ or $(x --1)$ does not exist, the zero flux boundary condition ($\partial N/\partial x = 0$) can be used to relate $N(x+h_x, t)$ to $N(x-h_x, t)$ (or vice versa) as shown in Eq. 13:

$$\frac{N(x+h_x,t) - N(x-h_x,t)}{2h_x} = 0 \Rightarrow N(x+h_x,t) = N(x-h_x,t). \quad (13)$$

The 3D implementation of Eq. 1 is shown below in Eq. 14:

$$\begin{aligned} \frac{N(x,y,z,t+h_t) - N(x,y,z,t)}{h_t} &= \left(\frac{\delta N(x,y,z,t)}{2h_x}\cdot\frac{\delta D(x,y,z)}{2h_x}\right) + D(x,y,z)\cdot\left(\frac{\delta^2 N(x,y,z,t)}{h_x^2}\right) \\ &+ \left(\frac{\delta N(x,y,z,t)}{2h_y}\cdot\frac{\delta D(x,y,z)}{2h_y}\right) + D(x,y,z)\cdot\left(\frac{\delta^2 N(x,y,z,t)}{h_y^2}\right) \\ &+ \left(\frac{\delta N(x,y,z,t)}{2h_z}\cdot\frac{\delta D(x,y,z)}{2h_z}\right) + D(x,y,z)\cdot\left(\frac{\delta^2 N(x,y,z,t)}{h_z^2}\right) \\ &+ k(x,y,z)\cdot N(x,y,z,t)\cdot\left(1 - \frac{N(x,y,z,t)}{\theta}\right). \end{aligned} \quad (14)$$

The derivation of the finite difference approximation of Eq. 2 is shown for the 1D implementation in Eqs. 15–17. Equation 2 is first rewritten in terms of the 1D stress in the x-direction (σ_x) in Eq. 15:

$$\nabla\cdot\sigma_x(x) - \lambda_f \nabla N(x,t) = 0. \quad (15)$$

σ_x is then replaced with Hooke's law for a linear elastic isotropic material ($\sigma_x = E\,\varepsilon_x$) in Eq. 16:

$$\nabla\cdot(E\varepsilon_x(x)) = \lambda_f \nabla N(x,t), \quad (16)$$

where E is Young's Modulus, and ε_x is equal to $\partial u/\partial x$. The divergence is then evaluated and the finite difference approximations are applied in Eq. 17:

$$\left(\frac{\delta E(x)}{2h_x}\frac{\delta u(x)}{2h_x} + E(x)\frac{\delta^2 u(x)}{h_x^2}\right) = \lambda_\mathrm{f}\frac{\delta N(x,t)}{2h_x}. \tag{17}$$

A similar approach as shown in Eqs. 15–17 can be followed to obtain the full 3D implementation of Eq. 2. Equations 18–20 show the finite difference approximation for the 3D implementation of Eq. 2. Equation 18 shows the x-direction component of Eq. 2:

$$\begin{aligned}&\left(\frac{2(1-v)}{1-2v}\right)\left(\frac{\delta G}{2h_x}\frac{\delta u}{2h_x}+G\frac{\delta^2 u}{{h_x}^2}\right)+\left(\frac{2v}{1-2v}\right)\left(\frac{\delta G}{2h_x}\frac{\delta v}{2h_y}+G\frac{\delta}{2h_x}\left(\frac{\delta v}{2h_y}\right)\right)\\&+\left(\frac{2v}{1-2v}\right)\left(\frac{\delta G}{2h_x}\frac{\delta w}{2h_z}+G\frac{\delta}{2h_x}\left(\frac{\delta w}{2h_z}\right)\right)+2\left(\frac{\delta G}{2h_x}\frac{\delta u}{2h_y}+G\frac{\delta}{2h_x}\left(\frac{\delta u}{2h_y}\right)\right)\\&+2\left(\frac{\delta G}{2h_x}\frac{\delta u}{2h_z}+G\frac{\delta}{2h_x}\left(\frac{\delta u}{2h_z}\right)\right)=\lambda_f\left(\frac{\delta N}{2h_x}\right),\end{aligned} \tag{18}$$

where u, v, and w represent tissue displacement in the x-, y-, and z-directions, respectively. Eq. 19 shows the y-direction component of Eq. 2:

$$\begin{aligned}&\left(\frac{2(1-v)}{1-2v}\right)\left(\frac{\delta G}{2h_y}\frac{\delta v}{2h_y}+G\frac{\delta^2 v}{{h_y}^2}\right)+\left(\frac{2v}{1-2v}\right)\left(\frac{\delta G}{2h_y}\frac{\delta u}{2h_x}+G\frac{\delta}{2h_y}\left(\frac{\delta u}{2h_x}\right)\right)\\&+\left(\frac{2v}{1-2v}\right)\left(\frac{\delta G}{2h_y}\frac{\delta w}{2h_z}+G\frac{\delta}{2h_y}\left(\frac{\delta w}{2h_z}\right)\right)+2\left(\frac{\delta G}{2h_y}\frac{\delta v}{2h_x}+G\frac{\delta}{2h_y}\left(\frac{\delta v}{2h_x}\right)\right)\\&+2\left(\frac{\delta G}{2h_y}\frac{\delta v}{2h_z}+G\frac{\delta}{2h_y}\left(\frac{\delta v}{2h_z}\right)\right)=\lambda_f\left(\frac{\delta N}{2h_y}\right).\end{aligned} \tag{19}$$

Equation 20 shows the z-direction component of Eq. 2:

$$\begin{aligned}&\left(\frac{2(1-v)}{1-2v}\right)\left(\frac{\delta G}{2h_z}\frac{\delta w}{2h_z}+G\frac{\delta^2 w}{{h_z}^2}\right)+\left(\frac{2v}{1-2v}\right)\left(\frac{\delta G}{2h_z}\frac{\delta u}{2h_x}+G\frac{\delta}{2h_z}\left(\frac{\delta u}{2h_x}\right)\right)\\&+\left(\frac{2v}{1-2v}\right)\left(\frac{\delta G}{2h_z}\frac{\delta v}{2h_y}+G\frac{\delta}{2h_z}\left(\frac{\delta v}{2h_y}\right)\right)+2\left(\frac{\delta G}{2h_z}\frac{\delta w}{2h_x}+G\frac{\delta}{2h_z}\left(\frac{\delta w}{2h_x}\right)\right)\\&+2\left(\frac{\delta G}{2h_z}\frac{\delta w}{2h_y}+G\frac{\delta}{2h_z}\left(\frac{\delta w}{2h_y}\right)\right)=\lambda_f\left(\frac{\delta N}{2h_z}\right).\end{aligned} \tag{20}$$

The unknown tissue displacements u, v, and w are solved by rewriting Eqs. 18–20 into a matrix system shown in Eq. 21:

$$[\mathbf{M}]\{\mathbf{U}\} = \lambda_\mathrm{f}\{\nabla\mathbf{N}\}, \tag{21}$$

where $[\mathbf{M}]$ is a square $3n \times 3n$ matrix of the finite difference coefficients, $\{\mathbf{U}\}$ is equal to $\{u_1, \cdots u_n, v_1, \cdots v_n, w_1, \cdots w_n\}^T$, where u_i, v_i, and w_i represent the displacement at the ith node in the x-, y-, and z-direction, respectively. $\{\nabla\mathbf{N}\}$ is equal to $\{\partial N_1/\partial x, \cdots \partial N_n/\partial x, \partial N_1/\partial y, \cdots \partial N_n/\partial y, \partial N_1/\partial z, \cdots \partial N_n/\partial z\}^T$, where $\partial N_i/\partial x$, $\partial N_i/\partial y$, and $\partial N_i/\partial z$ represent the gradient at the

ith node in the x-,y-, and z-direction, respectively. Rows 1 through n of [**M**]represent coefficients for Eq. 18, rows $n + 1$ through $2n$ of [**M**]represent the coefficients for Eq. 19, and $2n + 1$ through $3n$ of [**M**]represent the coefficients for Eq. 20. Rows 1 through n of $\{\mathbf{U}\}$ and $\{\nabla\mathbf{N}\}$ represent the x-direction components (u and $\partial N/\partial x$, respectively), rows $n + 1$ through $2n$ of $\{\mathbf{U}\}$ and $\{\nabla\mathbf{N}\}$ represent the y-direction components (v and $\partial N/\partial y$, respectively), and rows $2n + 1$ through $3n$ of $\{\mathbf{U}\}$ and $\{\nabla\mathbf{N}\}$ represent the z-direction components (w and $\partial N/\partial z$, respectively). [**M**] is built only once and can be factorized into lower and upper triangular matrices (refer to **Note 4** for further details on the construction and solving of Eq. 21). Equations 1 and 2 are solved using a three dimension in space (grid spacing: $250 \times 250 \times 1000$ μm), fully explicit in time (for Eq. 1, time step $= 0.01$ days) finite difference simulation. (Refer to **Note 5** for details on selecting an appropriate time step.) Equation 1 has no diffusive flux at the brain tissue boundaries (Neumann boundary condition [30]). Equation 2 has no tissue displacement in the Cartesian direction of the boundary (Dirichlet boundary condition), while displacement in the other Cartesian directions is unknown (slip condition [31]).

3.2.2 Forward Evaluation

A summary and example of the forward evaluation algorithm is presented in Fig. 3. The forward evaluation begins with solving the mechanical model (**steps 1** through **4** in Fig. 3). At the beginning of each iteration, the gradient of the current distribution of tumor cells, $\nabla N(x, y, z, t)$, is calculated and is assigned to $\{\nabla\mathbf{N}\}$ (**step 1** in Fig. 3). $\{\mathbf{U}\}$ is then solved for in Eq. 21 (**step 2** in Fig. 3). The strains (Eq. 4) and stresses (Eqs. 5 and 6) are calculated (**step 3** in Fig. 3). $\sigma_{\text{vm}}(x, y, z, t)$ is then used to update $D(x, y, z, t)$ (Eq. 7, **step 4** in Fig. 3). Finally, $D(x, y, z, t)$ is used in the evaluation of Eq. 1 to determine $N(x, y, z, t + 1)$ (**step 5** in Fig. 3). The forward evaluation of the model system is then repeated at each simulation time step.

3.3 Parameter Optimization and Tumor Growth Prediction

The optimal model parameters are determined using an iterative Levenberg-Marquardt [32, 33] weighted least squares optimization:

$$\left[J^T WJ + \alpha \cdot D_{J^T WJ}\right] \cdot \{\Delta\beta\} = J^T W\{N_{\text{meas}} - N_{\text{model}}(\beta)\}, \quad (22)$$

where J is the Jacobian matrix, W is a diagonal weighting matrix, α is a damping parameter, $D_{J^T WJ}$ is a diagonal matrix consisting of the diagonal elements of $J^T WJ$, $\{\Delta\beta\}$is as vector of updates to model parameters, $\{N_{\text{meas}}\}$ is a vector of the measured cell number, and $\{N_{\text{model}}(\beta)\}$ is a vector of the model described cell number using the current best set of parameters β. J is a (n (number of voxels) $\times$ nt (number of time points)) by p (the number of model parameters) matrix, W is a $(n \times nt) \times (n \times nt)$ matrix, has p components, and $\{N_{\text{meas}}\}$ has $(n \times nt)$ components. J can be estimated using

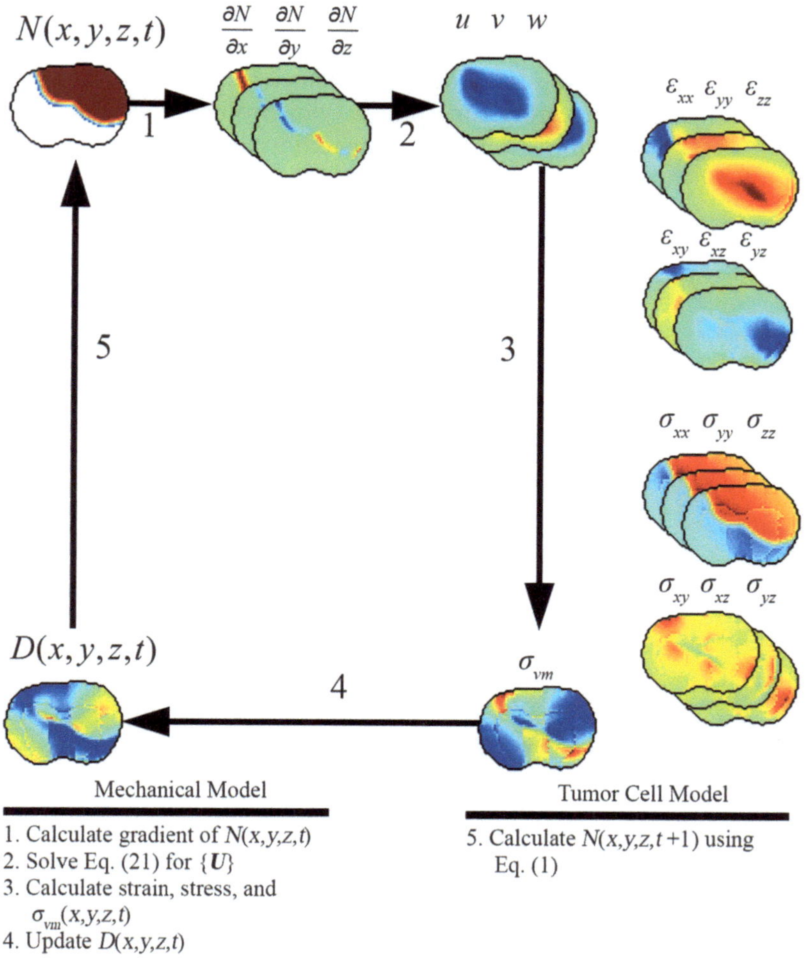

Fig. 3 Algorithm and example forward evaluation of mechanical and tumor cell model. The mechanical model is first solved to calculate the tissue displacement vector {**U**} due to $N(x, y, z, t)$, Eq. 21. {**U**} is then used to calculate strain, stress, and $\sigma_{\mathrm{vm}}(x, y, z, t)$. The new value of $D(x, y, z, t)$is calculated using Eq. 2 and $\sigma_{\mathrm{vm}}(x, y, z, t)$. Finally, $D(x, y, z, t)$ is used in Eq. 6 to calculate the value of $N(x, y, z, t + 1)$

numerical differentiation (refer to **Note 6** for further comments on J). For example, the J element at row i and column j, Eq. 23, represents the partial derivative of the model cell number at node i with respect to the jth model parameter and is calculated by individually perturbing model parameters as described below:

$$J_{i,j} = \frac{\partial N_i}{\partial \beta_j} = \frac{N_{\mathrm{model}}(i, \beta_{\mathrm{alt}}) - N_{\mathrm{model}}(i, \beta)}{\beta_{\mathrm{alt},j} - \beta_j}, \tag{23}$$

where $N_{\mathrm{model}}(i, \beta_{\mathrm{alt}})$ is the model cell number at the ith index of $\{N_{\mathrm{model}}\}$ using parameters β_{alt}, $N_{\mathrm{model}}(i, \beta)$ is the model cell number at the ith index of $\{N_{\mathrm{model}}\}$ using parameters β. β_{alt} is equal to β at all indices except for the jth index which is perturbed by a factor f(i.e., $\beta_{\mathrm{alt}\,,\,j} = f \times \beta_j$). (Note $f$should be a number close to but not equal to 1. In this work, we assign $f = 1.001$.) $W$is a square matrix with $n \times nt$ rows and columns. W weights the elements of J by the reciprocal of the total number of cells at each time point. This weighting is included to balance the influence of later time points to the earlier time points (which often have much fewer nonzero voxel measurements compared to the later time points). For $nt = 2$, $W_{i,i}$ is calculated using Eq. 24:

$$W_{i,i} = \begin{cases} i \leq n & \left(\sum_{j=1}^{j=n} (N_{\text{meas}}(j, t = 1)) \right)^{-1} \\ i > n \text{ and } i \leq 2n & \left(\sum_{j=1}^{j=n} (N_{\text{meas}}(j, t = 2)) \right)^{-1} \end{cases}. \quad (24)$$

Figure 4 summarizes the parameter optimization approach used to estimate model parameters $k(x, y, z)$ and D_0. The model is initially evaluated with a guess of the model parameters (**step 1** in

Parameter Optimization

1. Initial model evaluation with initial $\{\beta\}$ $N_{model}\{\beta\}$
2. Evaluate objective function with $\{\beta\}$: Error$\{\beta\}$

while-loop *stopping criteria not met*
  3. Build Jacobian
  4. Solve for $\{\Delta\beta\}$, calculate $\{\beta_{test}\}$
  5. Evaluate model with $\{\beta_{test}\}$
  6. Evaluate objective function with $\{\beta_{test}\}$: Error$\{\beta_{test}\}$

  7. if $Error\{\beta_{test}\} < Error\{\beta\}$
      $\{\beta\} = \{\beta_{test}\}$
      $\alpha = \alpha / 12$
    else $Error\{\beta_{test}\} < Error\{\beta\}$
      $\alpha = 3 \times \alpha$
    end

end

8. Report optimized parameters $\{\beta\}$

Fig. 4 Iterative parameter optimization approach. A schematic is shown above for the iterative parameter optimization algorithm using the Levenberg-Marquardt method [32, 33]. The model is first evaluated with an initial guess of model parameters, *line 1*. The objective function is then evaluated with the current set of model parameters, *line 2*. The optimal model parameters are then determined in an iterative "while-loop" which ceases when stopping criteria are met. At the beginning of each iteration, the Jacobian is built, *line 3*, and is used to solve for the new guess of model parameters, *line 4*. The model is then re-evaluated with the new model parameters, *line 5*, and the objective function is calculated, *line 6*. Finally, the error is compared to the previously observed lowest error to determine if the new parameter values are acceptable. The optimization ceases when the stopping criteria are met

Fig. 4). A guess of β is used to evaluate the objective function described in Eq. 25 (**step 2** in Fig. 4):

$$\text{Error} = \sum_{t=t_1}^{t_n}\left(\left(\sum_{i=1}^{i=n}(N_{\text{meas}}(i,t))\right)^{-1}\cdot\left(\sum_{i=1}^{i=n}(N_{\text{model}}(i,t,\beta)-N_{\text{meas}}(i,t))^2\right)\right). \tag{25}$$

The initial evaluation of Eq. 25 sets the current minimum error or Error(β). J, W, and $D_{J^T WJ}$ are then built (**step 3** in Fig. 4). The parameter update vector $\{\Delta\beta\}$ is then calculated using Eq. 22 and then added to $\{\beta\}$ for the current guess of model parameters $\{\beta_{\text{test}}\}$ (**step 4** in Fig. 4). The forward evaluation of the model is performed using model parameters $\{\beta_{\text{test}}\}$ (**step 5** in Fig. 4). Equation 25 is then re-calculated using $\{\beta_{\text{test}}\}$ (**step 6** in Fig. 4). The error evaluated using $\{\beta_{\text{test}}\}$ or Error(β_{test}) is compared to Error(β). If Error(β_{test}) is less than Error(β), $\{\beta_{\text{test}}\}$ is accepted (i.e., $\{\beta\} = \{\beta_{\text{test}}\}$) and α decreased by a factor of 12. If Error(β_{test}) is greater than Error(β), $\{\beta_{\text{test}}\}$ is rejected and α increased by a factor of 3. (Note, the factors that α is increased or decreased by (3 and 12 in this work) are often problem-specific and need to be empirically determined to improve convergence.) At this point, the stopping criteria are also evaluated. The stopping criterion can be a maximum number of iterations, a minimal threshold of error, or a minimal relative change in model error between successful iterations, or a minimal relative change in model [34] between successful iterations. In general, error will never reach zero for this type of system so selecting a stopping criteria that is sensitive to the relative change in error or parameter values will indicate convergence. The parameter optimization process continues by returning to **step 3** until the stopping criteria are met.

At the conclusion of the parameter optimization process, the optimized model parameters are used in a final forward evaluation of the model from t_i to t_f. The error between $N_{\text{model}}(x, y, z, t)$ and $N_{\text{meas}}(x, y, z, t)$is assessed at the time points not used in the parameter optimization t_{n+1} to t_f.

3.4 Summary and Outlook

In this chapter, a modeling and experimental framework was described which can be used to individualize a predictive biophysical model from an individual patient's imaging data. Clinically available imaging measurements from CE-MRI and DW-MRI were used to provide serial estimates of tumor cell number that were then used in an inverse problem to optimize model parameters for the measured tumor. These individually optimized model parameters could then be used to predict future growth or response. For example, acquiring data early in the course of a patient's therapy could be used to calibrate a patient-specific model that could then be used to predict the efficacy of the current treatment weeks or months before response is identifiable through standard criteria (e.g., the Response Evaluation Criteria in Solid Tumors [34]). For predicted non-responders, the calibrated model could potentially be used to evaluate other treatment regimens to adapt clinical care to improve the outcome on an individual patient basis. While this is a promising avenue for the future of clinical cancer care, further development of predictive biophysical models is needed to characterize patient response to a variety of available patient treatments [35].

4 Notes

1. When collecting a new dataset or evaluating this model in a different disease setting, model parameters should be measured or estimated on an individual basis. When this is not the case, however, model parameters should be assigned (or calculated) from literature values (e.g., G, ν, θ) obtained from experiments that most closely match the tumor or tumor location that is currently under investigation. For model parameters that cannot be measured experimentally or assigned from literature (e.g., λ_D, λ_f) can be assigned empirically based on results observed in a cohort. Sensitivity analysis (e.g., [36]) of the model system can also be used to help determine which model parameters require assignment on an individual basis and which model parameters may be assigned for the cohort.
2. The experimental time line may change depending on the particular cancer under investigation, its growth rate, and the initial size of the tumor. We selected day 10 to start our imaging experiments, as the tumors are approximately 20–40 mm^3 and typically extend over multiple imaging slices.
3. To calculate the physical carrying capacity (i.e., the maximum number of cells a space can contain) assumptions will need to be made about the overall tissue structure and cellular shape which can be verified through histological observations of the tissue. For the C6 line, we assumed that the tumor cell tissue was predominately composed of spherical tumor cells with a packing density and an average cell volume obtained from the literature [26, 27]. When comparing between the DW-MRI estimate of cellularity and the model predicted cellularity the precise values for packing density and average cell volume are not critical as long as the same carrying capacity is used in both the model and the ADC to cellularity calculation. However, when comparing to histological data, more care is required to match the average size, shape, and packing density of the tumor cells to what is observed in vivo. Packing density can be

calculated from Hematoxylin and Eosin (H&E) stained tissue sections by calculating the fraction of the H&E stained area over the total tumor ROI. The average cell area can then be calculated as the total occupied area (packing density multiplied by total ROI area) divided by the number of positive stained Hematoxylin cells. The average cell area can then be used to calculate an average cell radius and volume. In H&E stained sections obtained in one rat we calculated an average packing density of 0.764 ± 0.054% (mean ± 95% confidence interval) and an average volume of 982 ± 247 μm^3.

4. The coefficient matrix $[\mathbf{M}]$is a sparse and potentially very large $(3n \times 3n)$ matrix. To conserve memory and accelerate computational time, $[\mathbf{M}]$ can be represented by a sparse matrix $[\mathbf{M}_{\text{compact}}]$ which is an $nz \times 3$ matrix, where nz is the number of nonzero elements of $[\mathbf{M}]$, and the three columns represent the matrix nonzero entry, the entry's matrix row, and entry's matrix column entry, respectively. While many sparse matrix data formats exist, in this realization we used the format native to MATLAB. With respect to solution methods associated with sparse matrices, standardly some form of iterative approach would be adopted with an accompanying matrix precondition method to increase speed of calculation. In this realization, we employed one of the available MATLAB methods, namely, the bi-conjugate gradient stabilized method with an incomplete LU factorization as a preconditioner.
5. The simulation time step, h_t, is selected to maintain numerical stability for a range of diffusion coefficients for the parameter optimization process. To be stable, the product $D \cdot h_t\left(1/h_x^2 + 1/h_y^2 + 1/h_z^2\right)$ must be less than 1/2, or for isotropic dimensions the product$D \cdot h_t/h^2$must be less than 1/6. To be monotonic and stable, the product $D \cdot h_t\left(1/h_x^2 + 1/h_y^2 + 1/h_z^2\right)$ must be less than 1/4, or for isotropic dimensions the product $D \cdot h_t/h^2$must be less than 1/12.
6. Building or updating the Jacobian matrix, J, can be time intensive as the number of model parameters increases as Eq. 23 (and thus a full model evaluation) needs to be evaluated for each model parameter perturbation. Parallelized code can be used to simultaneously build several columns of J at a time, dramatically decreasing the computation time. For example, non-parallelized code takes approximately 13.1 min per 100 parameters, while parallelized code divided among 32 threads takes 0.7 min per 100 parameters. Alternatively, approaches such as Broyden's method [37] can be used to update J at each iteration while only building the full J matrix in the first iteration. Briefly, Broyden's method is a secant

method update that estimates J at the nth iteration based on the previous J, the difference between the model evaluation at the $(n-1)$ and $(n--2)$ iterations, and the difference between model parameters at the $(n-1)$ and $(n-2)$ iterations.

Acknowledgments

This work was supported through funding from CPRIT RR160005 and the National Cancer Institute U01CA174706, K25CA204599, and R01CA186193, from the National Institute of Neurological Disorders and Stroke R01NS049251 and the Vanderbilt-Ingram Cancer Center Support Grant (NIH P30CA68485).

References

1. Yankeelov TE, Quaranta V, Evans KJ, Rericha EC (2015) Toward a science of tumor forecasting for clinical oncology. Cancer Res 75(6):918–923
2. Atuegwu NC, Gore JC, Yankeelov TE (2010) The integration of quantitative multi-modality imaging data into mathematical models of tumors. Phys Med Biol 55(9):2429–2449
3. Atuegwu NC, Colvin DC, Loveless ME, Xu L, Gore JC, Yankeelov TE (2012) Incorporation of diffusion-weighted magnetic resonance imaging data into a simple mathematical model of tumor growth. Phys Med Biol 57 (1):225–240
4. Weis JA, Miga MI, Arlinghaus LR, Li X, Chakravarthy AB, Abramson V et al (2013) A mechanically coupled reaction-diffusion model for predicting the response of breast tumors to neoadjuvant chemotherapy. Phys Med Biol 58(17):5851–5866
5. Hormuth DA II, Weis JA, Barnes SL, Miga MI, Rericha EC, Quaranta V et al (2015) Predicting in vivo glioma growth with the reaction diffusion equation constrained by quantitative magnetic resonance imaging data. Phys Biol 12 (4):46006
6. Weis JA, Miga MI, Arlinghaus LR, Li X, Abramson V, Chakravarthy AB et al (2015) Predicting the response of breast cancer to neoadjuvant therapy using a mechanically coupled reaction-diffusion model. Cancer Res 75 (22):4697–4707
7. Baldock A, Rockne R, Boone A, Neal M, Bridge C, Guyman L et al (2013) From patient-specific mathematical neuro-oncology to precision medicine. Front Oncol 3:62
8. Corwin D, Holdsworth C, Rockne RC, Trister AD, Mrugala MM, Rockhill JK et al (2013) Toward patient-specific, biologically optimized radiation therapy plans for the treatment of glioblastoma. PLoS One 8(11):e79115
9. Hogea C, Davatzikos C, Biros G (2008) An image-driven parameter estimation problem for a reaction-diffusion glioma growth model with mass effects. J Math Biol 56(6):793–825
10. Liu Y, Sadowski SM, Weisbrod AB, Kebebew E, Summers RM, Yao J (2014) Patient specific tumor growth prediction using multimodal images. Med Image Anal 18 (3):555–566
11. Konukoglu E, Clatz O, Menze BH, Stieltjes B, Weber M-A, Mandonnet E et al (2010) Image guided personalization of reaction-diffusion type tumor growth models using modified anisotropic eikonal equations. IEEE Trans Med Imaging 29:77–95
12. Garg I, Miga MI (2008) Preliminary investigation of the inhibitory effects of mechanical stress in tumor growth. Proc SPIE 29:69182L-11
13. Venes D (2013) Taber's® cyclopedic medical dictionary, 22nd edn. F. A. Davis Company, Philadelphia, PA
14. DeAngelis LM (2001) Brain tumors. N Engl J Med 344(2):114–123
15. Helmlinger G, Netti PA, Lichtenbeld HC, Melder RJ, Jain RK (1997) Solid stress inhibits the growth of multicellular tumor spheroids. Nat Biotechnol 15(8):778–783
16. Padhani AR, Liu G, Mu-Koh D, Chenevert TL, Thoeny HC, Takahara T et al (2009) Diffusion-weighted magnetic resonance

imaging as a cancer biomarker: consensus and recommendations. Neoplasia 11(2):102–125
17. Yankeelov TE, Gore JC (2009) Dynamic contrast enhanced magnetic resonance imaging in oncology: theory, data acquisition, analysis, and examples. Curr Med Imaging Rev 3 (2):91–107
18. Barth R, Kaur B (2009) Rat brain tumor models in experimental neuro-oncology: the C6, 9L, T9, RG2, F98, BT4C, RT-2 and CNS-1 gliomas. J Neuro-Oncol 94(3):299–312
19. Hormuth DA II, Weis JA, Barnes SL, Miga MI, Rericha EC, Quaranta V, Yankeelov TE (2017). A mechanically-coupled reaction-diffusion model that incorporates intra-tumoral heterogeneity to predict in vivo glioma growth. J R Soc Interface 14:128
20. Barnes SL, Sorace AG, Loveless ME, Whisenant JG, Yankeelov TE (2015) Correlation of tumor characteristics derived from DCE-MRI and DW-MRI with histology in murine models of breast cancer. NMR Biomed 28 (10):1345–1356
21. Anderson AW, Xie J, Pizzonia J, Bronen RA, Spencer DD, Gore JC (2000) Effects of cell volume fraction changes on apparent diffusion in human cells. Magn Reson Imaging 18 (6):689–695
22. Guo Y, Cai Y-Q, Cai Z-L, Gao Y-G, An N-Y, Ma L et al (2002) Differentiation of clinically benign and malignant breast lesions using diffusion-weighted imaging. J Magn Reson Imaging 16(2):172–178
23. Sugahara T, Korogi Y, Kochi M, Ikushima I, Shigematu Y, Hirai T et al (1999) Usefulness of diffusion-weighted MRI with echo-planar technique in the evaluation of cellularity in gliomas. J Magn Reson Imaging 9(1):53–60
24. Humphries PD, Sebire NJ, Siegel MJ, Olsen ØE (2007) Tumors in pediatric patients at diffusion-weighted mr imaging: apparent diffusion coefficient and tumor cellularity. Radiology 245(3):848–854
25. Whisenant JG, Ayers GD, Loveless ME, Barnes SL, Colvin DC, Yankeelov TE (2014) Assessing reproducibility of diffusion-weighted magnetic resonance imaging studies in a murine model of HER2+ breast cancer. Magn Reson Imaging 32(3):245–249
26. Martin I, Dozin B, Quarto R, Cancedda R, Beltrame F (1997) Computer-based technique for cell aggregation analysis and cell aggregation in in vitro chondrogenesis. Cytometry 28 (2):141–146
27. Rouzaire-Dubois B, Milandri JB, Bostel S, Dubois JM (2000) Control of cell proliferation by cell volume alterations in rat C6 glioma cells. Pflugers Arch 440(6):881–888
28. Elkin BS, Ilankovan AI, Morrison B III (2011) A detailed viscoelastic characterization of the P17 and adult rat brain. J Neurotrauma 28:2235
29. Lee SJ, King MA, Sun J, Xie HK, Subhash G, Sarntinoranont M (2014) Measurement of viscoelastic properties in multiple anatomical regions of acute rat brain tissue slices. J Mech Behav Biomed Mater 29:213–224
30. Lynch D (2005) Numerical partial differential equations for environmental scientsits and engineers: a first practical course. Springer, New York, NY
31. Miga MI, Paulsen KD, Lemery JM, Eisner SD, Hartov A, Kennedy FE et al (1999) Model-updated image guidance: initial clinical experiences with gravity-induced brain deformation. IEEE Trans Med Imaging 10:866–874
32. Levenberg K (1944) A method for the solution of certain non-linear problems in least squares. Q J Appl Mathmatics II(2):164–168
33. Marquardt DW (1963) An algorithm for least-squares estimation of nonlinear parameters. J Soc Ind Appl Math 11(2):431–441
34. Eisenhauer EA, Therasse P, Bogaerts J, Schwartz LH, Sargent D, Ford R et al (2009) New response evaluation criteria in solid tumours: revised RECIST guideline (version 1.1). Eur J Cancer 45(2):228–247
35. Yankeelov TE, Atuegwu N, Hormuth DA, Weis JA, Barnes SL, Miga MI et al (2013) Clinically relevant modeling of tumor growth and treatment response. Sci Transl Med 5 (187):187ps9
36. Marino S, Hogue IB, Ray CJ, Kirschner DE (September 2008) A methodology for performing global uncertainty and sensitivity analysis in systems biology. J Theor Biol 254 (1):178–196
37. Broyden CG (1965) A class of methods for solving nonlinear simultaneous equations. Math Comput 19(92):577–593

Chapter 12

Profiling Tumor Infiltrating Immune Cells with CIBERSORT

Binbin Chen, Michael S. Khodadoust, Chih Long Liu, Aaron M. Newman, and Ash A. Alizadeh

Abstract

Tumor infiltrating leukocytes (TILs) are an integral component of the tumor microenvironment and have been found to correlate with prognosis and response to therapy. Methods to enumerate immune subsets such as immunohistochemistry or flow cytometry suffer from limitations in phenotypic markers and can be challenging to practically implement and standardize. An alternative approach is to acquire aggregative high dimensional data from cellular mixtures and to subsequently infer the cellular components computationally. We recently described CIBERSORT, a versatile computational method for quantifying cell fractions from bulk tissue gene expression profiles (GEPs). Combining support vector regression with prior knowledge of expression profiles from purified leukocyte subsets, CIBERSORT can accurately estimate the immune composition of a tumor biopsy. In this chapter, we provide a primer on the CIBERSORT method and illustrate its use for characterizing TILs in tumor samples profiled by microarray or RNA-Seq.

Key words Cancer immunology, Deconvolution, Support vector regression (SVR), Tumor infiltrating leukocytes (TILs), Tumor microenvironment, Tumor heterogeneity, Gene expression, Microarray, RNA-Seq, TCGA

1 Introduction

Neoplastic cells reside within a complex tumor microenvironment necessary for tumor growth and survival. Numerous non-neoplastic cell types including tumor-infiltrating leukocytes (TILs) comprise the tumor stroma. This immune infiltrate is often a heterogeneous mixture of immune cells that includes both innate and adaptive immune populations, and cell types associated with active (e.g., cytotoxic T lymphocytes) and suppressive (e.g., regulatory T cells, myeloid-derived suppressor cells) immune functions. The significance of TILs varies by cancer histology, with the presence of certain immune subsets often exhibiting a beneficial prognostic effect in one malignancy but a detrimental effect in another cancer type [1, 2]. The importance of TIL assessment continues to grow with the development of novel

Louise von Stechow (ed.), *Cancer Systems Biology: Methods and Protocols*, Methods in Molecular Biology, vol. 1711, https://doi.org/10.1007/978-1-4939-7493-1_12, © Springer Science+Business Media, LLC 2018

immunotherapeutic agents designed to target these cells. Recent studies have found that T lymphocyte subsets (e.g., CD8+) may predict response to existing and emerging immunotherapies, highlighting the importance of investigating tumor-associated immune cells as potential predictive biomarkers [3–5].

Measurement of the tumor immune infiltrate has traditionally been evaluated by histology on tissue sections and immune subsets have been inferred by immunohistochemistry of individual markers. However, immunophenotyping typically requires multiple parameters to accurately subset populations, and thus immunohistochemistry is unable to identify many immune populations and performs poorly at capturing functional phenotypes (e.g., activated vs. resting lymphocytes) [6]. Flow cytometry is an alternative method of quantifying immune infiltrates that enables simultaneous measurement of multiple parameters. However, this method requires prompt and careful processing of samples as well as tissue disaggregation, which may result in the loss of fragile cell types and the distortion of gene expression profiles. While flow cytometry can assess multiple markers, this number is still limited, potentially excluding markers that may better discriminate closely related cell populations [7].

In contrast, gene expression profiling of bulk tissues does not depend on surface markers and does not suffer from artifacts related to cellular dissociation. Samples can be readily processed and stored in a standardized fashion, mitigating issues that may confound data collected at different times and from different locations. Although previous studies of bulk tumor samples revealed a number of immune-enriched gene signatures with prognostic significance [8, 9], linking these signatures to specific TIL phenotypes has been challenging [10–14]. Methods for mathematically separating the bulk tumor gene expression profiles (GEP) into its component cell types can overcome this issue.

Several computational tools, including linear least-square regression (LLSR) [7], microarray microdissection with analysis of differences (MMAD) [15], and digital sorting algorithm (DSA) [16], have been applied to the deconvolution of complex GEP mixtures to infer cellular composition. Although these approaches are effective for enumerating highly distinct cell types in mixtures with minimal unknown content (e.g., lymphocytes, monocytes, and neutrophils in whole blood), they are sensitive to experimental noise, high unknown mixture content, and closely related cell types, limiting their utility for TIL assessment [17, 18].

CIBERSORT, a computational approach developed by our group, aims to address these challenges (*see* Fig. 1) [17]. Like other methods, CIBERSORT requires a specialized knowledgebase of gene expression signatures, termed a "signature matrix," for the deconvolution of cell types of interest. However, in contrast to previous efforts, CIBERSORT implements a machine learning

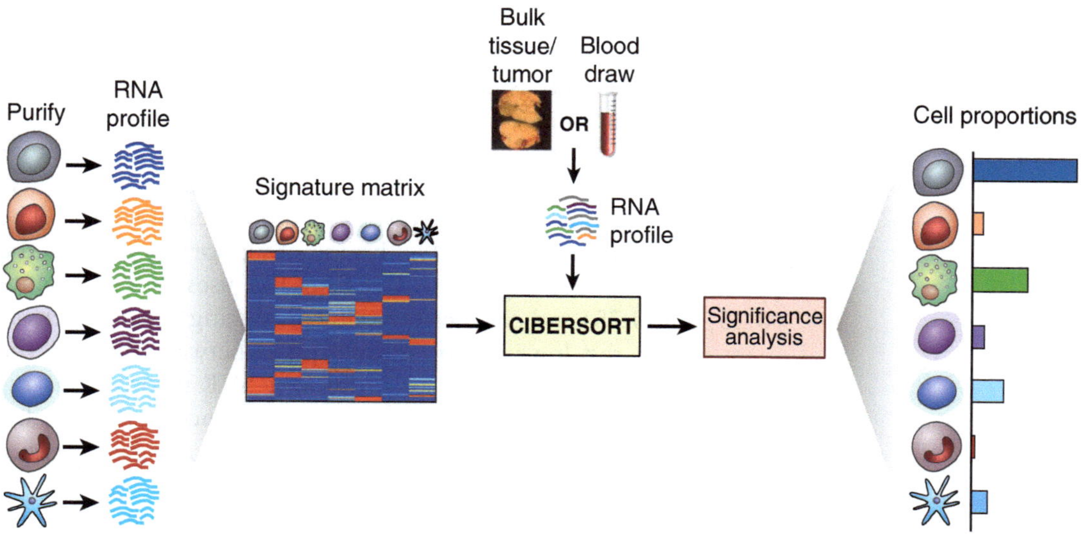

Fig. 1 Overview of CIBERSORT. As input, CIBERSORT requires a "signature matrix" comprised of barcode genes that are enriched in each cell type of interest. Once a suitable knowledgebase is created and validated, CIBERSORT can be applied to characterize cell type proportions in bulk tissue expression profiles. Although originally validated using a signature matrix containing 22 functionally defined human immune subsets (LM22) profiled by microarrays, CIBERSORT is a general framework that can be applied to diverse cell phenotypes and genomic data types, including RNA-Seq. To quantitatively capture deconvolution confidence, CIBERSORT calculates several quality control metrics, including a deconvolution *p*-value

approach, called support vector regression (SVR), that improves deconvolution performance through a combination of feature selection and robust mathematical optimization techniques (*see* Subheading 1.1 for details). In benchmarking experiments, CIBERSORT was more accurate than other methods in resolving closely related cell subsets and in mixtures with unknown cell types (e.g., solid tissues) [17]. Thus, CIBERSORT is a useful approach for high-throughput characterization of diverse cell types, such as TILs, from complex tissues. Here, we provide users with a practical roadmap for dissecting leukocyte content in tumor gene expression datasets with CIBERSORT.

1.1 CIBERSORT Model

A common objective of gene expression deconvolution algorithms is to solve the following system of linear equations for **f**:

$\mathbf{m} = \mathbf{f} \times \mathbf{B}$.

m: a vector consisting of a mixture GEP (input requirement).

f: a vector consisting of the fraction of each cell type in the signature matrix (unknown).

B: a "signature matrix" containing signature genes for cell subsets of interest (input requirement).

CIBERSORT differs from previous deconvolution methods in its application of a machine learning technique, ν-support vector regression (ν-SVR), to solve for **f** [19]. Briefly, SVR defines a hyperplane that captures as many data points as possible, given defined constraints, and reduces overfitting by only penalizing data points outside a certain error radius (termed support vectors) using a linear "epsilon-insensitive" loss function. The orientation of the hyperplane determines **f**. In the original description of CIBERSORT, the support vectors were genes selected from a signature matrix; however, the CIBERSORT algorithm is completely generalizable and can be applied to diverse genomic features [20]. The parameter ν determines the lower bound of support vectors and the upper bound of training errors. CIBERSORT uses a set of ν values (0.25, 0.5, 0.75) and chooses the value producing the best performance (i.e., the lowest root mean square between **m** and the deconvolution result $\mathbf{f} \times \mathbf{B}$). In addition, ν-SVR incorporates L_2-norm regularization, which minimizes the variance in the weights assigned to highly correlated cell types, thereby mitigating issues owing to multicollinearity.

CIBERSORT also allows users to create a custom signature matrix. Differentially expressed genes between cell types of interest are identified by a two-sided unequal variance *t*-test corrected for multiple hypothesis testing. A feature selection step is then performed to minimize the condition number, a matrix property that captures how well the linear system tolerates input variation and noise. For signature matrices comprised exclusively of immune cell types, there is an option to filter non-hematopoietic and cancer-specific genes to reduce the influence of non-immune cells on deconvolution results. By choosing features that minimize the condition number, CIBERSORT improves the stability of the signature matrix and further reduces the impact of multicollinearity. Additional details of the CIBERSORT method can be found in the original publication [17].

2 Materials

The general workflow for analyzing RNA admixtures with CIBERSORT consists of two key input files (*see* Fig. 1):

1. The "mixture file" is a single tab-delimited text file containing 1 or more GEPs of biological mixture samples (*see* Table 1). The first column contains gene names and should have "Name" (or similar) as a column header (i.e., in the space occupying column 1, row 1). Multiple samples may be analyzed in parallel, with the remaining columns (2, 3, etc.) dedicated to mixture GEPs, where each row represents the expression value for a given gene and the column header is the name of the mixture sample. Note that the mixture file and the signature matrix must share the same naming scheme for gene identifiers.
2. The "signature matrix" is a tab-delimited text file consisting of sets of "barcode genes" whose expression values collectively

Table 1
Format of input mixture files (tab separated plain text)

Gene_symbol (required)	Mixture 1	Mixture 2	...
Gene1			
Gene2			
...			

define unique gene expression signatures for each cell subset of interest. The file format is similar to the mixture file, with gene names in column 1. The remaining columns consist of signature GEPs from individual cell subsets. A validated leukocyte gene signature matrix (LM22) is available for the deconvolution of 22 functionally defined human hematopoietic subsets. LM22 was generated using Affymetrix HGU133A microarray data [17] and has been rigorously tested on Affymetrix HGU133 and Illumina Beadchip platforms. For the application of LM22 to RNA-Seq data, *see* **Note 1**.

Importantly, all expression data should be non-negative, devoid of missing values, and represented in non-log linear space. For Affymetrix microarrays, a custom chip definition file (CDF) is recommended (*see* Subheading 3.2.2) and should be normalized with MAS5 or RMA. Illumina Beadchip and single color Agilent arrays should be processed as described in the limma package. Standard RNA-Seq expression quantification metrics, such as fragments per kilobase per million (FPKM) and transcripts per kilobase million (TPM), are suitable for use with CIBERSORT.

In the sections below, we illustrate how CIBERSORT can be used to analyze complex tissues, whether profiled by microarray (Subheading 3.2) or RNA-Seq (Subheadings 3.3 and 4). We also provide instructions for custom signature matrix creation (Subheading 3.3). Although this protocol focuses on the deconvolution of gene expression data, CIBERSORT can be applied to other genomic data types, such as ATAC-Seq, provided that data from purified components are available on the same platform. Public genomic data repositories include the NIH Gene Expression Omnibus database (GEO, http://www.ncbi.nlm.nih.gov/geo/) and the NIH Genomic Data Commons (https://gdc.cancer.gov/). *See* Subheadings 3.3 and **Note 1** for more details.

All files necessary for this protocol, including R and Java implementations of CIBERSORT, can be downloaded through the links provided herein or from the CIBERSORT website (http://cibersort.stanford.edu). The following packages are required for the standalone R version: "e1071," "parallel," and "preprocessCore." The Java version additionally requires the following R packages: "Rserve" and "colorRamps." The "affy," "annotate," and "org.Hs.eg.db" R packages are required only if using the R script from the CIBERSORT website to process Affymetrix CEL files, as described in Subheading 3.2.2.

3 Methods

3.1 Installation

CIBERSORT can be run online (http://cibersort.stanford.edu/) or downloaded for local use, and is freely available for academic non-profit research. While the current R script can be used to run the CIBERSORT deconvolution engine, users wishing to create a custom signature matrix will need to use the website or the Java executable. To download and install the R dependencies described in Materials, run the following commands from an R terminal:

Within R

```
> install.packages('e1071') #R and Java versions.
> source(http://bioconductor.org/biocLite.R).
> biocLite('parallel') #R and Java versions.
> biocLite('preprocessCore') #R and Java versions.
> biocLite('Rserve') #Java version only.
> biocLite('colorRamps') #Java version only.
> biocLite('affy') # used to normalize Affymetrix CEL files (Subheading 3.2.2).
> biocLite('annotate') # used to annotate Affymetrix CEL files (Subheading 3.2.2).
> biocLite('org.Hs.eg.db') # used annotate human Affymetrix CEL files (Subheading 3.2.2).
```

3.2 Enumerating TIL Subsets with LM22

LM22 is a signature matrix file consisting of 547 genes that accurately distinguish 22 mature human hematopoietic populations isolated from peripheral blood or in vitro culture conditions, including seven T cell types, naive and memory B cells, plasma cells, NK cells, and myeloid subsets. LM22 was designed and extensively validated using gene expression microarray data, but is also applicable to RNA-Seq data for hypothesis generation (*see* **Note 1**). Here, we illustrate how to prepare Affymetrix microarray data for use with LM22, and how to run CIBERSORT with LM22 to characterize the leukocyte composition of prostate biopsies obtained from patients with prostate cancer and from healthy subjects. To follow the examples in this section, download GSE55945 CEL files from GEO (https://www.ncbi.nlm.nih.gov/geo/download/?acc=GSE55945&format=file). Processed data for GSE55945 can be downloaded from the CIBERSORT website.

3.2.1 General Tips for Mixture File Preparation

Gene expression data must be preprocessed as specified in Subheadings 2 and 3.2.2. Because LM22 uses HUGO gene symbols (e.g., *CD8A*, *MS4A1*, *CTLA4*, etc.), all mixture files need to possess matching HUGO identifiers. *See* **Note 2** for using non-HUGO gene symbols. Importantly, all expression values should be in non-log (i.e., linear) space with positive numerical values and no missing data. Not all signature matrix genes need to be present in the mixture expression data, but performance will improve with the presence of more signature genes.

3.2.2 Preparation of Affymetrix CEL Files

The CIBERSORT website provides an R script to convert Affymetrix CEL files, the raw data format for Affymetrix microarray experiments, into a tabular format that is ready for analysis with CIBERSORT (Menu>Download). All packages specified in the *Installation* section will need to be downloaded, along with a custom CDF from BrainArray (http://brainarray.mbni.med.umich.edu/Brainarray/Database/CustomCDF/20.0.0/entrezg.asp). The custom CDF must be compatible with the microarray platform used to profile the mixtures (e.g., for HGU133 Plus 2.0, download hgu133plus2hsentrezgcdf_20.0.0.tar.gz); the latest entrezg version is always recommended. Download the custom CDF and run the following terminal command to install the R library:

```
sudo R CMD INSTALL downloaded_customCDF_filename.tar.gz
```

The user is advised to run this step on a machine with root access or a self-contained R environment like RGui. Next, navigate to the directory containing raw Affymetrix CEL files (GSE55945 in this example) and run CEL_to_mixture.R, an R script that should be placed in the same folder as the CEL files. The script will output a correctly formatted CIBERSORT mixture file named: *NormalizedExpressionArray.customCDF.txt*. For this example, rename to "prostate_cancer.txt."

3.2.3 Running CIBERSORT

Before running CIBERSORT, all mixture files need to be uploaded (Menu > "Upload Files"). The user needs to select "Mixture" when uploading mixture files. After uploading the correctly formatted mixture file (e.g., prostate_cancer.txt) to the website, go to "Run CIBERSORT" under Menu (*see* Fig. 2). Select "LM22 (22 immune cell types)" for "Signature gene file." When clicking "Mixture file," the uploaded mixture file will be one of the options. Select "Run" after choosing both the mixture file of interest and a permutation number. At least 100 permutations are recommended to achieve statistical rigor.

To run CIBERSORT locally in R, navigate to the directory containing the CIBERSORT.R script, and run the following commands within the R terminal:

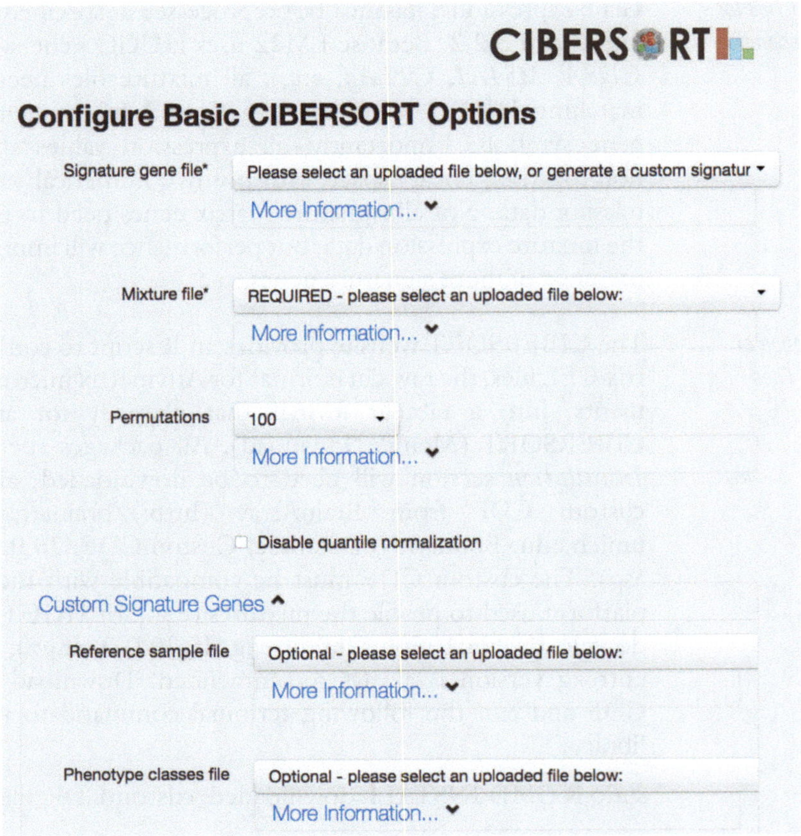

Fig. 2 CIBERSORT web interface. All the files except the LM22 gene signature need to be uploaded to the CIBERSORT website before proceeding to this page. When using LM22, the user will need to select the uploaded mixture file and specify "LM22 (22 immune cell types)" for the signature gene file. When creating custom gene signatures, a reference sample file and a phenotype classes file are required, and need to be uploaded to the webserver. For CIBERSORT to generate a meaningful *p*-value, we recommend at least 100 permutations; however, this parameter can be set to a small number for exploratory analyses

```
> source('CIBERSORT.R')
> results <- CIBERSORT('sig_matrix_file.txt','mixture_file.txt',
   perm=100, QN=TRUE)
```

Deconvolution output will be saved to a *results* object in R and written to disk as *CIBERSORT-results.txt* in the same directory.

In this example, *sig_matrix_file.txt* should be "LM22.txt" (obtain under Menu>Download); *mixture_file.txt* should be "prostate_cancer.txt"; *perm* is an integer number for the number of permutations; and *QN* is a Boolean value (TRUE or FALSE) for performing quantile normalization. QN is set to TRUE by default and recommended when the gene signature matrix is derived from several different studies or sample batches.

3.2.4 Interpretation of Results

Once the online analysis is complete, the website will output a stacked bar plot (*see* Fig. 3) and a heat map (*see* Fig. 4). The output

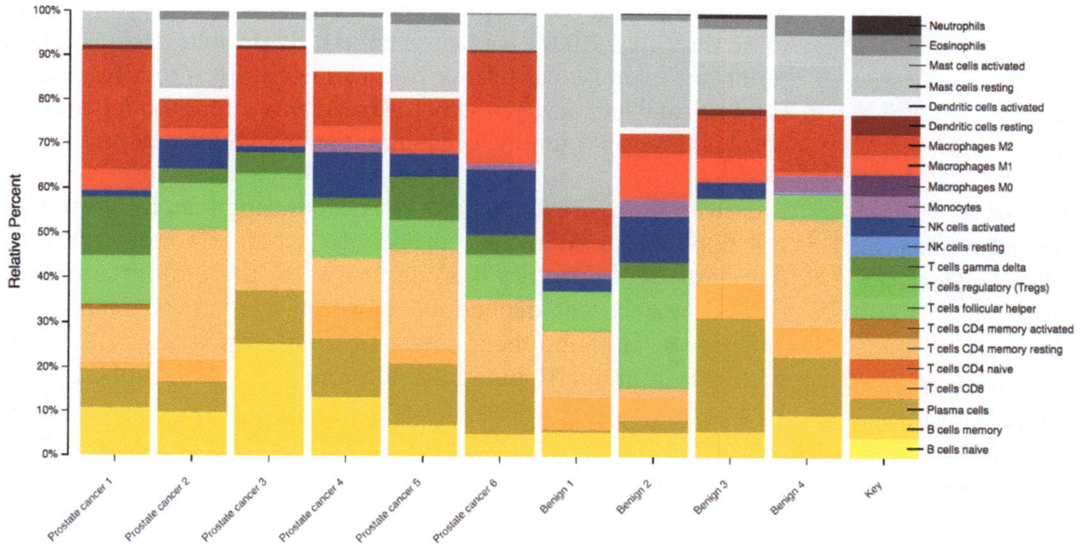

Fig. 3 Inferred composition of 22 immune cell subsets in malignant and normal prostate biopsies (related to Subheading 3.2). The results were generated using CIBERSORT and the built-in LM22 immune cell gene signature, and the stacked bar plot display was automatically generated by the CIBERSORT webserver

Input Sample	B cells	CD8 T cells	CD4 T cells	NK cells	Monocytes	Neutrophils	P-value	Pearson Correlation	RMSE
TCGA.EE.A29N.06A.12R.A18S...	0.341	0.108	0.186	0.007	0.358	0	0.000	0.569	0.822
TCGA.GN.A26A.06A.11R.A18T...	0.012	0.07	0.068	0.029	0.821	0	0.000	0.497	1.028
TCGA.EE.A2MR.06A.11R.A18S...	0.386	0.206	0.143	0	0.266	0	0.000	0.432	0.907
TCGA.ER.A2NG.06A.11R.A18T...	0.02	0.545	0	0.051	0.384	0	0.000	0.419	0.914
TCGA.FR.A8YE.06A.11R.A37K...	0.446	0.01	0.362	0.018	0.165	0	0.000	0.397	0.925
TCGA.EE.A3AF.06A.11R.A18S...	0.099	0.191	0	0.019	0.691	0	0.000	0.366	1.069
TCGA.ER.A19A.06A.21R.A18U...	0.051	0.249	0	0.048	0.653	0	0.000	0.365	1.044
TCGA.EE.A29G.06A.12R.A18T...	0.078	0	0.152	0.02	0.75	0	0.000	0.358	1.114
TCGA.GN.A4U5.01A.11R.A32P...	0	0.522	0	0.031	0.446	0	0.000	0.354	0.964
TCGA.ER.A193.06A.12R.A18S...	0.056	0.325	0	0.025	0.594	0	0.000	0.353	1.022
TCGA.EE.A2GL.06A.11R.A18S...	0.534	0.144	0.13	0	0.193	0	0.000	0.348	0.969
TCGA.XV.AAZV.01A.11R.A40A...	0	0.068	0.068	0.082	0.781	0	0.000	0.348	1.140
TCGA.EB.A3XC.01A.11R.A239...	0.069	0	0.298	0.063	0.57	0	0.000	0.346	1.017
TCGA.EE.A2MJ.06A.11R.A18S...	0.39	0.264	0	0.032	0.313	0	0.000	0.342	0.968
TCGA.EE.A2GP.06A.11R.A18S...	0.084	0	0.146	0.017	0.752	0	0.000	0.342	1.129

Fig. 4 Estimated proportions of six major leukocyte subsets (B cells, CD8 T cells, CD4 T cells, NK cells, monocytes/macrophages, neutrophils) in skin cutaneous melanoma tumor biopsies profiled by The Cancer Genome Atlas (TCGA). The results were determined using a custom RNA-Seq leukocyte signature matrix ("LM6," Subheading 3.3.3), and the heat map figure was generated by the CIBERSORT webserver

includes a p-value for the global deconvolution of each sample. A p-value threshold <0.05 is recommended. By default, deconvolution results are expressed as relative fractions normalized to 1 (e.g., fractions of total leukocyte content). Researchers interested in studying absolute levels of immune cells should refer to Subheading 3.6.

3.3 TIL Characterization with a Custom Signature Matrix

3.3.1 Generation of Expression Profiles for Custom Gene Signature Matrix Creation

A custom signature matrix can be created using data from purified cell populations. While the process to generate a custom matrix from expression profiles is straightforward, the performance of a custom matrix will depend on the quality of the data used to generate it. Immunophenotyping of leukocytes is a dynamic field with new immune populations continuing to be identified. Care should be taken in determining which immune "cell types" should be included in the signature matrix and which canonical markers should be used to isolate these populations. For example, it is clear that the population of "CD4-expressing T lymphocytes" encompasses heterogeneous populations with diverse functional phenotypes including naive, memory, Th1, Th2, Th17, T-regulatory cells, and T follicular helper cells. Replicates for each purified immune cell type are required to gauge variance in the expression profile (*see* **Note 4** for further details). The platform and methods used to generate data for the signature matrix ideally should be identical to that applied to the analysis of the mixture samples. *See* **Note 3** for analyzing murine data. While SVR is robust to unknown cell populations, performance can be adversely affected by genes that are highly expressed in a relevant unknown cell population (e.g., in the malignant cells) but not by any immune components present in the signature matrix. A simple option implemented in CIBERSORT to limit this effect is to remove genes highly expressed in non-hematopoietic cells or tumor cells. If expression data is available from purified tumor cells for the malignancy to be studied, this can be used as a guideline to filter other confounding genes from the signature matrix.

3.3.2 Input Data Preparation

The mixture input data format for custom signature gene matrix option is identical to the analysis with the LM22 signature gene matrix (Subheading 3.2.1). To generate the custom signature gene matrix, the user needs to provide a reference sample file containing the GEPs for each purified immune population of interest, and a phenotype class file assigning the profiles to each phenotypic type of immune cell to be included in the signature matrix. The expression data in the reference sample file should be in non-log (i.e., linear) space with genes listed in the rows and reference populations listed in columns. The phenotype class file lists the desired cell populations in the signature matrix listed in rows and the purified reference samples contained in the reference sample file listed in columns (refer to the CIBSERORT website manual for more details). These must be listed in the exact same order as the reference sample file. The cells are used to assign phenotypic classes to

Table 2
Format of input files to generate reference files and class files necessary for custom gene signatures (tab separated plain text)

Gene symbol (required)	Cell type Name1	Cell type Name1	Cell type Name1	Cell type Name2	Cell type Name2	Cell type Name2	...
Gene1							
Gene2							
...							

each purified reference sample. Importantly, all cell types should be represented by at least two replicates in order to identify genes with significantly differential expression (*see* **Note 4**).

For ease of use, we have created an R script to generate both intermediate files (the script is available from the CIBERSORT website). Gene expression data for each purified sample should be formatted similarly to the mixture input data (*see* Table 2) and each replicate of the same cell type must be labeled with the identical phenotypic class name. To run the script, execute the following command:

```
Rscript generate_ref_and_class.R your_input_mixture_file.txt
```

The script will produce two output files, both of which are required to build a signature matrix: *class_file.input.txt* (i.e., phenotype class file) and *reference_file.input.txt* (i.e., reference sample file).

3.3.3 Creating the Signature Matrix

In the following two sections, we describe how to create a custom leukocyte signature matrix and apply it to study cellular heterogeneity and TIL survival associations in melanoma tumors profiled by The Cancer Genome Atlas (TCGA). Readers can follow along by creating "LM6," a leukocyte RNA-Seq signature matrix comprised of six peripheral blood immune subsets (B cells, CD8 T cells, CD4 T cells, NK cells, monocytes/macrophages, neutrophils; GSE60424 [21]). Key input files are provided on the CIBERSORT website ("Menu>Download").

A custom signature file can be created by uploading the Reference sample file and the Phenotype classes file (Subheading 3.3.2) to the online CIBERSORT application (*see* Fig. 2) or can be created using the downloadable Java package. To build a custom gene signature matrix with the latter, the user should download the Java package from the CIBERSORT website and place all relevant files under the package folder. To link Java with R, run the following in R:

Within R:

```
> library(Rserve)
```

> Rserve(args="--no-save")

Command line:

> *java -Xmx3g -Xms3g -jar CIBERSORT.jar -M Mixture_file -P Reference_sample_file -c phenotype_class_file -f*

The last argument (-f) will eliminate non-hematopoietic genes from the signature matrix and is generally recommended for signature matrices tailored to leukocyte deconvolution. The user can also run this step on the website by choosing the corresponding reference sample file and phenotype class file (*see* Fig. 2). The CIBERSORT website will generate a gene signature matrix located under "Uploaded Files" for future download.

Following signature matrix creation, quality control measures should be taken to ensure robust performance (*see* "Calibration of in silico TIL profiling methods" in Newman et al.) [18]. Factors that can adversely affect signature matrix performance include poor input data quality, significant deviations in gene expression between cell types that reside in different tissue compartments (e.g., blood versus tissue), and cell populations with statistically indistinguishable expression patterns. Manual filtering of poorly performing genes in the signature matrix (e.g., genes expressed highly in the tumor of interest) may improve performance.

To benchmark our custom leukocyte matrix (LM6), we compared it to LM22 using a set of TCGA lung squamous cell carcinoma tumors profiled by RNA-Seq and microarray ($n = 130$ pairs). Deconvolution results were significantly correlated for all cell subsets shared between the two signature matrices ($p < 0.0001$). Notably, since LM6 was derived from leukocytes isolated from peripheral blood [21, 22], we restricted the CD4 T cell comparison to naive and resting memory CD4 T cells in LM22. Once validation is complete, a CIBERSORT signature matrix can be broadly applied to mixture samples as described in Subheading 3.3 (e.g., *see* Fig. 4).

3.4 Correlating TIL Levels with Clinical Outcomes

Associations with clinical indices and outcomes are commonly assessed using a log-rank test for binary variables and Cox proportional hazards regression for continuous variables. There are a number of freely available tools for such analyses. We typically use the R "survival" package or the python "lifelines" package. To illustrate TIL survival analysis in primary tumor samples, we applied LM6 (Subheading 3.3.3) to 473 TCGA skin cutaneous melanoma tumor samples profiled by RNA-Seq (*see* Fig. 4). We then analyzed the influence of estimated CD8 T cell levels on overall survival. Higher levels of CD8 T lymphocytes were associated with favorable overall survival in both dichotomous (Fig. 5) and continuous models ($p = 0.013$, Cox regression), consistent with previous studies [1, 2].

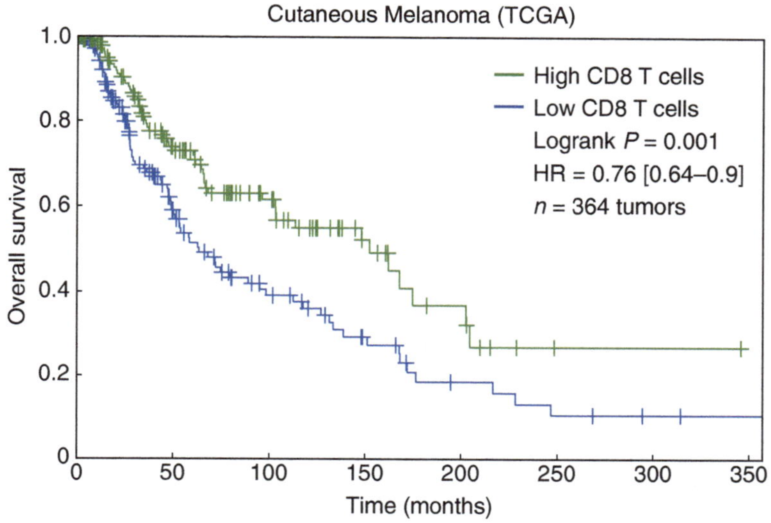

Fig. 5 Association between inferred tumor-infiltrating CD8 T cell content and overall survival in patients with skin cutaneous melanoma profiled by TCGA (related to Subheading 4). Estimated CD8 T cell levels were stratified by a median split, and the separation between survival curves was evaluated using a log-rank test. Only patients with available survival data and with a significant CIBERSORT *p*-value (<0.05) were considered for this analysis ($n = 364$). HR, hazard ratio. 95% confidence intervals for the hazard ratio are shown in *brackets*

3.5 Use of CIBERSORT to Infer Absolute TIL Levels

By default, CIBERSORT estimates the relative fraction of each cell type in the signature matrix, such that the sum of all fractions is equal to 1 for a given mixture sample. CIBERSORT can also be used to produce a score that quantitatively measures the overall abundance of each cell type (as described in "Analysis of deconvolution consistency" in Newman et al.) [17]. Briefly, the absolute immune fraction score is estimated by the median expression level of all genes in the signature matrix divided by the median expression level of all genes in the mixture. Using this metric coupled with LM22, we have found that CIBERSORT effectively captures overall immune content in RNA-Seq and microarray datasets when benchmarked against other methods. These include H&E staining and computational inference by ESTIMATE [23], a previously published method for determining overall immune content in tumor expression profiles.

Absolute results can be easily accessed from the CIBERSORT website by toggling the output between relative and absolute modes in the Results page (*see* online manual for details). When using the R script (Subheading 3.2.3), the user should download the latest version of the script and set "absolute=TRUE." For example:

```
results <- CIBERSORT('sig_matrix_file.txt','mixture_file.txt',
    perm=100, absolute=TRUE)
```

3.6 Conclusion

CIBERSORT is an in silico approach for characterizing cell subsets of interest in high-dimensional genomic data derived from bulk tissue samples. Given a validated signature matrix, CIBERSORT can profile compositional differences in a standardized manner, facilitating robust and reproducible analyses of cellular heterogeneity in both newly measured and archived genomic datasets, fresh/frozen tissue biopsies, and fixed clinical specimens. Since CIBERSORT is platform agnostic, it can be applied to diverse genomic data types other than mRNA, including DNA methylation, microRNA, proteomic, and chromatin accessibility profiles. CIBERSORT is therefore a versatile framework for tissue characterization, with applications for identifying predictive and prognostic cellular biomarkers, and novel therapeutic targets.

4 Notes

1. CIBERSORT is platform agnostic and can be applied to any genomic admixture that satisfies its mathematical model (Subheading 1.2), including mixtures profiled by RNA-Seq. Although LM22 was derived and originally validated using microarray data, we have observed significant correlations for most of LM22 populations on paired microarray/RNA-Seq TCGA datasets, suggesting that it is reasonable to apply LM22 to RNA-Seq data for hypothesis generation. Nevertheless, if significant subsets of genes within LM22 are not present in the RNA-Seq summarization, the deconvolution of the corresponding cell types may be adversely affected. To avoid such potential degradation of deconvolution, we strongly recommend including as many genes as possible within LM22 (e.g., components of BCR and TCR genes). Separately, it has been noted that the RNA-Seq mixture samples analyzed by the LM22 matrix will have a higher frequency of samples with p-values above 0.05. This is largely due to the differing dynamic range of RNA-Seq and microarray data, and may not accurately reflect the quality of the deconvolution results. Users should therefore exercise caution in interpreting cross-platform *p*-values. An RNA-Seq derived signature matrix analogous to LM22 is currently being developed with an expanded set of immune populations by the authors.
2. HUGO gene symbols are required as input when the LM22 signature matrix is used. However, CIBERSORT is not restricted to HUGO gene symbols, and users working with custom gene signatures can employ any set of unique alphanumeric identifiers, provided they are consistent between the signature matrix and the mixture file. When a user is not using HUGO gene symbols, the non-hematopoietic gene filtering

functions will not work since these lists are represented in HUGO format.

3. Applying the LM22 matrix to a murine tumor may be unreliable due to cross-species differences in immune biology. A user working with murine data should consider building a custom signature matrix with either publicly available data (e.g., ImmGen; https://www.immgen.org/) or in-house data.
4. The CIBERSORT model builds a gene signature matrix by minimizing gene expression variance within the same cell type and by maximizing variance between cell types; it is therefore important to use data replicates. Cell types should be isolated from the same tissue type or culture conditions, and biological replicates are recommended to help the model capture donor-to-donor variations. To increase statistical power, we recommend using three or more replicates for each cell subset.

Acknowledgments

We would like to thank David Steiner, M.D., Ph.D. for his assistance in generating the RNA-Seq derived signature matrix. This work is supported by grants from the Doris Duke Charitable Foundation (A.A.A.), the Damon Runyon Cancer Research Foundation (A.A.A.), the B&J Cardan Oncology Research Fund (A.A.A.), the Ludwig Institute for Cancer Research (A.A.A.), NIH grant 1K99CA187192-01A1 (A.M.N.), NIH grant PHS NRSA 5T32 CA09302-35 (A.M.N.), US Department of Defense grant W81XWH-12-1-0498 (A.M.N.), a grant from the Siebel Stem Cell Institute and the Thomas and Stacey Siebel Foundation (A.M.N.), an NIH/Stanford MSTP training grant (B.C.), and a PD Soros Fellowship (B.C.).

References

1. Fridman WH, Pagès F, Sautès-Fridman C, Galon J (2012) The immune contexture in human tumours: impact on clinical outcome. Nat Rev Cancer 12(4):298–306. https://doi.org/10.1038/nrc3245
2. Gentles AJ, Newman AM, Liu CL, Bratman SV, Feng W, Kim D, Nair VS, Xu Y, Khuong A, Hoang CD, Diehn M, West RB, Plevritis SK, Alizadeh AA (2015) The prognostic landscape of genes and infiltrating immune cells across human cancers. Nat Med 21 (8):938–945. https://doi.org/10.1038/nm.3909
3. Tumeh PC, Harview CL, Yearley JH, Shintaku IP, Taylor EJ, Robert L, Chmielowski B, Spasic M, Henry G, Ciobanu V, West AN, Carmona M, Kivork C, Seja E, Cherry G, Gutierrez AJ, Grogan TR, Mateus C, Tomasic G, Glaspy JA, Emerson RO, Robins H, Pierce RH, Elashoff DA, Robert C, Ribas A (2014) PD-1 blockade induces responses by inhibiting adaptive immune resistance. Nature 515 (7528):568–571. https://doi.org/10.1038/nature13954
4. Herbst RS, Soria JC, Kowanetz M, Fine GD, Hamid O, Gordon MS, Sosman JA, McDermott DF, Powderly JD, Gettinger SN, Kohrt HE, Horn L, Lawrence DP, Rost S, Leabman M, Xiao Y, Mokatrin A, Koeppen H, Hegde PS, Mellman I, Chen DS, Hodi FS (2014) Predictive correlates of

response to the anti-PD-L1 antibody MPDL3280A in cancer patients. Nature 515 (7528):563–567. https://doi.org/10.1038/nature14011

5. Ji RR, Chasalow SD, Wang L, Hamid O, Schmidt H, Cogswell J, Alaparthy S, Berman D, Jure-Kunkel M, Siemers NO, Jackson JR, Shahabi V (2012) An immune-active tumor microenvironment favors clinical response to ipilimumab. Cancer Immunol Immunother 61(7):1019–1031. https://doi.org/10.1007/s00262-011-1172-6
6. Tung JW, Heydari K, Tirouvanziam R, Parks DR, Herzenberg LA, Herzenberg LA (2007) Modern flow cytometry: a practical approach. Clin Lab Med 27(3):453. https://doi.org/10.1016/j.cII.2007.05.001
7. Abbas AR, Wolslegel K, Seshasayee D, Modrusan Z, Clark HF (2009) Deconvolution of blood microarray data identifies cellular activation patterns in systemic lupus erythematosus. PLoS One 4(7):e6098. https://doi.org/10.1371/journal.pone.0006098
8. Curtis C, Shah SP, Chin SF, Turashvili G, Rueda OM, Dunning MJ, Speed D, Lynch AG, Samarajiwa S, Yuan Y, Gräf S, Ha G, Haffari G, Bashashati A, Russell R, McKinney S, Langerød A, Green A, Provenzano E, Wishart G, Pinder S, Watson P, Markowetz F, Murphy L, Ellis I, Purushotham A, Børresen-Dale AL, Brenton JD, Tavaré S, Caldas C, Aparicio S, Group M (2012) The genomic and transcriptomic architecture of 2,000 breast tumours reveals novel subgroups. Nature 486(7403):346–352. https://doi.org/10.1038/nature10983
9. Ascierto ML, Kmieciak M, Idowu MO, Manjili R, Zhao Y, Grimes M, Dumur C, Wang E, Ramakrishnan V, Wang XY, Bear HD, Marincola FM, Manjili MH (2012) A signature of immune function genes associated with recurrence-free survival in breast cancer patients. Breast Cancer Res Treat 131 (3):871–880. https://doi.org/10.1007/s10549-011-1470-x
10. Mann GJ, Pupo GM, Campain AE, Carter CD, Schramm SJ, Pianova S, Gerega SK, De Silva C, Lai K, Wilmott JS, Synnott M, Hersey P, Kefford RF, Thompson JF, Yang YH, Scolyer RA (2013) BRAF mutation, NRAS mutation, and the absence of an immune-related expressed gene profile predict poor outcome in patients with stage III melanoma. J Invest Dermatol 133(2):509–517. https://doi.org/10.1038/jid.2012.283
11. Galon J, Angell HK, Bedognetti D, Marincola FM (2013) The continuum of cancer immunosurveillance: prognostic, predictive, and mechanistic signatures. Immunity 39 (1):11–26. https://doi.org/10.1016/j.immuni.2013.07.008
12. Galon J, Costes A, Sanchez-Cabo F, Kirilovsky A, Mlecnik B, Lagorce-Pagès C, Tosolini M, Camus M, Berger A, Wind P, Zinzindohoué F, Bruneval P, Cugnenc PH, Trajanoski Z, Fridman WH, Pagès F (2006) Type, density, and location of immune cells within human colorectal tumors predict clinical outcome. Science 313(5795):1960–1964. https://doi.org/10.1126/science.1129139
13. Tosolini M, Kirilovsky A, Mlecnik B, Fredriksen T, Mauger S, Bindea G, Berger A, Bruneval P, Fridman WH, Pagès F, Galon J (2011) Clinical impact of different classes of infiltrating T cytotoxic and helper cells (Th1, th2, treg, th17) in patients with colorectal cancer. Cancer Res 71(4):1263–1271. https://doi.org/10.1158/0008-5472.can-10-2907
14. Verhaak RG, Tamayo P, Yang JY, Hubbard D, Zhang H, Creighton CJ, Fereday S, Lawrence M, Carter SL, Mermel CH, Kostic AD, Etemadmoghadam D, Saksena G, Cibulskis K, Duraisamy S, Levanon K, Sougnez C, Tsherniak A, Gomez S, Onofrio R, Gabriel S, Chin L, Zhang N, Spellman PT, Zhang Y, Akbani R, Hoadley KA, Kahn A, Köbel M, Huntsman D, Soslow RA, Defazio A, Birrer MJ, Gray JW, Weinstein JN, Bowtell DD, Drapkin R, Mesirov JP, Getz G, Levine DA, Meyerson M, Network CGAR (2013) Prognostically relevant gene signatures of high-grade serous ovarian carcinoma. J Clin Invest 123(1):517–525. https://doi.org/10.1172/jci65833
15. Liebner DA, Huang K, Parvin JD (2014) MMAD: microarray microdissection with analysis of differences is a computational tool for deconvoluting cell type-specific contributions from tissue samples. Bioinformatics 30 (5):682–689. https://doi.org/10.1093/bioinformatics/btt566
16. Zhong Y, Wan YW, Pang K, Chow LM, Liu Z (2013) Digital sorting of complex tissues for cell type-specific gene expression profiles. BMC Bioinformatics 14:89. https://doi.org/10.1186/1471-2105-14-89
17. Newman AM, Liu CL, Green MR, Gentles AJ, Feng W, Xu Y, Hoang CD, Diehn M, Alizadeh AA (2015) Robust enumeration of cell subsets from tissue expression profiles. Nat Methods 12(5):453–457. https://doi.org/10.1038/nmeth.3337
18. Newman AM, Alizadeh AA (2016) High-throughput genomic profiling of tumor-infiltrating leukocytes. Curr Opin Immunol

41:77–84. https://doi.org/10.1016/j.coi.2016.06.006

19. Scholkopf B, Smola AJ, Williamson RC, Bartlett PL (2000) New support vector algorithms. Neural Comput 12(5):1207–1245
20. Corces MR, Buenrostro JD, Wu BJ, Greenside PG, Chan SM, Koenig JL, Snyder MP, Pritchard JK, Kundaje A, Gkeenleaf WJ, Majeti R, Chang HY (2016) Lineage-specific and single-cell chromatin accessibility charts human hematopoiesis and leukemia evolution. Nat Genet 48(10):1193–1203. https://doi.org/10.1038/ng.3646
21. Linsley PS, Speake C, Whalen E, Chaussabel D (2014) Copy number loss of the interferon gene cluster in melanomas is linked to reduced T cell infiltrate and poor patient prognosis. PLoS One 9(10):Artn E109760. https://doi.org/10.1371/Journal.Pone.0109760
22. Sleasman JW, Leon BH, Aleixo LF, Rojas M, Goodenow MM (1997) Immunomagnetic selection of purified monocyte and lymphocyte populations from peripheral blood mononuclear cells following cryopreservation. Clin Diagn Lab Immunol 4(6):653–658
23. Yoshihara K, Shahmoradgoli M, Martinez E, Vegesna R, Kim H, Torres-Garcia W, Trevino V, Shen H, Laird PW, Levine DA, Carter SL, Getz G, Stemke-Hale K, Mills GB, Verhaak RGW (2013) Inferring tumour purity and stromal and immune cell admixture from expression data. Nat Commun 4:Artn 2612. https://doi.org/10.1038/Ncomms3612

Chapter 13

Systems Biology Approaches in Cancer Pathology

Aaron DeWard and Rebecca J. Critchley-Thorne

Abstract

The complex network of the tissue system, in both pre-neoplastic tissues and tumors, demonstrates the need for a systems biology approach to cancer pathology, in which quantification of key tissue system processes is combined with informatics tools to produce actionable scores to aid clinical decision-making. A systems biology approach to cancer pathology enables integration of key system features that are relevant to diagnoses, patient outcomes, and responses to therapies. Key tissue system features relevant to cancer pathology include molecular and morphologic abnormalities in epithelia, cellular changes in the stroma such as immune infiltrates, and relationships between components of the system, such as interactions and spatial relationships between epithelial and stromal components, and also between specific immune cell subsets. Here, we describe a method for objective quantification of multiple epithelial and stromal biomarkers in the context of tissue architecture to generate a high dimensional tissue profile that can be used to build multivariable predictive models for cancer pathology.

Key words Biomarkers, Multiplexed immunofluorescence, Whole slide fluorescence imaging, Digital pathology, Quantitative image analysis, Cancer systems biology

1 Introduction

Current pathology methods for the assessment of pre-neoplastic and neoplastic tissues have been valuable for many decades, but are limited by subjectivity and observer variability. Digital slide scanning and algorithms for automated scoring of biomarkers stained by immunohistochemistry are gaining traction in clinical laboratories, which will improve workflows and reduce variability [1, 2]. However, the majority of current biomarkers used in cancer pathology testing are markers of epithelial cell processes and abnormalities. The complexity of the tissue system and the important roles of stromal components in the development and progression of cancer, and in the responses of cancer to therapies, highlight the need for a systems biology approach to cancer pathology [3, 4]. Assessment of key tissue system biomarkers can improve on the current diagnostic tools by creating multivariate profiles that capture key molecular and cellular features of the tissue

Louise von Stechow (ed.), *Cancer Systems Biology: Methods and Protocols*, Methods in Molecular Biology, vol. 1711, https://doi.org/10.1007/978-1-4939-7493-1_13, © Springer Science+Business Media, LLC 2018

environment, including relationships between biomarkers [3, 5]. Objective, reproducible measurement of multiple biomarkers per slide can be achieved via automated multiplexed immunofluorescence labeling of four or more biomarkers per slide coupled with standardized whole slide fluorescence scanning. Quantitative image analysis can be used to automatically extract an array of quantitative biomarker and morphologic features from whole slide tissue images, resulting in a multivariate tissue profile. Such profiles can be mined in samples from patient cohorts with clinicpathologic data to identify clinically relevant signatures, and to build diagnostic, prognostic, and predictive models. Here we describe methods for automated immunofluorescent labeling of multiple biomarkers in tissue slides, and standardized whole slide fluorescence imaging to produce tissue images with multiple registered channels of biomarker signal and morphology data. These composite images can be analyzed by tissue image analysis platforms to generate quantitative data on key tissue system features and processes. A systems biology approach to assess multiple epithelial and stromal biomarkers in Barrett's esophagus biopsies is discussed as an application of the method.

2 Materials

2.1 Multiplexed Immunofluorescence Slide Labeling Procedure

1. Slides prepared with 5 μm sections of formalin-fixed paraffin-embedded (FFPE) tissue (*see* **Note 1**).
2. BondRX autostainer (Leica BioSystems).
3. Bond Dewax Solution (Leica BioSystems).
4. 100% ethanol (reagent grade).
5. Bond Epitope Retrieval (ER) Solution 2 (Leica BioSystems).
6. Bond Wash Solution 10× (Leica BioSystems).
7. Image-iT® FX Signal Enhancer (Thermo Fisher).
8. Blocking buffer: Tris-buffered saline, 5% goat serum, 1% glycerol, 0.1% bovine serum albumin, 0.1% cold water fish skin gelatin, 0.04% sodium azide (*see* **Note 2**).
9. Primary antibody cocktails: primary antibodies in blocking buffer at a dilution predetermined by titration to produce optimal signal:noise (*see* **Notes 3** and **4**).
10. Secondary antibody cocktails: Alexa Fluor®-conjugated secondary antibodies raised in goat, specific to the species and species isotypes of the primary antibody cocktail (*see* **Notes 5** and **6**).
11. Hoechst 33342 or equivalent label that emits blue fluorescence when bound to double-stranded DNA.

12. Deionized water.
13. Prolong Gold Antifade Mountant (Thermo Fisher), or similar aqueous mounting medium containing components to protect against photobleaching.
14. Glass coverslips (#1.5).
15. Lens cleaner.
16. Clear nail polish.

2.2 Whole Slide Fluorescence Scanning

1. ScanScope® FL (Leica BioSystems) equipped with:
   (a) BrightLine® Pinkel quadband filter set optimized for DAPI, FITC, TRITC, & Cy5 (FF01-440/521/607/700-25).
   (b) BrightLine® single-band bandpass excitation filters FF01-387/11-25, FF01-485/20-25, FF01-560/25-25 and FF01-650/13-25 (Semrock).
   (c) X-Cite® exacte light source (Lumen Dynamics/Excelitas Technologies Corp.).
   (d) Light source calibration device (*see* **Note 7**) (Lumen Dynamics/Excelitas Technologies Corp.).
2. TetraSpeck Fluorescent Microspheres.
3. FocalCheck Fluorescent Microsphere.
4. ImageScope software (Leica BioSystems).

2.3 Quantitative Image Analysis

1. TissueCypher® Image Analysis Platform (Cernostics).
2. Image Processing Toolbox™ (MathWorks [6]).
3. ImageScope software (Leica BioSystems).

3 Methods

3.1 Multiplexed Immunofluorescence Slide Labeling Procedure

Program the BondRX autostainer to perform the following steps (application volume is 150 μL for all the reagent steps):

1. Bake slides (no reagent), incubation time 30 min, 60 °C.
2. Apply Bond Dewax solution, incubation time 30 min, 72 °C.
3. Reapply Bond Dewax solution, incubation time 0 s, 72 °C.
4. Reapply Bond Dewax solution, incubation time 0 s, ambient temperature (*see* **Note 8**).
5. Apply ethanol, incubation time 0 s, ambient temperature, repeat twice for total of three ethanol washes.
6. Apply Bond Wash (diluted to 1× with deionized water), incubation time 0 s, ambient temperature, repeat for total of three washes with Bond Wash.

7. Apply Bond ER Solution 2, incubation time 0 s, ambient temperature, repeat once.
8. Apply Bond ER Solution 2, incubation time 30 min, 98–100 °C (*see* **Note 9**).
9. Apply Bond ER Solution 2, incubation time 0 s, ambient temperature.
10. Apply Bond Wash, incubation time 0 s, ambient temperature, repeat for total of four washes with Bond Wash.
11. Apply Bond Wash, incubation time 0 s, ambient temperature.
12. Apply Image-iT FX, incubation time 30 min ambient temperature.
13. Apply Bond Wash, incubation time 0 s, ambient temperature, repeat for total of three washes with Bond Wash.
14. Apply blocking buffer, incubation time 30 min, ambient temperature.
15. Apply Bond Wash, incubation time 0 s, ambient temperature, repeat for a total of three washes.
16. Apply primary antibody cocktail, incubation time 60 min, ambient temperature.
17. Apply Bond Wash, incubation time 0 s–1 min, ambient temperature, repeat for total of 3–10 washes with Bond Wash (*see* **Note 10**).
18. Apply secondary antibody cocktail, incubation time 60 min, ambient temperature.
19. Apply Bond Wash, incubation time 0 s–1 min, ambient temperature, repeat for total of 3–10 washes with Bond Wash (*see* **Note 10**).
20. Apply Hoechst 33342, incubation time 3 min, ambient temperature.
21. Apply deionized water, incubation time 0 s, ambient temperature, repeat for a total of three washes.
22. Remove the slides immediately, wipe excess water from slides, allow sections to air dry at ambient temperature protected from light. Once dry mount the slides with aqueous mounting medium and allow curing for at least 12 h at room temperature protected from light (*see* **Note 11**).

3.2 Whole Slide Fluorescence Scanning

1. Calibrate the light source to steady absolute output using the X-Cite® XR2100 Power Meter. An output of 2.2 W will ensure adequate illumination for most imaging applications that can be maintained for 1500–2500 h of scanning depending on the initial attainable wattage of the bulb.

2. Follow the manufacturer's procedure for producing whole slide scans with four registered channels of image data, including:
   (a) Image Hoechst (or equivalent) in the FF01-387/11-25 (or equivalent) channel, Alexa Fluor 488 in the FF01-485/20-25 channel, Alexa Fluor 555 in the FF01-560/25-25 channel and Alexa Fluor 647 in the FF01-650/13-25 channel (*see* **Note 12**). The example images of the p16, AMACR, p53 biomarker panel and HIF-1alpha, CD45RO, CD1a biomarker panel in Barrett's esophagus biopsies are shown in Fig. 1.
   (b) Optimize exposure times on a test set of known negative, intermediate and high controls for each biomarker. Maintain consistent exposure times for all channels on all patient samples (*see* **Note 11**).
   (c) Review images for quality including focus, even illumination across imaging stripes/seams, artifacts, etc., and rescan if necessary to produce high-quality tissue images for quantitative image analysis.
   (d) Verify channel registration periodically using slides mounted with FocalCheck Fluorescent Microspheres (*see* **Note 13**).
   (e) Verify scanner precision periodically, after bulb calibration, and after the replacement of the bulb or light guide, using slides mounted with TetraSpeck Fluorescent Microspheres (*see* **Note 14**).

3.3 Quantitative Image Analysis

We utilize the TissueCypher® Image Analysis Platform, which includes a high performance file reading mechanism based on BigTiff format to decode raw image data, MatLab algorithms for segmenting low level tissue objects such as nuclei, cytoplasm, plasma membrane, and whole cells to allow feature collection at the cellular and subcellular level. It further contains higher order computer vision models for spatial quantification of biomarkers in tissue compartments, such as epithelium and lamina propria, as described by Prichard et al. [4]. There are multiple commercially available tools for quantitative analysis of digital tissue slide images, such as the Image Processing Toolbox™ that provides algorithms and functions for image processing, image analysis and development of algorithms for application-specific features.

Image analysis to create a multivariable tissue systems profile should include:

1. **Handling of image artifacts**. Artifacts such as bubbles, folds, fibers, and out of focus regions can be removed via manual annotation of images or algorithms [7, 8] (*see* **Note 15**).

Fig. 1 Representative images of multiplexed panels of tissue system biomarkers in Barrett's esophagus pinch biopsies. Sections of Barrett's esophagus pinch biopsies were fluorescently immunolabeled for the multiplexed panels of biomarkers described in **Notes 4** and **6**. Whole slide images were acquired at 20× magnification using the ScanScope FL. (**Panels a–d**) (**a**) HIF-1α-*green* (**b**) CD45RO-*red*, (**c**) CD1a-*yellow*, (**d**) HIF-1α-*green*, CD1a-*yellow* overlay demonstrating infiltration of the lamina propria by cells expressing HIF-1α, which indicates stromal angiogenesis, and also memory lymphocytes and dendritic cells. (**Panels e–h**) (**e**) HIF-1α-*green* (**f**) CD45RO-red, (**g**) CD1a-*yellow*, (**h**) HIF-1α-*green*, CD45RO-*red*, CD1a-*yellow* overlay, providing an additional example of infiltration of the lamina propria by cells expressing HIF-1α, memory lymphocytes and dendritic cells. (**Panels i–l**) (**i**) p16-*green*, (**j**) AMACR-*red*, (**k**) p53-*yellow*, (**l**) p16-*green*, AMACR-*red*, p53-*yellow* overlay showing loss of p16, focal overexpression of AMACR and overexpression of p53. (**Panels m–p**) (**m**) p16-*green*, (**n**) AMACR-*red*, (**o**) p53-*yellow*, (**p**) p16-*green*, AMACR-*red*, p53-*yellow* overlay showing normal/positive expression of p16, multi-focal overexpression of AMACR and loss of p53. Hoechst shown in *blue* in all panels

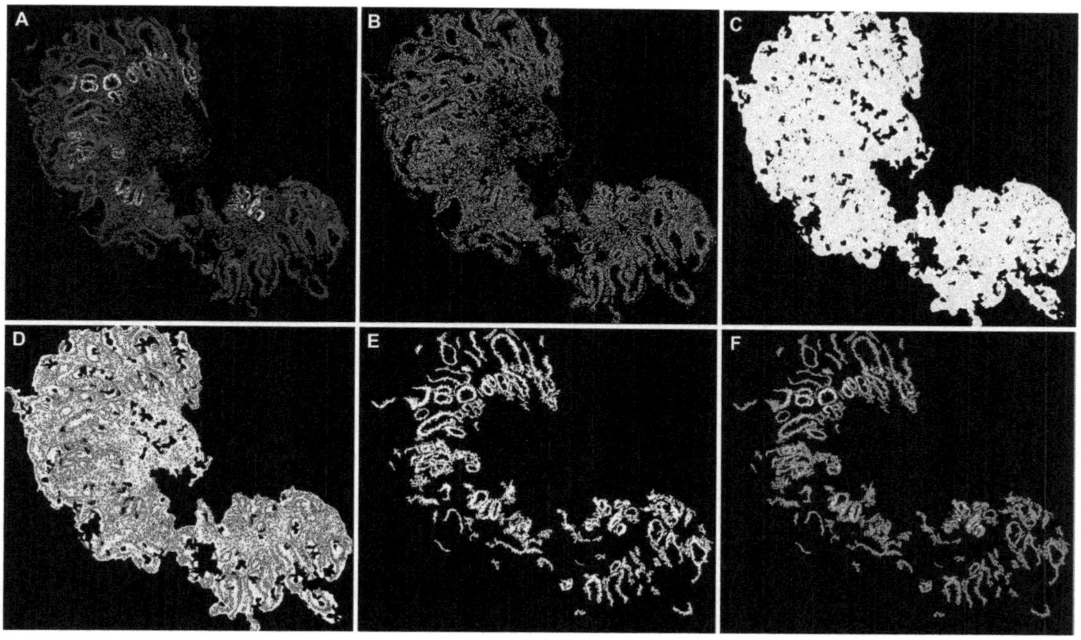

Fig. 2 Cellular object segmentation and tissue structure segmentation to enable quantitative, contextual feature measurements. The TissueCypher® Image Analysis Platform was used to detect a Barrett's esophagus biopsy and segment subcellular compartments and tissue objects. (**a**) Barrett's esophagus biopsy labeled for p16 (*green*), AMACR (*red*), p53 (*yellow*), and Hoechst (*blue*). (**b**) Segmentation of nuclei objects based on the Hoechst channel. (**c**) Segmentation of cell objects containing nuclei by first creating a distance map to which the watershed operation was applied, and then performing connected components labeling, as previously described [4]. (**d**) Segmentation of cytoplasm by subtracting the nuclei mask shown in Panel **b** from the cell mask shown in Panel **c**. (**e**) A nuclei cluster mask was produced via Gaussian smoothing of the Hoechst signal, rank order filter, image thresholding, morphological operations, and connected components labeling, as previously described [4]. (**f**) p53 signal (*yellow*) was measured within the segmented nuclei clusters

2. **Segmentation of low level objects such as nuclei (based on Hoechst signal), cytoplasm, plasma membrane, and whole cells**. Examples of object segmentation masks are shown in Fig. 2. Object segmentation allows collection of quantitative biomarker feature data at the cellular and subcellular levels, which in turn allows calculation of basic intensity measurements on biomarkers (mean, sum, standard deviation, moment, etc.), co-expression of multiple biomarkers, ratios of biomarkers between subcellular compartments, gating on subpopulations of cells with overexpression/lack of expression of multiple biomarkers, spatial arrangements of cells expressing 1 or more biomarkers, texture, nuclear morphology, etc. [4] (*see* **Note 16**). Examples of cell object-based features extracted from the biomarker panels described in this method include p53 mean intensity in nuclei objects, and nuclear area in cell objects with p16-loss and p53-overexpression. Both the features have been shown to have diagnostic significance in

distinguishing between Barrett's esophagus biopsies with high grade dysplasia versus non-dysplastic reactive atypia. This separation has prognostic significance in predicting risk of future progression in patients with Barrett's esophagus [4, 5]. Additional examples include CD45RO sum intensity in plasma membrane structures, using two-dimensional anisotropic diffusion, histogram equalization, and conversion to binary using the CD45RO signal [4], which has prognostic significance in Barrett's esophagus [5].

3. **Computer vision models for segmentation of tissue structures and components, such as epithelium, lamina propria, tumor nests, etc.** Computer vision models allow localization of biomarker signals to specific compartments and collection of feature data in the context of tissue architecture [4]. Examples of computer vision model/tissue structure-based features with diagnostic and/or prognostic significance in Barrett's esophagus include mean intensity of p53 in nuclei clusters. A nuclei cluster mask can be developed using the Hoechst (or equivalent dye) signal as we have previously described in detail [4]. An example nuclei cluster mask is shown in Fig. 2. Features derived from cell-based objects can also be localized to rectangular regions of tissue images to create microenvironment-based features that capture localized or focal biomarker abnormalities. Such features collected across whole slides can be summarized to quantify the cell-object biomarker features in, for example, the top scoring 5% of microenvironments on each slide. The size of the rectangular regions should be optimized to the specific application; we used regions of 161×161 pixels to capture focal overexpression of AMACR, and to detect clusters of stromal cells expressing HIF-1alpha in Barrett's esophagus biopsies, both of which have diagnostic and prognostic significance [4, 5].

4. **Statistical analysis of image-derived features.** Image analysis as described above generates multiple measurements per biomarker and when applied to multiplexed panels of biomarkers will generate a high dimensional feature data set. When performed on samples from an appropriately designed patient cohort with corresponding clinicopathologic data, the high dimensional data can be mined with the aid of bioinformatics to identify quantitative features relevant to diagnosis, prognosis, and responses to therapies. Combinations of relevant features can be used to build multivariable diagnostic, prognostic, and predictive models that integrate data on key tissue system processes to produce clinically actionable information. We have previously described an application of this approach in detail [5], in which 13,538 quantitative image analysis features extracted from 14 candidate protein-based biomarkers and Hoechst were mined in a training cohort of Barrett's esophagus patients with clinical outcome data in order to identify

prognostic features. A risk prediction model was built that integrated 15 of the prognostic features, which were derived from nine of the candidate protein biomarkers and Hoechst, into an individualized risk score that is correlated with risk of future progression to high grade dysplasia or esophageal adenocarcinoma. The pre-specified risk prediction model was validated on an independent, multi-institutional cohort of patients with Barrett's esophagus, demonstrating significant risk stratification of patients who progressed and patients who did not progress to high-grade dysplasia or EAC, and showing prognostic power that was independent of current clinical variables, including pathologic diagnosis provided by a gastrointestinal subspecialist [5].

Sample workflow to generate and apply quantitative feature data from AMACR, one of several representative biomarkers used for risk stratification in Barrett's Esophagus:

1. Open ImageScope software to view the digital image of a slide containing fluorescently labeled AMACR. Author annotations to remove artifacts and/or select regions of interest using the pen tool available within the software. For example, dust fibers brightly autofluoresce, and have the potential to be interpreted as positive signal. Annotating out a dust fiber will prevent its incorporation into subsequent image analysis.
2. Focal overexpression of AMACR in Barrett's Esophagus tissue is correlated with an increased risk of disease progression, whereas no/low expression is associated with a low risk of progression. An example image containing focal AMACR expression is shown in Fig. 1. The TissueCypher® Image Analysis Platform reads the annotated digital slide image, detects tissue fragments, and segments cell-based objects (Fig. 2). To quantify focal AMACR expression the software separates the whole image into 161 × 161 pixel tiles/microenvironments. The TissueCypher® Image Analysis software quantifies the fluorescence intensity of AMACR in the cell-based objects, e.g., plasma membrane objects, in each tile. The fluorescence intensity can be quantified as mean, sum, standard deviation, n^{th} percentile, etc. The top five tiles, or top 5% of tiles based on highest AMACR intensity, are averaged to generate a feature value for AMACR.
3. The feature derived from AMACR can be evaluated in a patient cohort with corresponding clinicopathologic data, e.g., a cohort including patients whose Barrett's esophagus progressed to esophageal adenocarcinoma (cases) and patients who Barrett's esophagus did not progress during surveillance (controls). Conditional logistic regression or Cox regression can be used to compare the feature in cases versus controls, and

to return a coefficient to weigh the feature in a univariate risk prediction analysis. The feature can also be entered into multivariable model building along with features derived from other biomarkers, as described above.

4 Notes

1. 5 μm is the optimal section thickness for tissue image object segmentation since it is thick enough to include an optimal number of whole nuclei, yet thin enough to avoid too many 3D overlaps. Prior to labeling store slides at 2–10 °C under vacuum to protect epitopes.
2. The serum type in the blocking buffer should match the species in which the secondary antibodies were generated. We use secondary antibodies raised in goat and thus use blocking buffer containing goat serum. The concentrations of blocking buffer components such as BSA, glycerol, cold water fish skin gelatin, and sodium azide can be titrated to minimize nonspecific labeling, depending on the antibodies and tissue-type used.
3. A range of primary antibody dilutions should be tested on tissues or cell line controls with known negative, intermediate, and high expression of the target biomarker. Fluorescence signal should be quantified in tissue areas/cells with positive expression (signal) and tissue areas/cells with negative/background or nonspecific labeling (noise). The signal:noise should be calculated for each dilution to determine the appropriate dilution for the specific application, which in our experience is the dilution that results in signal:noise $\geq$5. Use of in vitro diagnostic (IVD)-labeled antibodies will ensure lot-to-lot reproducibility. Even with IVD-labeled antibodies, new lots should be validated to ensure that the antibody specificity and signal:noise are equivalent between lots. We recommend using sections of FFPE cell lines on slides with at least 1 negative, 1 intermediate, and 1 high expressing cell line for each biomarker assessed. FFPE cell lines can also be utilized as batch controls in each run of patient slides being labeled, imaged, and analyzed. A method for the preparation of FFPE cell line controls has been previously described [9].
4. Primary antibodies within a single cocktail must be raised in different species or different species isotypes. For example: (a) Mouse IgG2a anti-p16 antibody, rabbit IgG anti-AMACR antibody, and mouse IgG2b anti-p53 antibody can be multiplexed within a cocktail to assess p16, AMACR, and p53 expression on a slide. (b) Rabbit IgG anti-HIF-1alpha

antibody, mouse IgG2a anti-CD45RO antibody, and mouse IgG1 anti-CD1a antibody can be multiplexed within a cocktail to assess CD1a, CD45RO, and HIF-1alpha on a slide.

5. Refer to the manufacturer's product information to ensure that the secondary antibodies have been highly cross-adsorbed to minimize cross-reactivity. Protein aggregates may form in the secondary antibody solutions. Therefore, the cocktail of fluorescently conjugated secondary antibodies should be centrifuged at high speed for 3 s prior to use. Only the supernatant should be applied to slides in order to minimize nonspecific labeling.
6. Alexa Fluor 488-conjugated goat anti-mouse IgG2a antibody, Alexa Fluor 555-conjugated goat anti-rabbit IgG, and Alexa Fluor 647-conjugated goat anti-mouse IgG2b can be prepared in a cocktail to detect the p16, AMACR, p53 antibody panel described in **Note 4a**. Alexa Fluor 488-conjugated goat anti-rabbit IgG, Alexa Fluor 555-conjugated goat anti-mouse IgG2a, and Alexa Fluor 647-conjugated goat anti-mouse IgG1 can be prepared in a cocktail to detect the HIF-1alpha, CD45RO, and CD1a antibody **Note 4b**. The fluorescently conjugated antibodies should be used at a dilution pre-determined by titration to produce optimal signal:noise. Dilutions ranging from 1:200–1:400 are used for the secondary antibody cocktails described here.
7. The light source should be equipped with a calibration device to ensure the consistent illumination necessary for quantitative image analysis of biomarkers and morphology.
8. Ambient laboratory temperature and humidity should be monitored and maintained within an established range. Variations in these environmental conditions will affect tissue processing and labeling steps that are performed under ambient conditions, which may increase intra- and inter-run imprecision.
9. The epitope retrieval temperature should be optimized for the specific panel of primary antibodies used. We use 100 °C for the panel containing p16, AMACR, p53 panel, and 98 °C for the HIF-1a, CD45RO, CD1a panel.
10. Longer washing incubation times and/or increased numbers of washes can be used to minimize nonspecific labeling where necessary. We use three 0 s washes post-primary antibody incubation for the p16, AMACR, p53 biomarker panel and ten 1 min washes for the HIF-1a, CD45RO, CD1a panel described here, and three 0 s washes post-secondary antibody incubation.
11. Proper mounting is essential to generation of high-quality tissue images suitable for image analysis. Mounting medium

such as Prolong Gold Antifade Mountant should be at room temperature prior to use. Care should be taken to avoid bubbles, fibers, and particulate matter in the mounting medium as these will result in image artifacts that will interfere with quantitative image analysis. Artifacts in tissue images should be removed via image annotation or algorithms prior to image analysis. Following mounting store slides horizontally at room temperature protected from light for at least 12 h to ensure proper curing of the mounting medium. Seal edges of coverslips with clear nail polish. Thoroughly clean the back of the slide and the outside of the coverslip with lens cleaner prior to slide scanning. Store fluorescently immunolabeled slides at 2–10 °C protected from light.

12. The quadband filter set and single-band bandpass excitation filters (described under Subheading 2.2) are calibrated to separate the DAPI, FITC, TRITC, Cy5, or equivalent fluorophores as recommended in this protocol (Hoechst 33342, Alexa Fluors 488, 555, and 647).
13. Correct image registration is necessary for quantitative image analysis involving measurement of biomarkers across different fluorescent channels, which is required to generate a systems profile of a tissue, which may include co-expression of biomarkers and spatial relationships between biomarkers that are imaged and quantified in different fluorescent channels.
14. Scanner precision should be monitored to ensure consistent excitation within imaging runs and between imaging runs. This is particularly important for clinical studies that can involve imaging of hundreds of patients over many months.
15. Whole slide digital images generated on the ScanScope FL scanner can be annotated to remove artifacts or select regions of interest using ImageScope software prior to reading into image analysis software.
16. Image analysis features should be normalized (centered or standardized) to correct for intra- and inter-run variability.

Acknowledgments

National Cancer Institute of the National Institutes of Health under Award Number R44CA192416. We thank Lia Reese, Bruce Campbell, and Kathleen Repa for technical assistance in the development and validation of the TissueCypher® methodology described in this chapter.

References

1. Pantanowitz L, Valenstein PN, Evans AJ, Kaplan KJ, Pfeifer JD, Wilbur DC, Collins LC, Colgan TJ (2011) Review of the current state of whole slide imaging in pathology. J Pathol Inf 2:36
2. Dennis J, Parsa R, Chau D, Koduru P, Peng Y, Fang Y, Sarode VR (2015) Quantification of human epidermal growth factor receptor 2 immunohistochemistry using the Ventana image analysis system: correlation with gene amplification by fluorescence in situ hybridization: the importance of instrument validation for achieving high (>95%) concordance rate. Am J Surg Pathol 39(5):624–631
3. Gough A, Lezon T, Faeder J, Chennubhotla C, Murphy R, Critchley-Thorne R, Taylor DL (2014) High content analysis and cellular and tissue systems biology: a bridge between cancer cell biology and tissue-based diagnostics. In: Mendelsohn J, Howley PM, Israel MA, Gray JW, Thompson CB (eds) The molecular basis of cancer 4th edition, 4th edn. Elsevier, New York
4. Prichard JW, Davison JM, Campbell BB, Repa KA, Reese LM, Nguyen XM, Li J, Foxwell T, Taylor DL, Critchley-Thorne RJ (2015) TissueCypher: a systems biology approach to anatomic pathology. J Pathol Inf 6:48
5. Critchley-Thorne RJ, Duits LC, Prichard JW, Davison JM, Jobe BA, Campbell BB, Repa KA, Reese LM, Li J, Diehl DL, Jhala NC, Ginsberg GG, DeMarshall M, Foxwell T, Zaidi AH, Taylor DL, Rustgi AK, Bergman JJ, Falk GW (2016) A novel tissue systems pathology test predicts progression in Barrett's esophagus patients. Cancer Epidemiol Biomark Prev 25 (6):958–968
6. MathWorks image processing toolbox. https://www.mathworks.com/products/image
7. Kothari S, Phan JH, Wang MD (2013) Eliminating tissue-fold artifacts in histopathological whole-slide images for improved image-based prediction of cancer grade. J Pathol Inf 4:22
8. Hang W, Phan JH, Bhatia AK, Cundiff CA, Shehata BM, Wang MD (2015) Detection of blur artifacts in histopathological whole-slide images of endomyocardial biopsies. Conf Proc IEEE Eng Med Biol Soc 2015:727–730
9. Dolled-Filhart M, McCabe A, Giltnane J, Cregger M, Camp RL, Rimm DL (2006) Quantitative in situ analysis of beta-catenin expression in breast cancer shows decreased expression is associated with poor outcome. Cancer Res 66 (10):5487–5494

Part V

Modeling Drug Responses in Cancer Cells

Chapter 14

Bioinformatics Approaches to Predict Drug Responses from Genomic Sequencing

Neel S. Madhukar and Olivier Elemento

Abstract

Fulfilling the promises of precision medicine will depend on our ability to create patient-specific treatment regimens. Therefore, being able to translate genomic sequencing into predicting how a patient will respond to a given drug is critical. In this chapter, we review common bioinformatics approaches that aim to use sequencing data to predict sample-specific drug susceptibility. First, we explain the importance of customized drug regimens to the future of medical care. Second, we discuss the different public databases and community efforts that can be leveraged to develop new methods for identifying new predictive biomarkers. Third, we cover the basic methods that are currently used to identify markers or signatures of drug response, without any prior knowledge of the drug's mechanism of action. We further discuss how one can integrate knowledge about drug targets, mechanisms, and predictive markers to better estimate drug response in a diverse set of samples. We begin this section with a primer on popular methods to identify targets and mechanism of action for new small molecules. This discussion also includes a set of computational methods that incorporate other drug features, which do not relate to drug-induced genetic changes or sequencing data such as drug structures, side-effects, and efficacy profiles. Those additional drug properties can aid in gaining higher accuracy for the identification of drug target and mechanism of action. We then progress to discuss using these targets in combination with disease-specific expression patterns, known pathways, and genetic interaction networks to aid drug choice. Finally, we conclude this chapter with a general overview of machine learning methods that can integrate multiple pieces of sequencing data along with prior drug or biological knowledge to drastically improve response prediction.

Key words Bioinformatics, Precision medicine, Drug response, Machine learning, Biomarkers

1 Introduction

One of the greatest challenges in the current paradigm of medicine is how to deal with patient heterogeneity—both across different diseases and even within patients diagnosed with the same disease. Over the past 50 years there have been many studies showing that patients with the same disease have completely different responses when treated with the same drug [1–3]. The prevailing hypothesis to explain the heterogeneous response is each patient's specific genetic profile. Precision medicine involves using this patient-

Louise von Stechow (ed.), *Cancer Systems Biology: Methods and Protocols*, Methods in Molecular Biology, vol. 1711, https://doi.org/10.1007/978-1-4939-7493-1_14, © Springer Science+Business Media, LLC 2018

specific genomic information to guide drug treatment, with the expectation that this will ultimately improve clinical outcomes [4]. With the decrease in sequencing costs over the past decade, it is now possible to obtain genomic information for patients prior to determining a specific treatment regimen. In addition, there has been an emergence of bioinformatics methods to interpret this sequencing data and come up with actionable strategies for precise drug choices. These methods not only allow for the identification of specific genetic traits that confer susceptibility or resistance to drug treatment, but can also combine genetic markers with gene ontologies and biological networks to predict precise response levels. In this chapter, we provide an overview of these bioinformatics methods, review the basic premises for each type of method, and discuss some of the current problems and future challenges that need to be solved. While we tend to focus on cancer, the databases and methods we described are often applicable to other diseases, as well.

2 Databases

In recent years, there have been a number of community efforts to generate and publicly release datasets that could be used to improve drug response prediction. Table 1 lists some of these datasets. In this review we will cover what we believe to currently be the best-suited and most popular public resources for aiding drug response prediction.

2.1 NCI60 Drug Sensitivity Database

The National Cancer Institute's (NCI) 60 cell line drug screen is a database of in vitro drug efficacies (either in terms of GI50, LD50, or TGI) for over 50,000 compounds screened against the NCI60 panel of cancer cell lines [5]. With 60 cancer cell lines from nine distinct tumor types—leukemia, colon, lung, central nervous system, renal, melanoma, ovarian, breast, and prostate—the NCI60 collection aims to provide information on a broad set of genetic conditions and tumor types. The NCI60 panel has itself been profiled using a variety of assays from genomic to gene expression and proteomics [6–9]. The profiling data can be used in conjunction with the Developmental Therapeutics Program's (DTP) drug screening database to identify genetic signatures indicative of a certain response pattern.

2.2 Cancer Cell Line Encyclopedia

The Cancer Cell Line Encyclopedia (CCLE) [10, 11] is a database of 947 different human cancer cell lines encompassing 36 different tumor types that have been genetically profiled—gene expression, copy number, mutations, etc. Furthermore, 24 known anticancer drugs were profiled against approximately 500 of these cell lines. Though the number of compounds profiled is smaller than the

Table 1
List of databases and abbreviations that are mentioned throughout the text of the chapter

Abbreviation	Full description	Website
GI50	Concentration of a compound that leads to a 50% inhibition of cell proliferation	
IC50	Concentration of a compound that leads to a 50% decrease in the desired activity	
LD50	Concentration of a compound that leads to 50% cell death	
TGI	Total growth inhibition	
GWAS	Genome-Wide Association Study	
SNP	Single-nucleotide polymorphism	
DREAM	Dialogue on reverse engineering assessment and methods	
NCI60-DTP	Drug screen of 60 cancer cell lines by the National Cancer Institute's (NCI) Developmental Therapeutics Program (DTP)	https://dtp.cancer.gov
CCLE	Cancer cell line encyclopedia	http://www.broadinstitute.org/ccle/home
CMap	Connectivity map	http://www.broadinstitute.org/cmap
GDSC	Genomics of drug sensitivity in cancer	http://www.cancerrxgene.org
TCGA	The Cancer Genome Atlas	http://cancergenome.nih.gov
GTEx	Genotype-tissue expression	http://www.gtexportal.org/home/

NCI60 drug screen, the greater number of cell lines tested allows for more precise identification of genetic predictors of sensitivity for the drugs measured.

2.3 Genomics of Drug Sensitivity in Cancer

Hosted by the Wellcome Trust Sanger Institute, the Genomics of Drug Sensitivity in Cancer (GDSC) database is a massive drug screen project similar to the NCI60 and CCLE. In their initial release, investigators screened a set of 138 known anti-cancer compounds against over 1000 different cancer cell lines (on average 525 cell lines tested per compound). Each cell line also was subjected to thorough expression and copy number profiling along with targeted mutation data for a set of 75 cancer genes. This dataset constitutes another great resource for the identification of genomic markers of drug responses.

2.4 Connectivity Map/LINCS

Released by the Broad Institute, the Connectivity Map (CMap) seeks to find connections between small molecules, physiological processes, and disease states [12]. Using mRNA expression (measured by DNA microarrays) as the "language" of cellular response, the CMap measures how a panel of cancer cell lines responds transcriptionally to a variety of different drug treatments. This approach had previously been successful in identifying drug mechanisms in yeast but had never been applied to cancer cells [13]. The investigators profiled four different cancer cell lines before and after treatment with a panel of more than 1000 small molecules. The LINCS database is an updated version of this profiling system with a much larger number of drugs and cell lines. This database makes use of the LINC1000 expression profiling system where the expression of 1000 key genes is measured and used to infer the global gene expression profile. From these transcriptional changes it is possible to explore a drug's mechanisms of action. These could be used to successfully repurpose drugs for specific diseases or genetic states [14, 15].

3 Identification of Genomic Markers of Drug Response

A key first step to any drug response prediction effort involves the identification of genomic markers that can impact efficacy. Identifying those markers makes response prediction a much simpler task. Once a polymorphism, gene expression pattern, or pathway has been identified, all new samples can simply be screened for that marker and, using known correlations with drug response, a prediction of drug susceptibility can be made. Here, we focus on a variety of approaches that can be used to identify genomic markers indicative of drug response.

3.1 Using Genome-Wide Associate Studies to Identify Polymorphisms Related to Drug Response

Genome-Wide Associate Studies (GWAS) have classically been used to detect genetic variations associated with specific disease phenotypes. However, in recent years, the use of GWAS has proved to be a powerful method to identify polymorphisms that can affect drug efficacy and toxicity [16]. Unlike approaches focusing on known drug targets or candidate gene lists, GWAS provides a hypothesis-free method that can systematically test a large number of variants [17, 18]. In order to run a GWAS one must provide a measure of response or toxicity for a large number of samples, as well as a thorough genotyping of each sample.

GWA studies typically fall into two main categories depending on whether the provided response measure is categorical (such as case/control, responder/non-responder, adverse reaction/no reactions, etc.) or quantitative (such as IC50 or a measure of side effect severity). Recently, there have been a series of developments improving the traditional GWAS, such as taking into account a

Table 2
Sample contingency table showing how we can use the number of responders with a certain SNP to test whether it is related to drug efficacy

	Responders	Non-responders
SNP present	90	15
SNP absent	10	485

gene's functional information [19], epistasis [20], or missing data [21]. Here, we review the basic premise of the categorical and quantitative GWA studies:

1. *Categorical*—The goal of a categorical GWAS is to identify SNPs that are highly predictive of which category a given sample will be assigned to. To begin with, samples are assigned to one of the two categories based on either their response to a given drug or the observation of a given adverse effect. For each observed SNP, we count the number of samples where that SNP is present (or absent). This data is then used to populate what is known as a contingency table. For instance, if in a dataset with 100 responders and 500 non-responders we observe 90 responders with a certain SNP and 15 non-responders with that same SNP, the resulting contingency table is shown in Table 2. A statistical test is then run on each contingency table to measure the deviation from the null-hypothesis, which assumes that there is no association between the SNP and categorical classes. The most common test used is either the chi-squared test (or the related Fishers exact test). This approach has successfully identified variants related to interferon beta [22] and anti-TNF treatment efficacy [23] as well as variants predictive of statin-induced myopathy [24].
2. *Quantitative*: Instead of using a contingency table test to detect significantly associated SNPs, a quantitative GWAS traditionally uses a generalized linear model (GLM), such as an Analysis of Variance (ANOVA)—a variant of a linear regression analysis—to identify SNPs that are highly correlated to the variable of interest (such as drug IC50) [25]. Though more complicated than the categorical case, there exist a number of public bioinformatics software packages such as PLINK [26] or SNPTEST [27] that can run quantitative GWAS and output a p-value for each polymorphism. While these analyses are less common for drug response prediction because of the difficulty in measuring quantitative response values, various groups have successfully used them to identify SNPs associated with susceptibility to chemotherapeutic drugs [28] or ACE inhibitors [29].

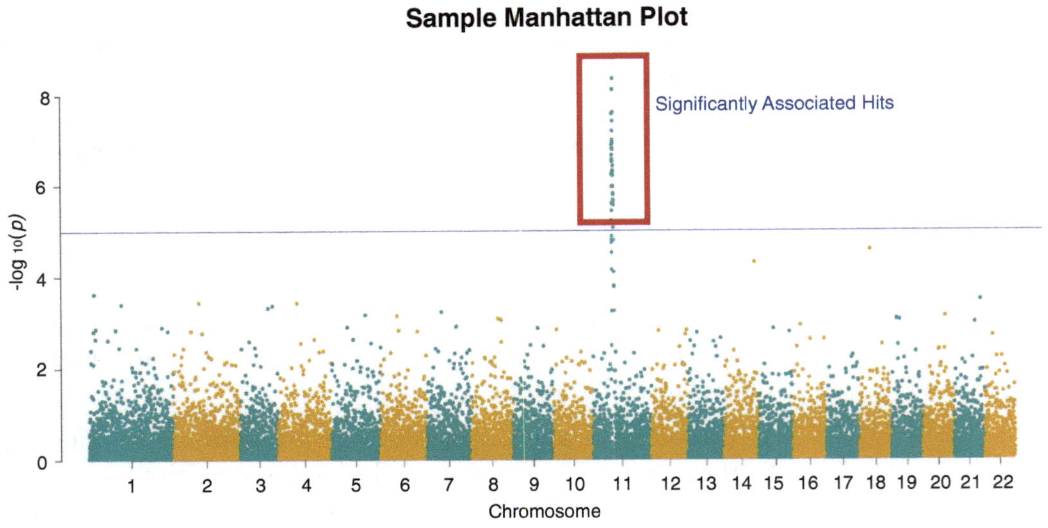

Fig. 1 Sample Manhattan plot showcasing how one can use the output of GWAS calculation to find SNPs related to drug efficacy. *Boxed hits* represent those that pass the significant *p* value cutoff and thus may be relevant to treatment response

Regardless of the type of GWAS used, the output is a set of *p*-values, one for each polymorphism tested. One important caveat is that all *p*-values must be corrected for multiple hypothesis testing (MHT) to account for the large number of statistical tests being performed. The most commonly used methods for MHT are the Bonferroni or Benjamini-Hochberg corrections. Adjusted *p*-values are then visualized using a Manhattan plot where the genomic position of each SNP is plotted against the negative log of its *p*-value (*see* Fig. 1). Using the Manhattan plot one can visually identify genomic regions or particular SNPs that are significantly associated with the given response feature.

3.2 Using Gene Expression to Find Response Signatures and Predict Response

While GWA studies aim to find a set of mutations or polymorphisms that are predictive of how a patient will respond to a drug, another popular approach is using gene expression data to find an expression signature associated with a positive (or negative) response. Different transcriptional profiles can often lead to different levels of drug efficacy, and differential expression analyses can help pinpoint the specific genes or pathways that drive the heterogeneous drug response and can be used to predict response levels.

The classic approach involves treating a cohort of mice or patients, or patient samples or cell lines with a given drug and measuring the degree of response in each sample. Similar to a GWAS, the response rate can be measured either categorically (responder/non-responder) or as a continuous variable. Using either sequencing data from before treatment or differential gene expression (comparing pre and post-treatment samples) one can

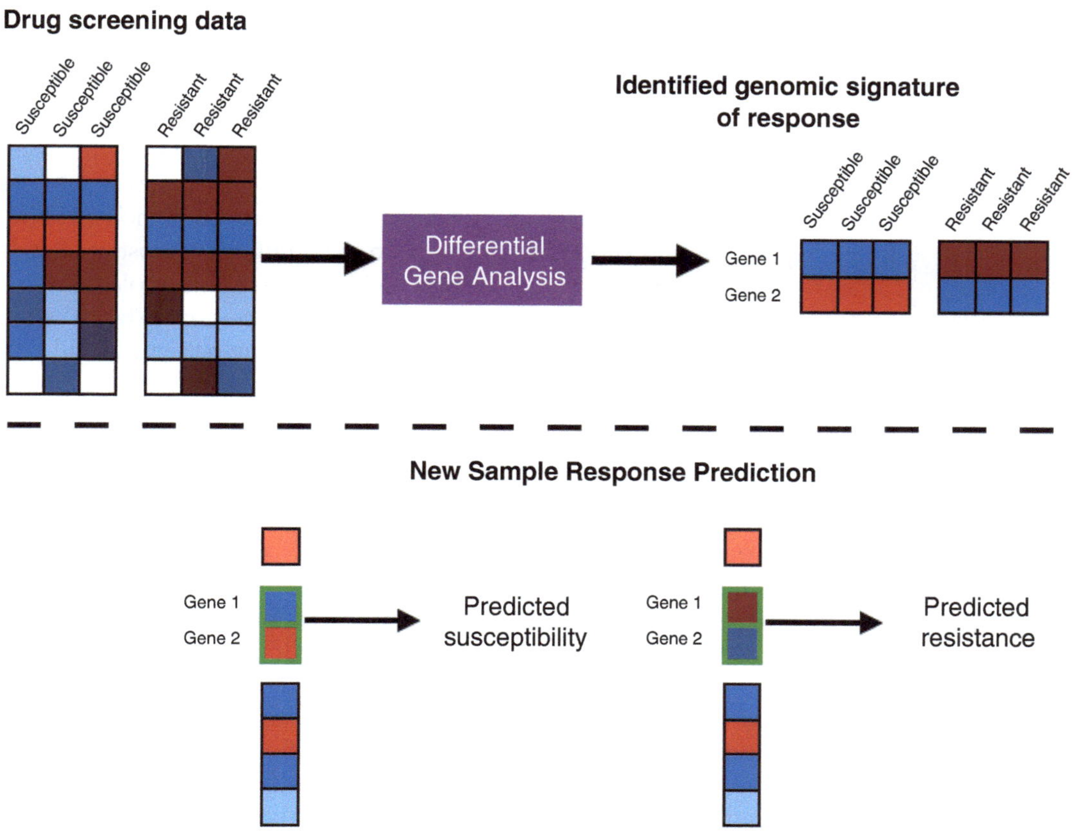

Fig. 2 Diagram on how gene expression patterns from responders and non-responders can be used to identify signatures related to response and how these can be used to better select new patients likely to respond

search for gene expression patterns that seems more prevalent in the samples that are susceptible (or resistant) to treatment (*see* Fig. 2). For instance, one would expect to see genes that confer drug resistance to be more highly expressed in samples where drug treatment shows a limited effect.

A number of methods exist for detecting differential expression across a set of samples. For microarray data oftentimes statistical tests such as an ANOVA would suffice, but packages such as limma [30] (*see* also Chapter 6 for an application of the limma package on phosphoproteomics data) use linear models that can help deal with more complicated experimental designs. For RNA-seq data the most popular methods include a limma-voom [31], DESeq2 [32], edgeR [33], and cufflinks (cuffdiff) [34]. DESeq2 and edgeR are currently considered the standard for differential expression analysis and both use similar underlying models (however with different dispersion estimates). However, in our experience we have found DESeq2 to be more conservative. One key difference between DESeq2/edgeR and limma-voom is that voom does not employ a negative binomial distribution and instead estimates the mean variance relationship. Therefore, voom may be a better choice if the input data differs strongly from a negative binomial distribution. Finally, one major difference between the cuffdiff pipeline and DESeq2 is that cuffdiff acts on the level of transcripts while DESeq2 uses gene counts as inputs. Additionally, Wright et al. [35] used a Bayesian predictor to automatically separate samples into subtypes based on their respective gene expression profiles, and used the output *p*-values to find the set of genes most predictive of subtype. This type of approach is useful for pooled sets of samples without knowledge of their subtype—for instance when one would like to determine if well-responding patients all fall into a certain disease subtype [36]. While initially tested on microarray data, this approach can be easily adapted to RNA-seq data and could generally be adapted to all types of predictive models.

3.3 Using Pathway Annotations and GSEA to Identify Differential Biological States

Often a differential gene expression analysis will have a set of genes as output, which has no obvious pattern or relevance to the type of drug being investigated. Additionally, it is quite common for a set of genes to be marked as significant in a differential gene expression analysis, but when experiments are done to perturb individual genes they seem to have little to no effect on drug response. In cases like these it is often helpful to translate the differentially expressed genes into a set of enriched biological pathways or gene sets. These can provide a broader explanation of a drug's mechanism of action and a clearer understanding on how to predict efficacy. This approach has previously been successful not only in drug response prediction, but also in the development of highly effective drugs. Overexpression of the mTOR pathway in lymphoma led to the development of inhibitors to specifically target genes in that pathway [37], and global activation of the epidermal growth factor receptor pathway was found to be predictive of erlotinib susceptibility in pancreatic cancer xenografts [38].

The basic technique to finding enriched pathways or canonical gene sets is to first annotate each gene based on the pathways/sets it falls into. A few popular resources for pathway and gene set annotation include: the Molecular Signatures Database (MSigDB) [39], Reactome [40, 41], the Kyoto Encyclopedia of Genes and Genomes (KEGG) [42], Gene Ontologies [43], and InnateDB [44, 45]. Reactome, KEGG, and InnateDB group genes based on their biochemical pathways (with InnateDB focusing on pathways relating to immunity), Gene Ontologies group genes based on their biological/molecular function or cellular localization, and MSigDB is a combination of all the aforementioned databases with custom sets of "hallmark" gene sets, or important genes involved in certain processes. Following annotation, a statistical test (such as the Fishers exact test) can be used to test whether a certain pathway is enriched for up (or down) regulated genes compared to what would be expected by random chance.

Another popular method for testing pathway enrichment is Gene Set Enrichment Analysis (GSEA) [46]. GSEA tests whether genes of a certain pathway/set are differentially expressed between the cases. It does this by computing an enrichment score for each gene set—increase in score if genes in set are differentially expressed, decrease in score if not—and using a number of permutations (number can be set by the user) it tests whether that enrichment score is significantly different than what would be expected by chance. Packaged with the MSigDB gene sets, GSEA has demonstrated success at identifying common biological pathways in independent lung cancer datasets while single-gene differential analyses could not [46].

4 Identifying Drug Targets and Mechanisms and Using Them to Improving Response

4.1 Computational Techniques to Identify Drug Targets and Mechanisms

For a small molecule in development the mechanisms of action and binding targets are often not fully understood. A number of computational methods exist that seek to predict targets for these orphan small molecules, based either on chemical structure or on its down-stream effects. These methods can broadly be divided into three categories:

1. Molecular dynamics: Using intricate mathematical models, molecular dynamics methods computationally simulate a drug's interaction with a given protein. To predict targets, an orphan small molecule is tested against a series of proteins to identify any with favorable binding results [47, 48]. However, this approach requires significant computation power, complex mathematical models, and full 3D structures for each queried protein—data that is often unavailable.
2. Ligand-based [49, 50]: Using a set of known protein binding partners for a given small molecule, ligand-based approaches apply machine learning techniques to find other proteins with high enough similarity to the known targets. The proteins with high degrees of similarity are predicted to be novel binding targets. However ligand-based methods often require a large number of known binding partners for each tested small molecule, and thus can mostly be used on drugs far enough in the drug development phase.
3. Downstream effect based: Recently, a number of methods emerged, which use the downstream effects of a small molecule (such as induced gene expression change [51] or side-effects [52]) to predict targets for orphan small molecules. The basic premise of these methods is to compare the effects of an orphan small molecule to the effects of drugs with known targets. If the

orphan molecule has an effect very similar to a drug with a known target, one would predict this known target to also be a target of the orphan small molecule. However, most current methods only utilize a small number of the available data sources and are thus not broadly applicable to all drug types. Our lab recently developed BANDIT, a novel computational method that integrates multiple different pieces of data on small molecules to predict specific binding targets and mechanisms [53]. When tested on a set of diverse drugs, BANDIT achieved an accuracy of approximately 90% at identifying known targets (validated using a standard cross validation setup), much higher than expected from other target prediction methods.

Another popular option is to focus on a drug's broad mechanism of action rather than its specific binding targets. One way to accomplish this is to observe how a given drug changes the transcriptional profile in a sample. For example, using gene expression data following cisplatin treatment, this type of analysis identified the p53 response and other pathways to be involved in cisplatin response [54]. This approach has become more practical with the emergence of public databases such as the Connectivity Map (CMap) [55]. From the CMap database, one can calculate fold change values for each gene after drug treatment. Using GSEA or other pathway enrichment methods, the fold change values can be converted into a set of pathway scores that reveal which pathways were enriched or mobilized. Though far less precise than specific target identification, this information is easier to obtain and could provide additional information on the context in which a given drug could be used.

4.2 Using Known Drug Targets To Predict Response

Assuming one can determine the mechanisms of action of a drug—either in terms of specific binding targets or broad knowledge on the biological pathways mobilized—the task of predicting efficacies is often much simpler. For example, if a drug's main mechanism of action is to target Protein A, then one would expect different efficacies in samples based on whether there is an amplification or deletion of Protein A. This type of reasoning also applies when there are mutations in a known drug target. Examples of this are treatments involving Gefitinib or Herceptin. Gefitinib is an anticancer small molecule known to target the EGFR kinase, and mutations in EGFR were found to predict sensitivity of samples to gefitinib treatment [56]. Herceptin, an antibody that targets HER2, was found to improve the outcomes of cancer patients with HER2 amplifications or activating mutations [57, 58]. Another example of this concept is vemurafenib—a small molecule that targets V600E BRAF mutation—that has been found to be selectively effective in cancer patients with this exact mutation, while having no beneficial effect on normal BRAF samples

[59–61]. These are just a few of the many examples showing how combining known drug targets with targeted sequencing can help detect instances of differential response.

However, it is also important to note that while the alterations of a drug's target are often predictive of efficacy, this is not always the case, even if the target itself serves as a biomarker [62]. Moreover, there are often cases where the predictive biomarker for a given drug is not the actual target, but rather another gene or set of genes involved in the same pathway or biological processes as drug's target. In cases like these sequencing could still prove to be a valuable tool, and we advise utilizing some of the other methods mentioned in this chapter. Drug target information could be used in combination with these methods to refine predictions and gain greater biological insights.

Sequencing-based approaches also can be very successful in positioning drugs for specific disease conditions—especially different cancer types. Using resources like the Cancer Genome Atlas (TCGA) [63] and Genotype-Tissue Expression (GTEx) project [64], one can find genes or pathways that are significantly upregulated in certain cancers or cancer types compared to either normal tissue samples or other cancer subtypes. Identifying such cancer-subtype-specific, upregulated signatures could highlight drugs known to target these signatures as particularly viable candidates for treatment. For instance, it was recently discovered that dopamine receptors were selectively upregulated in neoplastic stem cells in breast cancer. It was observed that thioridazine (a compound known to target dopamine receptors) was particularly effective against these cell populations [65].

4.3 Exploiting Genetic Interactions (SL/SDL)

One approach that has become increasingly popular is exploiting networks of synthetic lethality (SL) and synthetic dosage lethality (SDL) to predict drug efficacy. SL describes a specific type of genetic interactions involving two or more genes, where the loss of either gene individually is non-fatal, but the combined loss of all SL partner genes leads to a severe decrease in fitness or cell death. SDL describes a related genetic interaction where lethality is observed when one gene is lost while its SDL partner is overexpressed [66, 67]. Both SL and SDL interactions are highly relevant to cancer biology, as most cancers have both widespread losses and gains of certain genes. Exploiting these could drastically improve patient prognosis. For instance, if Gene A and Gene B are in an SL pair and Gene A is lost in a given cancer sample, then one would expect compounds targeting Gene B to have better responses in this sample (*see* Fig. 3).

To this end there have recently been many efforts to uncover underlying SL and SDL networks in cancer. Among the most successful efforts was the data mining synthetic lethality identification pipeline DAISY [68]. DAISY uses three distinct hypotheses to

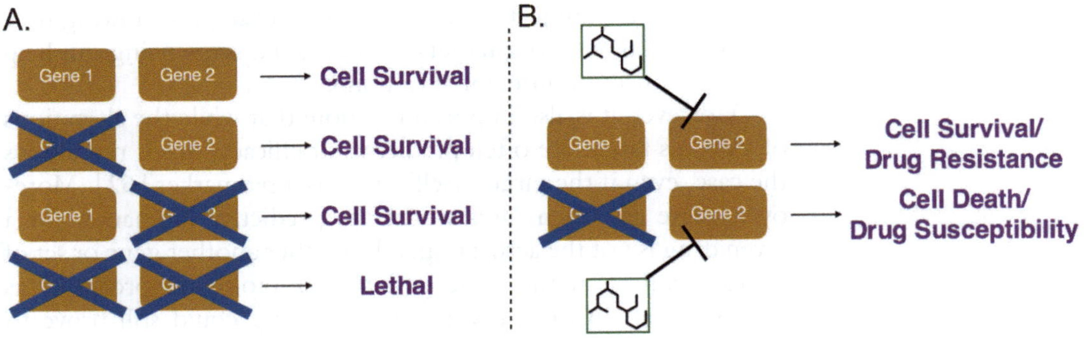

Fig. 3 (**a**) Diagram highlighting the concept of synthetic lethality and how known synthetic lethal relationships can be combined with genomic information to better predict drug response. (**b**) Using synthetic lethality to predict differential response

detect SL pairs (with the inverse hypotheses being used for SDL pair detection):

1. Genes in an SL pair will have significantly lower raters of co-mutation or co-loss.
2. Knockout/knockdown of a given gene will be more fatal in samples with under-expression or loss of its SL partner.
3. Genes in an SL pair are more likely to be co-expressed.

By scanning for gene pairs that fulfill all three hypotheses, DAISY predicted networks of SL and SDL interactions. It achieved an accuracy level of approximately 77% (measured by Area Under the Receiver Operating Curve) when compared to known SL interactions, demonstrating that DAISY could accurately infer SL and SDL genetic interactions. To translate this into predicting drug responses, the authors identified sample-specific exploitable interactions, or SDL interactions where one gene was overexpressed and SL interactions where one gene was lost. DAISY then identified drugs known to target the other gene in each exploitable interaction. For each drug DAISY ranked the most sensitive samples based on the number of exploitable interactions being targeted by each drug. They found that specific drugs were significantly more effective in cell lines predicted to be sensitive than those predicted to be resistant. Furthermore, the authors used a similar approach to predict the exact IC50 value for each drug across a set of cancer cell lines and observed a strong correlation between the predicted and observed values ($R = 0.721$). Taken together these results show how known genetic interactions (particularly SL and SDL interactions) can be combined with sequencing data to better predict drug sensitivities and inform treatment.

5 Machine Learning Approaches

In cases where identification of response biomarkers is too complex or the identified biomarkers do not reveal any underlying biological insight, machine learning approaches, which can combine sequencing data with information such as biological networks, are very powerful. The idea for employing machine learning approaches for drug response prediction is for the computational algorithm to learn how to combine a set of distinct features into a prediction of sensitivity. Most machine learning methods for drug sensitivity prediction are classified as supervised methods. Those supervised methods use a set of sequenced samples with known drug sensitivities to "train" the algorithm and determine how to combine features based on their predictive power (*see* Fig. 4). While the linear regression model discussed earlier can be considered the oldest form of machine learning, most popular methods currently utilize more advanced modeling to account for the complexity in genetic sequencing data. In fact, machine learning methods can often detect higher order genomic markers of drug response that other methods may have missed. One example is the use of machine

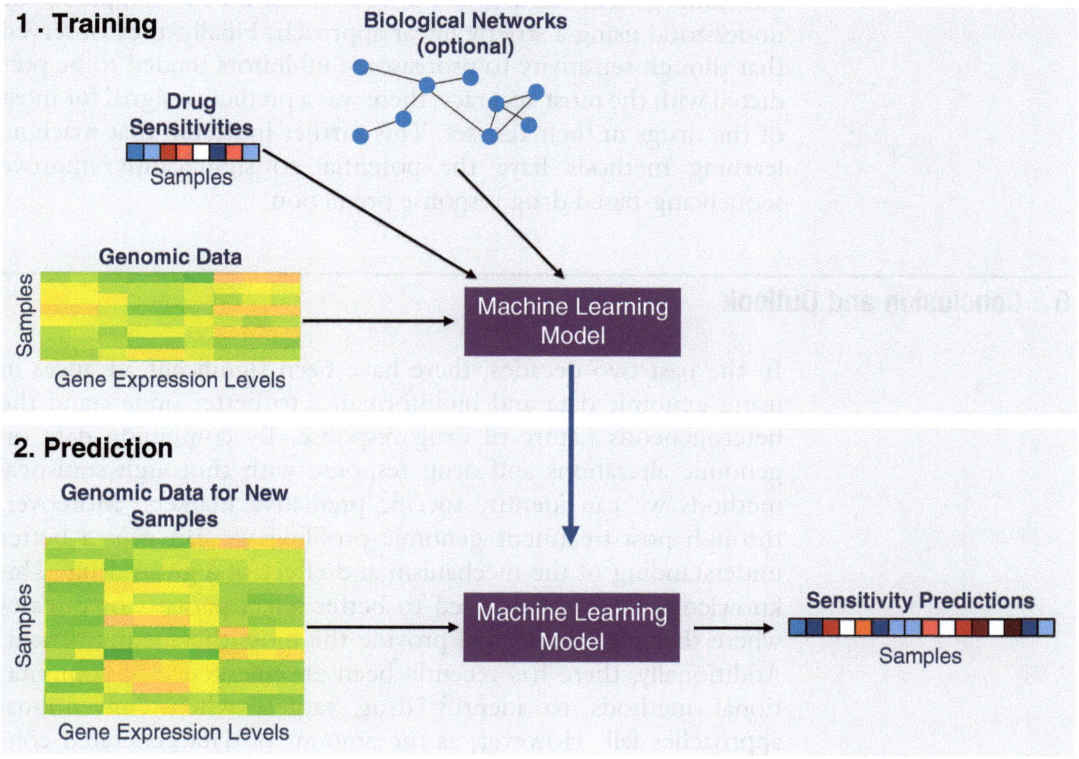

Fig. 4 Overview of how common machine-learning methods combine multiple data types to train a specific model that can be applied to new samples to predict sensitivity

learning to identify the EWS-FL11 translocation in Ewing's sarcoma as a marker of sensitivity to PARP inhibitors [69].

Many methods seek to improve their performance by including additional information on known biological networks, genetic interactions, or drug chemical properties. For instance, Menden et al. [70] found that including drug chemical information (such as weight and lipophilicity) with sequencing data improved the performance of both a neural network and random forest for sensitivity prediction. In collaboration with the NCI, the Dialogue on Reverse Engineering Assessment and methods (DREAM) project led a community effort to improve drug sensitivity predictions [71]. Through this effort, the NCI-DREAM consortium publicly released drug sensitivity data for a set of breast cancer cell lines along with thorough genetic, epigenetic, and proteomic sequencing data. Individual groups each submitted different sensitivity prediction methods and the NCI-DREAM consortium analyzed each method to identify any particular method features that led to higher accuracies. Interestingly, they found that the inclusion of annotated biological pathways was one of the two variables that significantly boosted performance [71]. Additionally, the consortium found that the top performing methods all utilized nonlinear modeling, indicating that in many cases the connections between individual genetic features and drug response are too complex to be understood using a strictly linear approach. Finally, they observed that though sensitivity to proteasome inhibitors tended to be predicted with the most accuracy, there was a predictive signal for most of the drugs in their test set. This further indicated that machine learning methods have the potential to significantly improve sequencing-based drug response prediction.

6 Conclusion and Outlook

In the past two decades, there have been significant advances in using genomic data and bioinformatics to better understand the heterogeneous nature of drug response. By combining data on genomic alterations and drug response with thorough statistical methods we can identify specific predictive markers. Moreover, through post-treatment genomic profiling we can gain a better understanding of the mechanism and effect of a given drug. This knowledge can then be used to better select patients or diseases where that mechanism will provide the most therapeutic benefit. Additionally, there has recently been an emergence of computational methods to identify drug targets when conventional approaches fail. However, as the amount of data generated continues to increase and drugs targeting new pathways are developed, we imagine that no single approach or method will provide high enough accuracy. Therefore, we expect the field to move toward using machine learning strategies that are able to integrate a variety of different data-types into a single predictive output. We are already seeing the creation of sophisticated methods for this purpose and anticipate this to only improve over the coming years. All together though we believe that the adoption of the methodology described in this chapter not only has the power to expand our understanding of pharmacology but can also significantly improve the current schema of patient treatment.

Acknowledgments

The authors would like to thank the Elemento Lab members and Natalie R. Davidson for their feedback and discussion. O.E. and N.M. are supported by the CAREER grant from National Science Foundation (DB1054964), NIH grant R01CA194547, the Starr CancerFoundation, as well as by startup funds from the Institute for Computational Biomedicine. Support for N.M. was also provided by the PhRMA Foundation Pre Doctoral Informatics Fellowship and by the Tri-Institutional Training Program in Computational Biology and Medicine.

References

1. Fry RC, Svensson JP, Valiathan C, Wang E, Hogan BJ, Bhattacharya S, Bugni JM, Whittaker CA, Samson LD (2008) Genomic predictors of interindividual differences in response to DNA damaging agents. Genes Dev 22 (19):2621–2626. https://doi.org/10.1101/gad.1688508
2. Rice SD, Heinzman JM, Brower SL, Ervin PR, Song N, Shen K, Wang DK (2010) Analysis of chemotherapeutic response heterogeneity and drug clustering based on mechanism of action using an in vitro assay. Anticancer Res 30 (7):2805–2811
3. Bosquet JG, Marchion DC, Chon H, Lancaster JM, Chanock S (2014) Analysis of chemotherapeutic response in ovarian cancers using publicly available high-throughput data. Cancer Res 74(14):3902–3912. https://doi.org/10.1158/0008-5472.CAN-14-0186
4. Sboner A, Elemento O (2016) A primer on precision medicine informatics. Brief Bioinform 17(1):145–153. https://doi.org/10.1093/bib/bbv032
5. Shoemaker RH (2006) The NCI60 human tumour cell line anticancer drug screen. Nat Rev Cancer 6(10):813–823. https://doi.org/10.1038/nrc1951
6. Abaan OD, Polley EC, Davis SR, Zhu YJ, Bilke S, Walker RL, Pineda M, Gindin Y, Jiang Y, Reinhold WC, Holbeck SL, Simon RM, Doroshow JH, Pommier Y, Meltzer PS (2013) The exomes of the NCI-60 panel: a genomic resource for cancer biology and systems pharmacology. Cancer Res 73 (14):4372–4382. https://doi.org/10.1158/0008-5472.Can-12-3342
7. Reinhold WC, Varma S, Sousa F, Sunshine M, Abaan OD, Davis SR, Reinhold SW, Kohn KW, Morris J, Meltzer PS, Doroshow JH, Pommier Y (2014) NCI-60 whole exome sequencing and pharmacological CellMiner analyses. PLoS One 9(7). https://doi.org/10.1371/journal.pone.0101670
8. Scherf U, Ross DT, Waltham M, Smith LH, Lee JK, Tanabe L, Kohn KW, Reinhold WC, Myers TG, Andrews DT, Scudiero DA, Eisen MB, Sausville EA, Pommier Y, Botstein D, Brown PO, Weinstein JN (2000) A gene expression database for the molecular pharmacology of cancer. Nat Genet 24(3):236–244. https://doi.org/10.1038/73439
9. Gholami AM, Hahne H, Wu ZX, Auer FJ, Meng C, Wilhelm M, Kuster B (2013) Global proteome analysis of the NCI-60 cell line panel. Cell Rep 4(3):609–620. https://doi.org/10.1016/j.celrep.2013.07.018

10. Barretina J, Caponigro G, Stransky N, Venkatesan K, Margolin AA, Kim S, Wilson CJ, Lehar J, Kryukov GV, Sonkin D, Reddy A, Liu M, Murray L, Berger MF, Monahan JE, Morais P, Meltzer J, Korejwa A, Jane-Valbuena J, Mapa FA, Thibault J, Bric-Furlong E, Raman P, Shipway A, Engels IH, Cheng J, Yu GK, Yu JJ, Aspesi P, de Silva M, Jagtap K, Jones MD, Wang L, Hatton C, Palescandolo E, Gupta S, Mahan S, Sougnez C, Onofrio RC, Liefeld T, MacConaill L, Winckler W, Reich M, Li NX, Mesirov JP, Gabriel SB, Getz G, Ardlie K, Chan V, Myer VE, Weber BL, Porter J, Warmuth M, Finan P, Harris JL, Meyerson M, Golub TR, Morrissey MP, Sellers WR, Schlegel R, Garraway LA (2012) The cancer cell line encyclopedia enables predictive modelling of anticancer drug sensitivity (483:603, 2012). Nature 492 (7428):290–290. https://doi.org/10.1038/nature11735
11. Stransky N, Ghandi M, Kryukov GV, Garraway LA, Lehar J, Liu M, Sonkin D, Kauffmann A, Venkatesan K, Edelman EJ, Riester M, Barretina J, Caponigro G, Schlegel R, Sellers WR, Stegmeier F, Morrissey M, Amzallag A, Pruteanu-Malinici I, Haber DA, Ramaswamy S, Benes CH, Menden MP, Iorio F, Stratton MR, McDermott U, Garnett MJ, Saez-Rodriguez J, Canc DS, Line CC, Inst B, Res NIB, Sensitivity GD, Hosp MG, Lab EMB, Inst EB, Inst WTS (2015) Pharmacogenomic agreement between two cancer cell line data sets. Nature 528(7580):84. https://doi.org/10.1038/nature15736
12. Lamb J, Crawford ED, Peck D, Modell JW, Blat IC, Wrobel MJ, Lerner J, Brunet JP, Subramanian A, Ross KN, Reich M, Hieronymus H, Wei G, Armstrong SA, Haggarty SJ, Clemons PA, Wei R, Carr SA, Lander ES, Golub TR (2006) The connectivity map: using gene-expression signatures to connect small molecules, genes, and disease. Science 313(5795):1929–1935. https://doi.org/10.1126/science.1132939
13. Hughes TR, Marton MJ, Jones AR, Roberts CJ, Stoughton R, Armour CD, Bennett HA, Coffey E, Dai HY, He YDD, Kidd MJ, King AM, Meyer MR, Slade D, Lum PY, Stepaniants SB, Shoemaker DD, Gachotte D, Chakraburtty K, Simon J, Bard M, Friend SH (2000) Functional discovery via a compendium of expression profiles. Cell 102(1):109–126. https://doi.org/10.1016/S0092-8674(00)00015-5
14. Gayvert KM, Dardenne E, Cheung C, Boland MR, Lorberbaum T, Wanjala J, Chen Y, Rubin MA, Tatonetti NP, Rickman DS, Elemento O (2016) A computational drug repositioning approach for targeting oncogenic transcription factors. Cell Rep 15(11):2348–2356. https://doi.org/10.1016/j.celrep.2016.05.037
15. Dudley JT, Deshpande T, Butte AJ (2011) Exploiting drug-disease relationships for computational drug repositioning. Brief Bioinform 12(4):303–311. https://doi.org/10.1093/bib/bbr013
16. Low SK, Takahashi A, Mushiroda T, Kubo M (2014) Genome-wide association study: a useful tool to identify common genetic variants associated with drug toxicity and efficacy in cancer pharmacogenomics. Clin Cancer Res 20(10):2541–2552. https://doi.org/10.1158/1078-0432.Ccr-13-2755
17. Zhou KX, Pearson ER (2013) Insights from genome-wide association studies of drug response. Annu Rev Pharmacol 53:299–310. https://doi.org/10.1146/annurev-pharmtox-011112-140237
18. McCarthy MI, Abecasis GR, Cardon LR, Goldstein DB, Little J, Ioannidis JPA, Hirschhorn JN (2008) Genome-wide association studies for complex traits: consensus, uncertainty and challenges. Nat Rev Genet 9 (5):356–369. https://doi.org/10.1038/nrg2344
19. Xu ZL, Taylor JA (2009) SNPinfo: integrating GWAS and candidate gene information into functional SNP selection for genetic association studies. Nucleic Acids Res 37: W600–W605. https://doi.org/10.1093/nar/gkp290
20. McKinney BA, Pajewski NM (2011) Six degrees of epistasis: statistical network models for GWAS. Front Genet 2:109. https://doi.org/10.3389/fgene.2011.00109
21. Howie B, Marchini J, Stephens M (2011) Genotype imputation with thousands of genomes. G3 (Bethesda) 1(6):457–470. https://doi.org/10.1534/g3.111.001198
22. Byun E, Caillier SJ, Montalban X, Villoslada P, Fernandez O, Brassat D, Comabella M, Wang J, Barcellos LF, Baranzini SE, Oksenberg JR (2008) Genome-wide pharmacogenomic analysis of the response to interferon beta therapy in multiple sclerosis. Arch Neurol Chicago 65(3):337–E332. https://doi.org/10.1001/archneurol.2008.47
23. Liu CY, Batliwalla F, Li WT, Lee A, Roubenoff R, Beckman E, Khalili H, Damle A, Kern M, Furie R, Dupuis J, Plenge RM, Coenen MJH, Behrens TW, Carulli JP, Gregersen PK (2008) Genome-wide association scan identifies candidate polymorphisms associated with differential response to anti-TNF treatment in rheumatoid arthritis. Mol

Med 14(9-10):575–581. https://doi.org/10.2119/2008-00056.Liu

24. Link E, Parish S, Armitage J, Bowman L, Heath S, Matsuda F, Gut I, Lathrop M, Collins R, Grp SC (2008) SLCO1B1 variants and statin-induced myopathy – a genomewide study. New Engl J Med 359(8):789–799
25. Bush WS, Moore JH (2012) Chapter 11: genome-wide association studies. PLoS Comput Biol 8(12). https://doi.org/10.1371/journal.pcbi.1002822
26. Purcell S, Neale B, Todd-Brown K, Thomas L, Ferreira MAR, Bender D, Maller J, Sklar P, de Bakker PIW, Daly MJ, Sham PC (2007) PLINK: a tool set for whole-genome association and population-based linkage analyses. Am J Hum Genet 81(3):559–575. https://doi.org/10.1086/519795
27. Burton PR, Clayton DG, Cardon LR, Craddock N, Deloukas P, Duncanson A, Kwiatkowski DP, McCarthy MI, Ouwehand WH, Samani NJ, Todd JA, Donnelly P, Barrett JC, Davison D, Easton D, Evans D, Leung HT, Marchini JL, Morris AP, Spencer CCA, Tobin MD, Attwood AP, Boorman JP, Cant B, Everson U, Hussey JM, Jolley JD, Knight AS, Koch K, Meech E, Nutland S, Prowse CV, Stevens HE, Taylor NC, Walters GR, Walker NM, Watkins NA, Winzer T, Jones RW, McArdle WL, Ring SM, Strachan DP, Pembrey M, Breen G, St Clair D, Caesar S, Gordon-Smith-K, Jones L, Fraser C, Green EK, Grozeva D, Hamshere ML, Holmans PA, Jones IR, Kirov G, Moskvina V, Nikolov I, O'Donovan MC, Owen MJ, Collier DA, Elkin A, Farmer A, Williamson R, McGuffin P, Young AH, Ferrier IN, Ball SG, Balmforth AJ, Barrett JH, Bishop DT, Iles MM, Maqbool A, Yuldasheva N, Hall AS, Braund PS, Dixon RJ, Mangino M, Stevens S, Thompson JR, Bredin F, Tremelling M, Parkes M, Drummond H, Lees CW, Nimmo ER, Satsangi J, Fisher SA, Forbes A, Lewis CM, Onnie CM, Prescott NJ, Sanderson J, Mathew CG, Barbour J, Mohiuddin MK, Todhunter CE, Mansfield JC, Ahmad T, Cummings FR, Jewell DP, Webster J, Brown MJ, Lathrop GM, Connell J, Dominiczak A, Marcano CAB, Burke B, Dobson R, Gungadoo J, Lee KL, Munroe PB, Newhouse SJ, Onipinla A, Wallace C, Xue MZ, Caulfield M, Farrall M, Barton A, Bruce IN, Donovan H, Eyre S, Gilbert PD, Hider SL, Hinks AM, John SL, Potter C, Silman AJ, Symmons DPM, Thomson W, Worthington J, Dunger DB, Widmer B, Frayling TM, Freathy RM, Lango H, Perry JRB, Shields BM, Weedon MN, Hattersley AT, Hitman GA, Walker M, Elliott KS, Groves CJ, Lindgren CM, Rayner NW, Timpson NJ, Zeggini E, Newport M, Sirugo G, Lyons E, Vannberg F, Brown MA, Franklyn JA, Heward JM, Simmonds MJ, Hill AVS, Bradbury LA, Farrar C, Pointon JJ, Wordsmith P, Gough SCL, Seal S, Stratton MR, Rahman N, Ban M, Goris A, Sawcer SJ, Compston A, Conway D, Jallow M, Bumpstead SJ, Chaney A, Downes K, Ghori MJR, Gwilliam R, Inouye M, Keniry A, King E, McGinnis R, Potter S, Ravindrarajah R, Whittaker P, Withers D, Easton D, Pereira-Gale J, Hallgrimsdottir IB, Howie BN, Su Z, Teo YY, Vukcevic D, Bentley D, Caulfield M, Mathew CG, Worthington J, Consortium WTCC, Syndicate BRGGS, Collaborat BCS (2007) Genome-wide association study of 14,000 cases of seven common diseases and 3,000 shared controls. Nature 447 (7145):661–678. https://doi.org/10.1038/nature05911
28. Gamazon ER, Huang RS, Cox NJ, Dolan ME (2010) Chemotherapeutic drug susceptibility associated SNPs are enriched in expression quantitative trait loci. Proc Natl Acad Sci U S A 107(20):9287–9292. https://doi.org/10.1073/pnas.1001827107
29. Chung CM, Wang RY, Chen JW, Fann CSJ, Leu HB, Ho HY, Ting CT, Lin TH, Sheu SH, Tsai WC, Chen JH, Jong YS, Lin SJ, Chen YT, Pan WH (2010) A genome-wide association study identifies new loci for ACE activity: potential implications for response to ACE inhibitor. Pharmacogenomics J 10 (6):537–544. https://doi.org/10.1038/tpj.2009.70
30. Diboun I, Wernisch L, Orengo CA, Koltzenburg M (2006) Microarray analysis after RNA amplification can detect pronounced differences in gene expression using limma. BMC Genomics 7. https://doi.org/10.1186/1471-2164-7-252
31. Ritchie ME, Phipson B, Wu D, Hu Y, Law CW, Shi W, Smyth GK (2015) limma powers differential expression analyses for RNA-sequencing and microarray studies. Nucleic Acids Res 43 (7):e47. https://doi.org/10.1093/nar/gkv007
32. Anders S, Huber W (2010) Differential expression analysis for sequence count data. Genome Biol 11(10). https://doi.org/10.1186/gb-2010-11-10-r106
33. Robinson MD, McCarthy DJ, Smyth GK (2010) edgeR: a bioconductor package for differential expression analysis of digital gene expression data. Bioinformatics 26 (1):139–140. https://doi.org/10.1093/bioinformatics/btp616

34. Trapnell C, Roberts A, Goff L, Pertea G, Kim D, Kelley DR, Pimentel H, Salzberg SL, Rinn JL, Pachter L (2012) Differential gene and transcript expression analysis of RNA-seq experiments with TopHat and Cufflinks. Nat Protoc 7(3):562–578. https://doi.org/10.1038/nprot.2012.016
35. Wright G, Tan B, Rosenwald A, Hurt EH, Wiestner A, Staudt LM (2003) A gene expression-based method to diagnose clinically distinct subgroups of diffuse large B cell lymphoma. Proc Natl Acad Sci U S A 100 (17):9991–9996. https://doi.org/10.1073/pnas.1732008100
36. Lam LT, Davis RE, Pierce J, Hepperle M, Xu Y, Hottelet M, Nong Y, Wen D, Adams J, Dang L, Staudt LM (2005) Small molecule inhibitors of IkappaB kinase are selectively toxic for subgroups of diffuse large B-cell lymphoma defined by gene expression profiling. Clin Cancer Res 11(1):28–40
37. Briones J (2009) Emerging therapies for B-cell non-Hodgkin lymphoma. Expert Rev Anticancer 9(9):1305–1316. https://doi.org/10.1586/Era.09.86
38. Jimeno A, Tan AC, Coffa J, Rajeshkumar NV, Kulesza P, Rubio-Viqueira B, Wheelhouse J, Diosdado B, Messersmith WA, Lacobuzio-Donahue C, Maitra A, Varella-Garcia M, Hirsch FR, Meijer GA, Hidalgo M (2008) Coordinated epidermal growth factor receptor pathway gene overexpression predicts epidermal growth factor receptor inhibitor sensitivity in pancreatic cancer. Cancer Res 68 (8):2841–2849. https://doi.org/10.1158/0008-5472.Can-07-5200
39. Liberzon A, Subramanian A, Pinchback R, Thorvaldsdottir H, Tamayo P, Mesirov JP (2011) Molecular signatures database (MSigDB) 3.0. Bioinformatics 27 (12):1739–1740. https://doi.org/10.1093/bioinformatics/btr260
40. Fabregat A, Sidiropoulos K, Garapati P, Gillespie M, Hausmann K, Haw R, Jassal B, Jupe S, Korninger F, McKay S, Matthews L, May B, Milacic M, Rothfels K, Shamovsky V, Webber M, Weiser J, Williams M, Wu G, Stein L, Hermjakob H, D'Eustachio P (2016) The reactome pathway knowledgebase. Nucleic Acids Res 44(D1):D481–D487. https://doi.org/10.1093/nar/gkv1351
41. Croft D, Mundo AF, Haw R, Milacic M, Weiser J, Wu G, Caudy M, Garapati P, Gillespie M, Kamdar MR, Jassal B, Jupe S, Matthews L, May B, Palatnik S, Rothfels K, Shamovsky V, Song H, Williams M, Birney E, Hermjakob H, Stein L, D'Eustachio P (2014) The reactome pathway knowledgebase. Nucleic Acids Res 42(Database issue): D472–D477. https://doi.org/10.1093/nar/gkt1102
42. Ogata H, Goto S, Sato K, Fujibuchi W, Bono H, Kanehisa M (1999) KEGG: Kyoto encyclopedia of genes and genomes. Nucleic Acids Res 27(1):29–34. https://doi.org/10.1093/nar/27.1.29
43. Ashburner M, Ball CA, Blake JA, Botstein D, Butler H, Cherry JM, Davis AP, Dolinski K, Dwight SS, Eppig JT, Harris MA, Hill DP, Issel-Tarver L, Kasarskis A, Lewis S, Matese JC, Richardson JE, Ringwald M, Rubin GM, Sherlock G (2000) Gene ontology: tool for the unification of biology. The gene ontology consortium. Nat Genet 25(1):25–29. https://doi.org/10.1038/75556
44. Breuer K, Foroushani AK, Laird MR, Chen C, Sribnaia A, Lo R, Winsor GL, Hancock RE, Brinkman FS, Lynn DJ (2013) InnateDB: systems biology of innate immunity and beyond—recent updates and continuing curation. Nucleic Acids Res 41(Database issue): D1228–D1233. https://doi.org/10.1093/nar/gks1147
45. Lynn DJ, Winsor GL, Chan C, Richard N, Laird MR, Barsky A, Gardy JL, Roche FM, Chan TH, Shah N, Lo R, Naseer M, Que J, Yau M, Acab M, Tulpan D, Whiteside MD, Chikatamarla A, Mah B, Munzner T, Hokamp K, Hancock RE, Brinkman FS (2008) InnateDB: facilitating systems-level analyses of the mammalian innate immune response. Mol Syst Biol 4:218. https://doi.org/10.1038/msb.2008.55
46. Subramanian A, Tamayo P, Mootha VK, Mukherjee S, Ebert BL, Gillette MA, Paulovich A, Pomeroy SL, Golub TR, Lander ES, Mesirov JP (2005) Gene set enrichment analysis: a knowledge-based approach for interpreting genome-wide expression profiles. Proc Natl Acad Sci U S A 102(43):15545–15550. https://doi.org/10.1073/pnas.0506580102
47. Li HL, Gao ZT, Kang L, Zhang HL, Yang K, Yu KQ, Luo XM, Zhu WL, Chen KX, Shen JH, Wang XC, Jiang HL (2006) TarFisDock: a web server for identifying drug targets with docking approach. Nucleic Acids Res 34:W219–W224. https://doi.org/10.1093/nar/gkl114
48. Rarey M, Kramer B, Lengauer T, Klebe G (1996) A fast flexible docking method using an incremental construction algorithm. J Mol Biol 261(3):470–489. https://doi.org/10.1006/jmbi.1996.0477
49. Butina D, Segall MD, Frankcombe K (2002) Predicting ADME properties in silico: methods and models. Drug Discov Today 7(11):

S83–S88. https://doi.org/10.1016/S1359-6446(02)02288-2

50. Nantasenamat C, Isarankura-Na-Ayudhya C, Prachayasittikul V (2010) Advances in computational methods to predict the biological activity of compounds. Expert Opin Drug Dis 5 (7):633–654. https://doi.org/10.1517/17460441.2010.492827
51. Wang KJ, Sun JZ, Zhou SF, Wan CL, Qin SY, Li C, He L, Yang L (2013) Prediction of drug-target interactions for drug repositioning only based on genomic expression similarity. PLoS Comput Biol 9(11):e1003315. https://doi.org/10.1371/journal.pcbi.1003315
52. Campillos M, Kuhn M, Gavin AC, Jensen LJ, Bork P (2008) Drug target identification using side-effect similarity. Science 321 (5886):263–266. https://doi.org/10.1126/science.1158140
53. Madhukar NS, Huang L, Khade P, Gayvert K, Giannakakou P, Elemento O (2015) Abstract B162: small molecule target prediction and identification of novel anti-cancer compounds using a data-driven bayesian approach. Mol Cancer Ther 14(12 Supplement 2):B162. https://doi.org/10.1158/1535-7163.targ-15-b162
54. Li J, Wood WH, Becker KG, Weeraratna AT, Morin PJ (2007) Gene expression response to cisplatin treatment in drug-sensitive and drug-resistant ovarian cancer cells. Oncogene 26 (20):2860–2872. https://doi.org/10.1038/sj.onc.1210086
55. Lamb J (2007) The connectivity map: a new tool for biomedical research. Nat Rev Cancer 7 (1):54–60. https://doi.org/10.1038/nrc2044
56. Paez JG, Janne PA, Lee JC, Tracy S, Greulich H, Gabriel S, Herman P, Kaye FJ, Lindeman N, Boggon TJ, Naoki K, Sasaki H, Fujii Y, Eck MJ, Sellers WR, Johnson BE, Meyerson M (2004) EGFR mutations in lung cancer: correlation with clinical response to gefitinib therapy. Science 304 (5676):1497–1500. https://doi.org/10.1126/science.1099314
57. Cappuzzo F, Bemis L, Varella-Garcia M (2006) HER2 mutation and response to trastuzumab therapy in non-small-cell lung cancer. N Engl J Med 354(24):2619–2621. https://doi.org/10.1056/NEJMc060020
58. Romond EH, Perez EA, Bryant J, Suman VJ, Geyer CE Jr, Davidson NE, Tan-Chiu E, Martino S, Paik S, Kaufman PA, Swain SM, Pisansky TM, Fehrenbacher L, Kutteh LA, Vogel VG, Visscher DW, Yothers G, Jenkins RB, Brown AM, Dakhil SR, Mamounas EP, Lingle WL, Klein PM, Ingle JN, Wolmark N (2005) Trastuzumab plus adjuvant chemotherapy for operable HER2-positive breast cancer. N Engl J Med 353(16):1673–1684. https://doi.org/10.1056/NEJMoa052122
59. Young K, Minchom A, Larkin J (2012) BRIM-1, -2 and -3 trials: improved survival with vemurafenib in metastatic melanoma patients with a BRAF(V600E) mutation. Future Oncol 8(5):499–507. https://doi.org/10.2217/fon.12.43
60. Chapman PB, Hauschild A, Robert C, Haanen JB, Ascierto P, Larkin J, Dummer R, Garbe C, Testori A, Maio M, Hogg D, Lorigan P, Lebbe C, Jouary T, Schadendorf D, Ribas A, O'Day SJ, Sosman JA, Kirkwood JM, Eggermont AM, Dreno B, Nolop K, Li J, Nelson B, Hou J, Lee RJ, Flaherty KT, GA MA, Group B-S (2011) Improved survival with vemurafenib in melanoma with BRAF V600E mutation. N Engl J Med 364(26):2507–2516. https://doi.org/10.1056/NEJMoa1103782
61. Bollag G, Tsai J, Zhang J, Zhang C, Ibrahim P, Nolop K, Hirth P (2012) Vemurafenib: the first drug approved for BRAF-mutant cancer. Nat Rev Drug Discov 11(11):873–886. https://doi.org/10.1038/nrd3847
62. Vogel CL, Cobleigh MA, Tripathy D, Gutheil JC, Harris LN, Fehrenbacher L, Slamon DJ, Murphy M, Novotny WF, Burchmore M, Shak S, Stewart SJ, Press M (2002) Efficacy and safety of trastuzumab as a single agent in first-line treatment of HER2-overexpressing metastatic breast cancer. J Clin Oncol 20 (3):719–726. https://doi.org/10.1200/JCO.2002.20.3.719
63. Brennan CW, Verhaak RGW, McKenna A, Campos B, Noushmehr H, Salama SR, Zheng SY, Chakravarty D, Sanborn JZ, Berman SH, Beroukhim R, Bernard B, Wu CJ, Genovese G, Shmulevich I, Barnholtz-Sloan J, Zou LH, Vegesna R, Shukla SA, Ciriello G, Yung WK, Zhang W, Sougnez C, Mikkelsen T, Aldape K, Bigner DD, Van Meir EG, Prados M, Sloan A, Black KL, Eschbacher J, Finocchiaro G, Friedman W, Andrews DW, Guha A, Iacocca M, O'Neill BP, Foltz G, Myers J, Weisenberger DJ, Penny R, Kucherlapati R, Perou CM, Hayes DN, Gibbs R, Marra M, Mills GB, Lander E, Spellman P, Wilson R, Sander C, Weinstein J, Meyerson M, Gabriel S, Laird PW, Haussler D, Getz G, Chin L, Network TR (2013) The somatic genomic landscape of glioblastoma. Cell 155(2):462–477. https://doi.org/10.1016/j.cell.2013.09.034
64. Lonsdale J, Thomas J, Salvatore M, Phillips R, Lo E, Shad S, Hasz R, Walters G, Garcia F, Young N, Foster B, Moser M, Karasik E, Gillard B, Ramsey K, Sullivan S, Bridge J,

Magazine H, Syron J, Fleming J, Siminoff L, Traino H, Mosavel M, Barker L, Jewell S, Rohrer D, Maxim D, Filkins D, Harbach P, Cortadillo E, Berghuis B, Turner L, Hudson E, Feenstra K, Sobin L, Robb J, Branton P, Korzeniewski G, Shive C, Tabor D, Qi LQ, Groch K, Nampally S, Buia S, Zimmerman A, Smith A, Burges R, Robinson K, Valentino K, Bradbury D, Cosentino M, Diaz-Mayoral N, Kennedy M, Engel T, Williams P, Erickson K, Ardlie K, Winckler W, Getz G, DeLuca D, MacArthur D, Kellis M, Thomson A, Young T, Gelfand E, Donovan M, Meng Y, Grant G, Mash D, Marcus Y, Basile M, Liu J, Zhu J, Tu ZD, Cox NJ, Nicolae DL, Gamazon ER, Im HK, Konkashbaev A, Pritchard J, Stevens M, Flutre T, Wen XQ, Dermitzakis ET, Lappalainen T, Guigo R, Monlong J, Sammeth M, Koller D, Battle A, Mostafavi S, McCarthy M, Rivas M, Maller J, Rusyn I, Nobel A, Wright F, Shabalin A, Feolo M, Sharopova N, Sturcke A, Paschal J, Anderson JM, Wilder EL, Derr LK, Green ED, Struewing JP, Temple G, Volpi S, Boyer JT, Thomson EJ, Guyer MS, Ng C, Abdallah A, Colantuoni D, Insel TR, Koester SE, Little AR, Bender PK, Lehner T, Yao Y, Compton CC, Vaught JB, Sawyer S, Lockhart NC, Demchok J, Moore HF (2013) The genotype-tissue expression (GTEx) project. Nat Genet 45(6):580–585. https://doi.org/10.1038/ng.2653

65. Sachlos E, Risueno RM, Laronde S, Shapovalova Z, Lee JH, Russell J, Malig M, McNicol JD, Fiebig-Comyn A, Graham M, Levadoux-Martin M, Lee JB, Giacomelli AO, Hassell JA, Fischer-Russell D, Trus MR, Foley R, Leber B, Xenocostas A, Brown ED, Collins TJ, Bhatia M (2012) Identification of drugs including a dopamine receptor antagonist that selectively target cancer stem cells. Cell 149(6):1284–1297. https://doi.org/10.1016/j.cell.2012.03.049

66. Madhukar NS, Elemento O, Pandey G (2015) Prediction of genetic interactions using machine learning and network properties. Front Bioeng Biotechnol 3(172). https://doi.org/10.3389/fbioe.2015.00172

67. Chan DA, Giaccia AJ (2011) Harnessing synthetic lethal interactions in anticancer drug discovery. Nat Rev Drug Discov 10(5):351–364. https://doi.org/10.1038/nrd3374

68. Jerby-Arnon L, Pfetzer N, Waldman YY, McGarry L, James D, Shanks E, Seashore-Ludlow B, Weinstock A, Geiger T, Clemons PA, Gottlieb E, Ruppin E (2014) Predicting cancer-specific vulnerability via data-driven detection of synthetic lethality. Cell 158 (5):1199–1209. https://doi.org/10.1016/j.cell.2014.07.027

69. Garnett MJ, Edelman EJ, Heidorn SJ, Greenman CD, Dastur A, Lau KW, Greninger P, Thompson IR, Luo X, Soares J, Liu Q, Iorio F, Surdez D, Chen L, Milano RJ, Bignell GR, Tam AT, Davies H, Stevenson JA, Barthorpe S, Lutz SR, Kogera F, Lawrence K, McLaren-Douglas A, Mitropoulos X, Mironenko T, Thi H, Richardson L, Zhou W, Jewitt F, Zhang T, O'Brien P, Boisvert JL, Price S, Hur W, Yang W, Deng X, Butler A, Choi HG, Chang JW, Baselga J, Stamenkovic I, Engelman JA, Sharma SV, Delattre O, Saez-Rodriguez J, Gray NS, Settleman J, Futreal PA, Haber DA, Stratton MR, Ramaswamy S, McDermott U, Benes CH (2012) Systematic identification of genomic markers of drug sensitivity in cancer cells. Nature 483 (7391):570–575. https://doi.org/10.1038/nature11005

70. Menden MP, Iorio F, Garnett M, McDermott U, Benes CH, Ballester PJ, Saez-Rodriguez J (2013) Machine learning prediction of cancer cell sensitivity to drugs based on genomic and chemical properties. PLoS One 8 (4). https://doi.org/10.1371/journal.pone.0061318

71. Costello JC, Heiser LM, Georgii E, Gonen M, Menden MP, Wang NJ, Bansal M, Ammad-ud-din M, Hintsanen P, Khan SA, Mpindi JP, Kallioniemi O, Honkela A, Aittokallio T, Wennerberg K, Collins JJ, Gallahan D, Singer D, Saez-Rodriguez J, Kaski S, Gray JW, Stolovitzky G, Community ND (2014) A community effort to assess and improve drug sensitivity prediction algorithms. Nat Biotechnol 32(12):1202–U1257. https://doi.org/10.1038/nbt.2877

Chapter 15

A Robust Optimization Approach to Cancer Treatment under Toxicity Uncertainty

Junfeng Zhu, Hamidreza Badri, and Kevin Leder

Abstract

The design of optimal protocols plays an important role in cancer treatment. However, in clinical applications, the outcomes under the optimal protocols are sensitive to variations of parameter settings such as drug effects and the attributes of age, weight, and health conditions in human subjects. One approach to overcoming this challenge is to formulate the problem of finding an optimal treatment protocol as a robust optimization problem (ROP) that takes parameter uncertainty into account. In this chapter, we describe a method to model toxicity uncertainty. We then apply a mixed integer ROP to derive the optimal protocols that minimize the cumulative tumor size. While our method may be applied to other cancers, in this work we focus on the treatment of chronic myeloid leukemia (CML) with tyrosine kinase inhibitors (TKI). For simplicity, we focus on one particular mode of toxicity arising from TKI therapy, low blood cell counts, in particular low absolute neutrophil count (ANC). We develop optimization methods for locating optimal treatment protocols assuming that the rate of decrease of ANC varies within a given interval. We further investigated the relationship between parameter uncertainty and optimal protocols. Our results suggest that the dosing schedule can significantly reduce tumor size without recurrence in 360 weeks while insuring that toxicity constraints are satisfied for all realizations of uncertain parameters.

Key words Robust optimization, Mixed integer optimization, Cancer treatment, Toxicity uncertainty

1 Introduction

An important problem in the study of cancer is the development of resistance to anti-cancer therapies. In particular, resistance-mediated treatment failure has been a problem for several blockbuster anti-cancer therapies [1, 2]. The problem of therapy resistance has been extensively studied from the perspective of evolutionary biology [3]. For example, in [4], the authors developed a stochastic model with experimental data to study the likelihood, composition, and diversity of pre-existing resistance. Their results show that there is at most one resistant clone present at the time of diagnosis for most patients. In another work [5], the authors constructed a stochastic model to study the timing of resistance-mediated treatment failure. They found that in the

Louise von Stechow (ed.), *Cancer Systems Biology: Methods and Protocols*, Methods in Molecular Biology, vol. 1711, https://doi.org/10.1007/978-1-4939-7493-1_15, © Springer Science+Business Media, LLC 2018

setting of treatment of non-small cell lung cancer with the targeted therapy Tarceva it is possible that treatment is discontinued too early.

Mathematical models of cancer evolution during treatment have the potential to be very useful in the creation of optimal treatment schedules. If one can construct computationally tractable mathematical models of cancer evolution under treatment, then it is possible to compare various treatment regimens and thereby search for the most effective regimen. A significant hurdle in the use of such models is parameter variability and uncertainty. In particular, one may have a computationally tractable mathematical model for tumor evolution during treatment, but finding an optimal treatment regimen for a patient requires knowing the model parameters for that patient. One possible solution to this problem is to estimate model parameters for a specific patient [6]. However, this approach is often hindered by a lack of sequential tumor size data for individual patients. An alternative approach is developing optimal treatment schedules that are robust to uncertainty in model parameters. In [7] we developed an approach for optimizing radiation therapy schedules in the presence of uncertain model parameters.

An exciting development of the past 15 years of cancer medicine is the development of new small molecule pharmaceutical agents that specifically target cancer cells [8, 9]. One stunning success has been in the treatment of chronic myeloid leukemia (CML) with the tyrosine kinase inhibitor (TKI) imatinib [10]. Since the launch of imatinib several other TKIs have been developed that are effective in the treatment of CML, e.g., dasatinib and nilotinib [11, 12]. In general, these drugs target the fusion protein BCR-ABL which results in the unchecked proliferation of CML cells [43]. While TKI therapy has been largely successful, a fraction of patients' experience treatment failure due to the evolution of mutated cancer cells that are resistant to the TKI they have been treated with. For example, in [13] researchers reported that the failure rate at 60 months for patients receiving imatinib was 17%. One possible method for reducing the risk of this evolved resistance is to treat patients with a variety of TKIs thereby reducing the risk of treatment failure due to a cell that is resistant to a specific TKI.

This leads to a challenging optimization problem where the goal is to decide on a sequence of TKI therapies that maximize patient outcomes. In our earlier work [14] we considered this problem, and worked with a mathematical model to study the evolution of CML and normal blood cells under treatment with a variety of TKIs. A potential roadblock for clinical implementation of our prior work is that model parameters are difficult to estimate accurately. Therefore, in the current work we consider an extension of [14] by allowing for uncertainty in model parameters.

The structure of the chapter is the following. In Subheading 2, we review general literature on optimization of cancer therapy in continuous time. In Subheading 3, we present our methodology for solving cancer treatment optimization problems with uncertain toxicity response. This is done in the context of treatment of CML with multiple TKI. In Subheading 4, we present numerical results in the setting of CML. We conclude the Chapter by summarizing our innovative methodology and providing insight into clinical management.

2 General Models

The general statement of an optimization problem in cancer therapy consists of the objective function, the control system of cell dynamics, and the toxicity constraints. Optimal control theory is widely used in the design of treatment protocol problems. The general form for a continuous-time cancer optimization problem can be described as follows:

$$\min J \tag{1a}$$

$$\text{s.t. } \dot{x}(t) = f(x, y) \tag{1b}$$

$$\dot{y}(t) = g(x, y) \tag{1c}$$

$$\widehat{f}(x(t), y(t)) \leq 0 \tag{1d}$$

$$\tilde{f}(x(t), y(t)) = 0 \tag{1e}$$

$$x_{\min} \leq x(t) \leq x_{\max} \tag{1f}$$

$$y_{\min} \leq y(t) \leq y_{\max} \tag{1g}$$

where J is the objective function and is determined by the intended outcome of the therapy, $x(t) = (x_1(t), x_2(t), \ldots, x_{n-1}(t), x_n(t))$ is the state vector which represents the population of n different types of cells at time t, e.g., normal, wild type, or mutant cells, and $y(t) = (y_1(t), y_2(t), \ldots, y_{l-1}(t), y_l(t))$ is the control vector which represents the l control types such as drug dosages, treatment methods (i.e., chemotherapy, radiation therapy, TKIs) or which drug will be applied during the treatment. The equation $\dot{x}(t) = f(x, y)$ is a differential equation governing the cell dynamics. The equation $\dot{y}(t) = g(x, y)$ is a differential equation that governs the drug levels in the system as a function of cell population, i.e., the relationship between drug dosage and tumor sizes with respect to time t. Equations 1d and 1e indicate that the $x(t)$ and $y(t)$ may be constrained by inequality and equality constraints, and Eqs. 1f and 1g indicate the lower and upper bounds for $x(t)$ and $y(t)$, respectively.

2.1 Objective Functions

The role of the objective function in Eq. 1 is to specify the desired outcome of the course of anti-cancer therapy. The simplest form of an objective function is to minimize the tumor population at the end of treatment [15], i.e.,

$$J = C(T) \tag{2}$$

where $C(T)$ is the tumor cell population at time t and T is a given constant parameter indicating the length of treatment period. Although objective functions of the form (Eq. 2) are easy to implement, they suffer from the drawback that they allow for large tumor populations during treatment. To deal with this shortcoming, Murray et al. [16] minimized the total tumor cell population over the interval $[0, T]$ while limiting the side effects of therapy. In particular, they consider the objective function

$$J = \int_0^T (\alpha_1 C(t) + \alpha_2 S_e(t))\mathrm{d}t$$

where $S_e(t)$ is a function modeling side effects. It can be a function of dosage [17], or loss of body weight [18], and α_1 and α_2 are weighting values for the cumulative tumor population and normal tissues toxicity, respectively. Note that if one chooses parameter α_2 as zero, then the goal is to minimize the cumulative tumor population over the time frame $[0, T]$.

2.2 Tumor Growth Models

Most optimization models of cancer therapy assume that tumor growth can be accurately modeled by a set of differential equations (usually ordinary differential equations). Some important questions to consider when building these kinds of models are how the tumor cells grow, how they interact, and how they are affected by anti-cancer therapy. The simplest tumor growth model assumes that all tumor cells proliferate with constant cell cycle duration which results in an exponential growth model:

$$\dot{x}(t) = \lambda x(t)$$

where $x(t)$ is the tumor size at time t, and λ is a constant related to the net-growth rate of the tumor. By using a single parameter, an exponential growth model can capture some key features of the beginning phase of tumor growth. However, the prediction of tumor size based on the exponential growth model does not match well with clinical datasets, since the exponential model will give unreasonably large values over a long time. In particular, limited nutrient availability for large tumors makes the exponential growth an inappropriate model for tumor growth [19]. To overcome this drawback researchers often use models such as logistic or Gompertz models, where the growth rate decays as the tumor population increases [19]. Thus, as t increases, tumor size converges to a maximal volume, the so-called carrying capacity, denoted by K. The logistic growth model is defined based on a linear reduction in the tumor growth which is proportional to the tumor size [20]:

$$\dot{x}(t) = \lambda x(t)\left(1 - \frac{x(t)}{K}\right)$$

Like the logistic growth model, the Gompertz growth model assumes that decreasing growth rate is due to competition for the nutrients in a more densely populated tumor

$$\dot{x}(t) = \lambda x(t) \ln \frac{K}{x(t)} \tag{3}$$

In [20, 21], the authors propose a modified Gompertz model, which incorporates drug concentration. The dynamics of drug concentration are modeled by the following equation:

$$\dot{v}(t) = u(t) - \beta v(t)$$

where $v(t)$ is the drug concentration at time t and $u(t)$ is a piecewise continuous function in time that indicates the rate of drug infusion. Drug concentration falls by a fraction of $\beta v(t)$ over the time dt. The authors assume that the net-growth of a tumor cell population comes from two sources: tumor growth due to cell proliferation and tumor shrinkage due to drug administration. The tumor growth is modeled by a general Gompertz model, i.e., Eq. 3. For modeling cell death, they make two more assumptions: the tumor size linearly decreases $x(t)$, and tumor killing stops if drug concentration drops below v_{th}. In summary, the tumor cell kill is given by the function:

$$L(x(t), v(t)) = k(v(t) - v_{\text{th}})H(v(t) - v_{\text{th}})x(t)$$

where k is the proportion of tumor cells killed per unit time per unit drug concentration, and H is the Heaviside step function which is a discontinuous function whose value is zero for negative argument and one for positive argument [22]:

$$H(v(t) - v_{\text{th}}) = \begin{cases} 0; \text{if } v(t) < v_{\text{th}} \\ 1; \text{if } v(t) \geq v_{\text{th}} \end{cases}$$

The cell dynamics are described as

$$\dot{x}(t) = \gamma x(t) \ln\left(\frac{K}{x(t)}\right) - L(x(t), v(t))x(t)$$

In [23], the authors propose a model to describe the dynamics of acute myeloblastic leukemia (AML). The two cell types considered in this model are normal and leukemic hematopoietic cells. The authors assume that the leukemic population inhibits the growth of normal cells, and that drug treatment can kill both leukemic and normal cells. The models are described in the following:

$$\dot{L}(t) = g\left(\log\frac{L_A}{L}\right)L - fL - Kv(t)L$$

$$\dot{N}(t) = a\left(\log\frac{N_A}{N}\right)N - bN - cNL - hu(t)N + G(t)$$

where $L(t)$ and $N(t)$ denote the population of leukemic and normal cells at time t, respectively. Parameters g and a represent the birth rates of leukemic and normal cells, respectively, f and b are the death rates of leukemic and normal cells, respectively. The parameter c is the degree of inhibition exercised by the leukemic cells over the normal cells, while L_A and N_A are the carrying capacities of leukemic cells and normal cells, respectively. Parameters k and h represent the drug's effect on killing of both leukemic and normal cells. Finally, $G(t)$ is the regrowth rate of normal cells due to the infusion and action of recombinant hemolytic growth factors.

A four-compartment model is proposed to explain the kinetics of the molecular response to imatinib [24]. There are three different cell types in the model: normal cells, wild-type leukemic cells, and mutant leukemic cells. For each cell type, the authors considered four layers: stem cells (SC), progenitor cells (PC), differentiated cells (DC), and terminally differentiated cells (TC). Wild-type leukemic cells can acquire mutations that confer resistance to imatinib at rate μ. The authors assume that imatinib only decreases the birth rates of mutant PC and DC. The basic model is given by

$$\text{SC}: \quad \dot{x}_0 = [\lambda(x_0) - d_0]x_0 \quad \dot{y}_0 = \left[r_y(1-\mu) - d_0\right]y_0 \quad \dot{z}_0 = (r_z - d_0)z_0 + r_y y_0 \mu$$

$$\text{PC}: \quad \dot{x}_1 = a_x x_0 - d_1 x_1 \quad \dot{y}_1 = a_y y_0 - d_1 y_1 \quad \dot{z}_1 = a_z z_0 - d_1 z_1$$

$$\text{DC}: \quad \dot{x}_2 = b_x x_1 - d_2 x_2 \quad \dot{y}_2 = b_y y_1 - d_2 y_2 \quad \dot{z}_2 = b_z z_1 - d_2 z_2$$

$$\text{TC}: \quad \dot{x}_3 = c_x x_2 - d_3 x_3 \quad \dot{y}_3 = c_y y_2 - d_3 y_3 \quad \dot{z}_3 = c_z z_2 - d_3 z_3$$

where x_0, x_1, x_2, and x_3 indicate the populations of normal SC, PC, DC, and TC, respectively. y_0, y_1, y_2, and y_3 indicate the populations of wild-type leukemic SC, PC, DC, and TC, respectively. z_0, z_1, z_2, and z_3 indicate the populations of mutant leukemic SC, PC, DC, and TC, respectively. The rate constants are given by a, b, and c with appropriate indices between normal, wild type, and mutant leukemic cells. d_0, d_1, d_2, and d_3 indicate the death rates of SC, PC, DC, and TC, respectively. λ is a decreasing function describing the homeostasis of normal SC. r_y and r_z are the birth rates of sensitive leukemic and resistant leukemic SC, respectively. In our previous work [14] an optimization problem was designed based on an extension of this model that considered multiple possible drugs (imatinib, dasatinib, and nilotinib). The goal of the optimization problem was to identify a sequence of drug exposures that led to a minimal leukemic cell burden at the end of a fixed time interval.

2.3 Toxicity Modeling

In cancer therapy, the goal is to achieve a maximal reduction in tumor burden while keeping toxic side effects within acceptable levels. Mathematical modeling can be used to understand the relationship between toxic side effects and treatment administration. The control variables may be drug dosages or selections during the course of therapy. Some existing models ignore the toxicity effects by assuming that patients can tolerate the side effects during treatment [25, 26]. However, it is often the case that patients are required to go off drug for a period (drug holiday) due to severe side effects such as grade 3–4 neutropenia [27]. Taking toxicity into account brings an important phenomenon into the model and allows for greater confidence when proposing treatment schedules for the clinical setting.

One approach for modeling toxicity of cancer therapy is a statistical approach that takes into account patient factors such as immune system performance, loss of body weight, and side effects experienced by patients [27–29]. For example, Sokal et al. [28] developed a model to calculate the risk of drug toxicity during treatment as a function of patient's age and the number of platelet and blast cells:

$$r = \exp(0.0116 \times (\text{age} - 43.4) + (\text{spln} - 7.51) + 0.188 \\ \times [(\text{pc}/700)2 - 0.563] + 0.0887 \times (\text{bc} - 2.10))$$

where spln represents the spleen size, pc is the platelet count, and bc is the number of blast cells. If $r < 0.8$, patient is in a low risk protocol. If $0.8 \leq r \leq 1.2$, patient is in an intermediate risk protocol, and if $r > 1.2$, patient is in a high-risk protocol.

In [20], the authors proposed a mathematical model for the prevention of excessive side effects in cancer chemotherapy. First, the drug concentration at the cancer site at any time should be less than a positive constant value $v_{\max}$

$$0 \leq v(t) \leq v_{\max}$$

second, the total cumulative toxicity obtained by taking the integral of drug concentration over the course of treatment should be less than a positive constant value v_{cum}

$$\int_{0}^{T} v(s)\mathrm{d}s \leq v_{\mathrm{cum}}$$

and third, it is also possible to limit the cumulative toxicity over a window of time which is shorter than the total treatment time T under threshold v_{di}

$$\int_{0}^{t+\mathrm{d}t} v(s)\mathrm{d}s \leq v_{\mathrm{di}}$$

The above toxicity constraints are widely applied in cancer optimization problems.

In our previous work [14], we propose a discrete optimization model for studying optimal treatment regimens of CML. We defined cytotoxic regimens as schedules resulting in low absolute neutrophil count (ANC) values in patients at any time during the therapy. We have built a simple mathematical model for the evolution of ANC levels under a variety of therapies and then used this model to monitor the dynamics of the patient's ANC in response to each therapy protocol to ensure that the resulting toxicity in the patient falls within acceptable ranges.

3 Robust Optimization for Patients with CML

In this section, we introduce our original work that develops a dynamical model to study the optimal treatment protocol under toxicity uncertainty in the context of a specific cancer type, CML. CML is a cancer of the blood and bone marrow that is normally caused by the oncogene BCR-ABL [30]. The treatment of CML was transformed by the development of imatinib which is a selective inhibitor of the chimeric protein Bcr-Abl (product of the oncogene BCR-ABL). Initial clinical trials showed that the use of imatinib for the treatment of CML resulted in rapid response in the majority of patients [9, 31]. Despite a positive effect, around 20% of patients who were treated with imatinib do not achieve a complete cytogenetic response (CCR) [44]. One possible cause of this is the presence of imatinib resistant CML cells. In another study of BCR-ABL mutations in CML patients, researchers report that mutations were detected in 195/467 (41%) patients [32]. Several new inhibitors, such as nilotinib and dasatinib, have been developed to obtain an increased potency and a broader range of activity against the known imatinib-resistant mutants [33]. Nilotinib has a 20–30-fold increase in potency over imatinib, while dasatinib shows 100–300-fold higher potency than imatinib in vitro [34]. Overall, these three drugs are promising in the treatment of CML. An important issue that also needs to be considered is that side effects arise due to drug toxicity, including low blood cell count, fever, heart problems, as well as a number of other adverse events [35–37]. Different patients may suffer different side effects and even for the same patient, due to the change in health condition over time, side effects may vary over the course of treatment. This complexity of the side effects induced by TKI therapy makes the scheduling of treatment for CML challenging.

3.1 Nominal Problem Formulation

In this section, we first introduce a series of ordinary differential equations (ODE) that describes the dynamics of normal stem cells, wild-type CML cells, and mutant CML cells in response to combination therapy. Then we explain how the toxicity associated with treatment protocols quantified by monitoring ANC values during treatment. Next, we propose a deterministic optimization problem to find the best schedule of multiple therapies based on the evolution of CML cells according to our ordinary differential equation model. The resulting optimization problem is nontrivial due to the presence of ordinary different equation constraints and integer variables. We explain how the nominal problem can be solved efficiently.

3.1.1 CML Dynamics

We use ODEs to describe the dynamics of stem cells for CML patients over a given time period of M weeks. There are three different types of stem cells: normal stem cells (NSC), wild-type stem cells (WSC), and mutant stem cells (MSC). Let $I = \{1, 2, 3, \ldots, n\}$ be the set of stem cell types, where types 1, 2, and i denote NSC, WSC, and type $(i-2)$ MSC $(3 \leq i \leq n)$, respectively. Let $J = \{0, 1, 2, 3\}$ be the set of drugs used to treat CML, where drug 0, 1, 2, and 3 denote a drug holiday, nilotinib, dasatinib, and imatinib, respectively. Let $M = \{1, 2, 3, \ldots, M\}$ be the set of treatment periods and $x_{i(t)}$ the abundance of NSC, WSC, and MSC at time t for $i \in I$, respectively. In this project, we assume that $\Delta t = 7$ days. If drug $j \in J$ is taken for week m, the cell dynamics are modeled as below:

$$\dot{x}_1(t) = \left(b_1^j \psi_{x_1} - d\right) x_1, \qquad t \in [m\Delta t, (m+1)\Delta t], m \in M \setminus \{M\}, \quad (4a)$$

$$\dot{x}_2(t) = \left(b_2^j (1 - (n-2)\mu) \psi_{x_2} - d\right) x_2, \qquad t \in [m\Delta t, (m+1)\Delta t], m \in M \setminus \{M\}, \quad (4b)$$

$$\dot{x}_i(t) = \left(\left(b_i^j \psi_{x_2} - d\right) x_i + \mu b_2^j \psi_{x_2} x_2\right), \qquad t \in [m\Delta t, (m+1)\Delta t], m \in M \setminus \{M\}, 3 \leq i \leq n, \quad (4c)$$

Here, we assume that the birth rates of the NSC, WSC, and MSC are drug specific, but drugs do not affect the death rates of stem cells and all the stem cells have the same death rate d. The division rates of NSC, WSC, and MSC under drug j are b_1^j, b_2^j, and b_i^j per week, respectively. MSC are mutated from WSC with a mutation rate μ. The competition between normal and leukemic stem cells is modeled by the density dependence function ψ_{x_i}, where $\psi_{x_i} = 1/\left(1 + p_i \sum_{k=1}^{n} x_k(t)\right)$. These functions ensure that the total number of normal and leukemic stem cells remains constant once the system reaches a steady state [38]. We set the constants $p_1 = \left(\frac{b_1^0}{d_1} - 1\right)/K_1$ and $p_2 = \left(\frac{b_2^0}{d_1} - 1\right)/K_2$, where K_1 and K_2 are the equilibrium abundance of NSC and WSC. In the equilibrium system of NSC, we further assume only NSC is present. In the equilibrium system of WSC, we assume only WSC is present. We also assume that $p_2 = p_i\ (3 \le i \le n)$.

Note that in this model we focus solely on the stem cell layer since our earlier work [14] thoroughly investigated the structure of optimal schedules in a hierarchical population model, i.e., model with multiple layers.

3.1.2 Toxicity Modeling in Nominal Optimization Problem

In our previous work [14], we developed a model to quantify the ANC levels in patients during the course of therapy. Here we review this model. We assume the patient's ANC level decreases at rate $d_{\text{anc},j}$ per week taking drug j, for $j = 1, 2, 3$. During drug holiday, ANC increases at rate $-d_{\text{anc},0}$ per week but never exceeds the normal level $\text{ANC}_{\text{normal}}$. At the same time, ANC should stay above an acceptable threshold level L_{anc}. The ANC levels are modeled as:

$$y^{m+1} = \min\left(y^m - \sum_{j \in J} d_{\text{anc},j} z^{m,j},\ \text{ANC}_{\text{normal}}\right)$$

$$y^m \ge L_{\text{anc}}$$

where y^m is the ANC value at week m. $z^{m,j}$, the binary decision variables, indicate whether drug j is taken in week m or not, for each $j = 0, 1, 2, 3$ and $m = 0, 1, \ldots, M - 1$.

3.1.3 Nominal Optimization Problem

Assume that the initial population for each cell type is known. The goal of the nominal problem is to develop a treatment protocol to minimize the cumulative leukemic cell number over a given planning period subject to the toxicity constraints. The drug used in each treatment cycle is determined by the weekly treatment decision. Within each week, the dosing regimen stays identical on a day-to-day basis. The cumulative leukemic cell numbers at time t are modeled by the total number of WSC and MSC which is $\sum_{m \in M,\ i \in I \setminus \{1\}} x_{i,m}$, where $x_{i,m} = x_i(m\Delta t)$.

The nominal optimization problem can be formulated as a mixed-integer optimization problem with ODE constraints: details are provided in Appendix 1.

3.2 Robust Problem Formulation

A challenge of utilizing this optimization procedure in the clinical setting is that parameters such as birth rates, death rates, and toxicity decreasing rates in model (Eq. 7) may vary among patients.

Even for the same patient, due to changes in health status, these parameters may vary over time. By modeling the uncertainty in (Eq. 7), we investigate how parametric uncertainty affects the optimal solution. Specifically, we consider the uncertainty of drug toxicity in the model. The problem is formulated as a mixed integer robust optimization problem. Our objective function is to minimize the cumulative leukemic cell number over a fixed period. The goal of the study is to investigate how the parameter uncertainty affects the optimal solution.

3.2.1 Toxicity Uncertainty

We primarily focus on uncertainty in the rate at which the ANC level decreases under the different treatment options. In particular, we assume

$$d^{m}_{\mathrm{anc},j} = L^{j} + \widehat{C}^{j}\eta^{m,j} \tag{5}$$

where L^j is the lower bound of ANC decrease rates under drug j; $\widehat{C}^j$ a positive constant value and $\eta^{m,j}$ is an unknown random variable between 0 and 1, which is used to capture the uncertainty of drug toxicity. First note that we can relax constraint (Eq. 8g) with the following inequality:

$$y^{m+1} \le \widehat{y}^{m} - \sum_{j\in J}\left(L^{j} + \widehat{C}^{j}\eta^{m,j}\right)z^{m,j} = \widehat{y}^{m} - \sum_{j\in J} L^{j} z^{m,j} - \sum_{j\in J}\widehat{C}^{j}\eta^{m,j}z^{m,j}.$$

In robust optimization, we are interested in finding the best solution that is feasible for all realizations of uncertain parameters and we do not allow any violation of the toxicity constraint for any parameters taking values in the sets (Eq. 5). Therefore, the robust counterpart of the nominal problem associated with uncertainty sets defined in Eq. 5 is found by solving

$$\min \sum_{m\in M,\quad i\in I\backslash\{1\}} x_{i,m} \tag{6a}$$

$$\text{s.t.}\ \sup\left\{ y^{m+1} - \widehat{y}^{m}\sum_{j\in J} L^{j} z^{m,j} + \sum_{j\in J}\widehat{C}^{j}\eta^{m,j}z^{m,j} | \eta^{m,j}\in[0,1]\right\} \le 0 \tag{6b}$$

$$\begin{gathered}(8b),(8c),(8d),(8e),(8f),(8h),\\(8i),(8j),(8k),(8l),(8m),(8n)\end{gathered} \tag{6c}$$

Further mathematical details on the solution and derivation of the robust optimization problem can be found in Appendix 2.

4 Example of Applying Modeling Methodology to CML Treatment

In this section, first we will describe the dataset and parameters that were used in our numerical experiments, then the dynamics of the CML cells under three mono-therapies will be simulated. Next, the solution to the nominal and robust optimum drug schedule under toxicity constraints will be explored. At the end of this section the sensitivity of the optimal solution to model parameters and the effect of the robust optimization on treatment outcome are studied.

4.1 Parameter Selection

For our model, we assume there are two BCR-ABL-mutant cell types that are Y253F and F317L. We consider patients harboring three different levels of the BCR-ABL mutant cells before the start of therapy: low, medium, and high. The corresponding initial cell populations are given in Table 1. The parameter settings for birth rates and death rates in our model (Eq. 7) are given below. Based on [38], we set death rate d to be 0.003. The net-growth rate of NSC is assumed to be 0.005. The net-growth rates of WSC $\left(b_2^j\right)$ under drug holiday and mono-therapies are 0.008 and 0.002, respectively. We assume that the net-growth rates of MSC under holiday $\left(b_i^0,\ \ i = 3,\ \ 4\right)$ are the same as b_2^0 which is 0.008. b_i^j for $i = 3\ ,4$ and $j \geq 1$ are estimated based on the work presented in [39] which studied the in vivo mutational selectivity profile for mono-therapies. We consider two mutant cell types in the model, i.e., Y 253F and F317L. For Y 253F, the estimated values of b_3^j are 0.0088, −0.0097, and 0.0101 under nilotinib, dasatinib, and imatinib, respectively. For F317L, the estimated values of b_4^j are −0.0228, 0.0509, and −0.0079 under nilotinib, dasatinib, and imatinib, respectively. The mutation rate of WSC is 10^{-7} [24]. We assume the equilibrium abundance of NSC (K_1) and WSC (K_2) are 10^7 and 2×10^7, respectively.

For toxicity constraints, we assume the patient's normal ANC level is $U_{\text{anc}} = 3000/\text{mm}^3$ANC and its ANC cannot fall below $L_{\text{anc}} = 1000/\text{mm}^3$. We assume that the patient's initial ANC is $3000/\text{mm}^3$. Based on the median time of grade 3 or 4 episode of neutropenia, we estimated the weekly decrease rates of ANC as $d_{\text{anc}\ ,\ 1} = -\ 145.8333/\text{mm}^3$ under nilotinib [40], $d_{\text{anc}\ ,\ 2} = -\ 125/\text{mm}^3$ under dasatinib [41], and $d_{\text{anc}\ ,\ 3} = -\ 56.4516/\text{mm}^3$ under imatinib [10]. We assume that the ANC of a patient increases by $d_{\text{anc}\ ,\ 0} = 500/\text{mm}^3$ during a drug holiday, before it reaches the normal level $3000/\text{mm}^3$. In this project, we consider two types of uncertainties: $\widehat{C}^j = 0.2 \times L^j$ and $\widehat{C}^j = 0.3 \times L^j$.

Table 1
Initial cell population conditions

	NSC	WSC	Y253F	F317L
Low	9.00×10^6	9.00×10^5	1.00×10^4	1.00×10^4
Medium	9.00×10^6	9.00×10^5	1.00×10^5	1.00×10^5
High	9.00×10^6	9.00×10^5	3.00×10^6	3.00×10^6

4.2 Cell Dynamics Simulations

In this part, we present the dynamics of stem cells with the preexisting BCR-ABL mutation Y 253F and F317L under monotherapies. As reported, Y 253F is highly resistant to imatinib, lightly resistant to nilotinib and sensitive to dasatinib; F317L is highly resistant to dasatinib, and sensitive to imatinib and nilotinib. For this simulation pattern, we expect that all monotherapies will fail eventually because of the presence of the mutant cells and their differentiated responses to drugs. The initial levels of NSC, WSC, Y 253F, and F317L are $9E+06$, $9E+05$, $1E+05$, and $1E+05$, respectively.

Figure 1 plots the cell dynamics over 420 weeks (around 8 years) for six treatment protocols: nilotinib, dasatinib, and imatinb mono-therapy, all of which are performed with and without drug holiday. As F317L is resistant to dasatinib, the population of F317L explodes around week 50 when administering dasatinib [42]. On the other hand, we note that the population of Y 253F is well controlled. In Fig. 2, we only look at the performances of imatinb and nilotinib mono-therapy. The population of Y 253 increases over time, but the population size of F317L decreases in both the cases. Those results indicate that drug combination may be more effective for treating patients with multiple mutant cell types.

Next, we discuss the results for nominal and robust optimization problems. We first report the recurrence time of the optimal schedule and mono therapies assuming that the toxicity parameters are known, i.e., the nominal problem. In addition, we investigate the recurrence time of the resulting optimal schedule when

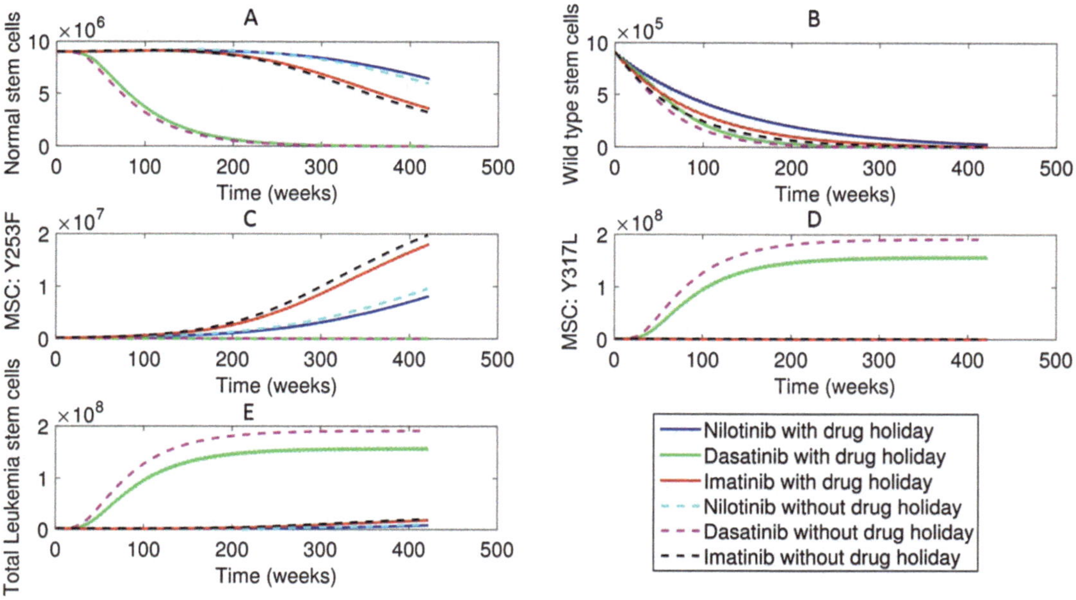

Fig. 1 **(a–e)** Cell dynamics under mono-therapies for 420 weeks (mutant cell types: Y 253F and F317L)

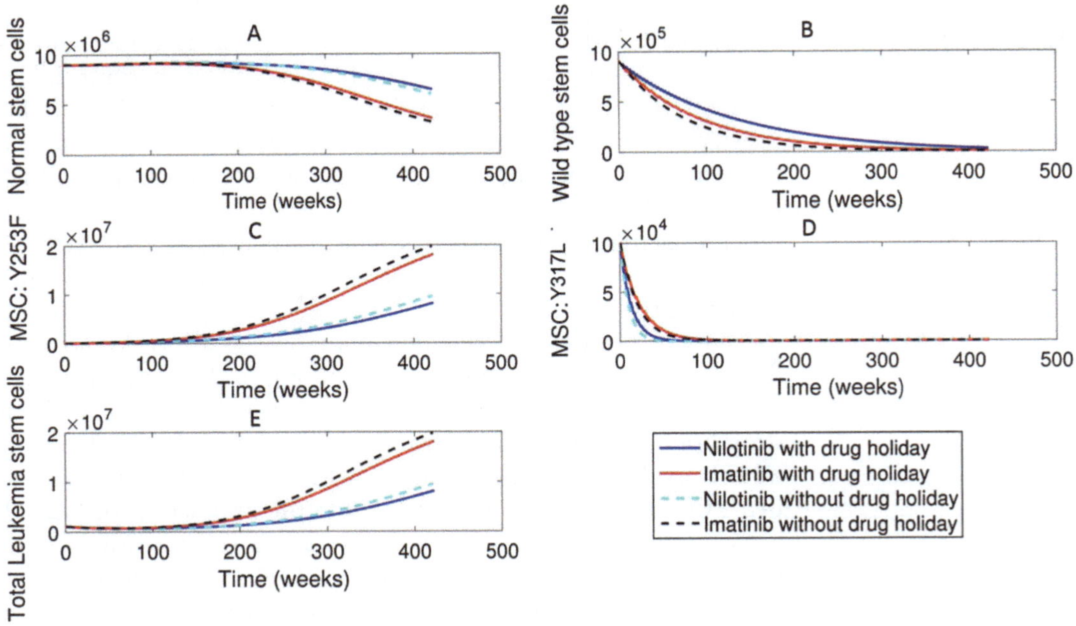

Fig. 2 (**a–e**) Cell dynamics under mono-therapies (without dasatinib) for 420 weeks (mutant cell types: Y 253F and F317L)

perturbing model parameters. Finally, we solve the robust optimization problem under different uncertainty settings and initial conditions.

4.3 Nominal Optimal Treatment Plans

In this section, we are interested in the recurrence time for the two scenarios: mono-therapy and the nominal optimized therapies that are achieved by solving the model presented in Appendix 1 for 360 weeks. The recurrence time is defined as the time at which the tumor cell population returns to its size at the start of treatment. The initial conditions for NSC, WSC, Y 253F, and F317L are $9E+06$, $5E+05$, $3E+05$, and $3E+05$, respectively. The nominal optimal treatment plans are given in Fig. 3. The cell growth is shown in Fig. 4. Since F317L is highly resistant to dasatinib, we show the dynamics of tumor growth under dasatinib only for 50 weeks. The results are summarized in Table 2. Under the optimal schedule, the tumor size keeps decreasing, and thus there is no recurrence time. We thus denote recurrence time by NA. Under nilotinib, the tumor size reaches its minimal size at week 88, then reaches the initial population size at week 183, and doubles its size at week 261. Under imatinib, the tumor size reaches the minimal size at week 63, reaches the initial population size at week 130, and doubles its size at week 185. Under dasatinib, the tumor size keeps increasing.

We also performed a sensitivity analysis on the nominal optimal solution (shown in Fig. 3) with respect to the birth rates of mutant

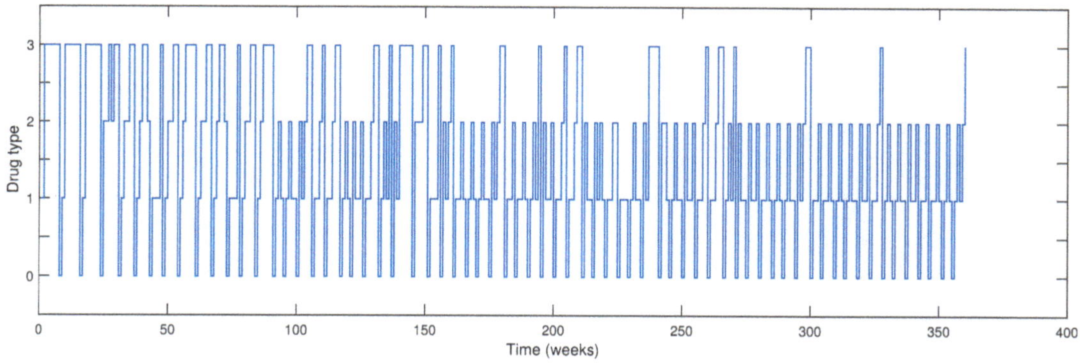

Fig. 3 Optimal solution of the nominal problem for 360 weeks (mutant cell types: Y 253F and F317L). Digits 0, 1, 2, and 3 represent drug holiday, nilotinib, dasatinib, and imatinib

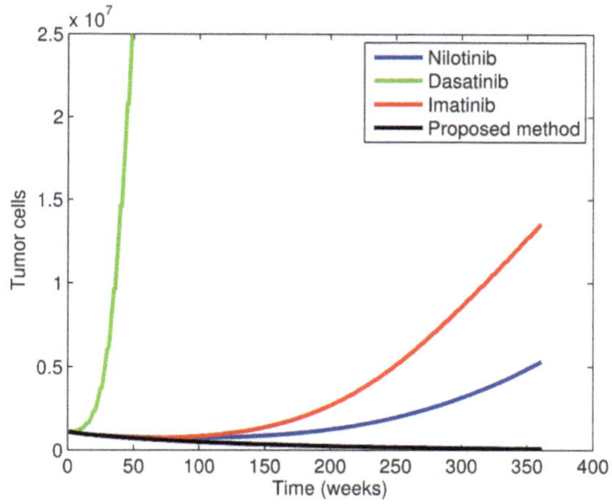

Fig. 4 Cell dynamics under mono-therapies and Optimal solution of the nominal problem for 360 weeks (mutant cell types: Y 253F and F317L)

Table 2
Recurrence time for multiple mutants

	To minimal	Recurrence time	Double the size
Nilotinib	88	183	261
Dasatinib	1	1	20
Imatinib	63	130	185
Drug combination	*NA*	*NA*	*NA*

cells (b_i^j for $i = 3\ ,\ 4$ and $j = 1\ ,\ 2\ ,\ 3$). We are interested in how the recurrence time under schedule (shown in Fig. 3) changes as we vary the birth rates of mutant cells. A 360-week simulation is run to study the behavior of recurrence time. We consider two scenarios.

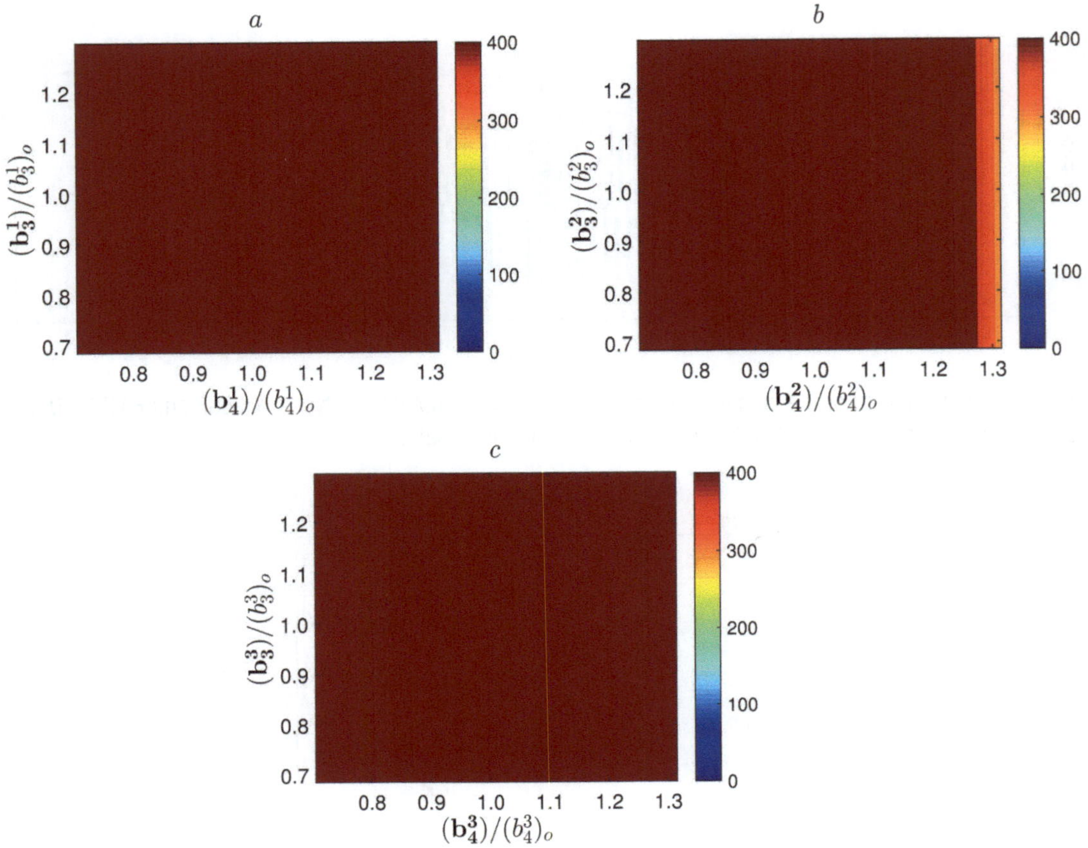

Fig. 5 The recurrence time with respect to the birth rate changes of Y253F and F317L under one drug when the treatment protocols are fixed as the nominal optimal solution. (**a–c**) show the recurrence time when birth rates of Y253F and F317L vary under nilotinib, dasatinib, and imatinib, respectively

Scenario one is that the birth rates of both mutant cells types vary under only one drug, while the birth rates of mutant cells stay constant under the other two drugs, i.e., if the drug affecting birth rates is nilotinib, then b_3^1 and b_4^1 are set to be uniformly distributed on [0.7, 1.3], while b_3^2, b_4^2, b_3^3, and b_4^3 are fixed. The other scenario is that the birth rates of one mutant cell type change under all drugs, whereas the birth rates of the second mutant cell type stay constant, i.e., the birth rates of Y 253F change under all three drugs, while the birth rates of F317L stay the same under all three drugs.

Figure 5 shows the results for scenario one. The colors indicate different recurrence time as indicated by the colorbar, i.e., blue, green, and red corresponding to a recurrence time of 0, 150, and 360, respectively. The original birth rate of type $(i - 2)$ mutant cell under drug j, $\left(b_i^j\right)_o$, is given in Subheading 4.1. The varied birth rates of type $(i - 2)$ cell under drug j are represented by $\boldsymbol{b}_i^j$. The ratio, $\boldsymbol{b}_i^j/\left(b_i^j\right)_o$ is set to be uniformly distributed on [0.7, 1.3]. The results in Fig. 5a, c indicate that the tumor size is below the initial tumor size after 360 weeks using the proposed method. The results in Fig. 5b show that if $\boldsymbol{b}_4^2/\left(b_4^2\right)_o$ is >1.27, recurrence happens before the end of treatment. Recall that $\left(b_4^2\right)_o$, a positive value, is the original net growth rate of F317L under dasatinib. As we increase $\boldsymbol{b}_4^2/\left(b_4^2\right)_o$, $\boldsymbol{b}_4^2$ increases which causes F317L grows faster under dasatinib. However, overall we see that the optimal schedule (shown in Fig. 3) is largely robust to changes in the birth rates of the mutant cells.

Figure 6 shows the results for scenario two where birth rates of F317L vary. For better visualization purposes, we fix the birth rates under one drug while varying the birth rates under the other two, and show the results of the recurrence time. Figure 6a indicates that the ratio of F317L birth rate under nilotinib (drug 1) is fixed at $\boldsymbol{b}_4^1/\left(b_4^1\right)_o = 0.7$ and the ranges of $\boldsymbol{b}_4^2/\left(b_4^2\right)_o$ and $\boldsymbol{b}_4^3/\left(b_4^3\right)_o$ are uniformly distributed on [0.7, 1.3]. Columns 1, 2, and 3 show the recurrence time of tumor for fixed ratio of $\boldsymbol{b}_4^j/\left(b_4^j\right)_o$ set at 0.7, 1.0, and 1.3, respectively. Figures in rows 1 (a, b, c), 2 (d, e, f), and 3 (g, h, i) correspond to $j = 1\ ,2\ ,3$, respectively. The figures in the first row show that the increase in $\boldsymbol{b}_4^1/\left(b_4^1\right)_o$ is less likely to cause the tumor reaching the initial size at the end of treatment. The reason is that F317L is highly sensitive to drug 1 which is nilotinib. As we increase the ratio of $\boldsymbol{b}_4^1/\left(b_4^1\right)_o$, the growth rate of F317L is reduced. From Fig. 6a, we can see that under the extreme case $\left(\boldsymbol{b}_4^1/\left(b_4^1\right)_o = 0.7, \boldsymbol{b}_4^2/\left(b_4^2\right)_o = 1.3, \text{and } \boldsymbol{b}_4^3/\left(b_4^3\right)_o = 0.7\right)$, recurrences happen around week 150. The result is consistent with the recurrence time reported in Fig. 6f, g. There is no recurrence when $\boldsymbol{b}_4^2/\left(b_4^2\right)_o \leq 1$, but if $\boldsymbol{b}_4^2/\left(b_4^2\right)_o = 1.3$, recurrence happens in almost half of the cases. Since F317L is sensitive to both nilotinib and imatinib, the results in Fig. 6g–i are similar to the ones in Fig. 6a–c. The difference is that there is still a chance for tumor recurrence when $\boldsymbol{b}_4^3/\left(b_4^3\right)_o = 1.3$, because nilotinib is applied more often compared to imatinib in the nominal optimal solution.

4.4 Robust Optimal Treatment Plans

As we discuss in Appendix 2, protection levels (Γ^m) adjust the robustness of the proposed model against the conservation level of the solution. In this part, we first compare the robust optimal solutions under different protection levels, which are provided in Appendix 3 for two monotherapies (imatnib and nilotinib). Figure 7 shows the dynamics of tumor growth for 30 weeks under nilotinib, imtinib, and optimal solutions with different protection levels (Fig. 8). For this simulation, we assume $\widehat{C}^j = 0.2 \times L^j$, and initial population sizes are $9E + 06$, $9E + 05$, $1E + 05$, and $1E + 05$, for NSC, WSC, Y 253F, and F317L, respectively. It is interesting to note that the tumor sizes under the proposed methods at week 30 are lower than those predicted for either of the

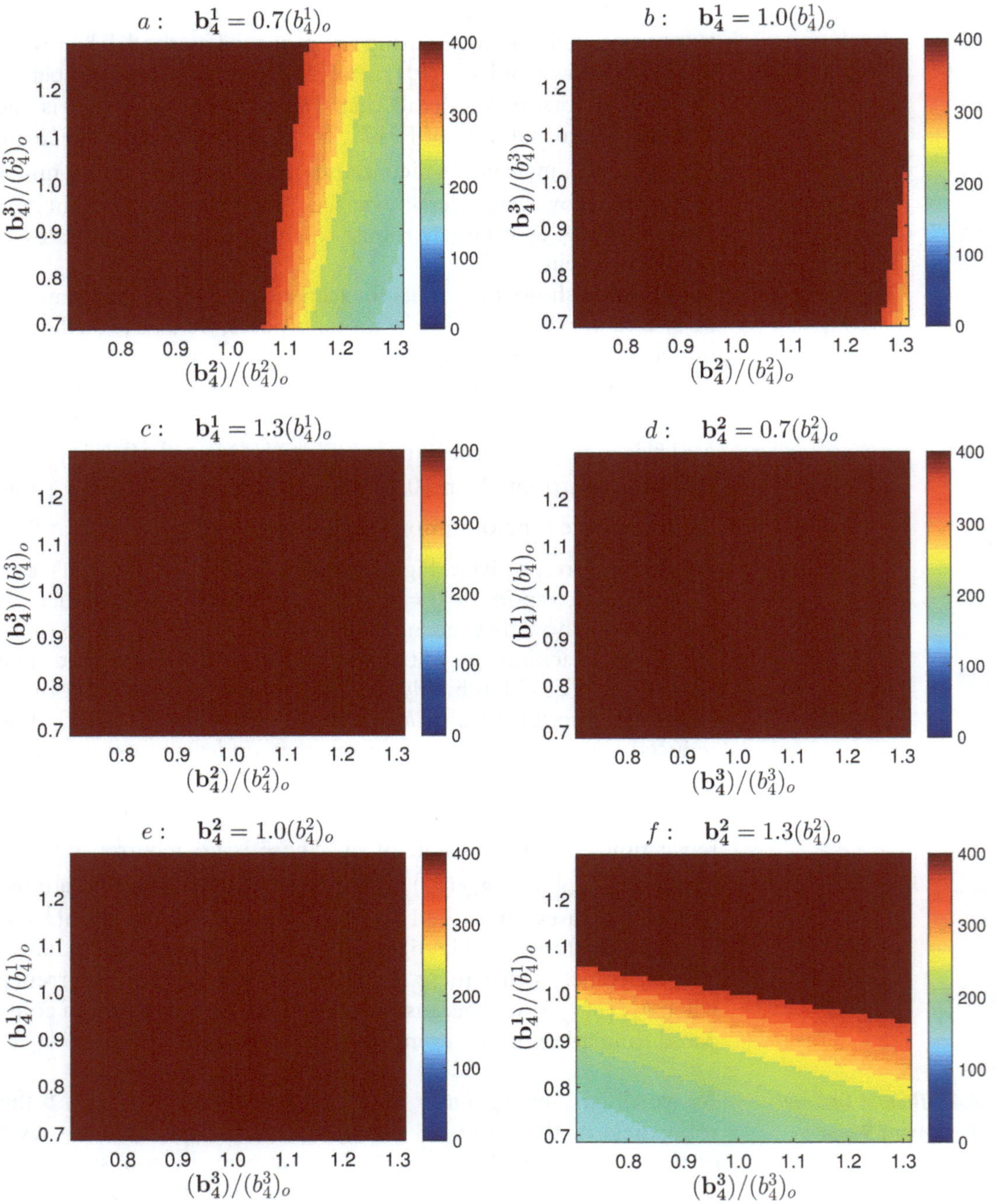

Fig. 6 (a–i) The recurrence time with respect to the varied birth rates of F317L under three drugs when the treatment protocols are fixed as the nominal optimal solution. Rows: constant birth rate set under nilotnib, dasatnib, and imatinib, respectively. Columns: constant birth rate ratio set at 0.7, 1.0 and 1.3, respectively

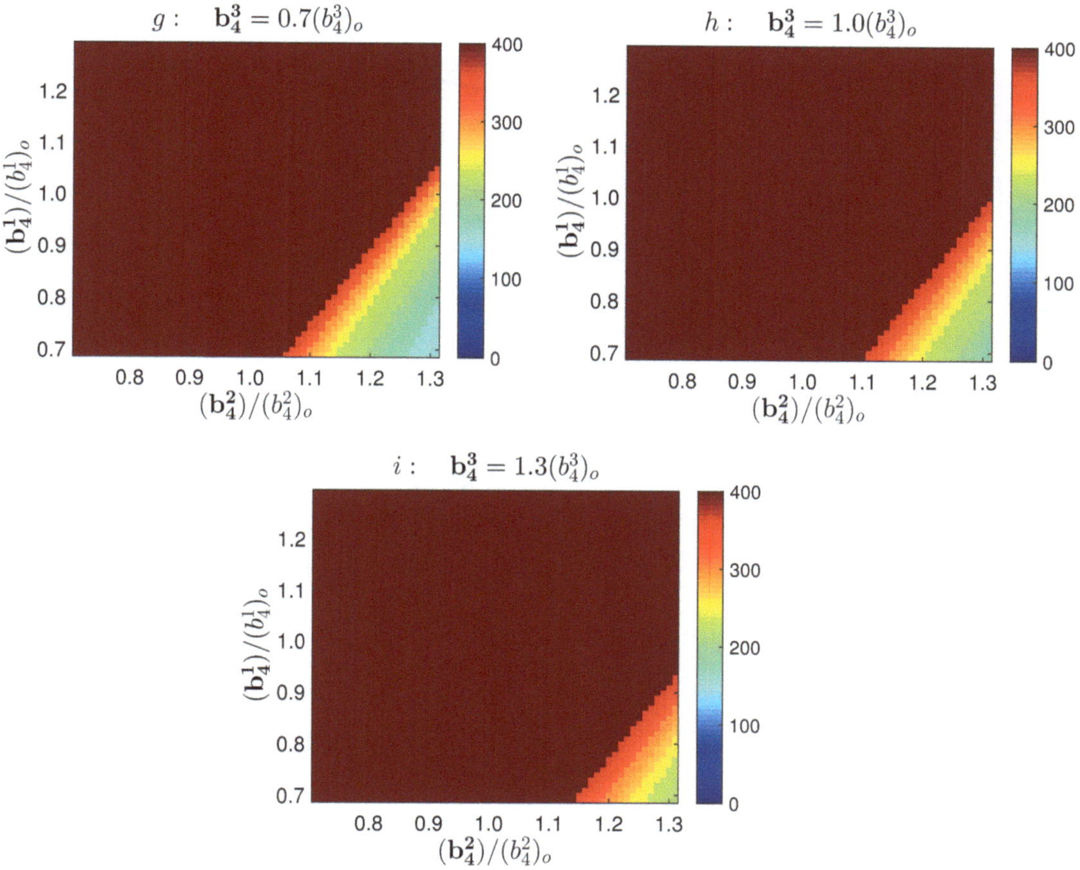

Fig. 6 (continued)

mono-therapies even though our objective function aims to minimize the cumulative tumor sizes.

Table 3 shows the robust optimal solutions under $\widehat{C}^j = 0.2 \times L^j$ for patients with initial tumor sizes at low, medium, and high levels (Table 3). For example, if we take $\widehat{C}^j = 0.2 \times L^j$, for patients with an initial tumor size at low level, under zero protection level, $\Gamma = 0$, the optimal value is 2.41632×10^7. However, with full protection, $\Gamma = 3$, the optimal value is increased by 0.541% to 2.42939×10^7. For patients with an initial tumor size at medium level, under zero protection level, the optimal value is 2.9189×10^7. With full protection, the optimal value is increased by 0.7044% to 2.9395×10^7. For patients with an initial tumor size at high level, under zero protection level, the optimal value is 2.7933×10^7. With full protection, the optimal value is increased by 1.3519% to 2.8311×10^7.

Figure 9 shows the increments of optimal values under different protection levels for patients with initial tumor sizes at low, medium, and high levels when assuming $\widehat{C}^j = 0.2 \times L^j$. The

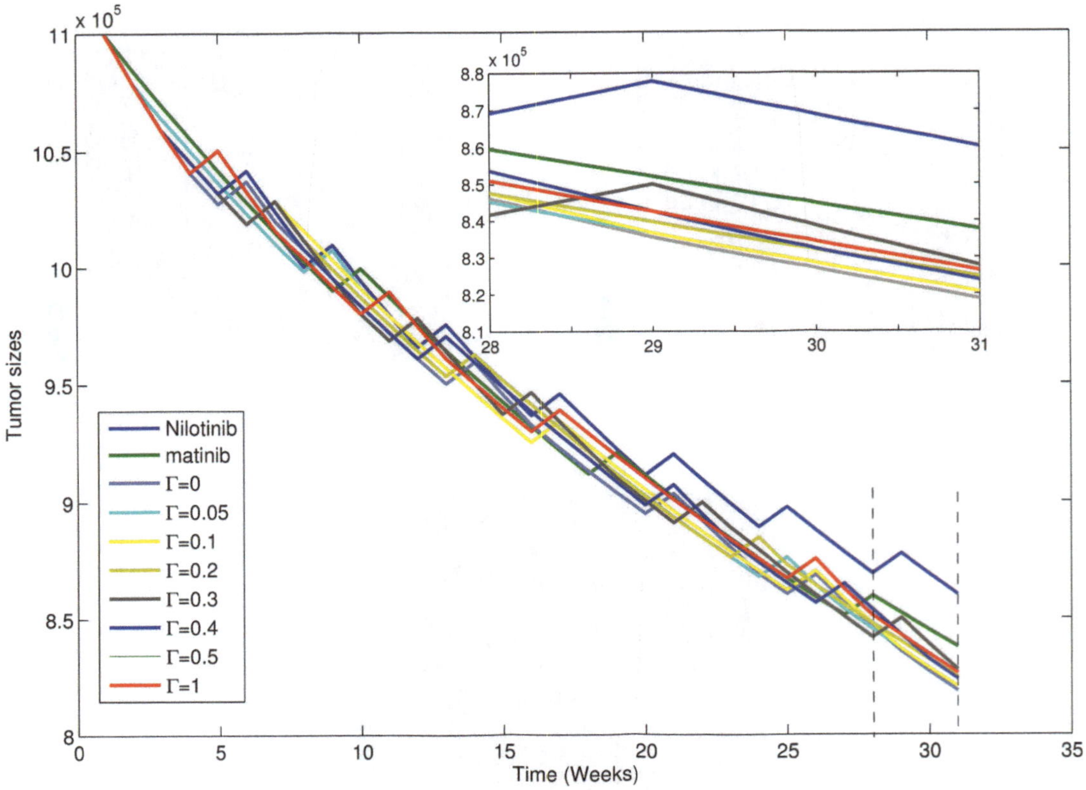

Fig. 7 Cell dynamics under mono-therapies (without dasatinib) and optimal solutions for 30 weeks (mutant cell types: Y 253F and F317L)

increments are calculated by: $\frac{\Upsilon_\Gamma^* - \Upsilon_0^*}{\Upsilon_0^*}$, where Υ_0^* and Υ_Γ^* are the optimal values of the nominal and robust optimization problems under different protection levels, respectively. It is interesting to note that the optimal value of the objective function increases as we increase the protection level of robust solutions.

Next, we consider how the optimal treatment protocols are affected by protection levels Γ and initial conditions of tumor size. Figure 10a–c show the optimal treatment protocols for $\widehat{C}^j = 0.2 \times L^j$ and initial tumor size at low (a), medium (b), and high (c) levels. For initial tumor size at low level, as wild-type cells dominate the total tumor size at the beginning of treatment, it is efficient to reduce tumor size by taking the drug with the lowest toxicity, which is drug 3. Recall that drug 0, 1, 2, and 3 represent drug holiday, nilotinib, dasatinib, and imatinib. At the end of treatment, as the number of mutant cells increases, it is necessary to switch to dasatinib, which can reduce the number of Y 253F cells efficiently. As we increase the protection level, more drug holidays are needed, i.e., for unprotected optimal solutions ($\Gamma = 0$), the third break happens at the end of treatment, week 30, however for

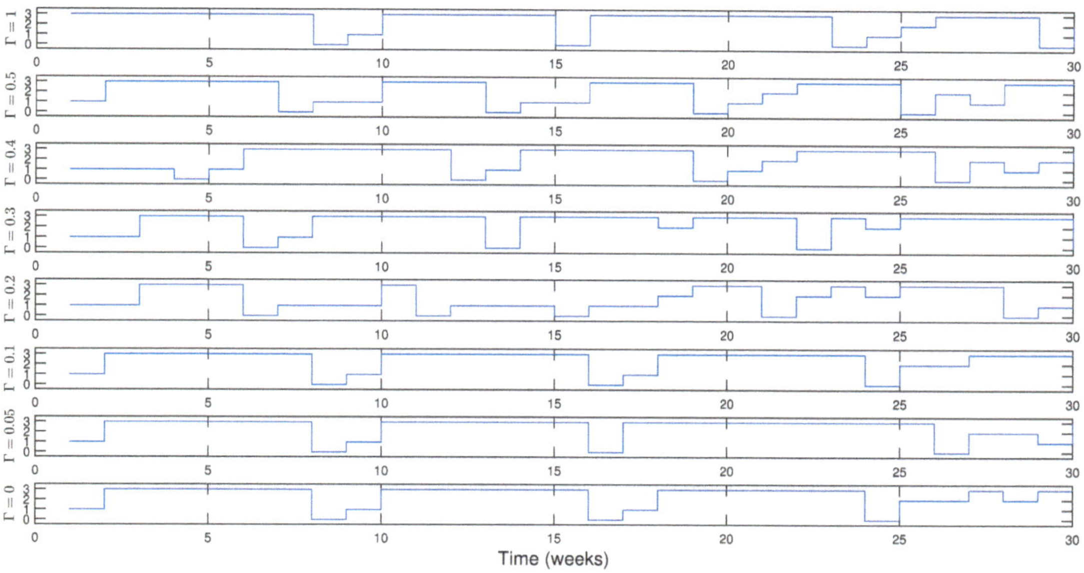

Fig. 8 Robust optimal solutions under $\widehat{C}^j = 0.2 \times L^j$ for 30 weeks. The initial conditions for NSC, WSC, $Y253F$, and $F317L$ are $9E+06$, $9E+05$, $1E+05$, and $1E+05$, respectively.

Table 3
Robust solution for $\widehat{C}^j = 0.2 \times L^j$: Multiple mutants

Γ	(a): Low level Optimal value ($\times 10^7$)	Increment (%)	(b): Medium level Optimal value ($\times 10^7$)	Increment (%)	(c): High level Optimal value ($\times 10^7$)	Increment (%)
0	2.41632	0	2.9166	0	2.7933	0
0.05	2.41632	0	2.9192	0.0877	2.7957	0.0842
0.1	2.41750	0.049	2.9203	0.1250	2.7957	0.0842
0.2	2.41900	0.110	2.9261	0.3240	2.8004	0.2542
0.3	2.41948	0.131	2.9262	0.3287	2.8035	0.3626
0.4	2.42093	0.191	2.9265	0.3379	2.8061	0.4560
0.5	2.42367	0.304	2.9311	0.4940	2.8084	0.5386
1	2.42939	0.541	2.9403	0.8098	2.8288	1.2707

conservative solutions ($\Gamma = 3$), it happens at week 25. Furthermore, as the protection level increases, nilotinib and dasatinib are more frequently used in the treatment to guarantee that optimal solutions do not violate the toxicity constraint under different realizations of model parameters.

For patients with an initial tumor size at medium level, as the mutant cell population increases, nilotinib and dasatinib appear

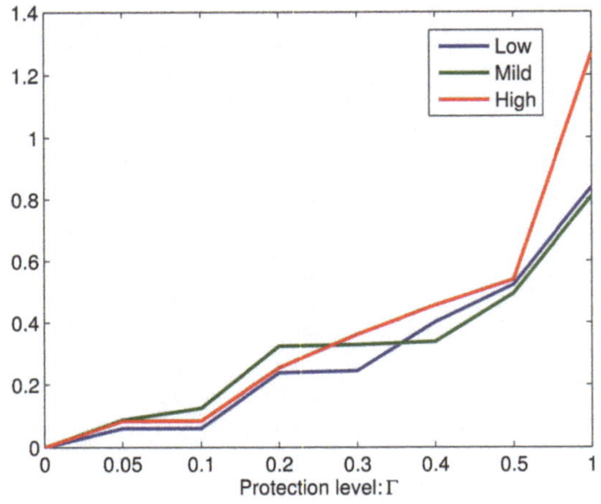

Fig. 9 Tumor size increments of optimal values under Γ with three different initial conditions

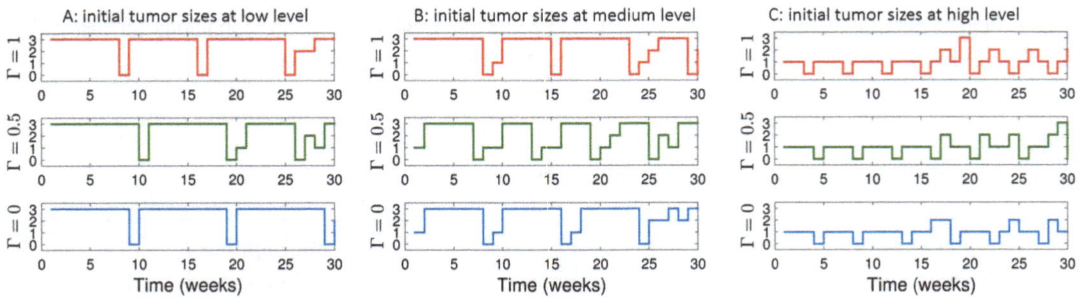

Fig. 10 Optimal solutions under $\widehat{C}^j = 0.2 \times L^j$ with three different initial conditions: (**a**) initial tumor size at low level; (**b**) initial tumor size at medium level; (**c**) initial tumor size at high level

more often in the optimal solution during the course of therapy. Similarly, patients with an initial tumor size at medium level also need to take longer breaks as we increase the protection level of the toxicity constraint.

For patients with initial tumor size at high level, the populations of mutant cells dominate the tumor sizes. For the first 15 weeks, nilotinib is delivered to reduce the population size of F317L. Note that the population size of Y 253F keeps increasing during the first 15 weeks due to its resistance to nilotinib. To control the size of Y 253F, dasatinib is administrated during the last 15 weeks.

To investigate the effects of the size of the uncertainty ranges on the optimal solutions, we perform simulation studies to see how the structure of optimal schedules changes in the context of various uncertainty ranges. Table 4 shows the robust optimal values under $\widehat{C}^j = 0.3 \times L^j$ for patients with an initial tumor size at low, medium, and high levels. Figure 11 shows the optimal solutions

Table 4
Robust solution for $\widehat{C}^j = 0.3 \times L^j$: Multiple mutants

Γ	(a): Low level Optimal value (×10⁷)	Increment (%)	(b): Medium level Optimal value (×10⁷)	Increment (%)	(c): High level Optimal value (×10⁷)	Increment (%)
0	2.4163	0	2.9167	0	2.7933	0
0.05	2.4178	0.0604	2.9197	0.1036	2.7957	0.0842
0.1	2.4178	0.0604	2.9220	0.1840	2.8000	0.2397
0.2	2.4221	0.2386	2.9251	0.2913	2.8050	0.4176
0.3	2.4222	0.2448	2.9281	0.3912	2.8102	0.6044
0.4	2.4260	0.4018	2.9313	0.5033	2.8107	0.6217
0.5	2.4289	0.5223	2.9341	0.5980	2.8219	1.0237
1	2.4366	0.8394	2.9524	1.2242	2.8443	1.8250

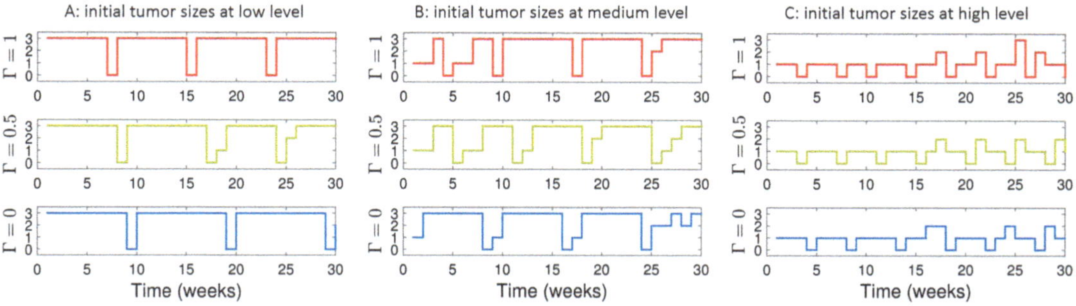

Fig. 11 Optimal solutions under $\widehat{C}^j = 0.3 \times L^j$ with three different initial conditions: (**a**) initial tumor size at low level; (**b**) initial tumor size at medium level; (**c**) initial tumor size at high level

for patients with an initial tumor size at low, medium, and high levels, respectively. These results are similar to those of $\widehat{C}^j = 0.2 \times L^j$. Hence, we can conclude that the structure of the optimal solution is only mildly sensitive to the size of the uncertainty range.

Next, we focus on comparing the differences that resulted from the uncertainty ranges $\left(\widehat{C}^j = 0.3 \times L^j \text{ and } \widehat{C}^j = 0.3 \times L^j\right)$. From Fig. 12, we observe that: for patients with an initial tumor size at low, medium, and high levels, the larger the toxicity uncertainty ranges, the larger the optimal value.

The idea of imposing protection levels on robust optimization is to use conservative constraints that guarantee no toxic side effects occur. Here, we compare the performance of nominal solutions versus robust optimization solutions in terms of objective function and toxic side effects. We do this by randomly generating ANC

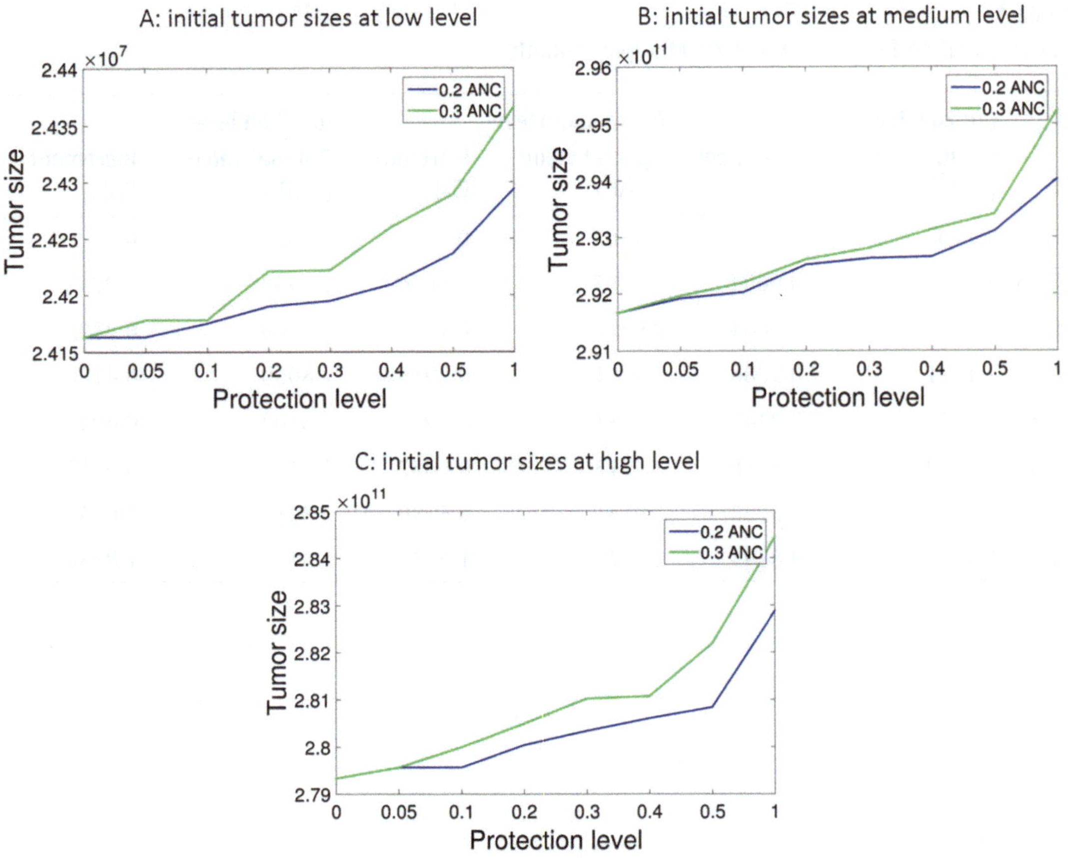

Fig. 12 The effects of uncertainty ranges under three different initial conditions: (**a**) initial tumor size at low level; (**b**) initial tumor size at medium level; (**c**) initial tumor size at high level

decay rates and comparing performance of robust and nominal optimal solutions for the generated decay rates. We do this repeatedly and look at the average increase in leukemic cell burden that results when using the robust optimal schedule, we also look at the fraction of times we have toxic side effects when using the nominal optimal solution. This process of repeatedly generating random variables and averaging results is known as Monte-Carlo simulation. To summarize, we use Monte-Carlo simulation to understand how much greater the cumulative tumor size is under robust optimization to guarantee that patients will not show toxic side effects, and how much more toxicity (in terms of decreasing ANC) patients will suffer if they are treated with the nominal therapy. Two sets of simulations are conducted, corresponding to different uncertainty ranges ($\widehat{C}^j = 0.2 \times L^j$ and $\widehat{C}^j = 0.3 \times L^j$). For both simulations, the initial populations for NSC, WSC, Y 253F, and F317L are $9E+06$, $5E+05$, $3E+05$, and $3E+05$, respectively. The ANC decrease rate $d^m_{\mathrm{anc},j}$ under drug j at mth week is randomly generated by assuming $\eta^{m,j}$ is uniformly distributed on $[0, 1]$. If

Table 5
Price of robust optimization

Γ	$\widehat{C}^j = 0.2 \times L^j$ Increments in OBJ (%)	Toxicity invalidation (%)	$\widehat{C}^j = 0.3 \times L^j$ Increments in OBJ (%)	Toxicity invalidation (%)
1	1.2707	0	1.8250	0
0.5	0.5386	49.98	1.0237	48.64
0.4	0.4560	61.80	0.6217	61.89
0.3	0.3626	69.18	0.6044	72.31
0.2	0.2542	79.07	0.4176	82.67
0.1	0.0842	94.08	0.2397	96.53
0.05	0.0842	94.08	0.0842	97.38
0	0	100	0	100

the ANC value is $<L_{\text{anc}}$ during the simulation, then the patient received a toxic side effect and the simulation is considered infeasible. The fraction of cases that are infeasible due to toxic side effects is calculated by the total number of infeasible cases divided by the total number of cases (10^6). The results are shown in Table 5. Recall that $\Gamma = 0$ is equivalent to the nominal problem. For both simulations, as we increase Γ the objective value increases, while the probability of toxicity violation decreases. The optimal solution obtained by ROP seems to yield an interesting tradeoff between the two objectives of minimizing cumulative tumor population and the infeasibility of toxicity constraints beyond which allowing for more risky regimens, i.e., using smaller Γ, does not lead to any significant gain in objective function. In particular, if $\widehat{C}^j = 0.2 \times L^j$, then it appears that around $\Gamma = 0.2$ there is a sharp change in the fraction of runs that lead to toxic side effects and a significant increase in objective value.

5 Conclusion

The major focus of this chapter was to introduce a mathematical model for identifying optimal anti-cancer treatment strategies in the presence of parameter uncertainty. These methods have great potential for designing and understanding optimal anti-cancer treatments. Our general framework is to build a differential equation model for the relevant cancer and normal cell populations undergoing a particular treatment. For many differential equations, it is necessary to develop a linear approximation via a linear regression model. We next develop a mathematical model for the most relevant toxicities in the treatment we are studying. With these mathematical models in place we are able to build a mathematical optimization model for identifying optimal treatment schedules. In order to account for patient variability, we make our model robust to inter-patient heterogeneity in the rate at which side effects develop. We can then use software solvers to identify treatment schedules that are optimal and robust. In this chapter, we applied this methodology to study the treatment of chronic myeloid leukemia (CML) with a variety of possible tyrosine kinase inhibitors (TKI).

There are several areas for improvement in our method. First, we assume the drug dosages are constant, and therefore ignore the possible benefits or risks of varying doses. Second, we assume that there is no drug present in the patient from the previous treatment when we switch to a new drug. In order to better characterize this residual term detailed analysis of drug-drug interaction would be necessary. Third our method only accounts for inter-patient variability in toxicity terms and not in cancer cell growth or death rates. This is an important aspect of inter-patient variability that we plan to further pursue. Finally, our method requires that we approximate the governing differential equations with a model that is linear in the state. This prevents us from finding the true optimal solution; furthermore, this approximation can be problematic in systems that exhibit strongly nonlinear behaviors.

Acknowledgments

JZ was supported by NSF grant DMS-1224362. HB was supported by NSF grant CMMI-1362236. KL was supported by NSF grants CMMI-1362236 and CMMI-1552764.

Appendix 1: Nominal Optimization Problem

The nominal optimization problem can be formulated as a MIOP as below:

$$\min \sum_{m\in M,\quad i\in I\setminus\{1\}} x_{i,m} \tag{7a}$$

$$\text{s.t.}\quad \dot{x}_1(t)=\sum_{j=0}^{3} z^{m,j}\left(b_1^j\psi_{x_1}-d\right)x_1,\quad t\in[m\Delta t,(m+1)\Delta t],\ m\in M\setminus\{M\} \tag{7b}$$

$$\dot{x}_2(t)=\sum_{j=0}^{3} z^{m,j}\left(b_2^j(1-(n-2)\mu)\psi_{x_2}-d\right)x_2, t\in[m\Delta t,(m+1)\Delta t],\ m\in M\setminus\{M\} \tag{7c}$$

$$\dot{x}_i(t)=\sum_{j=0}^{3} z^{m,j}\left(\left(b_i^j\psi_{x_2}-d\right)x_i+\mu b_2^j\psi_{x_2}x_2\right),\ t\in[m\Delta t,(m+1)\Delta t], m\in M\setminus\{M\}, 3\le \text{i}\le n \tag{7d}$$

$$\sum_{j\in J} z^{m,j} = 1, \qquad m\in M\backslash\{M\}, \tag{7e}$$

$$y^{m+1} = \hat{y}^m - \sum_{j\in J} d_{\text{anc},j} z^{m,j}, \qquad m\in M\backslash\{M\}, \tag{7f}$$

$$\hat{y}^m = \min(y^m, \text{ANC}_{\text{normal}}), \qquad m\in M, \tag{7g}$$

$$L_{anc} \le \hat{y}^m, \qquad m\in M, \tag{7h}$$

$$z^{m,j}\in\{0,1\}, \;\; m\in M\backslash\{M\}, j\in J \tag{7i}$$

where $x(0)$, y^0 are given. In Eqs. 7b, 7c, and 7d, the dynamics of NSC, WSC, and MSC are described, respectively. Equations 7e, and 7i indicate that during each week, only one type of drug or no drug is allowed. Equations 7f, 7g, and 7h describe the toxicity constraints.

As discussed in the previous work [26], the ODEs can be approximated by linear functions:

$$\min \sum_{m\in M,\;\; i\in I\backslash\{1\}} x_{i,m}$$

$$\text{s.t.} \quad x_{i,m+1} = \sum_{j=0}^{3} z^{m,j}\left(C_{i,0}^{j} + \sum_{k=1}^{n} C_{i,k}^{j} x_{k,m}\right), \;\; t\in[m\Delta t, (m+1)\Delta t], m\in M\backslash\{M\}$$

$$\sum_{j\in J} z^{m,j} = 1, \qquad m\in M\backslash\{M\}$$

$$y^{m+1} = \hat{y}^m - \sum_{j\in J} d_{\text{anc},j} z^{m,j}, \qquad m\in M\backslash\{M\}$$

$$\hat{y}^m = \min(y^m, \text{ANC}_{\text{normal}}), \qquad m\in M$$

$$L_{\text{anc}} \le \hat{y}^m, \qquad m\in M$$

$$z^{m,j}\in\{0,1\}, \;\; m\in M\backslash\{M\}, j\in J$$

where $x(0)$, y^0 are given.

There are two types of nonlinear terms here: $z^{m,j}x_{i,m}$ and $\hat{y}^m = \min(y^m, \text{ANC}_{\text{normal}})$.

To linearize $z^{m,j}x_{i,m}$, we introduce a new variable

$$0 \le v_i^{m,j} \le U_i z^{m,j}$$

$$-U_i\left(1 - z^{m,j}\right) \le v_i^{m,j} - x_{i,m} \le U_i\left(1 - z^{m,j}\right)$$

To linearize $\hat{y}^m = \min(y^m, \text{ANC}_{\text{normal}})$, we introduce a binary variable p^m

$$\hat{y}^m \geq \text{ANC}_{\text{normal}} - U_y(1 - p^m),$$
$$\hat{y}^m \geq y^m - U_y p^m,$$
$$\hat{y}^m \leq y^m,$$
$$\hat{y}^m \leq \text{ANC}_{\text{normal}},$$
$$p^m \in \{0, 1\}.$$

The nominal problem can be transformed into a MILP as

$$\min \sum_{m \in M, \quad i \in I \setminus \{1\}} x_{i,m} \tag{8a}$$

$$\text{s.t. } x_{i,m+1} = \sum_{j=0}^{3} \left(z^{m,j} C_{i,0}^{j} + \sum_{k=1}^{n} C_{i,k}^{j} v_k^{m,j} \right), \quad t \in [m\Delta t, (m+1)\Delta t], m \in M \setminus \{M\}, \tag{8b}$$

$$0 \leq v_i^{m,j} \leq U_i z^{m,j}, \tag{8c}$$

$$-U_i\left(1 - z^{m,j}\right) \leq v_i^{m,j} - x_{i,m}, \tag{8d}$$

$$v_i^{m,j} - x_{i,m} \leq U_i\left(1 - z^{m,j}\right), \tag{8e}$$

$$\sum_{j \in J} z^{m,j} = 1, \quad m \in M \setminus \{M\}, \tag{8f}$$

$$y^{m+1} = \hat{y}^m - \sum_{j \in J} d_{\text{anc},j} z^{m,j}, \quad m \in M \setminus \{M\}, \tag{8g}$$

$$\hat{y}^m \geq \text{ANC}_{\text{normal}} - U_y(1 - p^m), \tag{8h}$$

$$\hat{y}^m \geq y^m - U_y p^m, \tag{8i}$$

$$\hat{y}^m \leq y^m, \tag{8j}$$

$$\hat{y}^m \leq \text{ANC}_{\text{normal}}, \tag{8k}$$

$$p^m \in \{0, 1\}, \tag{8l}$$

$$L_{\text{anc}} \leq \hat{y}^m, \quad m \in M, \tag{8m}$$

$$z^{m,j} \in \{0, 1\}, \quad m \in M \setminus \{M\}, j \in J, \tag{8n}$$

where $x(0)$, y^0 are given.

Appendix 2: ROP Model

In this section, we describe the mathematical details of the robust problem. We introduce parameters Γ^m that take values in the bounded intervals $[0, |V^m|]$, where V^m is the index sets of parameters with uncertainty. Γ^m is not necessarily an integer. The role of parameters Γ^m is to adjust the robustness of the proposed model against the conservation level of solution, thus it is called protection level. The motivation of Γ^m is that it is unlikely that all the parameters with uncertainty vary at the same time and reach the maximal uncertainty. In other words, the model assumes that there exists only a subset of the parameter drift that influence the solution. More specifically, it assumes that there are up to $\lfloor \Gamma^m \rfloor$ of uncertainty parameters which are allowed to deviate from their nominal values, and the toxicity decreasing rate $d^m_{\mathrm{anc},j}$ changes by at most $(\Gamma^m - \lfloor \Gamma^m \rfloor)\widehat{C}^j$, where $\lfloor \Gamma^m \rfloor$ is the greatest integer $\leq \Gamma^m$. Note that, if we choose $\Gamma^m = 0$, we completely ignore the influence of parameter uncertainty and are using the nominal values of the uncertain parameters, and if we choose $\Gamma^m = V^m$, then all the uncertain parameters are subjected to deviate from their nominal values. In this project, the maximum value of Γ^m is 3 since there are three parameters with uncertainty. Note however that only one drug is chosen for each period, parameter uncertainty of the other two drugs will not affect the robust optimal solution. Thus, the robust optimal solution under $\Gamma^m = 1$ is exactly the same as the ones under $\Gamma^m > 1$. The proposed robust counterpart of problem (Eq. 6) is as follows:

$$\min \sum_{m\in M,\quad i\in I\setminus\{1\}} x_{i,m} \tag{9a}$$

$$\text{s.t.}\quad y^{m+1} - \widehat{y}^m + \sum_{j\in J} L^j z^{m,j} + \max_{C_m^{\mathrm{RO}}} \left\{ \sum_{j\in S^m} \widehat{C}^j z^{m,j} + (\Gamma^m - \lfloor \Gamma^m \rfloor)\widehat{C}^{t^m} z^{m,t^m} \right\}, \tag{9b}$$

$$(8b), (8c), (8d), (8e), (8f), (8h), (8i), (8j), (8k), (8l), (8m), (8n) \tag{9c}$$

where $C_m^{\mathrm{RO}} = \{S^m \cup \{t^m\} | S^m \subseteq V^m, |S^m| \leq \lfloor \Gamma^m \rfloor, t^m \in V^m \setminus S^m\}$, S^m is the index sets of uncertain parameters which are allowed to deviate from their nominal values. According to the method developed in [20], the maximization problem in Eq. 9b is equivalent to the following auxiliary problem:

$$\max \sum_{j\in J} \widehat{C}^j \eta^{m,j} z^{m,j} \tag{10a}$$

$$\text{s.t.}\quad 0 \leq \eta^{m,j} \leq 1, \quad j\in J, \tag{10b}$$

$$\sum_{j\in J} \eta^{m,j} \leq \Gamma^m, \tag{10c}$$

Equation 10c indicates that the total variation of the parameters cannot exceed some threshold Γ^m. Notice that problem (Eq. 11) is bounded. It is clear that $\eta^{m,j} = 0$ is a feasible solution of (Eq. 11). By strong duality, the optimal objective value of problem (Eq. 11) is the same as the optimal objective value of its dual problem. It is easy to check that the dual problem can be written as

$$\max q^m \Gamma^m + \sum_{j \in J} p^{m,j} \tag{11a}$$

$$\text{s.t.} \quad -\widehat{C}^j z^{m,j} + \mathrm{q}^{\mathrm{m}} + p^{m,j} \geq 0, \tag{11b}$$

$$q^m \geq 0, \tag{11c}$$

$$p^{m,j} \geq 0, \tag{11d}$$

Thus, the optimal solution of our robust problem can be obtained by solving the MILP:

$$\min \sum_{m \in M, \quad i \in I \setminus \{1\}} x_{i,m} \tag{12a}$$

$$s.t. \quad x_{i,m+1} = \sum_{j=0}^{3} \left(z^{m,j} C^j_{i,0} + \sum_{k=1}^{n} C^j_{i,k} v^{m,j}_k \right), \quad t \in [m\Delta t, (m+1)\Delta t], \, m \in M \setminus \{M\}, \tag{12b}$$

$$0 \leq v^{m,j}_i \leq U_i z^{m,j}, \tag{12c}$$

$$-U_i\left(1 - z^{m,j}\right) \leq v^{m,j}_i - x_{i,m}, \tag{12d}$$

$$v^{m,j}_i - x_{i,m} \leq U_i\left(1 - z^{m,j}\right), \tag{12e}$$

$$\sum_{j \in J} z^{m,j} = 1, \quad m \in M \setminus \{M\}, \tag{12f}$$

$$y^{m+1} \leq \widehat{y}^m - \sum_{j \in J} L^j z^{m,j} - q^m \Gamma^m - \sum_{j \in J} p^{m,j}, \quad m \in M \setminus \{M\}, \tag{12g}$$

$$\widehat{y}^m \geq \mathrm{ANC}_{\mathrm{normal}} - U_y(1 - p^m), \tag{12h}$$

$$\widehat{y}^m \geq y^m - U_y p^m, \tag{12i}$$

$$\widehat{y}^m \leq y^m, \tag{12j}$$

$$\widehat{y}^m \leq \mathrm{ANC}_{\mathrm{normal}}, \tag{12k}$$

$$p^m \in \{0, 1\}, \tag{12l}$$

$$L_{\mathrm{anc}} \leq \widehat{y}^m, \quad m \in M, \tag{12m}$$

$$z^{m,j} \in \{0,1\}, \quad m \in M \setminus \{M\}, j \in J, \tag{12n}$$

$$-\widehat{C}^j z^{m,j} + q^m + p^{m,j} \geq 0, \tag{12o}$$

$$q^m \geq 0, \tag{12p}$$

$$p^{m,j} \geq 0, \tag{12q}$$

$$x(0), y^1, \widehat{y}^1 \text{ are given}, \quad p^1 = 0 \tag{12r}$$

Appendix 3: Robust Optimal Solutions for 30 Weeks

In this section, we summarize all the robust optimal solutions discussed in Subheading. 4.4 (*see* Figs. 13, 14, 15, 16, and 17).

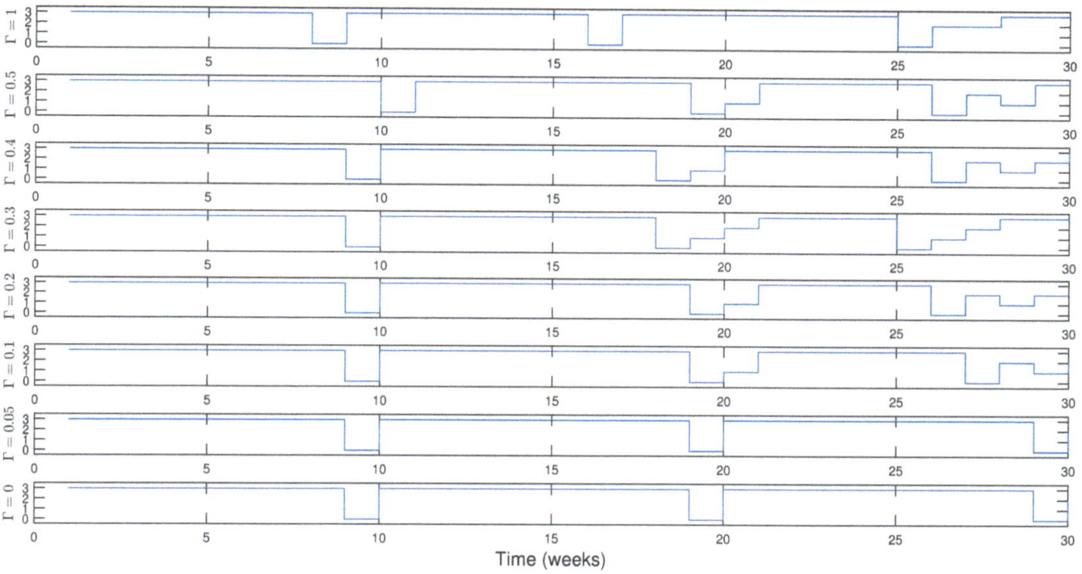

Fig. 13 Robust optimal solutions under $\widehat{C}^j = 0.2 \times L^j$ for 30 weeks. The initial conditions for NSC, WSC, *Y253F*, and *F317L* are $9E+06$, $9E+05$, $1E+04$, and $1E+04$, respectively

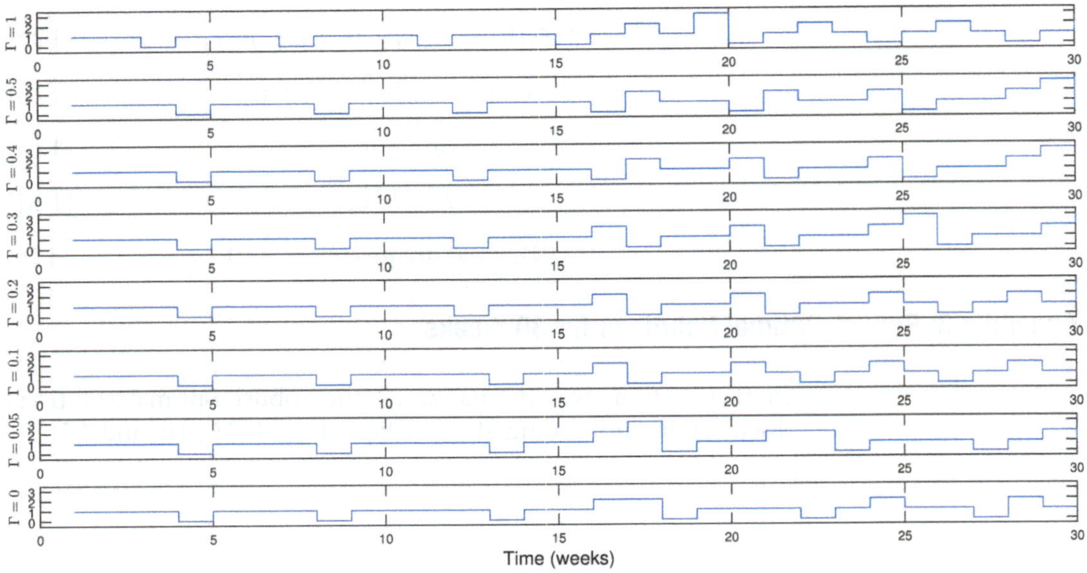

Fig. 14 Robust optimal solutions under $\widehat{C}^j = 0.2 \times L^j$ for 30 weeks. The initial conditions for NSC, WSC, *Y253F*, and *F317L* are $9E+06$, $5E+05$, $3E+05$, and $3E+05$, respectively

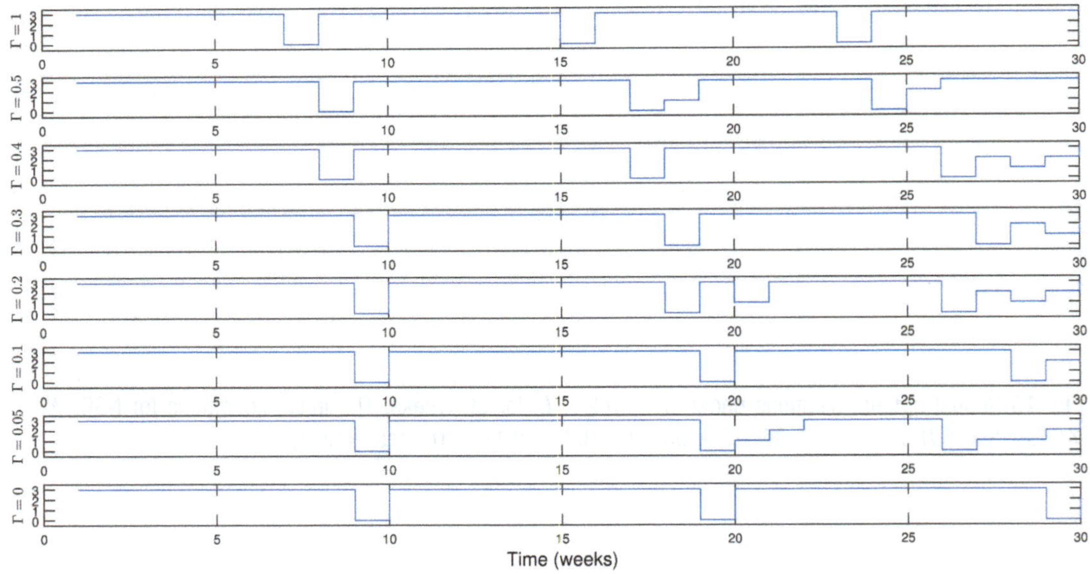

Fig. 15 Robust optimal solutions under $\widehat{C}^j = 0.3 \times L^j$ for 30 weeks. The initial conditions for NSC, WSC, *Y253F*, and *F317L* are $9E+06$, $9E+05$, $1E+04$, and $1E+04$, respectively

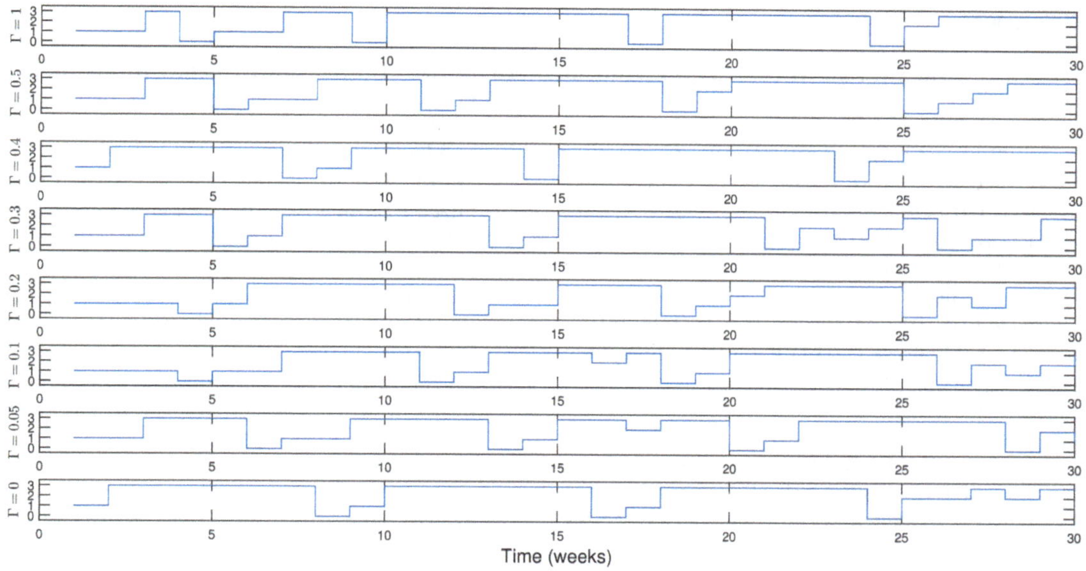

Fig. 16 Robust optimal solutions under $\widehat{C}^j = 0.3 \times L^j$ for 30 weeks. The initial conditions for NSC, WSC, $Y253F$, and $F317L$ are $9E+06$, $9E+05$, $1E+05$, and $1E+05$, respectively

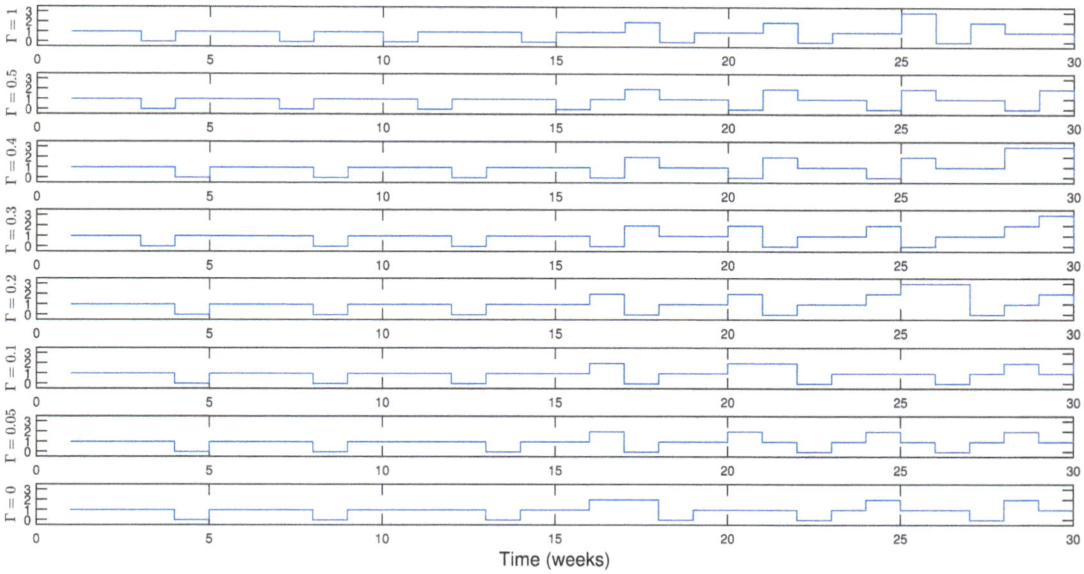

Fig. 17 Robust optimal solutions under $\widehat{C}^j = 0.3 \times L^j$ for 30 weeks. The initial conditions for NSC, WSC, $Y253F$, and $F317L$ are $9E+06$, $5E+05$, $3E+05$, and $3E+05$, respectively

References

1. Shi Z, Peng XX, Kim IW et al (2007) Erlotinib (Tarceva, OSI-774) antagonizes ATP-binding cassette subfamily B member 1 and ATP-binding cassette subfamily G member 2-mediated drug resistance. Cancer Res 67:1101220
2. Paraiso KH, Xiang Y, Rebecca VW et al (2011) PTEN loss confers BRAF inhibitor resistance to melanoma cells through the suppression of BIM expression. Cancer Res 71:27502760
3. Foo J, Michor F (2014) Evolution of acquired resistance to anti-cancer therapy. J Theor Biol 355:10
4. Leder K, Foo J, Skaggs B et al (2011) Fitness conferred by BCR-ABL kinase domain mutations determines the risk of pre-existing resistance in chronic myeloid leukemia. PLoS One 6(11):e27682. https://doi.org/10.1371/journal.pone.0027682
5. Foo J, Leder K (2013) Dynamics of cancer recurrence. Annals Appl Probab 23 (4):1437–1468
6. Swanson KR, Bridge C, Murray JD et al (2003) Virtual and real brain tumors: using mathematical modeling to quantify glioma growth and invasion. J Neurol Sci 216(1):1–10
7. Badri H, Watanabe Y, Leder K (2015) Optimal radiotherapy dose schedules under parametric uncertainty. Phys Med Biol 61(1):338
8. Zhou C, Wu YL, Chen G et al (2011) Erlotinib versus chemotherapy as first-line treatment for patients with advanced EGFR mutation-positive non-small-cell lung cancer (OPTIMAL, CTONG-0802): a multicentre, open-label, randomised, phase 3 study. Lancet Oncol 12(8):735–742
9. Druker BJ, Talpaz M, Resta DJ et al (2001) Efficacy and safety of a specific inhibitor of the BCR-ABL tyrosine kinase in chronic myeloid leukemia. N Engl J Med 344:1031–1037
10. Kantarjian H, Sawyers C, Hochhaus A et al (2002) Hematologic and cytogenetic responses to imatinib mesylate in chronic myelogenous leukemia. N Engl J Med 346:645–652
11. Cortes JE, Jones D, O'Brien S et al (2010) Results of dasatinib in patients with early chronic-phase chronic myeloid leukemia. J Clin Oncol 28(3):398–404
12. Giles FJ, Abruzzese E, Rosti G et al (2010) Nilotinib is active in chronic and accelerated phase chronic myeloid leukemia following failure of imatinib and dasatinib therapy. Leukemia 24:1299–1301
13. O'Hare T, Eide CA, Deininger MWN (2007) Bcr-Abl kinase domain mutations, drug resistance, and the road to a cure for chronic myeloid leukemia. Blood 110:2242–2249
14. He Q, Zhu JF, Dingli D et al (2016) Optimized treatment schedules for chronic myeloid leukemia. PLoS Comput Biol 12:e1005129
15. Harrold JM, Parker RS (2009) Clinically relevant cancer chemotherapy dose scheduling via mixedinteger optimization. Comput Chem Eng 33(12):2042–2054
16. Murray JM (1990) Some optimal control problems in cancer chemotherapy with a toxicity limit. Math Biosci 100(1):49–67
17. Murray JM (1990) Optimal control for a cancer chemotherapy problem with general growth and loss functions. Math Biosci 98:273–287
18. Hadjiandreou MM, Mitsis GG (2014) Mathematical modeling of tumor growth, drug-resistance, toxicity, and optimal therapy design. IEEE Trans Biomed Eng 61(2):415–425
19. Laird AK (1964) Dynamics of tumour growth. Br J Cancer 18(3):490–502
20. Martin RB (1992) Optimal control drug scheduling of cancer chemotherapy. Automatica 28:11131123
21. Floares A Neural networks control of drug dosage regimens in cancer chemotherapy. SAIA, Cluj-Napoca, Transilvania
22. Weisstein, Eric W. Heaviside step function. MathWorld
23. Afenya EK (2001) Recovery of normal hemopoiesis in disseminated cancer therapy-a model. Math Biosci 172
24. Michor F, Hughes TP, Iwasa Y et al (2005) Dynamics of chronic myeloid leukaemia. Nature 435:1267–1270
25. Bozic I, Reiter JG, Allen B et al (2013) Evolutionary dynamics of cancer in response to targeted combination therapy. Elife 2:e00747
26. Nanda S, Moore H, Lenhart S (2007) Optimal control of treatment in a mathematical model of chronic myelogenous leukemia. Math Biosci 210:143
27. O'Brien S, Berman E, Borghaei H et al (2009) NCCN clinical practice guidelines in oncology: chronic myelogenous leukemia. J Natl Compr Canc Netw 7(9):984–1023
28. Sokal JE, Cox EB, Baccarani M et al (1984) Prognostic discrimination in "good-risk" chronic granulocytic leukemia. Blood 63:789–799

29. Hasford J, Pfirrmann M, Hehlmann R et al (1998) A new prognostic score for survival of patients with chronic myeloid leukemia treated with interferon alfa. Writing Committee for the Collaborative CML Prognostic Factors Project Group. J Natl Cancer Inst 90:850–858
30. Scheijen B, Griffin JD (2002) Tyrosine kinase oncogenes in normal hematopoiesis and hematological disease. Oncogene 21:3314
31. Deininger MW, O'Brien S, Ford JM et al (2003) Practical management of patients with chronic myeloid leukemia receiving imatinib. J Clin Oncol 21(8):1637–1647
32. Katia BBP, Israel B, Carla B et al (2015) BCR-ABL mutations in Chronic Myeloid Leukemia treated with tyrosine kinase inhibitors and impact on survival. Cancer Invest 33:451–458
33. Ravin JG, Hagop K, Susan O et al (2009) The use of nilotinib or dasatinib after failure to 2 prior tyrosine kinase inhibitors: long-term follow-up. Blood 114(20):4361
34. Wei G, Rafiyath S, Liu D (2010) First-line treatment for chronic myeloid leukemia: dasatinib, nilotinib, or imatinib. J Hematol Oncol 3:47
35. Cornelison M, Jabbour EJ, Welch MA (2012) Managing side effects of tyrosine kinase inhibitor therapy to optimize adherence in patients with chronic myeloid leukemia: the role of the midlevel practitioner. J Support Oncol 10 (1):14–24
36. Conchon M, Freitas CM, Rego MA et al (2011) Dasatinib - clinical trials and management of adverse events in imatinib resistant/ intolerant chronic myeloid leukemia. Rev Bras Hematol Hemoter 33(2):131–139
37. Marin D (2012) Initial choice of therapy among plenty for newly diagnosed chronic myeloid leukemia. Hematology Am Soc Hematol Educ Program 1:115–121
38. Foo J, Drummond MW, Clarkson B et al (2009) Eradication of chronic myeloid leukemia stem cells: a novel mathematical model predicts no therapeutic benefit of adding G-CSF to imatinib. PLoS Comput Biol 5(9): e1000503
39. Gruber FX, Ernst T, Porkka K et al (2012) Dynamics of the emergence of dasatinib and nilotinib resistance in imatinib-resistant CML patients. Leukemia 26:172–177
40. Cortes JE, Jones D, O'Brien S et al (2010) Nilotinib as front-line treatment for patients with chronic myeloid leukemia in early chronic phase. J Clin Oncol 28(3):392–397
41. Radich JP, Kopecky KJ, Appelbaum FR et al (2012) A randomized trial of dasatinib 100 mg versus imatinib 400 mg in newly diagnosed chronic-phase chronic myeloid leukemia. Blood 120(19):3898–3905
42. Deininger M, Mauro M, Matloub Y et al (2008) Prevalence of T315I, dasatinib-specific resistant mutations (F317L, V299L, and T315A), and nilotinib-specific resistant mutations (P-loop and F359) at the time of imatinib resistance in chronic-phase chronic myeloid leukemia (CP-CML). Blood 112:3236
43. Sawyers C (2004) Targeted cancer therapy. Nature 432:294–297
44. Cortes J, Talpaz M, O'Brien S et al (2005) Molecular responses in patients with chronic myelogenous leukemia in chronic phase treated with imatinib mesylate. Clin Cancer Res 11:3425

Chapter 16

Modeling of Interactions between Cancer Stem Cells and their Microenvironment: Predicting Clinical Response

Mary E. Sehl and Max S. Wicha

Abstract

Mathematical models of cancer stem cells are useful in translational cancer research for facilitating the understanding of tumor growth dynamics and for predicting treatment response and resistance to combined targeted therapies. In this chapter, we describe appealing aspects of different methods used in mathematical oncology and discuss compelling questions in oncology that can be addressed with these modeling techniques. We describe a simplified version of a model of the breast cancer stem cell niche, illustrate the visualization of the model, and apply stochastic simulation to generate full distributions and average trajectories of cell type populations over time. We further discuss the advent of single-cell data in studying cancer stem cell heterogeneity and how these data can be integrated with modeling to advance understanding of the dynamics of invasive and proliferative populations during cancer progression and response to therapy.

Key words Breast cancer, Cancer stem cell, Mathematical model, Optimal therapy design

1 Introduction

Mathematical modeling of cancer stem cells has proven useful in several important areas of translational cancer research. Those include, for example: understanding evolutionary dynamics of clonal populations and prediction of therapeutic resistance [1–3]; understanding tumor growth dynamics [4]; inferring the evolutionary dynamics that occur during cancer initiation and progression [5]; understanding the dynamics of stem cell state transitions and estimation of dedifferentiation rates [6, 7]; understanding the complex regulatory pathways that modulate stem cell behavior; and predicting clinical responses to combination therapies targeting the cancer stem cell niche [7].

Based on predictions from modeling, clinical oncologists are able to optimize dosing, frequency, and duration of therapies (e.g., dose dense treatments in adjuvant breast cancer therapies), which increase efficacy and minimize side effects, leading to improved

Louise von Stechow (ed.), *Cancer Systems Biology: Methods and Protocols*, Methods in Molecular Biology, vol. 1711, https://doi.org/10.1007/978-1-4939-7493-1_16, © Springer Science+Business Media, LLC 2018

outcomes [8–10]. Statistical modeling has also proven valuable in selecting prognostic and predictive markers in clinical trials in translational oncology [11]. There are many opportunities where modeling can expand its contribution to translational oncology. As single-cell transcriptomics and epigenetic data become more readily available and methods of simulation become more sophisticated, multiscale modeling will permit the integration of data that will inform models and improve predictions, which will ultimately lead to more effective therapies. In this chapter, we will review the methods that are currently being used in mathematical oncology, and suggest areas where modeling could further be applied in cancer stem cell systems biology research.

2 Compelling Research Questions in (Cancer) Stem Cell Research That Can Be Addressed with Mathematical Modeling

2.1 Single-Cell Gene Expression and Epigenetic Data: How to Extract Information to Best Inform Models?

Single-cell sequencing and transcriptomics on a genome-wide level has advanced greatly in recent years. Statistical methods have been developed to analyze single-cell data in order to characterize tumor heterogeneity [12, 13], demonstrate clonal evolution [14], and infer phylogenetic relationships and ordering of mutations [15].

Genetic and epigenetic patterns that emerge during the processes of stem cell quiescence, activation, and differentiation can be captured using single-cell analysis. Intra-tumoral heterogeneity creates a challenge for the study of the interconnecting molecular events that guide these processes. Single-cell gene expression analysis has been used to explore cell heterogeneity in breast cancer and unravel gene expression variation in both cell line and patient-derived xenograft samples [16, 17]. By examining expression levels of 96 genes from pathways involved in cell self-renewal, adhesion, and differentiation, three different patterns of expression in these genes were observed in single cells obtained from cell lines and from patient-derived xenograft samples. These patterns correspond to three distinct cell populations: epithelial Breast cancer stem cells (BCSCs), mesenchymal BCSCs, and non-stem cancer cells. Applying these methods to populations of circulating tumor cells will allow for the characterization of cell types within a patient at diagnosis and in response to treatment.

Whole transcriptome RNA-sequencing is used to transcriptional events that are continuously changing within a cell over time. Changes that are observed using this technology include alternative gene spliced transcripts, post-transcriptional modifications, gene fusion, mutations, and alterations in gene expression. Additionally, whole genome bisulfite sequencing is used to generate genome-wide analysis of DNA methylation. As these technologies become available within single-cell studies, sophisticated methods will need to be developed to analyze these data and distinguish relevant patterns from inherent noise that is anticipated within a single cell over time. Signaling pathways can be reconstructed from genome, transcriptome, and proteome data [18, 19]. While statistical inference has been successful in studying these components individually, combining information from each level is essential for understanding the system as a whole [20]. As our understanding of cellular networks improves, these results can be integrated with dynamic modeling approaches to estimate rates of stem cell state transitions and to identify regulatory nodes. As samples from circulating tumor cells from patients exposed to therapeutic combinations become available, these methods could be used to sort cell populations and track responses of each population to therapy.

2.2 Modeling Cell-Cell Interactions Between Cancer Stem Cells and Their Microenvironment

Because tumors consist of many cell types that interact with each other, as well as with the numerous cell types that are present in the tumor microenvironment, models that account for these interactions are required. Evolutionary game theory has been useful in modeling these interactions [1]. Models based on evolutionary game theory have been employed to examine mechanisms of growth control under conditions of competing resources [21], and have predicted the evolution of cooperation among tumor cells [22].

The breast cancer stem cell microenvironment consists of a number of diverse cell types including more differentiated tumor cells, stromal cells, endothelial cells, and immune cells. These cells interact with each other through a number of signaling mechanisms involving cytokines, growth factors, and other signaling molecules, such as miRNAs [23–29].

Under normal conditions, the stem cell niche regulates how stem cells participate in tissue generation, maintenance, and repair, preventing stem cell depletion and overpopulation. The interaction between these normal, tissue-specific stem cells and their niche is required for balanced tissue maintenance, and aberrant function of the niche may contribute to malignant transformation.

The cancer stem cell niche plays an important role in the regulation of tumor growth, and metastasis as well as in modulating therapeutic response. Here, we will describe the cellular elements of the breast cancer stem cell niche.

Breast cancer stem cells, exist in either a proliferative, epithelial state characterized by expression of ALDH as well as epithelial markers such as E-cadherin, or in a quiescent, invasive, mesenchymal state, characterized by expression of CD44 as well as additional mesenchymal markers such as vimentin, N-cadherin, Twist, and Slug and Snail [30]. When a BCSC is in the proliferative state, it can undergo symmetric self-renewal, or asymmetric self-renewal, giving rise to one identical copy of itself and one **bipotent progenitor cell** [31, 32]. Alternatively, it can undergo symmetric

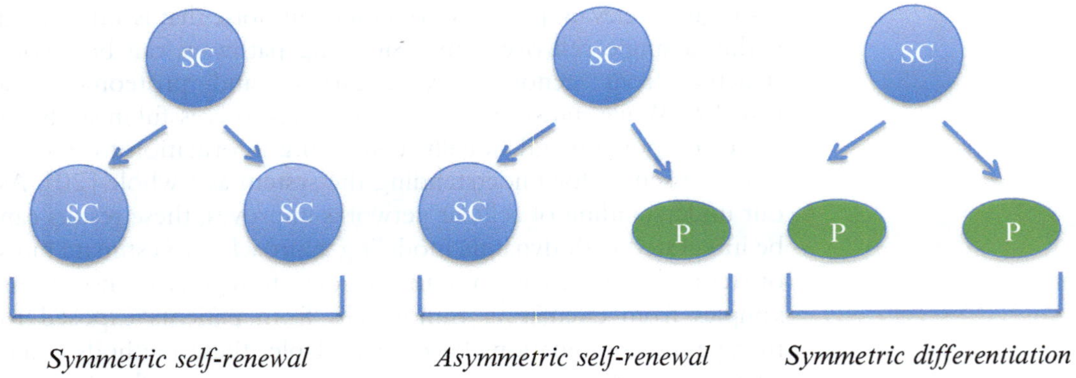

Fig. 1 Types of stem cell division. A stem cell or stem-like cell can undergo symmetric self-renewal, giving rise two identical copies of themselves, or asymmetric self-renewal, giving rise to one identical copy of itself and one partially differentiated progenitor cell. It can also undergo symmetric differentiation, in which it gives rise to two partially differentiated daughter cells

differentiation generating two bipotent progenitors (*see* Fig. 1). Mathematical modeling has shown that slight disruption in the balance between symmetric self-renewal, asymmetric self-renewal, and symmetric differentiation can lead to Gompertzian growth kinetics in tumors [7]. The bipotent progenitors give rise to either **luminal cells** or **basal cells**. These differentiated cancer cells comprise the bulk of the tumor, and currently most cancer treatment modalities are focused on this population. Other cells that are present in the stem cell microenvironment include **mesenchymal stem cells** that give rise to and maintain the stroma, **endothelial cells** that reside in the tumor vasculature and various elements of the immune system. In fact, recent studies have indicated that myeloid-derived suppressor cells (MDSCs) are able to directly stimulate BCSC self-renewal through the activation of the Notch pathway [33].

All of these cell types, as well as the microenvironmental signaling pathways that guide their interaction, need to be considered in a multiscale model of the breast cancer stem cell niche. These models may be helpful in predicting patient responses to combinatorial therapies targeting angiogenesis, for promoting activation of cancer stem cells that are quiescent, and the prevention of invasion.

2.3 Relevance of Spatial Factors?

Spatial organization is a key factor for growth and tissue renewal during development and regeneration of healthy tissues [34]. It was first observed in the germ stem cell niche of *Drosophila melanogaster* that during cell division, the mitotic spindle is aligned with support cells of the niche so that the daughter cell that remains within the niche retains stem cell identity, whereas the daughter cell that is displaced outside the niche (away from self-renewal signals) initiates differentiation [35]. These oriented divisions have also been observed in mammalian epithelia. For example, the position of a stem cell within a hair follicle predicts whether it is likely to remain committed, generate precursors, or progress to a different fate [34]. Another example is that of stratified epithelial cells. Alignment of the stem cell niche along rigid basal lamina leads to regular morphologies, whereas alignment along a freely moving basal lamina leads to distorted epithelial morphologies [36].

The dynamics of the stem cell niche have been well described in the hematopoietic system.

Mathematical models designed to explore the mechanisms by which stem cells communicate with the niche, as well as the fact that cancer arises as a results of failure of this communication, have shown that coupled lineages allow for more controlled regulation of total blood cell numbers than uncoupled lineages and respond better to random perturbation to maintain homeostatic equilibrium [37].

In a model of the breast cancer stem cell niche, it would be ideal to also consider spatial effects. Spatial stochastic models have been used to study cancer initiation and progression [38] as well as mutational heterogeneity [39]. Spatial models have the potential to be helpful for the optimization of therapies targeting the stem cell niche.

2.4 Do Hypoxic Microenvironments Promote Late Recurrence?

The vasculature of tumors is very important in determining how nutrients and drugs are delivered to tumor cells. Recent evidence from mouse xenograft studies demonstrates that hypoxia, mediated by hypoxia-inducible factor 1α, drives the stem/progenitor cell enrichment, and activates the Akt/β-catenin cancer stem cell regulatory pathway [40]. Hypoxia stimulates ALDH+ epithelial BCSCs, located in the interior hypoxic zones of breast tumors, while the invasive mesenchymal cells are located on the leading edge of the tumor. Models that take into consideration the fractal geometric properties of tumor vascular networks, as well as the spatial gradients in resources and metabolic states, have been used to predict metabolic rates of tumors and derive universal growth curves to predict growth dynamics in response to targeted treatments [41]. Extensions of these growth equations including necrotic, quiescent, and proliferative states have been used to understand growth trajectories across tumor types. This type of modeling may be ideally suited to answer questions related to the growth of stem cell compartments in response to hypoxia, and for the selection of combined, targeted treatments for the eradication of both quiescent and proliferative BCSCs. Another potential option would be to use recent updates to stochastic simulation methods that include spatial effects. Introducing the spatial aspects of the stem cell niche into simulation is required to answer questions related to hypoxic regulation of BCSC behavior.

2.5 Integration of Immunotherapy with Molecularly Targeted and Cytotoxic Therapies

The advent of immunotherapy has led to a dramatic shift in the treatment and survival of several tumors, such as melanoma, renal cell carcinoma, lung cancer, and Hodgkin lymphoma [42–49]. Approximately one-quarter of patients with triple negative breast cancer respond to immunotherapy [50]. Immunotherapy is particularly successful in aggressive malignancies, where the percentage of tumor-initiating cells is high. For example, in melanoma the majority of tumor cells have capacity for self-renewal [51]. These tumors were the first where immunotherapy was shown to be successful. Immunotherapy, informed by mathematical modeling, may have a greater chance of leading to durable remissions [52].

Successful immunotherapy should target stem-like cells as well as bulk tumor cells. Mathematical modeling can be helpful in predicting the variable response to immunotherapy based on different proportions of cell types comprising a tumor. These models are especially relevant in the adjuvant setting, where tumor growth and invasion are driven by a small number of cells on a longer time scale, and where considerably more time and resources are required to directly observe survival outcomes in relation to therapy. If immunotherapy is successful in activating the immune system to target the stem cell compartment, it should eventually lead to eradication of the tumor. However, the required duration of therapy required to observe an appreciable change in bulk tumor size is unknown. Stochastic models can be used to predict extinction times of the cell populations comprising the tumor, allowing the estimation of the treatment duration required to eradicate cancer cells [53]. Models should also take into account the potential costs of immunotherapy, including autoimmune side effects. These models would allow selection of the optimal treatment dosing and duration that would have the best the chance of tumor eradication while minimizing the risk of side effects.

Another area in immunotherapy where mathematical modeling may prove useful is in determining optimal combinations of therapies. A branching process model has been used to predict success of combination therapy under assumptions of mutations conferring resistance [54]. In models combining cytotoxic chemotherapy, vaccine therapy, CTLA4 and PD-1 inhibitors, and drugs targeting the BRAF and MEK pathways and other molecular pathways [55, 56], it will be important to model dosing and effectiveness in order to address the need to minimize potentially debilitating side effects, including autoimmune processes as well as the development of secondary malignancies.

3 Mathematical Modeling and Simulation Tools in Translational Oncology

In silico experiments can be used in concert with cell line experiments, animal xenograft model studies, and patient-oriented translational studies to complement and improve cancer stem cell research. Signaling networks in the cancer stem cell microenvironment are complex and much work is needed to understand the regulatory dynamics of this system. While gene knock-out experiments allow the delineation of the importance of each individual molecular component of the cancer stem cell microenvironment, the combination of mathematical modeling with laboratory research allows studying of the emergent properties and provides a framework for elucidating the integrative dynamics of this complex system [57–60].

Given the levels of complexity of the cancer stem cell niche, selection of the most appropriate mathematical model remains challenging. We will describe a variety of mathematical modeling approaches and situations where specific methods can address this challenge by providing important biological insights.

3.1 Defining the Model

The breast cancer stem cell niche is a complex system comprised of cancer stem cells, and the surrounding cells and molecular signals that govern the behavior of the stem cells. Multiple overlapping feedback loops regulate whether a cancer stem cell undergoes self-renewal, quiescence, differentiation, or apoptosis. The niche also regulates the rare event of partially differentiated breast epithelial cancer cells undergoing dedifferentiation into a stem-like state.

The scope of a model is defined by the *reactant species* involved, and by the *reactions* or *events* that take place. Examples of species involved in the breast cancer stem cell niche include cancer stem cells (quiescent and invasive versus proliferative), progenitor cells, differentiated luminal and basal cells, endothelial cells, mesenchymal cells, immune cells as well as the elements of signaling pathways, which regulate the transitions and interactions between these cell types [26, 61]. Those signaling pathway elements include cytokines (e.g., IL-6, IL-8, TGF-β, BMPs), receptors (e.g., HER2 and CXCR1), and intracellular signals, including protein kinases (e.g., Akt), transcription factor proteins (e.g., Lin28, IκB, Stat3), microRNA precursors (e.g., let-7), and microRNAs (e.g., mir-93) [23–29]. The reactions of a model describe the important events that change the abundance of reactant species. Examples of reactions in the breast cancer stem cell niche include stem cell self-renewal, quiescence, differentiation, and apoptosis. In general, a model should be kept as simple as possible, adding sufficient complexity to address the biological principles involved.

Figure 2 shows a simplified model of the state transitions that occur between the proliferative epithelial (MET) state of breast cancer stem cells (BCSCs) and their invasive quiescent mesenchymal (EMT) state (for illustration, a small number of species and reactions have been included here). The species include cell types (EMT and MET states of the BCSCs) and the factors (cytokines and intracellular signaling molecules) that regulate transitions

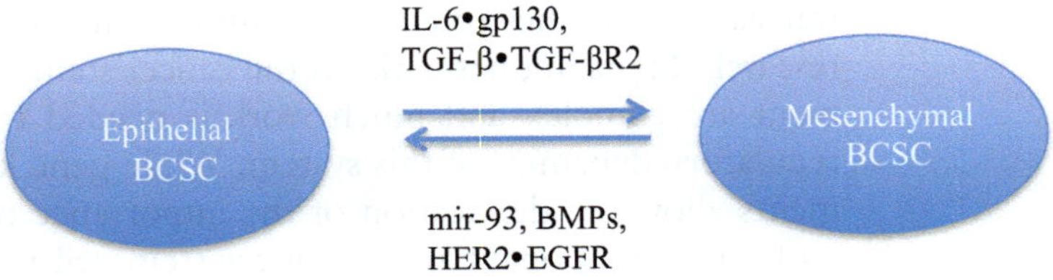

Species: epithelial BCSC, mesenchymal BCSC, IL-6 and its receptor (gp130), TGF-β and its receptor (TGF-βR2), mir-93, BMP, HER2 and its receptor (EGFR).

Reactions:

Receptor binding and dissociation

IL-6 + gp130 → IL-6 • gp130	IL-6 • gp130 → IL-6 + gp130
TGF-β + TGF-βR2 → TGF-β • TGF-βR2	TGF-β • TGF-βR2 → TGF-β + TGF-βR2
HER2 + EGFR → HER2 • EGFR	HER2 • EGFR → HER2 + EGFR

Stem cell state transitions

MET + IL-6 • gp130 → EMT + IL-6 • gp130

MET + TGF-β • TGF-βR2 → EMT + TGF-β • TGF-βR2

EMT + HER2 • EGFR → MET + HER2 • EGFR

EMT + mir-93 → MET + mir-93

EMT + BMP → MET + BMP

Fig. 2 Schematic of microenvironmental signals governing BCSC state transitions. In this simplified model of the BCSC niche, we identify the species involved, including cell types (the proliferative epithelial BCSCs and the quiescent mesenchymal BCSC populations) and cytokines and intracellular signals that regulate transition between these two states. The reactions included in our model directly or indirectly play a role in regulating the BCSC state transitions

between these two states. Reaction types include receptor binding and dissociation, as well as the state transitions. A more biologically complete model also includes the regulatory feedback loops that exist within this system, such as IL-6 activation of the Akt/Stat/NFκB pathway leading to increased transcription of IL-6, and the interaction of Lin-28/Let-7 and HER2 leading to activation of β-catenin driving self-renewal of epithelial BCSCs. The inclusion of such regulatory feedback loops thus enables the model to more closely simulate responses to environmentally stressful conditions.

3.2 Deterministic Versus Stochastic Models

Deterministic models can provide insight into many important aspects of microenvironmental signaling, including the understanding of dynamic control (as revealed by time-course studies), the impact of cellular cross-talk and identification of control points, and an indication of possible target points for treatment, as well as the exploration of dose-response relationships [62, 63]. There are several software packages in Matlab® and Mathematica® that enable investigators to explore nonlinear dynamics of complex systems based on a series of reaction rate equations. Examples of these include the Systems Biology Toolbox [40] in Matlab®, and ReactionKinetics [64] in Mathematica®.

An example of a reaction rate equation, applied to our simplified model of stem cell state transitions, describes the rate of change in E over time, the concentration of epithelial BCSCs and the rate of change in M over time, the concentration of mesenchymal BCSCs:

$$\frac{\mathrm{d}E}{\mathrm{d}t} = -(k_1y_1 + k_2y_2)\,E + \left(k_3y_3 + k_4y_4 + k_5y_5\right) \tag{1}$$

$$\frac{\mathrm{d}M}{\mathrm{d}t} = -\left(k_3y_3 + k_4y_4 + k_5y_5\right)\,M + (k_1y_1 + k_2y_2)\,E \tag{2}$$

where y_1 through y_5 are the concentrations of the microenvironmental factors (IL-6 • gp130, TGF-β • TGF-βR2, mir-93, BMP, and HER) that interact with the two cellular species, and k_1 through k_5 are rate constants describing the impact of the interaction of sets of species. In this simple model, we note that the rates of change over time for E and M are related as follows:

$$\frac{\mathrm{d}E}{\mathrm{d}t} = -\frac{\mathrm{d}M}{\mathrm{d}t} \quad . \tag{3}$$

If symmetric self-renewal, a process that results in an increase in the number of BCSCs was to be added into this mathematical model, as well as apoptosis, which decreases the number of BCSCs the system of equations would be:

$$\frac{\mathrm{d}E}{\mathrm{d}t} = -(k_1y_1 + k_2y_2)\,E + \left(\beta - \delta + k_3y_3 + k_4y_4 + k_5y_5\right)M \tag{4}$$

$$\frac{\mathrm{d}M}{\mathrm{d}t} = -\left(k_3y_3 + k_4y_4 + k_5y_5\right)M + (k_1y_1 + k_2y_2)E \tag{5}$$

where β and δ are the rates of symmetric self-renewal and apoptosis, respectively. In this case, the rate of change in epithelial and mesenchymal BCSCs would be equal only when $\beta = \delta$.

3.3 Visualizing the Model

Petri nets are diagrams that are used in systems biology to describe transitions and interactions that occur in complex systems [65]. In these graphs, boxes represent the occurrence of transitions, ovals represent species, and directed arcs delineate which reactant species enter the reaction (i.e., arrow flows from species to reaction) and products that are produced during the reaction (i.e., arrow flows from the reaction to the species). Figure 3 shows the petri net

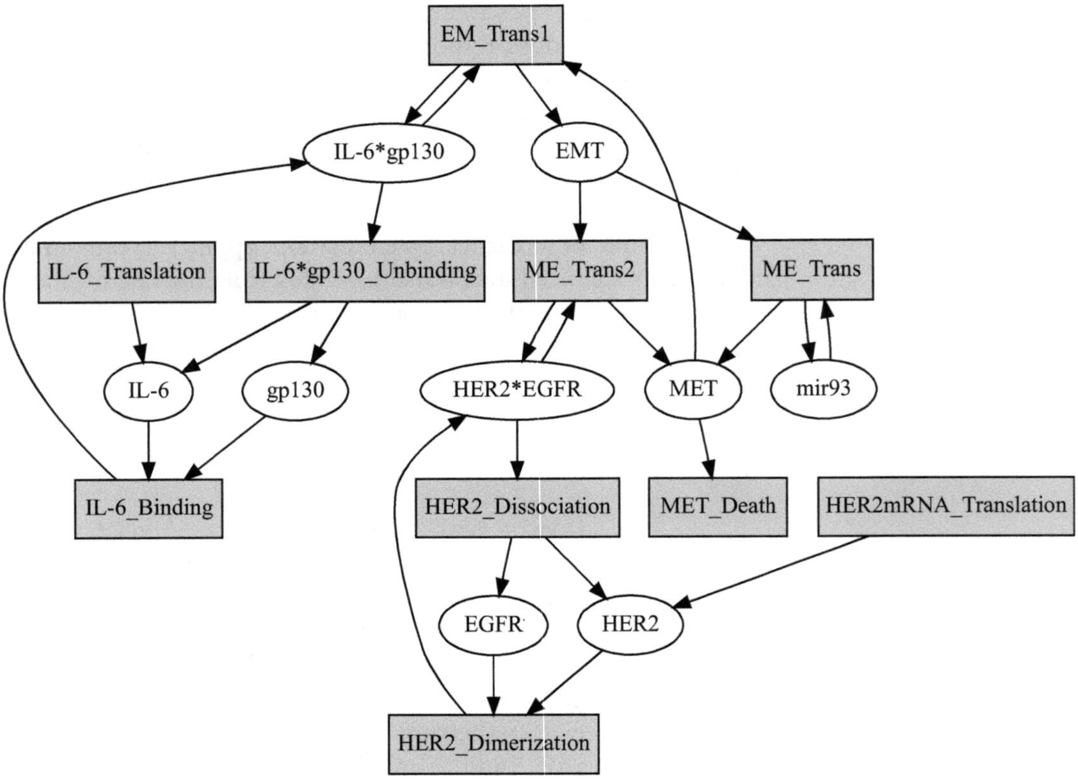

Fig. 3 Petri net generated by the simplified model of factors regulating transitions between proliferative and quiescent BCSC states. The Petri net demonstrates the interconnectivity of the model, defining its reactant species (*ovals*) and the transitions and events (*boxes*) that relate them to each other

generated by our simplified model, which describes the state transitions between the quiescent and proliferative BCSC states. While Petri nets are based on strong mathematical foundation, they are also helpful for use as a visual communication aid to understand system behavior. The Petri net graphs in Fig. 3 were made using the GraphViz package in the Julia language.

3.4 Stochastic Simulation

In certain situations, stochastic models provide additional information when approaching scientific questions in mathematical oncology. Rare events, such as mutation and extinction, can be accounted for with stochastic models, as can random fluctuations in species counts that may greatly impact the population dynamics of the system. Using probabilistic models, one is able to calculate how frequently a population would become extinct under a given condition or treatment, as well as the required duration of therapy that would be needed to eradicate a stem cell population [58].

As the system gains increasing layers of complexity, more sophisticated models are required and these models become difficult to solve analytically. In this scenario, stochastic simulation techniques are helpful in studying niche dynamics where there can

Table 1
Propensity and stoichiometric change for two example reactions

Reaction	Propensity	Stoichiometric matrix
MET + IL-6 • gp130 → EMT + IL-6 • gp130	$c_1 * x_1 * x_2$	$\nu_{11} = -1, \nu_{12} = 0, \nu_{13} = +1$
EMT + HER2 • EGFR → MET + HER2 • EGFR	$c_2 * x_3 * x_4$	$\nu_{21} = +1, \nu_{23} = -1, \nu_{24} = 0$

be large numbers of species and reactions and multiple overlapping feedback loops. In the stochastic reaction kinetics framework, a propensity must be specified for each reaction, as well as the net change in count of each species. For our simplified example of the stem cell state transitions, the species counts for epithelial BCSCs, mesenchymal BCSC, IL-6 • gp130, and HER2 • EGFR would be x_1 to x_4, respectively, allowing the calculation of the propensity of each reaction as well as the stoichiometric change in each species for each reaction. Table 1 shows the propensity and stoichiometric change for two example reactions.

Stochastic simulation algorithms proceed by updating the state vector, which consists of particle counts for each of the reactant species, after each reaction (or set of reactions) is allowed to fire. In the stochastic simulation algorithm [66] the counts are updated after each reaction fire. As a result, this algorithm is the most accurate, but also the slowest. Approximate algorithms, such as the τ-leaping algorithm, leap over a set of reactions, in which the mean number of times a given reaction fires during the interval is given by the product of its propensity and the length of the leap interval [44]. While these methods increase computational speed, they can compromise accuracy in situations where the propensity is abruptly changing. An update to the τ-leaping algorithm, the step-anticipation τ-leaping algorithm, allows the user to anticipate the change in propensity during the leap and leads to improved accuracy without compromising speed [67]. Outputs from stochastic simulation include full distributions of cell counts, as well as trajectories of cell counts over time. Figure 4 shows the full distributions (panel A) and mean trajectories (panel B) for epithelial BCSCs while varying the rate of symmetric self-renewal of epithelial BCSCs. Full distributions may be advantageous over the mean trajectory when one is interested in the frequency with which a population of cancer stem cells falls below a threshold of detectability or when investigating how frequently that population is eradicated in response to therapy.

3.5 3D Simulation and Agent-Based Modeling

Agent-based modeling is a microscale approach that combines elements of game theory, complex systems, emergence, and evolutionary programming to simulate the actions and interactions of individual cells and collective groups of cells to assess their effects

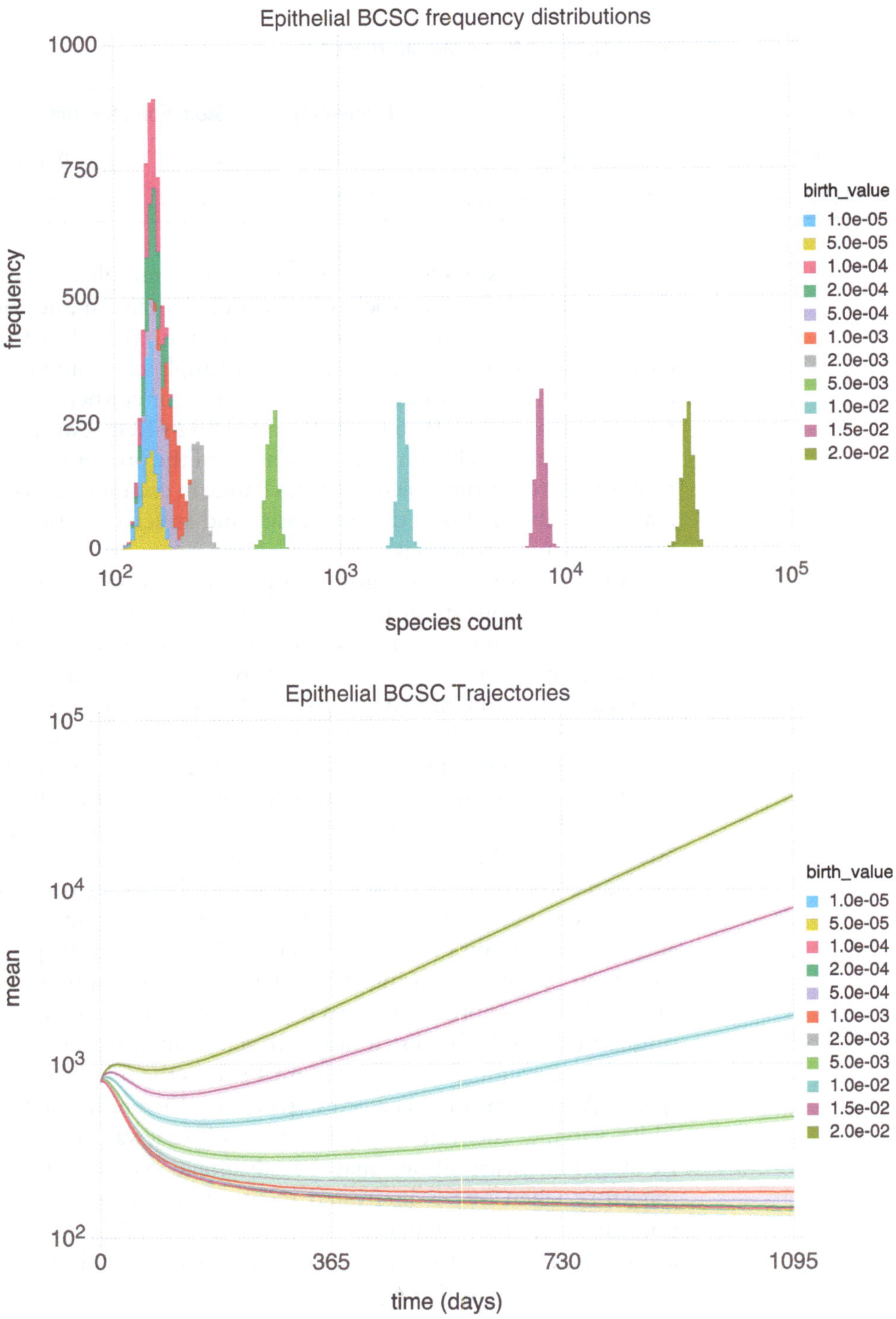

Fig. 4 Sample output from stochastic simulation of stem cell state transitions. The first panel shows the full distribution of epithelial BCSC cell counts over 1000 simulations for a fixed period of time. For slower birth rates, BCSC cell populations reach smaller final counts. In the second panel, the average trajectories of epithelial-like BCSC populations are shown. When the birth rate is faster, BCSC cell counts initially diminish in response to therapy but later increase over time

on the system. They are particularly useful in accounting for details of smaller levels of systems and the prediction of the appearance of complex phenomena that occur at a higher level. Open source simulation packages have recently become available that allow simulation of the behavior of millions of cells in three-dimensional tissues. These methods have been applied to patient-calibrated models of ductal carcinoma in situ to predict clinical progression [68]. While these methods currently do not distinguish stem cell states, it would be a useful extension of this approach to predict bulk tumor response when cancer stem cells are therapeutically targeted.

3.6 How to Integrate Models with Data

Mathematical modeling combined with experimental techniques in single-cell expression and epigenetic analysis represent a powerful combination to understand the dynamics of the cancer stem cell niche. An iterative approach is employed, where experimental data are used to validate models and further inform mathematical modeling parameters, and modeling predictions are used to guide experiments and suggest new ones [69]. In situations where known molecular mechanisms represented in the model are sufficient to account for physiologic or cell biological phenomena, the model can be used to explore the emergent system properties. When there are additional phenomena not explained by molecular mechanisms, the model could suggest new experiments to identify additional molecular mechanisms to explain these phenomena.

We anticipate that single-cell genomic and transcriptomic profiling will advance our understanding of intra-tumoral heterogeneity of cancer stem cells, the role of circulating CSC populations during cancer development and tumor progression and in the response to treatment. As our understanding of cellular interactions within the tumor and its tissue microenvironment advances, we will be able to design novel therapies that will more effectively target the tumor microenvironment. The ability to track the evolution of the cancer stem cell compartments in circulating tumor cells in response to therapy will be particularly helpful in the adjuvant setting where eradication of cancer stem cells is most critical.

Acknowledgments

Thanks are given to Jill Granger for manuscript review and editing. This work was supported by grants RO1 CA101860 and R35 CA129765, NIH/NCATS UCLA CTSI Grant KL2TR000122, and by the Breast Cancer Research Foundation

References

1. Nowak M (2006) Evolutionary dynamics: exploring the equations of life. Harvard University Press, Canada
2. Michor F (2008) Mathematical models of cancer stem cells. J Clin Oncol 26:2854–2861
3. Foo J, Michor F (2014) Evolution of acquired resistance to anti-cancer therapy. J Theor Biol 355:10–20
4. Weekes SL, Barker B, Bober S, Cisneros K, Cline J, Thompson A, Hlatky L, Hahnfeldt P, Enderling HA (2014) multicompartment mathematical model of cancer stem cell-driven tumor growth dynamics. Bull Math Biol 76:762–782
5. Beerenwinkel N, Schwarz RF, Gerstung M, Markowetz F (2014) Cancer evolution: mathematical models and computational inference. Syst Biol 0:1–24
6. Gupta PB, Fillmore CM, Jiang G, Shapira SD, Tao K, Kuperwasser C, Lander ES (2011) Stochastic state transitions give rise to phenotypic equilibrium in populations of cancer cells. Cell 146:633–644
7. Sehl ME, Shimada M, Landeros A, Lange K, Wicha MS (2015) Modeling of cancer stem cell state transitions predicts therapeutic response. PLoS One 10:e0135797
8. Norton L (2005) Conceptual and practical implications of breast tissue geometry: toward a more effective, less toxic therapy. Oncologist 10:370–381
9. Baldock AL, Rockne RC, Boone AD, Neal ML, Hawkins-Daarud A, Corwin DM, Bridge CA, Guyman LA, Trister AD, Mrugala MM, Rockhill JK, Swanson KR (2013) From patient-specific mathematical neuro-oncology to precision medicine. Front. Oncologia 3:62
10. Withers HR, Taylor JMG, Maciejewski B (1988) Treatment volume and tissue tolerance. Int J Radiat Oncol Biol Phys 14:751–759
11. Simon R, Altman DG (1994) Statistical aspects of prognostic factor studies in oncology. Br J Cancer 69:979–985
12. Almendro V, Cheng Y-K, Randles A, Itzkovitz S, Marusyk A, Ametller E, Gonzalez-Farre X, Munoz M, Russnes HG, Helland A, Rye IH, Borresen-Dale AL, Maruyama R, van Oudenaarden A, Dowsett M, Jones RL, Reis-Filho J, Gascon P, Goenen M, Michor F, Polyak K (2014) Inference of tumor evolution during chemotherapy by computational modeling and in situ analysis of genetic and phenotypic cellular diversity. Cell Rep 6:514–527
13. Trinh A, Rye IH, Almendro V, Helland A, Russnes HG, Markowetz F (2014) Goifish: a system for the quantification of single cell heterogeneity from ifish images. Genome Biol 15:442
14. Hou Y, Song L, Zhu P, Zhang B, Tao Y, Xu X, Li F, Wu K, Liang J, Shao D, Wu H, Ye X, Ye C, Wu R, Jian M, Chen Y, Xie W, Zhang R, Chen L, Liu X, Yao X, Zheng H, Yu C, Li Q, Gong Z, Mao M, Yang X, Yang L, Li J, Wang W, Lu Z, Gu N, Laurie G, Bolund L, Kristiansen K, Wang J, Yang X, Wang J (2012) Single-cell exome sequencing and monoclonal evolution of a JAK2-negative myeloproliferative neoplasm. Cell 148:873–885
15. Kim KI, Simon R (2014) Using single-cell sequencing data to model the evolutionary history of a tumor. BMC Bioinformatics 15:27
16. Azizi E, Fouladdel S, Deol YS, Bender J, McDermott S, Jiang H, Sehl M, Clouthier SG, Nagrath S, Wicha MS. Exploring cancer stem cells heterogeneity via single cell multiplex gene expression analysis. Abstract 1943. Proceedings: AACR 106th Annual Meeting 2015; April 5–9th, 2014; San Diego, CA.
17. Azizi E, Jiagge EM, Fouladdel S, Wong S, Dziubinski ML, Sehl M, Kyani A, Li J, Jiang H, Luther TK, Clouthier SG, McDermott SP, Carpten J, Newman LA, Merajver SD, Wicha M. Single cell multiplex gene expression analysis to unravel heterogeneity of PDX samples established from tumors of breast cancer patients with different ethnicity. Abstract 4834. Proceedings: AACR 106th Annual Meeting 2015; April 18–22, 2015; Philadelphia, PA.
18. Hwang D, Smith JJ, Leslie DM, Weston AD, Rust AG, Ramsey S, de Atauri P, Siegel AF, Bolouri H, Aitchison JD, Hood L (2005) A data integration methodology for systems biology: experimental verification. Proc Natl Acad Sci U S A 102:17302–17307
19. Yeang CH, Ideker T, Jaakkola T (2004) Physical Network Models. J Comput Biol 11:243–262
20. Markowetz F, Sprang R (2007) Inferring cellular networks – a review. BioMed Central Bioinformatics 8(Suppl 6):S5
21. Gatenby RA, Vincent TL (2003) An evolutionary model of carcinogenesis. Cancer Res 63:6212–6220
22. Axelrod R, Axelrod DE, Pienta KJ (2006) Evolution of cooperation among tumor cells. Proc Natl Acad Sci U S A 103:13474–13479

23. Korkaya H, Kim GI, Davis A, Malik F, Henry NL, Ithimakin S, Quraishi AA, Tawakkol N, D'Angelo R, Paulson AK, Chung S, Luther T, Paholak HJ, Liu S, Hassan KA, Zen Q, Clouthier SG, Wicha MS (2012) Activation of an IL6 inflammatory loop mediates trastuzumab resistance in HER2+ breast cancer by expanding the cancer stem cell population. Mol Cell 47:570–584
24. Korkaya H, Liu S, Wicha MS (2011) Breast cancer stem cells, cytokine networks, and the tumor microenvironment. J Clin Invest 121:3804–3809
25. Korkaya H, Liu S, Wicha MS (2011) Regulation of cancer stem cells by cytokine networks: attacking cancer's inflammatory roots. Clin Cancer Res 17:6125–6129
26. Liu S, Ginestier C, SJ O, Clouthier SG, Patel SH, Monville F, Korkaya H, Heath A, Dutcher J, Kleer CG, Jung Y, Dontu G, Taichman R, Wicha MS (2011) Breast cancer stem cells are regulated by mesenchymal stem cells through cytokine networks. Cancer Res 71:614–624
27. Liu S, Clouthier SG, Wicha MS (2012) Role of microRNAs in the regulation of breast cancer stem cells. J Mammary Gland Biol Neoplasia 17:15–21
28. Deng L, Shang L, Bai S, Chen J, He X, Martin-Trevino R, Chen S, Li XY, Meng X, Yu B, Wang X, Liu Y, McDermott SP, Ariazi AE, Ginestier C, Ibarra I, Ke J, Luther T, Clouthier SG, Xu L, Shan G, Song E, Yao H, Hannon GJ, Weiss SJ, Wicha MS, Liu S (2014) MicroRNA100 inhibits self-renewal of breast cancer stem-like cells and breast tumor development. Cancer Res 74:6648–6660
29. Liu S, Patel SH, Ginestier C, Ibarra I, Martin-Trevino R, Bai S, McDermott SP, Shang L, Ke J, SJ O, Heath A, Zhang KJ, Korkaya H, Clouthier SG, Charafe-Jauffret E, Birnbaum D, Hannon GJ, Wicha MS (2012) MicroRNA93 regulates proliferation and differentiation of normal and malignant breast stem cells. PLoS Genet 8:e1002751
30. Liu S, Cong Y, Wang D, Sun Y, Deng L, Liu Y, Martin-Trevino R, Shang L, McDermott SP, Landis MD, Hog S, Adams A, D'Angelo R, Ginestier C, Charafe-Jauffret E, Clouthier SG, Birnbaum D, Wong ST, Zhan M, Chang JC, Wicha MS (2013) Breast cancer stem cell transition between epithelial and mesenchymal states reflective of their normal counterparts. Stem Cell Rep 2:78–91
31. Morrison SJ, Kimble J (2006) Asymmetric and symmetric stem-cell divisions in development and cancer. Nature 441:1068–1074
32. Cicalese A, Bonizzi G, Pasi CE, Faretta M, Ronzoni S, Giulini B, Brisken C, Minucci S, Di Fiore PP, Pelicci PG (2009) The tumor suppressor p53 regulates polarity of self-renewing divisions in mammary stem cells. Cell 138:1083–1095
33. Peng D, Tanikawa T, Li W, Zhao L, Vatan L, Szeliga W, Wan S, Wei S, Wang Y, Liu Y, Staroslawska E, Szubstarski F, Rolinski J, Grywalska E, Stanisławek A, Polkowski W, Kurylcio A, Kleer C, Chang AE, Wicha M, Sabel M, Zou W, Kryczek I (2016) Myeloid-derived suppressor cells endow stem-like qualities to breast cancer cells through IL6/STAT3 and NO/NOTCH cross-talk signaling. Cancer Res 76:3156–3165
34. Rompolas P, Mesa KR, Greco V (2013) Spatial organization within a niche as a determinant of stem cell fate. Nature 402:513–518
35. Jones DL, Wagers AJ (2008) No place like home: anatomy and function of the stem cell niche. Nat Rev Mol Cell Biol 9:11–21
36. Ovadia J, Nie Q (2013) Stem cell niche structure as an inherent cause of undulating epithelial morphologies. Biophys J 104:237–246
37. Szekely T, Burrage K, Mangel M, Bonasall MB (2014) Stochastic dynamics of interacting haematopoietic stem cell niche lineages. PLoS Comput Biol 10:e1003794
38. Komarova NL (2006) Spatial stochastic models for cancer initiation and progression. Bull Math Biol 68:1573–1599
39. Komarova NL (2007) Loss- and gain-of-function mutations in cancer: mass-action, spatial and hierarchical models. J Stat Phys 128:413–446
40. Conley SJ, Gheordunescu E, Kakarala P, Newman B, Korkaya H, Heath AN, Clouthier SG, Wicha MS (2012) Antiangiogenic agents increase breast cancer stem cells via the generation of tumor hypoxia. Proc Natl Acad Sci U S A 109:1784–1789
41. Savage VM, Herman AB, West GB, Leu K (2013) Using fractal geometry and universal growth curves as diagnostics for comparing tumor vasculature and metabolic rate with healthy tissue and for predicting responses to drug therapies. Discr Cont Dyn Syst Ser B 18:1077–1108
42. Pardoll DM (2012) The blockade of immune checkpoints in cancer immunotherapy. Nat Rev Cancer 12:252–264
43. Hodi FS, O'Day SJ, DF MD, Weber RW, Sosman JA, Haanen JB, Gonzalez R, Robert C, Schadendorf D, Hassel JC, Akerley W, van den Eertwegh AJ, Lutzky J, Lorigan P, Vaubel JM, Linette GP, Hogg D, Ottensmeier CH,

Lebbé C, Peschel C, Quirt I, Clark JI, Wolchok JD, Weber JS, Tian J, Yellin MJ, Nichol GM, Hoos A, Urba WJ (2010) Improved survival with ipilimumab in patients with metastatic melanoma. N Engl J Med 363:711–723

44. Mellman I, Coukos G, Dranoff G (2011) Cancer immunotherapy comes of age. Nature 480:480–489
45. Luke JJ, Flaherty KT, Ribas A, Long GV. Targeted agents and immunotherapies: optimizing outcomes in melanoma. Nat Rev Clin Oncol. 2017 14 463
46. Ribas A, Hamid O, Daud A, Hodi FS, Wolchok JD, Kefford R, Joshua AM, Patnaik A, Hwu WJ, Weber JS, Gangadhar TC, Hersey P, Dronca R, Joseph RW, Zarour H, Chmielowski B, Lawrence DP, Algazi A, Rizvi NA, Hoffner B, Mateus C, Gergich K, Lindia JA, Giannotti M, Li XN, Ebbinghaus S, Kang SP, Robert C (2016) Association of Pembrolizumab With Tumor Response and Survival Among Patients With Advanced Melanoma. JAMA 315:1600–1609
47. Garon EB, Rizvi NA, Hui R, Leighl N, Balmanoukian AS, Eder JP, Patnaik A, Aggarwal C, Gubens M, Horn L, Carcereny E, Ahn MJ, Felip E, Lee JS, Hellmann MD, Hamid O, Goldman JW, Soria JC, Dolled-Filhart M, Rutledge RZ, Zhang J, Lunceford JK, Rangwala R, Lubiniecki GM, Roach C, Emancipator K, Gandhi L (2015) KEYNOTE-001 Investigators. Pembrolizumab for the treatment of non-small-cell lung cancer. N Engl J Med 372:2018–2028
48. Ansell SM, Lesokhin AM, Borrello I, Halwani A, Scott EC, Gutierrez M, Schuster SJ, Millenson MM, Cattry D, Freeman GJ, Rodig SJ, Chapuy B, Ligon AH, Zhu L, Grosso JF, Kim SY, Timmerman JM, Shipp MA, Armand P (2015) PD-1 blockade with nivolumab in relapsed or refractory Hodgkin's lymphoma. N Engl J Med 372:311–319
49. Motzer RJ, Escudier B, McDermott DF, George S, Hammers HJ, Srinivas S, Tykodi SS, Sosman JA, Procopio G, Plimack ER, Castellano D, Choueiri TK, Gurney H, Donskov F, Bono P, Wagstaff J, Gauler TC, Ueda T, Tomita Y, Schutz FA, Kollmannsberger C, Larkin J, Ravaud A, Simon JS, LA X, Waxman IM, Sharma P (2015) CheckMate 025 Investigators. Nivolumab versus everolimus in advanced renal-cell carcinoma. N Engl J Med 373:1803–1813
50. Leisha A. Emens, Fadi S. Braiteh, Philippe Cassier, Jean-Pierre Delord, Joseph Paul Eder, Marcella Fasso, Yuanyuan Xiao, Yan Wang, Luciana Molinero, Daniel S. Chen and Ian Krop. Abstract 2859: Inhibition of PD-L1 by MPDL3280A leads to clinical activity in patients with metastatic triple-negative breast cancer (TNBC). Proceedings: AACR 106th Annual Meeting 2015; April 18–22, 2015; Philadelphia, PA
51. Quintana E, Shackleton M, Sabel MS, Fullen DR, Johnson TM, Morrison SJ (2008) Efficient tumour formation by single human melanoma cells. Nature 456:593–598
52. Walker R, Enderling H (2015) From concept to clinic: mathematically informed immunotherapy. Curr Probl Cancer 40:68–83
53. Sehl M, Zhou H, Sinsheimer JS, Lange KL (2011) Extinction models for cancer stem cell therapy. Math Biosci 234(2):132–146
54. Robert L, Ribas A, Hu-Lieskovan S (2016) Combining targeted therapy with immunotherapy. Can 1+1 equal more than 2? Semin Immunol 28:73–80
55. Hu-Lieskovan S, Robert L, Homet Moreno B, Ribas A (2014) Combining targeted therapy with immunotherapy in BRAF-mutant melanoma: promise and challenges. J Clin Oncol 32:2248–2254
56. Lu H, Clauser KR, Tam WL, Fröse J, Ye X, Eaton EN, Reinhardt F, Donnenberg VS, Bhargava R, Carr SA, Weinberg RAA (2014) breast cancer stem cell niche supported by juxtacrine signalling from monocytes and macrophages. Nat Cell Biol 16:1105–1117
57. Bozic I, Reiter JG, Allen B, Antal T, Chatterjee K, Shah P, Moon YS, Yaqubie A, Kelly N, Le DT, Lipson EJ, Chapman PB, Diaz LA Jr, Vogelstein B, Nowak MA (2013) Evolutionary dynamics of cancer in response to targeted combination therapy. Elife 2:e00747
58. Sehl ME, Sinsheimer JS, Zhou H, Lange KL (2009) Differential destruction of stem cells: implications for targeted cancer stem cell therapy. Cancer Res 69(24):9481–9489
59. Rodriguez-Brenes IA, Komarova NL, Wodarz D (2011) Evolutionary dynamics of feedback escape and the development of stem-cell-driven cancers. Proc Natl Acad Sci U S A 108:18983–18988
60. Behar M, Barken D, Werner SL, Hoffmann A (2013) The dynamics of signaling as a pharmacological target. Cell 155:448–461
61. Sun Z, Komarova NL (2012) Stochastic modeling of stem-cell dynamics with control. Math Biosci 240:231–240
62. Mitchell S, Tsui R, Hoffmann A (2015) Studying NF-kB signaling with mathematical models. Methods Mol Biol 1280:647–661
63. Schmidt H, Jirstrand M (2006) Systems Biology Toolbox for MATLAB: a computational

platform for research in systems biology. Bioinformatics 22:514–515

64. Nagy AL, Papp D, Toth J (2012) ReactionKinetics-- a mathematica package with applications. Chem Eng Sci 83:12–23
65. Peterson JL (1981) Petri net theory and the modeling of systems. Prentice-Hall, Englewood Cliffs, NJ
66. Gillespie DT (1977) Exact stochastic simulation of coupled chemical reactions. J Phys Chem 81:2340–2361
67. Gillespie DT, Petzold LR (2003) Improved leap-size selection for accelerated stochastic simulation. J Chem Phys 119:8229–8234
68. Macklin P, Edgerton ME, Thompson AM, Cristini V (2012) Patient-calibrated agent-based modeling of ductal carcinoma in situ (DCIS): from microscopic measurements to macroscopic predictions of clinical progression. J Theor Biol 301:122–140
69. Enderling H (2013) Unveiling stem cell kinetics: prime time for integrating experimental and computational models. Front Oncol 3:291

Chapter 17

Methods for High-throughput Drug Combination Screening and Synergy Scoring

Liye He, Evgeny Kulesskiy, Jani Saarela, Laura Turunen, Krister Wennerberg, Tero Aittokallio, and Jing Tang

Abstract

Gene products or pathways that are aberrantly activated in cancer but not in normal tissue hold great promises for being effective and safe anticancer therapeutic targets. Many targeted drugs have entered clinical trials but so far showed limited efficacy mostly due to variability in treatment responses and often rapidly emerging resistance. Toward more effective treatment options, we will need multi-targeted drugs or drug combinations, which selectively inhibit the viability and growth of cancer cells and block distinct escape mechanisms for the cells to become resistant. Functional profiling of drug combinations requires careful experimental design and robust data analysis approaches. At the Institute for Molecular Medicine Finland (FIMM), we have developed an experimental-computational pipeline for high-throughput screening of drug combination effects in cancer cells. The integration of automated screening techniques with advanced synergy scoring tools allows for efficient and reliable detection of synergistic drug interactions within a specific window of concentrations, hence accelerating the identification of potential drug combinations for further confirmatory studies.

Key words Drug combinations, High-throughput screening, Experimental design, Synergy scoring, Computational modeling

1 Introduction

A pressing challenge in the development of personalized cancer medicine is to understand how to make the most out of genomic information from a patient when evaluating treatment options. Over the past decade, there has been an extensive effort to sequence cancer genomes in large patient cohorts, sparking expectations to identify novel targets for more effective and selective treatment opportunities. These sequencing efforts have revealed a remarkable degree of genetic heterogeneity between and within tumors, which partly explains why the traditional "one-size-fits-all" anticancer treatment strategies have often produced disappointing outcomes in clinical trials [1]. On the other hand,

Louise von Stechow (ed.), *Cancer Systems Biology: Methods and Protocols*, Methods in Molecular Biology, vol. 1711, https://doi.org/10.1007/978-1-4939-7493-1_17, © Springer Science+Business Media, LLC 2018

functional studies using high-throughput drug screening allowed linking cancer genomic vulnerabilities to targeted drug responses [2–4]. However, complex genetic and epigenetic changes may lead to re-activation of multiple compensatory pathways and to emergence of treatment-resistant subpopulations (so-called cancer clonal evolution).

Therefore, to reach effective and sustained clinical responses, one often needs multi-targeted drugs or drug combinations, which selectively inhibit multiple pathways in cancer cells [5, 6]. To facilitate discovery of effective drug combinations, preclinical studies often rely on drug combination screening in cancer cell models. Those serve as a starting point to prioritize the most promising hits for further experimental investigation and therapy optimization. Many of the existing drug combination studies, however, focus on conventional chemotherapeutic drugs tested in a panel of cell lines, for which the drug combination effects might not easily translate into treatment options in the clinic (*see* (7)). In contrast, primary cell cultures that are derived from patients have shown tremendous potential that could enable the rapid assessment of novel drugs or drug combinations at the individual level [8]. To facilitate clinical translation, we have established at FIMM an Individualized Systems Medicine (ISM) drug combination platform. The ISM platform combines genomics, drug testing, and computational tools to predict drug responses for individual cancer patients. The ISM platform has successfully been used to functionally profile primary leukemia, ovarian cancer, and prostate cancer patient samples ex vivo so that the drug responses can be translated to the in vivo setting [9–12].

The advances in high-throughput drug combination screening have enabled the assaying of a large collection of chemical compounds, generating dynamic dose-response profiles that allow us to quantify the effect of drug combinations at an unprecedented level. A drug combination is usually classified as synergistic, antagonistic, or non-interactive. This classification is based on the deviation of the observed drug combination response from the expected effect of non-interaction (the null hypothesis). To quantify the degree of drug synergy, several models have been proposed, such as those based on the Highest single agent model (HSA) [13], the Loewe additivity model (Loewe) [14], and the Bliss independence model (Bliss) [15]. These existing drug synergy scoring models, together with their software implementations, were initially proposed for low-throughput experiments. In those experiments a limited number of drugs were combined with a fixed level of response, e.g., at their IC50 concentrations. For example, CompuSyn has become a popular tool to calculate a combination index (CI) using the Loewe additivity model [16]. However, CompuSyn allows only for manual input of one drug combination at a time, which makes it less efficient for analyzing multiple drug combinations, particularly when the drug combinations are tested under various concentrations, in a so-called dose-response matrix design.

To facilitate the data analysis of high-throughput drug combination screens, more recent tools have been made available as R implementations (https://www.R-project.org). For example, mixlow is an R package which utilizes a nonlinear mixed-effects model to calculate the CI [17]. However, mixlow works only for an experimental design where the ratio of two drugs in a combination is fixed over all tested concentrations. Therefore, it may not be directly applicable for a dose-response matrix design, where the ratios of two drugs vary. Another R package, called drc, provides an URSA (universal response surface approach) model, which is more suitable for dose-response matrix data [18]. URSA extends the Loewe model by considering the response surfaces over all the tested concentrations. In contrast to the CI, which is defined at a fixed response level, the URSA model provides a summarized drug interaction score from the whole dose-response matrix. However, the URSA implementation in the drc package often leads to fitting errors when the dose responses fail to comply with the model assumptions. To evaluate the appropriateness of URSA, one needs to trace back to its underlying theoretical paper [19]. The Bliss model has also been extended recently by incorporating the response surface concept, similar as in the URSA model, based on which a contour plot of a Bliss interaction index can be constructed [20]. We have recently developed a response surface model, called Zero Interaction Potency (ZIP), which combines the Loewe and the Bliss models, and proposed a delta score to characterize the synergy landscape over the full dose-response matrix [21].

Here, we describe an experimental-computational drug combination analysis pipeline that has been widely used in Finland and elsewhere to test and score effects of drug combinations in cancer cells [22–24]. The pipeline includes both an experimental protocol for dose-response matrix drug combination assays, as well as computational tools to facilitate the plate design and synergy modeling. The pipeline is applicable not only to cancer cell lines but also to patient-derived cancer samples for individualized drug combination optimization. With the increasing size of our compound library, including compounds that target all the known cancer survival pathways, the drug combination discovery can now be targeted toward more personalized anticancer treatment. We first describe the experimental protocol including a computer program, called FIMMcherry, which enables efficient production and visualization of combination assay plates, the output of which can be directly exported to the robotic system for automated dispensing. To address the lack of tailored software tools for high-throughput drug combination scoring, we here report a new R-package, SynergyFinder, which provides efficient implementations for all the popular synergy scoring models, including HSA, Loewe, Bliss, and ZIP.

This implementation provides the lab users with more flexibility to explore their drug combination data. We expect that the use of SynergyFinder will greatly improve the interpretation of the drug combination results and may eventually lead to the standardization of preclinical drug combination studies.

2 Materials

2.1 Cell Culture

1. Established cancer cell lines can be purchased from multiple vendors (*see* **Note 1**).
2. Patient-derived samples are obtained with permission from Finnish biobanks, hospitals, and clinical collaborators [2].
3. Cell media, serum and supplements recommended by cell line providers.
4. Trypsin-EDTA.
5. HyQTase.
6. CellTox Green Cytotoxicity reagent (Promega).
7. CellTiter-Glo or CellTiter-Glo 2.0 reagent (Promega).
8. 384-well tissue culture treated sterile assay plates.
9. MicroClime Environmental Lids.
10. Beckman Coulter Biomek FX^P for dispensing primary cells, which tend to grow as aggregates.
11. Plate reader.

2.2 Drug Combination Plate Design

1. FIMMcherry software (*see* **Note 2**).
2. Source plate file in text format.
3. Drug combination file in text format.
4. Compound library (*see* **Note 3**, Fig. 1).
5. Labcyte Echo 550 acoustic dispenser for dispensing compounds in precise volume with high accuracy (2.5 nL).
6. Storage pods.

2.3 Phenotypic Readouts

1. CellTox Green Cytotoxicity Assay.
2. CellTiter-Glo or CellTiter-Glo 2.0 Assay.
3. MultiFlo FX Multi-Mode Dispenser with RAD module or Multidrop Combi Reagent Dispenser for dispensing growth media, CellTiter-Glo reagents and seeding cells.
4. Plate shaker.
5. PHERAstar FS or Cytation 5 Cell Imaging Multi-Mode plate readers for CellTox Green (fluorescence) and CellTiter-Glo (luminescence) detection on 384-well plates.

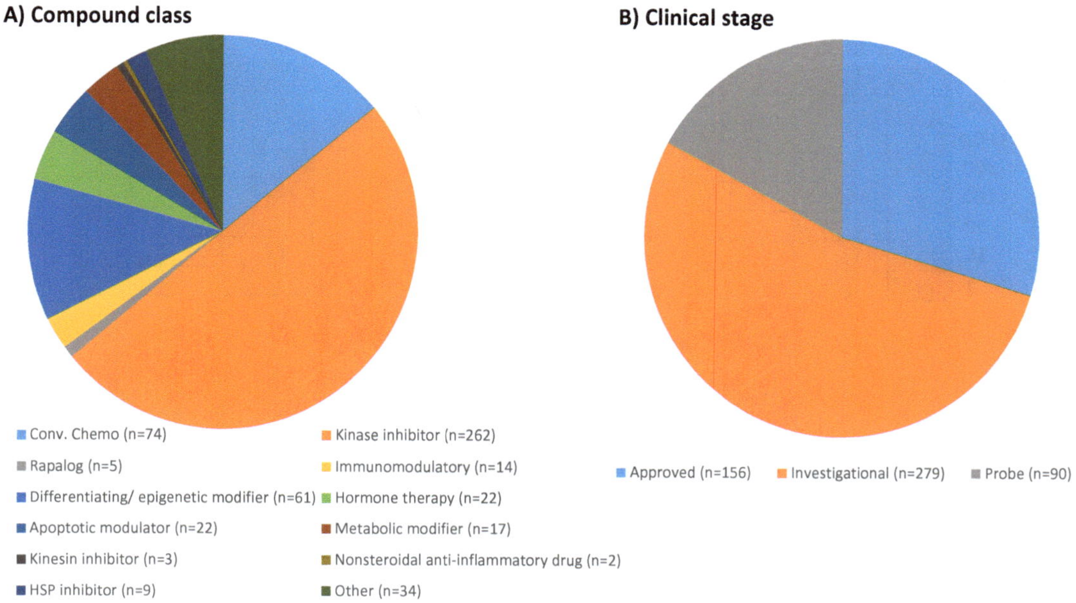

Fig. 1 An overview of the FIMM oncology compound collection. The drug combination platform enables the testing of pairwise drug combinations from 525 small-molecular anticancer compounds that cover mainly kinase inhibitors and other signal transduction modulators. About half of the compounds comprised in the library are either FDA-approved or being evaluated in clinical trials at different stages

2.4 Software Tools for Data Analysis

1. R.
2. Bioconductor.
3. SynergyFinder package (*see* **Notes 4–6**).
4. csv file that describes a drug combination dataset.

3 Methods

The drug combination analysis pipeline starts from sample preparation and compound selection, based on which an automated plate design program called FIMMCherry is utilized. The drug sensitivity and resistance is then profiled in the plate by cell viability, cytotoxicity, and other readouts. The resulting dose-response matrix data is analyzed with the SynergyFinder R package for the detection of synergistic drug combinations (*see* Fig. 2).

3.1 Cell Culture

1. Dissociate cells by adding 0.05% trypsin-EDTA or HyQTase to achieve a single-cell suspension.
2. Titrate cells to define optimal density within exponential growth (log phase). Seed cells in twofold serial dilution starting from 16,000 cells/well on 384-well plates. For most cell lines, the optimal cell number is in the range of 500–2000 cells/well.

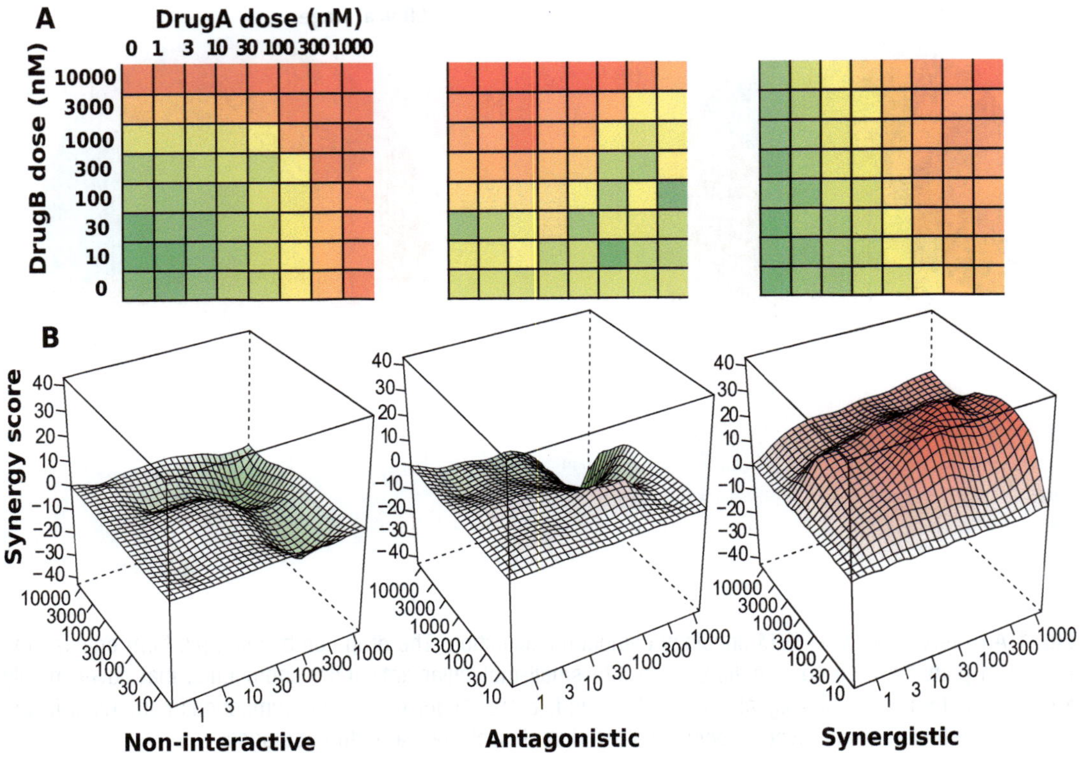

Fig. 2 An overview of the drug combination data analysis. (**a**) A typical high-throughput drug combination screen utilizes a dose-response matrix design where all possible dose combinations for a drug pair can be tested. Colors in the dose-response matrices show different levels of phenotypic responses of the cancer cell with *red* indicating stronger inhibition and *green* indicating lower inhibition. (**b**) Depending on the interaction pattern models derived from the dose-response matrices, a drug combination can be classified as non-interactive, antagonistic, or synergistic

3. Cell toxicity and viability detection after 72 h of incubation using CellTox Green and CellTiter-Glo reagents. Add 5 μL of culture medium to pre-drugged 384-well assay plates using MultiFlo FX Multi-Mode Dispenser with RAD module or Multidrop Combi Reagent Dispenser and shake the plates for 15 min. If toxicity measurement is performed, include 1:2000 dilution of CellTox Green reagent. Seed cells at optimal density to pre-drugged assay plates using MultiFlo FX Multi-Mode Dispenser with RAD module or Multidrop Combi Reagent Dispenser in 20 μL of culture medium. Culture cells for 72 h at 37 °C in the presence of 5% CO_2. Measure the amount of dead cells, stained by the CellTox Green reagent, using a plate reader with fluorescence mode. For viability measurement, add 25 μL of CellTiter-Glo reagent to assay plates using MultiFlo FX Multi-Mode Dispenser with RAD module or Multidrop Combi Reagent Dispenser. Shake the plates for 5 min and subsequently

spin the plates at 218 × *g* for 5 min. Measure the CTG signal in the assay wells using a plate reader with luminescence mode.

4. MicroClime Environmental Lids are used to minimize edge effect and to keep concentrations of solutions constant.

3.2 Drug Combination Plate Design

We utilize a combination plate layout where six compound pairs can be accommodated on one 384-well plate. A given pair of drugs is combined in a series of one blank and seven half-log dilution concentrations, resulting in an 8 × 8 dose matrix. To be able to transfer the compounds according to this matrix format, a pick list defining the source and destination plate locations and transfer volumes for the compounds is needed. An in-house program, called FIMMCherry, has been developed to automatically generate these rather complex pick lists effortlessly (*see* **Note 7**).

Two tab-delimited text files are needed as input:

1. A source plate file provides information of the compound stocks (compound identification, available concentration ranges, source plate identification, and well identification).
2. A drug combination file containing the selected compound pairs.

After loading the input files, FIMMCherry will show the layout of the plates accordingly (Fig. 3). A pick list that is compatible with the Labcyte Echo dispenser is then created by the program for compound dispensing. The Labcyte Echo 550 acoustic dispenser transfers liquid from source wells to destination wells in a non-contact fashion in 2.5 nL droplets. The pick list generated above is compatible with the Echo Cherry Pick software without further modifications to produce the pre-drugged assay plates [10].

1. The compounds are dissolved in DMSO except for 19 drugs (e.g., platinum drugs) with poor DMSO solubility or stability that are instead dissolved in water. All 525 compounds are transferred in five doses on eight 384-well plates.
2. The pre-dosed plates are stored in Storage Pods under nitrogen gas at room temperature for up to 1 month.
3. For quality control, a regular quality check-up of our compound library is performed which includes the testing of the compounds with four assay-ready cell lines (DU4475, HDQ-P1, IGROV-1, and MOLM-13) every 2 months. Following the time-dependent reproducibility of the drug responses allows us to precisely detect any changes in the compound stability and activity.

3.3 Viability Readouts

1. Transfer 5 μL of media with CellTox Green Cytotoxicity reagent into a 384-well containing the pre-diluted compound library (*see* **Note 8**).

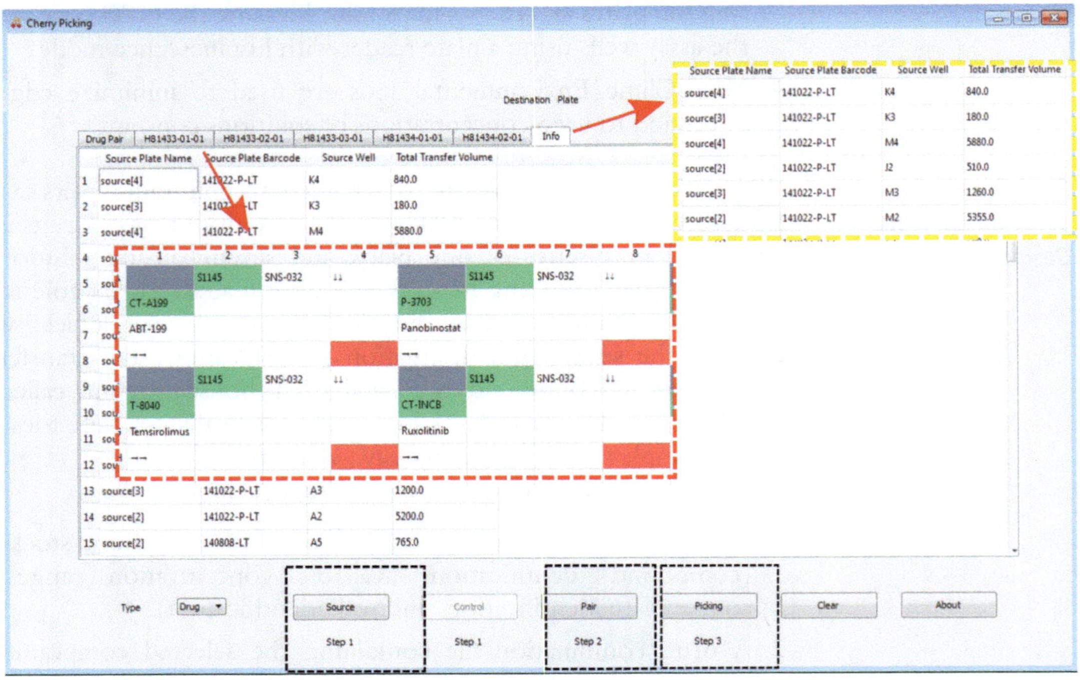

Fig. 3 Drug combination plate design using FIMMCherry. The graphical user interface contains a virtual plate enabling an interactive way of designing the plate. After loading the input files including the source, the control, and drug pair information (the *black inset boxes*), the selected drug combinations and their dose ranges will be listed in the "Drug Pair" tab, for which an echo file will be generated for acoustic dispensing. Each plate can be visualized in a separate tab and will be named by its plate identifier (the *red inset box*). The "Info" tab shows the liquids consumption in the source plates (the *yellow inset box*)

2. Shake the plate on the plate shaker at 450 rpm for 5 min for proper drug dissolving.
3. Transfer a single-cell suspension in 20 μL of media to a 384-well plate. Final dilution of CellTox Green reagent should be 1:2000 in 25 μL.
4. Incubate the cells in the plates for 72 h.
5. Shake the plates on the plate shaker at 500 rpm for 30 s. Read fluorescence in the plates using a plate reader for CellTox Green Cytotoxicity detection.
6. Transfer 25 μL of CellTiter-Glo reagent to the plate.
7. Shake the plates on the plate shaker at 450 rpm for 5 min and spin the plate at 218 × *g* for 5 min.
8. Read luminescence in the plates for detecting cell viability using a plate reader.

3.4 Synergy Scoring: Installation of the SynergyFinder R-package

1. Download and install R (https://www.R-project.org).
2. Download and install Bioconductor (https://www.bioconductor.org/).
3. Install the SynergyFinder package by typing in the R console as below:

```
> source("https://www.bioconductor.org/biocLite.R")
> biocLite("synergyfinder")
```

4. Load the package:

```
> library(synergyfinder)
```

3.5 Synergy Scoring: Input Data

1. A single csv file that describes a drug combination dataset is provided as input. The csv file is in a list format and must contain the following columns:
   - **BlockID**: the identifier for a drug combination. If multiple drug combinations are present, e.g., in the standard 384-well plate where six drug combinations are fitted, then the identifiers for each of them must be unique.
   - **Row and Col**: the row and column indexes for each well in the plate.
   - **DrugCol**: the name of the drug on the columns in a dose-response matrix.
   - **DrugRow**: the name of the drug on the rows in a dose-response matrix.
   - **ConcCol and ConcRow**: the concentrations of the column drugs and row drugs in combination.
   - **ConcUnit**: the unit of concentrations. It is typically nM or μM.
   - **Response**: the effect of drug combinations at the concentrations specified by ConcCol and ConcRow. The effect must be normalized to %inhibition of cell viability or proliferation based on the positive and negative controls. For a well-controlled experiment, the range of the response values is expected from 0 to 100. However, missing values or extreme values are allowed. For input data where the drug effect is represented as %viability, the program will internally convert it to %inhibition value by 100-%viability.
2. We provide example input data in the R package, which is extracted from a recent drug combination screen for treatment of diffuse large B-cell lymphoma (DLBCL) [7]. The example input data contains two representative drug combinations

(*ibrutinib and ispinesib* and *ibrutinib and canertinib*) for which the %viability of a cell line TMD8 was assayed using a 6 by 6 dose matrix design. The example data in the required list format can be loaded and reshaped to a dose-response matrix format for further analysis by typing:

```
> data("mathews_screening_data")
> dose.response.mat <- ReshapeData(mathews_screening_data,
data.type = "viability")
```

3. The "data.type" parameter specifies the type of drug response, which can be either "viability" or "inhibition." We will use these example data to illustrate the main functions of SynergyFinder below. More documentation of the input and output parameters for each function can be accessed by typing:

```
> help('ReshapeData')
```

3.6 Synergy Scoring: Input Data Visualization

1. The input data can be visualized using the function PlotDoseResponse by typing:

```
> PlotDoseResponse(dose.response.mat)
```

2. The function fits a four-parameter log-logistic model to generate the dose-response curves for the single drugs based on the first row and first column of the dose-response matrix. The drug combination responses are also plotted as heatmaps. From those, one can assess the therapeutic significance of the combination, e.g., by identifying the concentrations at which the drug combination can lead to a maximal effect on the inhibition of cancer cell survival/proliferation (*see* Fig. 4). The PlotDoseResponse function also provides a high-resolution pdf file by adding the "save.file" parameter:

```
> PlotDoseResponse(dose.response.mat, save.file = TRUE)
```

3. The pdf file will be saved under the current work directory with the syntax: "drug1.drug2.dose.response.blockID.pdf."

3.7 Synergy Scoring: Drug Synergy Scoring (See Notes 9 and 10)

1. The current SynergyFinder package provides the synergy scores of four major reference models, including HSA, Loewe, Bliss, and ZIP. In a drug combination experiment where drug 1 at dose x_1 is combined with drug 2 at dose x_2, the effect of such a combination is y_c as compared to the monotherapy effect $y_1(x_1)$ and $y_2(x_2)$. To be able to quantify the degree of drug interactions, one needs to determine the deviation of y_c from the

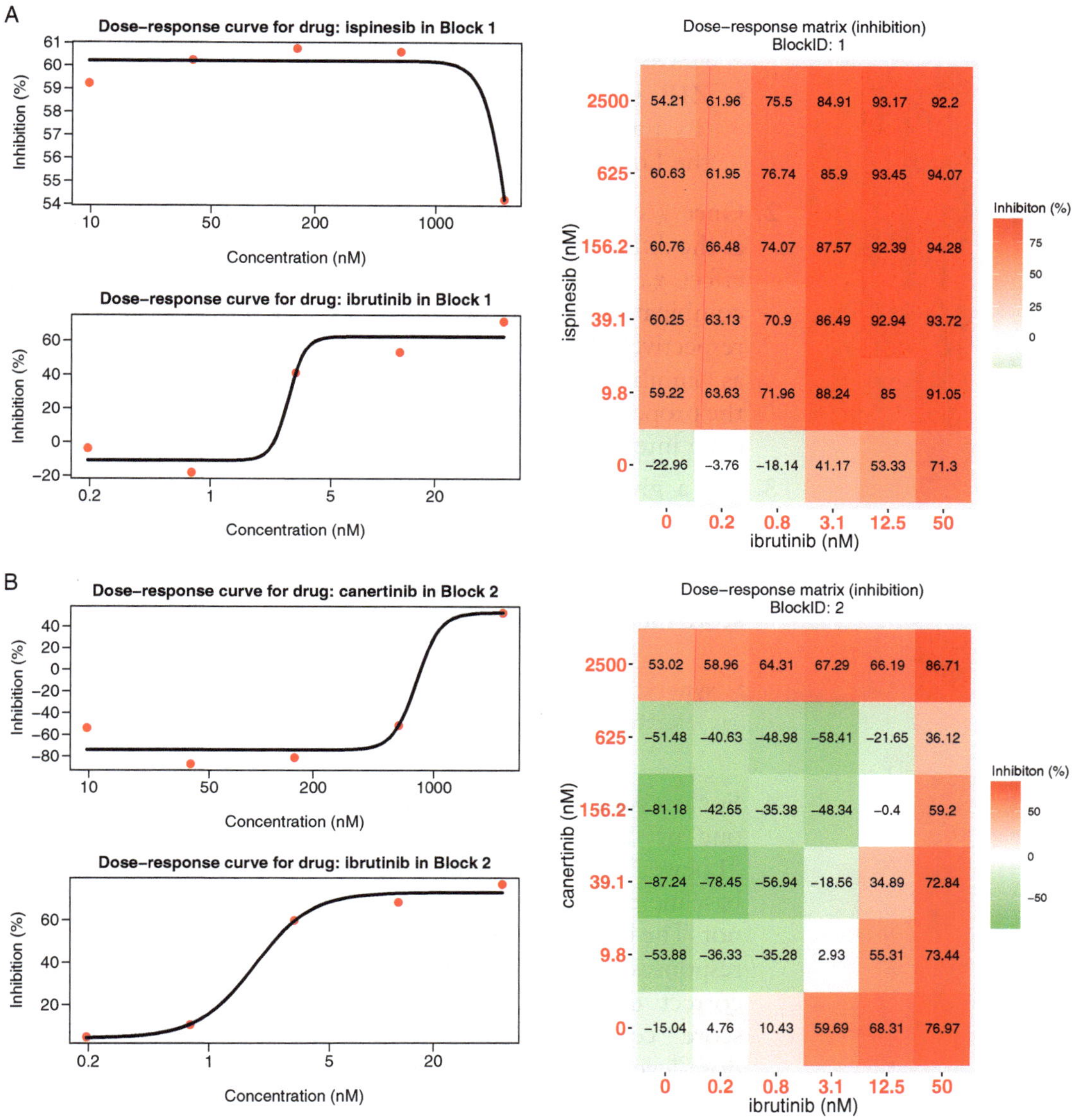

Fig. 4 Plots for single-drug dose-response curves and drug combination dose-response matrices. (**a**) The ibrutinib and ispinesib combination. (**b**) The ibrutinib and canertinib combination. *Left panel*: single drug dose-response curves fitted with the commonly-used 4-parameter log-logistic (4PL) function. *Right panel*: the raw dose-response matrix data is visualized as a heatmap

expected effect y_e of non-interaction, which is calculated in different ways with the individual reference models.

- **HSA**: y_e is the effect of the highest monotherapy effect, i.e., $y_e = \max(y_1, y_2)$.
- **Loewe**: y_e is the effect that would be achieved if a drug was combined with itself, i.e., $y_e = y_1(x_1 + x_2) = y_2(x_1 + x_2)$.

- **Bliss**: y_e is the effect that would be achieved if the two drugs are acting independently of the phenotype, i.e., $y_e = y_1 + y_2 - y_1 y_2$.
- **ZIP**: y_e is the effect that would be achieved if the two drugs do not potentiate each other, i.e., both the assumptions of the Loewe model and the Bliss model are met.

2. Once y_e can be determined, the synergy score can be calculated as the difference between the observed effect y_c and the expected effect y_e. Depending on whether $y_c > y_e$ or $y_c < y_e$ the drug combination can be classified as synergistic or antagonist, respectively. Furthermore, as the input data has been normalized as %inhibition, the synergy score can be directly interpreted as the proportion of cellular responses that can be attributed to the drug interactions.
3. For a given dose-response matrix, one needs to first choose which reference model to use and then apply the CalculateSynergy function to calculate the corresponding synergy score at each dose combination. For example, the ZIP-based synergy score for the example data can be obtained by typing:

```
> synergy.score <- CalculateSynergy(data = dose.response.
mat, method = "ZIP", correction = TRUE)
```

4. For assessing the synergy scores with the other reference models, one needs to change the "method" parameter to "HSA," "Loewe," or "Bliss." The "correction" parameter specifies if a baseline correction is applied on the raw dose-response data or not. The baseline correction utilizes the average of the minimum responses of the two single drugs as a baseline response to correct the negative response values. The output "synergy. score" contains a score matrix of the same size to facilitate a dose-level evaluation of drug synergy as well as a direct comparison of the synergy scores between two reference models.

3.8 Synergy Scoring: The Drug Interaction Landscape

1. The synergy scores are calculated across all the tested concentration combinations, which can be visualized as either a two-dimensional or a three-dimensional interaction surface over the dose matrix. The landscape of such a drug interaction scoring is very informative when identifying the specific dose regions where a synergistic or antagonistic drug interaction occurs. The height of the 3D drug interaction landscape is normalized as the % inhibition effect to facilitate a direct comparison of the degrees of interaction among multiple drug combinations. In addition, a summarized synergy score is provided by averaging over the whole dose-response matrix. To visualize the drug interaction landscape, one can utilize the PlotSynergy function as below (*see* Fig. 5):

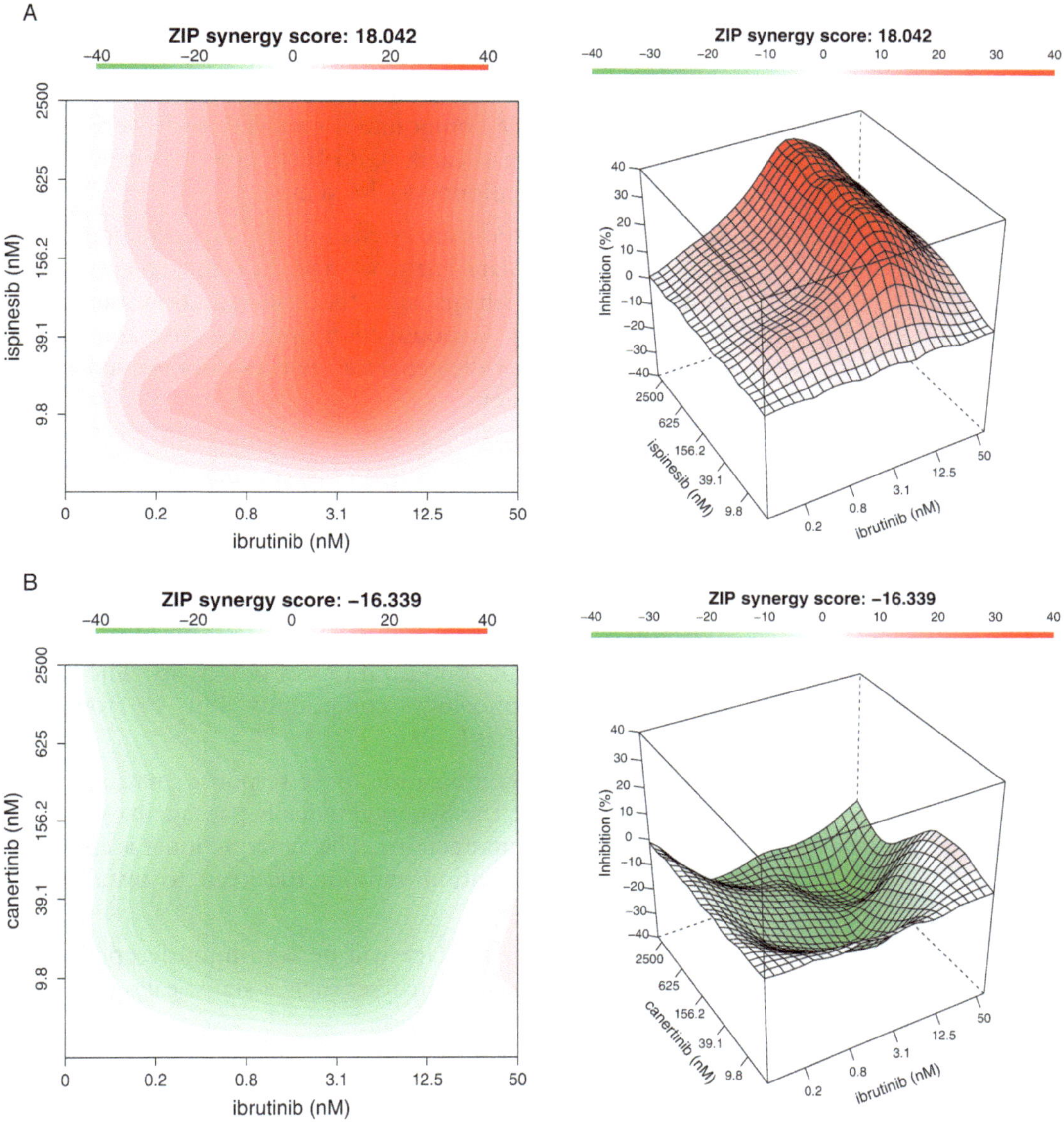

Fig. 5 The drug interaction landscapes based on the ZIP model. (**a**) The ibrutinib and ispinesib combination. (**b**) The ibrutinib and canertinib combination

```
> PlotSynergy(synergy.score, type = "all", save.file = TRUE)
```

2. The "type" parameter specifies the visualization type of the interaction surface as 2D, 3D, or both.

4 Notes

1. Examples of cell lines include four cell lines that are used for quality check of the compound library: DU4475 (breast cancer), HDQ-P1 (breast cancer), IGROV-1 (ovarian cancer), and MOLM-13 (acute monocytic leukemia).
2. Specific software tools are needed in the experimental design stage and in the data analysis stage. For the 384-well plate design, once the drugs and the concentration ranges are selected, we use the in-house cherry-picking program, FIMM-cherry, to automatically generate the echo files needed for the Labcyte Access system.
3. The FIMM oncology collection contains both FDA/EMA-approved drugs and investigational compounds (*see* Fig. 1). The collection is constantly evolving and the current FO4B version contains 525 compounds with concentrations ranging typically between 1 and 10,000 nM. For some compounds, the concentration range is adjusted upward (e.g., platinum drugs, 100,000 nM) or downward (e.g., rapalogs, 100 nM) to better match their relevant concentrations of bioactivity. The full list of the FIMM oncology compounds can be found in Supplementary Table 1.
4. When the drug combination dose-response matrix data is ready, we then use the SynergyFinder R-package to score and visualize the drug interactions. The SynergyFinder is also available as a web-application without the need to install the R environment.
5. The SynergyFinder package will be continuously updated for including more rigorous analyses such as statistical significance, effect size, and noise detection.
6. Availability: The source code for the FIMMCherry program is available at github (https://github.com/hly89/FIMM-Cherry). The SynergyFinder R package for drug combination data analysis is available at CRAN and Bioconductor.
7. FIMMCherry is a desktop GUI application, which is developed using Python (https://www.python.org/) and Qt application development framework (https://www.qt.io/). The integration of Python and Qt allows FIMMCherry to run on all the major computer platforms including Windows, Linux, and Mac OS X.
8. We have not seen problems in cell proliferation rate or other major effects when using the reagent. The reagent is stable at least 72 h in the cell culture and the cells dying at the beginning of the 72 h incubation are still stained after 72 h.

Table 1
The FIMM oncology compound collection

DRUG_ NAME	Mechanism targets	Class explained	High phase approval status	Alias	Trade names	Supplier	Supplier Ref	Solvent	High Conc. (nM)
SN-38	Active metabolite of irinotecan. Topoisomerase I inhibitor	A. Conv. Chemo	(approved)	BR-36613, 7 -Ethyl-10-hydroxy camptothecine		ChemieTek	CT-SN38	DMSO	10000
Idarubicin	Topoisomerase II inhibitor	A. Conv. Chemo	Approved		Zavedos, Idamycin	Sigma-Aldrich	I1656	DMSO	1000
Auranofin	Antirheumatic agent	A. Conv. Chemo	Approved			Sigma-Aldrich	A6733	DMSO	2500
Plicamycin	RNA synthesis inhibitor	A. Conv. Chemo	Approved	Mithramycin A		Santa Cruz Biotechnology	sc-200909-5	DMSO	10000
Bortezomib	Proteasome inhibitor (26S subunit)	A. Conv. Chemo	Approved	MS-341	Velcade, Cytomib	National Cancer Institute	NSC 681239-L/9	DMSO	1000
Clofarabine	Antimetabolite; Purine analog	A. Conv. Chemo	Approved		Evoltra, Clolar	National Cancer Institute	NSC 606869-X/4	DMSO	10000
Lomustine	Alkylating nitrosourea compound	A. Conv. Chemo	Approved	CCNU, CeeNU		National Cancer Institute	NSC 79037-R/12	DMSO	10000
Vincristine	Mitotic inhibitor. Vinca alkaloid microtubule depolymerizer	A. Conv. Chemo	Approved			Selleck	S1241	DMSO	1000
Vinorelbine	Mitotic inhibitor. Vinca alkaloid microtubule depolymerizer	A. Conv. Chemo	Approved			Selleck	S4269	DMSO	10000
Altretamine	Formaldehyde release, alkylating agent	A. Conv. Chemo	Approved			National Cancer Institute	NSC 13875-O/97	DMSO	10000
Vinblastine	Mitotic inhibitor. Vinca alkaloid microtubule depolymerizer	A. Conv. Chemo	Approved			National Cancer Institute	NSC 49842-J44	DMSO	1000
Chlorambucil	Nitrogen mustard alkylating agent	A. Conv. Chemo	Approved			National Cancer Institute	NSC 3088-N/6	DMSO	10000
Dacarbazine	Alkylating agent	A. Conv. Chemo	Approved			National Cancer Institute	NSC 45388-R/74	DMSO	10000
Cyclophosphamide	Alkylating agent	A. Conv. Chemo	Approved			Selleck	S1217	DMSO	40000

(continued)

Table 1 (continued)

DRUG_ NAME	Mechanism targets	Class explained	High phase approval status	Alias	Trade names	Supplier	Supplier Ref	Solvent	High Conc. (nM)
Cytarabine	Antimetabolite, interferes with DNA synthesis	A. Conv. Chemo	Approved	Ara-C		National Cancer Institute	NSC 63878-P/19	DMSO	10000
Fluorouracil	Antimetabolite	A. Conv. Chemo	Approved	5-fluorouracil, 5-FU		National Cancer Institute	NSC 19893-G/4	DMSO	10000
Ifosfamide	Nitrogen mustard alkylating agent	A. Conv. Chemo	Approved			National Cancer Institute	NSC 109724-X/4	DMSO	10000
Melphalan	Nitrogen mustard alkylating agent	A. Conv. Chemo	Approved			Sigma-Aldrich	M2011	AQ	12500
Mitoxantrone	Topoisomerase II inhibitor	A. Conv. Chemo	Approved			National Cancer Institute	NSC 279836-C/2	DMSO	1000
Paclitaxel	Mitotic inhibitor, taxane microtubule stabilizer	A. Conv. Chemo	Approved		Taxol	National Cancer Institute	NSC 125973-L/68	DMSO	1000
Procarbazine	Alkylating agent	A. Conv. Chemo	Approved			National Cancer Institute	NSC 77213-K/6	DMSO	10000
Topotecan	Topoisomerase I inhibitor. Camptothecin analog	A. Conv. Chemo	Approved			National Cancer Institute	NSC 609699-Y/16	DMSO	10000
Temozolomide	Alkylating agent	A. Conv. Chemo	Approved			National Cancer Institute	NSC 362856-R/31	DMSO	100000
Mechlorethamine	Nitrogen mustard alkylating agent	A. Conv. Chemo	Approved	Nitrogen mustard	Mustargen	Sigma-Aldrich	122564	DMSO	100000
Mitotane	Antineoplastic agent	A. Conv. Chemo	Approved			National Cancer Institute	NSC 38721-U/3	DMSO	10000
Allopurinol	Xanthine oxidase inhibitor	A. Conv. Chemo	Approved		Zyloprim	National Cancer Institute	NSC 1390-R/3	DMSO	10000
Busulfan	Alkylating antineoplastic agent	A. Conv. Chemo	Approved			Sigma-Aldrich	B2635	DMSO	100000
Hydroxyurea	Antineoplastic agent	A. Conv. Chemo	Approved		Myelostat	Sigma-Aldrich	H8627	DMSO	100000
Mercaptopurine	Antimetabolite	A. Conv. Chemo	Approved	6-mercaptopurine, 6-MP		National Cancer Institute	NSC 755-Z/13	DMSO	10000

Thioguanine	Antimetabolite; Purine analog	A. Conv. Chemo	Approved	6-thioguanine, 6-TG		National Cancer Institute	NSC 752-W/47	DMSO	10000
Carmustine	Alkylating agent	A. Conv. Chemo	Approved	BCNU		National Cancer Institute	NSC 409962-T/3	DMSO	10000
Thio-TEPA	Alkylating agent	A. Conv. Chemo	Approved			Sigma-Aldrich	T6069	DMSO	50000
Pipobroman	Alkylating agent	A. Conv. Chemo	Approved			National Cancer Institute	NSC 25154-X/2	DMSO	10000
Raltitrexed	DHFR/GARFT/thymidylate synthase inhibitor	A. Conv. Chemo	Approved	ICI-D 1694	Tomudex	Medchemexpress	HY-10821	DMSO	1000
Irinotecan	Topoisomerase I inhibitor. Camptothecin prodrug analog	A. Conv. Chemo	Approved		Camptosar	LC Laboratories	I-4122	DMSO	10000
Nelarabine	Nucleoside analog, DNA, RNA synth inhibitor	A. Conv. Chemo	Approved		Arranon, Atriance	SequoiaResearch Products	SRP003328n	DMSO	10000
Docetaxel	Mitotic inhibitor, taxane microtubule stabilizer	A. Conv. Chemo	Approved		Taxotere, Docecad	LC Laboratories	D-1000	DMSO	1000
Pentostatin	Antimetabolite; Purine analog	A. Conv. Chemo	Approved	Deoxycoformycin		National Cancer Institute	NSC 218321-O/48	DMSO	10000
Estramustine	Alkylating agent	A. Conv. Chemo	Approved			Sigma-Aldrich	E0407	AQ	10000
Floxuridine	Antimetabolite; Analog of 5-fluorouracil	A. Conv. Chemo	Approved	5-fluorodeoxyuridine		National Cancer Institute	NSC 27640-Z/31	DMSO	10000
Gemcitabine	Antimetabolite; Nucleoside analog	A. Conv. Chemo	Approved		Gemsar, Gemzar	National Cancer Institute	NSC 613327-S/2	DMSO	1000
Teniposide	Topoisomerase II inhibitor	A. Conv. Chemo	Approved			National Cancer Institute	NSC 122819-I/52	DMSO	10000
Dactinomycin	RNA and DNA synthesis inhibitor	A. Conv. Chemo	Approved	Actinomycin D		National Cancer Institute	NSC 3053-Y/14	DMSO	1000
Streptozocin	Alkylating glucosamine-nitrosourea agent	A. Conv. Chemo	Approved			National Cancer Institute	NSC 37917-V/5	DMSO	10000
Cladribine	Antimetabolite; Purine analog	A. Conv. Chemo	Approved		Leustatin	National Cancer Institute	NSC 105014-F/2	DMSO	1000
Mitomycin C	Antineoplastic anatibiotic; DNA crosslinker	A. Conv. Chemo	Approved			National Cancer Institute	NSC 26980-J/65	DMSO	10000

(continued)

Table 1 (continued)

DRUG_ NAME	Mechanism targets	Class explained	High phase approval status	Alias	Trade names	Supplier	Supplier Ref	Solvent	High Conc. (nM)
Carboplatin	Platinum-based antineoplastic agent	A. Conv. Chemo	Approved			Selleck	S1215-2	AQ	100000
Cisplatin	Platinum-based antineoplastic agent	A. Conv. Chemo	Approved			Selleck	S1166-2	AQ	100000
Oxaliplatin	Platinum-based antineoplastic agent	A. Conv. Chemo	Approved			Selleck	S1224-2	AQ	100000
Uracil mustard	Alkylating agent	A. Conv. Chemo	Approved	Uracil mustard		National Cancer Institute	NSC 34462-Q/2	DMSO	10000
Daunorubicin	Topoisomerase II inhibitor	A. Conv. Chemo	Approved			National Cancer Institute	NSC 82151-A/44	DMSO	1000
Etoposide	Topoisomerase II inhibitor	A. Conv. Chemo	Approved			National Cancer Institute	NSC 141540-H/184	DMSO	10000
Doxorubicin	Topoisomerase II inhibitor	A. Conv. Chemo	Approved		Adriamycin	Sigma-Aldrich	D1515	DMSO	1000
Valrubicin	Topoisomerase II inhibitor	A. Conv. Chemo	Approved		Valstar	National Cancer Institute	NSC 246131-R/4	DMSO	5000
Ixabepilone	Mitotic inhibitor. Epothilone microtubule stabilizer.	A. Conv. Chemo	Approved	azaepothilone B	Ixempra	National Cancer Institute	NSC 747973-W/3	DMSO	1000
Fludarabine	Antimetabolite; Purine analog	A. Conv. Chemo	Approved		Fludara	National Cancer Institute	NSC 312887-C/7	DMSO	10000
Bleomycin	Glycopeptide antibiotic; causes DNA breaks	A. Conv. Chemo	Approved			Selleck	S1214	DMSO	10000
Carfilzomib	Proteasome inhibitor (20S subunit)	A. Conv. Chemo	Approved		Kyprolis	ChemieTek	CT-CARF	DMSO	1000
Bendamustine	Nitrogen mustard alkylating agent	A. Conv. Chemo	Approved			National Cancer Institute	NSC 138783	DMSO	10000
Capecitabine	5-FU prodrug	A. Conv. Chemo	Approved			LC Laboratories	C-2799	DMSO	10000
Chloroquine	Antimalaria agent; chemo/radio sensitizer	A. Conv. Chemo	Approved			Sigma-Aldrich	C6628	AQ	100000

Omacetaxine	Protein synthesis inhib (80 S ribosome)	A. Conv. Chemo	Approved	Homoharringtonine	Synribo	Santa Cruz Biotechnology	sc-202652	DMSO	10000
Amsacrine	DNA intercalation, Topo II inhibitor	A. Conv. Chemo	Approved	Acridinyl anisidide		Santa Cruz Biotechnology	sc-214540	DMSO	10000
Cabazitaxel	Taxane microtubule stabilizer, antimitotic	A. Conv. Chemo	Approved	XRP6258	Jevtana	Medchemexpress	HY-15459	DMSO	1000
Oprozomib	proteasome (20 S) inhibitor	A. Conv. Chemo	Investigational (Ph 1)	PR-047		ChemieTek	CT-OPRO	DMSO	2500
ABT-751	Mitotic inhibitor. Colchicine site binding microtubule depolymerizer.	A. Conv. Chemo	Investigational (Ph 2)			Selleck	S1165	DMSO	10000
Indibulin	Mitoric inhibitor. Microtubule depolymerizer	A. Conv. Chemo	Investigational (Ph 2)			Tocris Biosciences	3728	DMSO	10000
Aldoxorubicin	Topo II, albumin	A. Conv. Chemo	Investigational (Ph 2)			Medchemexpress	HY-16261	DMSO	1000
Amonafide	Topoisomerase II inhibitor / DNA intercalator	A. Conv. Chemo	Investigational (Ph 3)	Xanafide, Quinamed		Selleck	S1367	DMSO	10000
Patupilone	Mitotic inhibitor, epothilone microtubule stabilizer	A. Conv. Chemo	Investigational (Ph 3)	Epothilone B, EpoB		LC Laboratories	E-5500-2	DMSO	1000
Pixantrone	topoisomerase II inhibitor	A. Conv. Chemo	Investigational (Ph 3)			Medchemexpress	HY-13727A	AQ	10000
Camptothecin	Topoisomerase I inhibitor	A. Conv. Chemo	Probe			Tocris Biosciences	1100	DMSO	5000
8-chloro-adenosine	Nucleoside analog, RNA synthesis inhibitor	A. Conv. Chemo	Probe	8-chloro-adenosine		Northwestern University	NWU 8CL	DMSO	50000
8-amino-adenosine	Nucleoside analog, RNA synthesis inhibitor	A. Conv. Chemo	Probe	8-amino-adenosine		Northwestern University	NWU 8NH2	DMSO	50000
Hydroxyfasudil	ROCK, PKA, PKG, PRK inhibitor	B. Kinase inhibitor	Active metabolite of approved drug			Santa Cruz Biotechnology	sc-202176	DMSO	37500
Gefitinib	EGFR inhibitor	B. Kinase inhibitor	Approved		Iressa	LC Laboratories	G-4408	DMSO	10000
Imatinib	Abl, Kit, PDGFRB inhibitor	B. Kinase inhibitor	Approved		Gleevec, Glivec	LC Laboratories	C-5508	DMSO	10000

(continued)

Table 1
(continued)

DRUG_ NAME	Mechanism targets	Class explained	High phase approval status	Alias	Trade names	Supplier	Supplier Ref	Solvent	High Conc. (nM)
Erlotinib	EGFR inhibitor	B. Kinase inhibitor	Approved	OSI-774	Tarceva	National Cancer Institute	NSC 718781-R/4	DMSO	10000
Lapatinib	HER2, EGFR inhibitor	B. Kinase inhibitor	Approved	GW2016	Tykerb, Tyverb	LC Laboratories	L-4804	DMSO	1000
Palbociclib	CDK inhibitor (Cdk4/6)	B. Kinase inhibitor	Approved		Ibrance	Selleck	S1116	DMSO	10000
Afatinib	EGFR inhibitor	B. Kinase inhibitor	Approved		Gilotrif, Giotrif	Selleck	S1011	DMSO	1000
Crizotinib	ALK, c-Met inhibitor	B. Kinase inhibitor	Approved		Xalkori	Selleck	S1068	DMSO	1000
Ponatinib	Broad TK inhibitor	B. Kinase inhibitor	Approved		Iclusig	Selleck	S1490	DMSO	1000
Trametinib	MEK1/2 inhibitor	B. Kinase inhibitor	Approved	JTP-74057	Mekinist	ChemieTek	CT-GSK112	DMSO	250
Ruxolitinib	JAK1&2 inhibitor	B. Kinase inhibitor	Approved		Jakafi, Jakavi	ChemieTek	CT-INCB	DMSO	10000
Nilotinib	Abl inhibitor	B. Kinase inhibitor	Approved		Tasigna	LC Laboratories	N-8207	DMSO	10000
Vemurafenib	B-Raf(V600E) inhibitor	B. Kinase inhibitor	Approved	RG7204, RO5185426	Zelboraf	ChemieTek	CT-P4032	DMSO	10000
Vandetanib	VEGFR,EGFR, RET inhibitor	B. Kinase inhibitor	Approved		Caprelsa	LC Laboratories	V-9402	DMSO	1000
Dasatinib	Abl, Src, Kit, EphR... Inhibitor	B. Kinase inhibitor	Approved		Sprycel	LC Laboratories	D-3307	DMSO	1000
Tofacitinib	JAK3, JAK2(V617F) inhibitor	B. Kinase inhibitor	Approved	tasocitinib	Xeljanz, Jakvinus	LC Laboratories	T-1377	DMSO	5000
Axitinib	VEGFR, PDGFR, KIT inhibitor	B. Kinase inhibitor	Approved		Inlyta	LC Laboratories	A-1107	DMSO	10000
Bosutinib	Abl, Src inhibitor	B. Kinase inhibitor	Approved		Bosulif	LC Laboratories	B-1788	DMSO	10000
Pazopanib	VEGFR inhibitor	B. Kinase inhibitor	Approved		Votrient	LC Laboratories	P-6706	DMSO	10000
Sorafenib	B-Raf, FGFR-1, VEGFR-2 & -3, PDGFR-beta, KIT, and FLT3 inhib	B. Kinase inhibitor	Approved		Nevaxar	LC Laboratories	S-8502	DMSO	1000
Sunitinib	Broad TK inhibitor	B. Kinase inhibitor	Approved		Sutent	LC Laboratories	S-8803	DMSO	1000

Regorafenib	B-Raf, c-Kit, VEGFR2 inhibitor	B. Kinase inhibitor	Approved		Stivarga	Selleck	S1178	DMSO	10000
Cabozantinib	VEGFR2, Met, FLT3, Tie2, Kit and Ret inhibitor	B. Kinase inhibitor	Approved	XL184	Cometriq	ChemieTek	CT-XL184	DMSO	1000
Ibrutinib	Btk inhibitor	B. Kinase inhibitor	Approved	CRA-032765	Imbruvica	Selleck	S2680	DMSO	1000
Dabrafenib	B-Raf(V600E) inhibitor	B. Kinase inhibitor	Approved		Tafinlar	ChemieTek	CT-DABR	DMSO	2500
Ceritinib	ALK inhibitor	B. Kinase inhibitor	Approved	LDK378	Zykadia	Selleck	S7083	DMSO	2500
Fasudil	ROCK, PKA, PKG, PRK inhibitor, prodrug	B. Kinase inhibitor	Approved (Japan)	HA-1077		LC Laboratories	H-2330	DMSO	50000
Alectinib	ALK (incl gatekeeper mut) inhib	B. Kinase inhibitor	Approved (Japan)		Alecensa	ChemieTek	CT-CH542	DMSO	1000
Idelalisib	PI3K inhibitor, p110δ-selective	B. Kinase inhibitor	Approved (US)	CAL-101	Zydelig	ChemieTek	CT-CAL101	DMSO	10000
Nintedanib	VEGFR, PDGFR, FGFR inhibitor	B. Kinase inhibitor	Approved (US)	Indetanib	Vargatef, Ofev	Selleck	S1010	DMSO	10000
Lenvatinib	VEGFR inhibitor	B. Kinase inhibitor	Approved (US)		Lenvima	Selleck	S1164	DMSO	2500
CUDC-101	HDAC & EGFR, Her2 inhibitor	B. Kinase inhibitor	Investigational (Ph 1)			Selleck	S1194	DMSO	10000
PF-00477736	Chk1 inhibitor	B. Kinase inhibitor	Investigational (Ph 1)			Axon Medchem	Axon 1379-2	DMSO	10000
AZD7762	Chk1 inhibitor	B. Kinase inhibitor	Investigational (Ph 1)			Axon Medchem	Axon 1399	DMSO	1000
AZD8055	mTOR inhibitor	B. Kinase inhibitor	Investigational (Ph 1)			ChemieTek	CT-A8055-3	DMSO	10000
Doramapimod	p38MAPK inhibitor	B. Kinase inhibitor	Investigational (Ph 1)			Axon Medchem	Axon 1358	DMSO	10000
Bryostatin 1	PKC activator	B. Kinase inhibitor	Investigational (Ph 1)			Santa Cruz Biotechnology	sc-201407-4	DMSO	100
EMD1214063	c-Met inhibitor	B. Kinase inhibitor	Investigational (Ph 1)			ChemieTek	CT-EMD063	DMSO	1000
AZD1480	JAK1/2, FGFR inhibitor	B. Kinase inhibitor	Investigational (Ph 1)			ChemieTek	CT-A1480	DMSO	1000
Tamatinib	Syk inhibitor	B. Kinase inhibitor	Investigational (Ph 1)			Selleck	S2194-2	DMSO	10000

(continued)

Table 1 (continued)

DRUG_ NAME	Mechanism targets	Class explained	High phase approval status	Alias	Trade names	Supplier	Supplier Ref	Solvent	High Conc. (nM)
TAK-733	MEK1/2 inhibitor	B. Kinase inhibitor	Investigational (Ph 1)			Selleck	S2617	DMSO	1000
Omipalisib	PI3K/mTOR inhibitor	B. Kinase inhibitor	Investigational (Ph 1)			Selleck	S2658	DMSO	1000
TAK-901	Aurora B inhibitor	B. Kinase inhibitor	Investigational (Ph 1)			Selleck	S2718	DMSO	1000
NVP-BGJ398	FGFR inhibitor	B. Kinase inhibitor	Investigational (Ph 1)	BGJ398		ChemieTek	CT-BGJ398	DMSO	1000
INK128	mTOR inhibitor	B. Kinase inhibitor	Investigational (Ph 1)	INK128		ChemieTek	CT-INK128	DMSO	1000
ZSTK474	PI3K gamma selective inhibitor	B. Kinase inhibitor	Investigational (Ph 1)			LC Laboratories	Z-1066	DMSO	10000
AZD2014	mTOR inhibitor, ATP-competitive	B. Kinase inhibitor	Investigational (Ph 1)			Selleck	S2783	DMSO	10000
GSK2636771	PI3K beta selective inhibitor	B. Kinase inhibitor	Investigational (Ph 1)			ChemieTek	CT-GSK263	DMSO	10000
Rebastinib	Allosteric ABL, FLT3, TIE2, TRKA inhibitor	B. Kinase inhibitor	Investigational (Ph 1)			ChemieTek	CT-DCC20	DMSO	1000
BMS-911543	JAK2 inhibitor	B. Kinase inhibitor	Investigational (Ph 1)			ChemieTek	CT-BMS911	DMSO	10000
LY-294002	PI3K inhibitor	B. Kinase inhibitor	Investigational (Ph 1)			LC Laboratories	L-7962	DMSO	100000
ASP3026	ALK inhibitor	B. Kinase inhibitor	Investigational (Ph 1)			ChemieTek	CT-ASP302	DMSO	10000
PF-03758309	PAK inhibitor	B. Kinase inhibitor	Investigational (Ph 1)			ChemieTek	CT-PF0375	DMSO	10000
AZD-8330	MEK1/2 inhibitor	B. Kinase inhibitor	Investigational (Ph 1)	ARRY-424704		ChemieTek	CT-A8330	DMSO	10000
BMS-599626	Pan-HER inhibitor	B. Kinase inhibitor	Investigational (Ph 1)	AC-480		ChemieTek	CT-BMS59	DMSO	10000

LY-2874455	FGFR inhibitor	B. Kinase inhibitor	Investigational (Ph 1)		Axon Medchem	Axon 1981	DMSO	1000
SGI-1776	PIM kinase inhibitor	B. Kinase inhibitor	Investigational (Ph 1)		Selleck	S2198	DMSO	10000
AT7519	CDK1, 2, 4, 6 and 9 inhibitor	B. Kinase inhibitor	Investigational (Ph 1)		Selleck	S1524	DMSO	10000
TAK-960	PLK1 inhibitor	B. Kinase inhibitor	Investigational (Ph 1)		Santa Cruz Biotechnology	sc-364631	DMSO	2500
Lucitanib	FGFR1, VEGFR inhibitor	B. Kinase inhibitor	Investigational (Ph 1)	CO-3810, S 80881	Axon Medchem	Axon 1942	DMSO	10000
AMG-208	MET inhibitor	B. Kinase inhibitor	Investigational (Ph 1)		Selleck	S1316	DMSO	2500
AMG-900	pan-Aurora inhibitor	B. Kinase inhibitor	Investigational (Ph 1)		Selleck	S2719	DMSO	1000
ARRY-380	HER2 inhibitor	B. Kinase inhibitor	Investigational (Ph 1)		Selleck	S2752	DMSO	2500
GSK-1070916	AURb, AURc inhibitor	B. Kinase inhibitor	Investigational (Ph 1)		Selleck	S2740	DMSO	1000
GSK-461364	PLK1 inhibitor	B. Kinase inhibitor	Investigational (Ph 1)		Selleck	S2193	DMSO	10000
NVP-INC280	MET inhibitor	B. Kinase inhibitor	Investigational (Ph 1)	INC280, INCB-28060	Selleck	S2788	DMSO	1000
OSI-930	KIT, VEGFR inhibitor	B. Kinase inhibitor	Investigational (Ph 1)		Selleck	S1220	DMSO	2500
Palomid-529	AKT, MTOR, PI3K inhibitor	B. Kinase inhibitor	Investigational (Ph 1)	P529	Selleck	S2238	DMSO	10000
PF-00562271	FAK inhibitor	B. Kinase inhibitor	Investigational (Ph 1)		Selleck	S2672	DMSO	10000
PF-03814735	AURa, AURb inhibitor	B. Kinase inhibitor	Investigational (Ph 1)		Selleck	S2725	DMSO	10000
Gedatolisib	PI3K/mTOR inhibitor	B. Kinase inhibitor	Investigational (Ph 1)	PKI-587	Selleck	S2628	DMSO	1000
SNS-314	AURa, AURb inhibitor	B. Kinase inhibitor	Investigational (Ph 1)		Selleck	S1154	DMSO	1000
TAK-285	HER2 inhibitor	B. Kinase inhibitor	Investigational (Ph 1)		Selleck	S2784	DMSO	2500

(continued)

Table 1 (continued)

DRUG_NAME	Mechanism targets	Class explained	High phase approval status	Alias	Trade names	Supplier	Supplier Ref	Solvent	High Conc. (nM)
MLN-8054	AURa AURb FLT3 KIT (PDGFR)	B. Kinase inhibitor	Investigational (Ph 1)			Selleck	S1100	DMSO	10000
KW-2449	AURa AURb FLT3 inhibitor	B. Kinase inhibitor	Investigational (Ph 1)			Selleck	S2158	DMSO	2500
KRN-633	VEGFR inhibitor	B. Kinase inhibitor	Investigational (Ph 1)			Selleck	S1557	DMSO	2500
PHA-793887	CDK inhibitor	B. Kinase inhibitor	Investigational (Ph 1)			Selleck	S1487	DMSO	10000
AZD-6482	PI3Kbeta-selective inhibitor	B. Kinase inhibitor	Investigational (Ph 1)			Selleck	S1462	DMSO	2500
GSK-1059615	PI3K/mTOR inhibitor	B. Kinase inhibitor	Investigational (Ph 1)	GSK-615		Selleck	S1360	DMSO	10000
CYC-116	Aurora and VEGFR2 inhibitor	B. Kinase inhibitor	Investigational (Ph 1)			Selleck	S1171	DMSO	10000
CP-724714	EGFR ERBB2 inhibitor	B. Kinase inhibitor	Investigational (Ph 1)			Selleck	S1167	DMSO	10000
SGX-523	MET inhibitor	B. Kinase inhibitor	Investigational (Ph 1)			Selleck	S1112	DMSO	5000
JNJ-38877605	MET inhibitor	B. Kinase inhibitor	Investigational (Ph 1)			Selleck	S1114	DMSO	10000
GSK-690693	AKT, PKA, PKC inhibitor	B. Kinase inhibitor	Investigational (Ph 1)			Selleck	S1113	DMSO	10000
OSU-03012	PDPK1 inhibitor	B. Kinase inhibitor	Investigational (Ph 1)	AR-12		Selleck	S1106	DMSO	25000
NVP-AEW541	IGF1R inhibitor	B. Kinase inhibitor	Investigational (Ph 1)	AEW541		Selleck	S1034	DMSO	10000
PF-04217903	MET inhibitor	B. Kinase inhibitor	Investigational (Ph 1)			Selleck	S1094	DMSO	2500
AZD-1080	GSK3 inhibitor	B. Kinase inhibitor	Investigational (Ph 1)	AZ-11548415		Selleck	S7145	DMSO	10000

RG-7603	PI3K inhibitor, pan-class I	B. Kinase inhibitor	Investigational (Ph 1)	GDC-0349	Selleck	S8040	DMSO	2500
MK-8776	CHEK1 inhibitor	B. Kinase inhibitor	Investigational (Ph 1)	SCH-900776	Selleck	S2735	DMSO	2500
CH-5132799	PI3K inhibitor, pan-class I	B. Kinase inhibitor	Investigational (Ph 1)	PA-799	Selleck	S2699	DMSO	10000
AZD-5438	CDK1,2,9 inhibitor	B. Kinase inhibitor	Investigational (Ph 1)		Selleck	S2621	DMSO	10000
Silmitasertib	CSNK2A1 inhibitor	B. Kinase inhibitor	Investigational (Ph 1)		Selleck	S2248	DMSO	10000
Mubritinib	ERBB2 inhibitor	B. Kinase inhibitor	Investigational (Ph 1)		Selleck	S2216	DMSO	1000
AZD-8186	PI3Kbeta inhibitor	B. Kinase inhibitor	Investigational (Ph 1)		Active Biochem	A-1610	DMSO	1000
XL019	JAK2 inhibitor	B. Kinase inhibitor	Investigational (Ph 1)		Selleck	S7036	DMSO	10000
Bentamapimod	JNK inhibitor	B. Kinase inhibitor	Investigational (Ph 1)		Medchemexpress	HY-14761	DMSO	10000
AZD1208	PIM1, 2, 3 kinase inhibitor	B. Kinase inhibitor	Investigational (Ph 1)		Medchemexpress	HY-15604	DMSO	10000
BGB324	Axl inhibitor	B. Kinase inhibitor	Investigational (Ph 1)	R 428	Axon Medchem	Axon 1946	DMSO	10000
CEP-37440	ALK inhibitor	B. Kinase inhibitor	Investigational (Ph 1)		ChemieTek	CT-CEP374	DMSO	5000
AT13148	p70S6K, PKA, ROCK (AKT) inhibitor	B. Kinase inhibitor	Investigational (Ph 1)		Medchemexpress	HY-16071	DMSO	10000
Cerdulatinib	JAK, SYK inhibitor	B. Kinase inhibitor	Investigational (Ph 1)		Medchemexpress	HY-15999	DMSO	10000
TEW-7197	TGF-β receptor ALK4/ ALK5 inhibitor	B. Kinase inhibitor	Investigational (Ph 1)	EW-7197	Selleck	S7530	DMSO	2500
GDC-0994	ERK inhibitor	B. Kinase inhibitor	Investigational (Ph 1)		Medchemexpress	HY-15947-2	DMSO	10000
Merestinib	Met inhibitor	B. Kinase inhibitor	Investigational (Ph 1)		Medchemexpress	HY-15514A	DMSO	1000
VS-4718	FAK inhibitor	B. Kinase inhibitor	Investigational (Ph 1)	PND-1186	Chemietek	CT-VS4718	DMSO	10000

(continued)

Table 1 (continued)

DRUG_NAME	Mechanism targets	Class explained	High phase approval status	Alias	Trade names	Supplier	Supplier Ref	Solvent	High Conc. (nM)
LY3009120	pan-RAF inhibitor	B. Kinase inhibitor	Investigational (Ph 1)	DP-4978		Medchemexpress	HY-12558	DMSO	10000
BI 2536	PLK1 inhibitor	B. Kinase inhibitor	Investigational (Ph 2)			Selleck	S1109	DMSO	1000
AT9283	Aurora A & B, Jak2, Flt, Abl inhibitor	B. Kinase inhibitor	Investigational (Ph 2)			Selleck	S1134	DMSO	1000
Danusertib	Aurora, Ret, TrkA, FGFR-1 inhibitor	B. Kinase inhibitor	Investigational (Ph 2)			Selleck	S1107	DMSO	10000
Foretinib	MET, VEGFR2 inhibitor	B. Kinase inhibitor	Investigational (Ph 2)	XL880, EXEL-2880		Selleck	S1111	DMSO	1000
SNS-032	CDK inhibitor	B. Kinase inhibitor	Investigational (Ph 2)	BMS-387032		Selleck	S1145	DMSO	10000
Alvocidib	CDK inhibitor	B. Kinase inhibitor	Investigational (Ph 2)	Flavopiridol, HMR-1275		Selleck	S1230	DMSO	10000
Pimasertib	MEK1/2 inhibitor	B. Kinase inhibitor	Investigational (Ph 2)	MSC1936369B		Selleck	S1475	DMSO	10000
Motesanib	VEGFR, PDGFR, Ret, Kit inhibitor	B. Kinase inhibitor	Investigational (Ph 2)			Selleck	S1032	DMSO	10000
PF-04691502	PI3K/mTOR inhibitor	B. Kinase inhibitor	Investigational (Ph 2)			ChemieTek	CT-PF1502	DMSO	10000
MK1775	Wee1 inhibitor	B. Kinase inhibitor	Investigational (Ph 2)			Axon Medchem	Axon 1494	DMSO	10000
BMS-754807	IGF1R inhibitor	B. Kinase inhibitor	Investigational (Ph 2)			ChemieTek	CT-BMS75	DMSO	10000
OSI-027	mTOR inhibitor	B. Kinase inhibitor	Investigational (Ph 2)			ChemieTek	CT-O027	DMSO	10000
Refametinib	MEK1/2 inhibitor	B. Kinase inhibitor	Investigational (Ph 2)	RDEA119		ChemieTek	CT-R119	DMSO	10000
MK-2206	AKT inhibitor	B. Kinase inhibitor	Investigational (Ph 2)			ChemieTek	CT-MK2206	DMSO	1000

Linsitinib	IGF1R, IR inhibitor	B. Kinase inhibitor	Investigational (Ph 2)	ASP7487	ChemieTek	CT-O906	DMSO	10000
Tandutinib	FLT3, PDGFR, KIT inhibitor	B. Kinase inhibitor	Investigational (Ph 2)	CT53518	LC Laboratories	T-7802	DMSO	1000
Pictilisib	PI3K inhibitor, pan-class I	B. Kinase inhibitor	Investigational (Ph 2)	RG-7321	LC Laboratories	G-9252	DMSO	10000
Seliciclib	CDK2/7/9 inhibitor	B. Kinase inhibitor	Investigational (Ph 2)	Roscovitine	LC Laboratories	R-1234	DMSO	10000
Dactolisib	mTOR/(PI3K) inhibitor	B. Kinase inhibitor	Investigational (Ph 2)		LC Laboratories	N-4288	DMSO	1000
Quizartinib	FLT3 inhibitor	B. Kinase inhibitor	Investigational (Ph 2)		ChemieTek	CT-AC220	DMSO	1000
Gandotinib	JAK2 inhibitor	B. Kinase inhibitor	Investigational (Ph 2)		Selleck	S2179	DMSO	10000
Sotrastaurin	PKC inhibitor	B. Kinase inhibitor	Investigational (Ph 2)	AEB071	Axon Medchem	Axon 1635-2	DMSO	10000
UCN-01	PKCbeta, PDK1, Chk, Cdk2 inhibitor	B. Kinase inhibitor	Investigational (Ph 2)	7-Hydroxy staurosporine	Sigma-Aldrich	U6508-4	DMSO	10000
Tivantinib	MET inhibitor	B. Kinase inhibitor	Investigational (Ph 2)		ChemieTek	CT-ARQ197	DMSO	1000
RAF265	C-Raf inhibitor	B. Kinase inhibitor	Investigational (Ph 2)	CHIR-265	Selleck	S2161	DMSO	1000
Rabusertib	Chk1 inhibitor	B. Kinase inhibitor	Investigational (Ph 2)	IC-83	Selleck	S2626	DMSO	1000
Galunisertib	TGF-B/Smad inhibitor	B. Kinase inhibitor	Investigational (Ph 2)		Selleck	S2230	DMSO	1000
Buparlisib	PI3K inhibitor, pan-class I	B. Kinase inhibitor	Investigational (Ph 2)	BKM-120	Selleck	S2247	DMSO	10000
Apitolisib	PI3K/mTOR inhibitor	B. Kinase inhibitor	Investigational (Ph 2)		ChemieTek	CT-G0980	DMSO	10000
AZD4547	FGFR inhibitor	B. Kinase inhibitor	Investigational (Ph 2)		ChemieTek	CT-A4547	DMSO	1000
Sonolisib	PI3K inhibitor, pan-class I. Irreversible	B. Kinase inhibitor	Investigational (Ph 2)	DJM-166	Active Biochem	PX-866	DMSO	10000

(continued)

Table 1 (continued)

DRUG_NAME	Mechanism targets	Class explained	High phase approval status	Alias	Trade names	Supplier	Supplier Ref	Solvent	High Conc. (nM)
Binimetinib	MEK1/2 inhibitor	B. Kinase inhibitor	Investigational (Ph 2)	MEK162, ARRY-438162, ARRY-162		ChemieTek	CT-A162	DMSO	1000
KX2-391	non-ATP competitive Src inhibitor	B. Kinase inhibitor	Investigational (Ph 2)			Selleck	S2700	DMSO	10000
Fostamatinib	Syk inhibitor	B. Kinase inhibitor	Investigational (Ph 2)			Selleck	S2206-3	AQ	2500
Momelotinib	JAK1 & 2 inhibitor	B. Kinase inhibitor	Investigational (Ph 2)	CYT1138		ChemieTek	CT-CYT387	DMSO	10000
Ralimetinib	p38MAPK inhibitor	B. Kinase inhibitor	Investigational (Ph 2)	CP868569		Selleck	S1494	DMSO	10000
Crenolanib	PDGFRA and PDGFRB inhibitor	B. Kinase inhibitor	Investigational (Ph 2)			Selleck	S2730	DMSO	10000
GDC-0068	AKT inhibitor	B. Kinase inhibitor	Investigational (Ph 2)	RG7440		ChemieTek	CT-G0068	DMSO	10000
Alpelisib	PI3Kalpha inhibitor	B. Kinase inhibitor	Investigational (Ph 2)	BYL719		Selleck	S2814	DMSO	2500
Baricitinib	JAK inhibitor	B. Kinase inhibitor	Investigational (Ph 2)	INCB28050		Selleck	S2851	DMSO	2500
AZD-5363	AKT inhibitor	B. Kinase inhibitor	Investigational (Ph 2)			ChemieTek	CT-A5363	DMSO	10000
SAR302503	JAK2-selective inhibitor	B. Kinase inhibitor	Investigational (Ph 2)	TG-101348		ChemieTek	CT-TG101	DMSO	10000
Bafetinib	Abl, Lyn inhibitor	B. Kinase inhibitor	Investigational (Ph 2)	NS-187		Selleck	S1369	DMSO	1000
Tideglusib	GSK3 inhibitor	B. Kinase inhibitor	Investigational (Ph 2)			Selleck	S2823	DMSO	3000
Rigosertib	Ras-Raf interaction inhibitor	B. Kinase inhibitor	Investigational (Ph 2)		Estybon, Novonex	Selleck	S1362	DMSO	10000
Milciclib	CDK2 inhibitor	B. Kinase inhibitor	Investigational (Ph 2)			Selleck	S2751	DMSO	10000

Duvelisib	PI3K inhibitor	B. Kinase inhibitor	Investigational (Ph 2)	INK1197	Selleck	S7028	DMSO	500
Icotinib	EGFR inhibitor	B. Kinase inhibitor	Investigational (Ph 2)		Selleck	S2922	DMSO	10000
Amuvatinib	Broad spectrum TK inhib	B. Kinase inhibitor	Investigational (Ph 2)		Selleck	S1244	DMSO	10000
Pelitinib	EGFR inhibitor	B. Kinase inhibitor	Investigational (Ph 2)	WAY-EKB 569	Selleck	S1392	DMSO	2500
Telatinib	VEGFR, KIT, PDGFR inhibitor	B. Kinase inhibitor	Investigational (Ph 2)		Selleck	S2231	DMSO	10000
Triciribine	AKT inhibitor	B. Kinase inhibitor	Investigational (Ph 2)	Pentaazacentopthylene, Tricyclic nucleoside	Selleck	S1117-2	DMSO	100000
Tozasertib	pan-Aurora inhibitor	B. Kinase inhibitor	Investigational (Ph 2)	MK-0457	Selleck	S1048	DMSO	10000
Varlitinib	EGFR HER2 inhibitor	B. Kinase inhibitor	Investigational (Ph 2)		Selleck	S2755	DMSO	10000
Golvatinib	MET, VEGFR2 inhibitor	B. Kinase inhibitor	Investigational (Ph 2)		Selleck	S2859	DMSO	2500
Copanlisib	PI3K alpha, beta selective inhibitor	B. Kinase inhibitor	Investigational (Ph 2)		Selleck	S2802-2	DMSO w/ 10mM TFA	1000
Sapitinib	Pan-HER inhibitor	B. Kinase inhibitor	Investigational (Ph 2)		Selleck	S2192	DMSO	1000
NVP-AEE788	EGFR, VEGFR, ABL, SRC inhibitor	B. Kinase inhibitor	Investigational (Ph 2)	AEE788, GNF-PF-5343	Selleck	S1486	DMSO	2500
NVP-BGT226	PI3K/mTOR inhibitor	B. Kinase inhibitor	Investigational (Ph 2)	BGT226	Selleck	S2749	DMSO	1000
BMS-777607	Met, Axl, Ron and Tyro3 inhibitor	B. Kinase inhibitor	Investigational (Ph 2)		Selleck	S1561	DMSO	2500
Abemaciclib	CDK4 and 6 inhibitor	B. Kinase inhibitor	Investigational (Ph 2)		Selleck	S7158	DMSO	2500
VX 745	p38MAPK inhibitor	B. Kinase inhibitor	Investigational (Ph 2)		Tocris Biosciences	3915	DMSO	10000
XL-647	EGFR, ERBB2, VEGFR, EPHB4	B. Kinase inhibitor	Investigational (Ph 2)		Santa Cruz Biotechnology	sc-364659	DMSO	1000

(continued)

Table 1 (continued)

DRUG_NAME	Mechanism targets	Class explained	High phase approval status	Alias	Trade names	Supplier	Supplier Ref	Solvent	High Conc. (nM)
PD184352	MEK1/2 inhibitor	B. Kinase inhibitor	Investigational (Ph 2)			Selleck	S1020	DMSO	10000
ENMD-2076	pan-Aurora inhibitor	B. Kinase inhibitor	Investigational (Ph 2)			Selleck	S1181	DMSO	10000
MK-2461	MET inhibitor	B. Kinase inhibitor	Investigational (Ph 2)			Selleck	S2774	DMSO	10000
PD0325901	MEK1/2 inhibitor	B. Kinase inhibitor	Investigational (Ph 2)			Selleck	S1036	DMSO	1000
PH-797804	p38MAPK inhibitor	B. Kinase inhibitor	Investigational (Ph 2)			Selleck	S2726	DMSO	1000
TAK-715	p38MAPK inhibitor	B. Kinase inhibitor	Investigational (Ph 2)			Selleck	S2928	DMSO	10000
TG100-115	PI3K gamma/delta inhibitor	B. Kinase inhibitor	Investigational (Ph 2)			Selleck	S1352	DMSO	10000
CEP-32496	BRAF inhibitor	B. Kinase inhibitor	Investigational (Ph 2)			Selleck	S8015	DMSO	10000
GDC-0623	MEK1/2 inhibitor	B. Kinase inhibitor	Investigational (Ph 2)			Active Biochem	A-1181	DMSO	2500
Talmapimod	p38MAPK alpha selective inhibitor	B. Kinase inhibitor	Investigational (Ph 2)			Axon Medchem	Axon 1671	DMSO	10000
Encorafenib	B-RAF(V600E)	B. Kinase inhibitor	Investigational (Ph 2)	LGX818		Selleck	S7108	DMSO	1000
Tanzisertib	JNK1, 2, 3 inhibitor	B. Kinase inhibitor	Investigational (Ph 2)			Medchemexpress	HY-15495	DMSO	10000
BMS863233	Cdc7 inhibitor	B. Kinase inhibitor	Investigational (Ph 2)			Selleck	S7547	AQ	10000
Entospletinib	SYK inhibitor	B. Kinase inhibitor	Investigational (Ph 2)			Selleck	S7523	DMSO	5000
Voxtalisib	mTOR/PI3K inhibitor	B. Kinase inhibitor	Investigational (Ph 2)	XL765		ChemieTek	CT-XL765c	DMSO	10000

Pilaralisib	PI3K inhibitor. Pan-class I	B. Kinase inhibitor	Investigational (Ph 2)	XL147		Medchemexpress	HY-16526	DMSO	2500
Uprosertib	AKT inhibitor	B. Kinase inhibitor	Investigational (Ph 2)			Medchemexpress	HY-15965	DMSO	10000
Filgotinib	JAK1-selective inhibitor	B. Kinase inhibitor	Investigational (Ph 2)			Selleck	S7605	DMSO	10000
Afuresertib	AKT1-selective inhibitor	B. Kinase inhibitor	Investigational (Ph 2)			Medchemexpress	HY-15966A	DMSO	1000
PF-06463922	ALK, ROS1 inhibitor	B. Kinase inhibitor	Investigational (Ph 2)			Selleck	S7536	DMSO	1000
SLx-2119	ROCK2 inhibitor	B. Kinase inhibitor	Investigational (Ph 2)	KD025		Medchemexpress	HY-15307	DMSO	5000
Poziotinib	pan-HER inhibitor	B. Kinase inhibitor	Investigational (Ph 2)	NOV120101		Medchemexpress	HY-15730	DMSO	1000
Spebrutinib	BTK inhibitor	B. Kinase inhibitor	Investigational (Ph 2)	AVL-292		Medchemexpress	HY-18012	DMSO	1000
Ulixertinib	ERK inhibitor	B. Kinase inhibitor	Investigational (Ph 2)	VRT752271		Chemietek	CT-VRT752	DMSO	10000
Prexasertib	Chk1 inhibitor	B. Kinase inhibitor	Investigational (Ph 2)			Medchemexpress	HY-18174A	DMSO	10000
Vatalanib	VEGFR-1 & -2 inhibitor	B. Kinase inhibitor	Investigational (Ph 3)	PTK 787, ZK222584		LC Laboratories	V-8303	DMSO	10000
Orantinib	KDR, FGFR, PDGFR inhibitor	B. Kinase inhibitor	Investigational (Ph 3)	SU6668		Selleck	S1470	DMSO	10000
Selumetinib	MEK1/2 inhibitor	B. Kinase inhibitor	Investigational (Ph 3)	ARRY-142886		Selleck	S1008	DMSO	10000
Dovitinib	FGFR inhibitor	B. Kinase inhibitor	Investigational (Ph 3)	TKI258		Selleck	S1018	DMSO	10000
Perifosine	AKT/PI3K inhibitor	B. Kinase inhibitor	Investigational (Ph 3)			Selleck	S1037	AQ	2500
Cediranib	KDR/Flt/VEGFR inhibitor	B. Kinase inhibitor	Investigational (Ph 3)		Recentin	Selleck	S1017	DMSO	1000
Tivozanib	VEGFR1, 2, 3, c-Kit, PDGFRB inhibitor	B. Kinase inhibitor	Investigational (Ph 3)	KRN951		ChemieTek	CT-AV951	DMSO	10000
AZD1152-HQPA	Aurora B inhibitor	B. Kinase inhibitor	Investigational (Ph 3)			ChemieTek	CT-A1152H	DMSO	1000

(continued)

Table 1 (continued)

DRUG_ NAME	Mechanism targets	Class explained	High phase approval status	Alias	Trade names	Supplier	Supplier Ref	Solvent	High Conc. (nM)
Alisertib	Aurora A inhibitor	B. Kinase inhibitor	Investigational (Ph 3)			ChemieTek	CT-M8237	DMSO	10000
Lestaurtinib	FLT3, JAK2, TrkA, TrkB, TrkC inhibitor	B. Kinase inhibitor	Investigational (Ph 3)			LC Laboratories	L-6307	DMSO	1000
Saracatinib	Src, Abl inhibitor	B. Kinase inhibitor	Investigational (Ph 3)			LC Laboratories	S-8906	DMSO	10000
Canertinib	pan-HER inhibitor	B. Kinase inhibitor	Investigational (Ph 3)	PD 183805		LC Laboratories	C-1201	DMSO	10000
Enzastaurin	PKCbeta inhibitor	B. Kinase inhibitor	Investigational (Ph 3)			LC Laboratories	E-4506	DMSO	10000
Masitinib	KIT inhibitor	B. Kinase inhibitor	Investigational (Ph 3)			LC Laboratories	M-7007	DMSO	10000
Midostaurin	PKC, PKA, S6K and EGFR inhibitor	B. Kinase inhibitor	Investigational (Ph 3)	PKC412, CGP 41251		LC Laboratories	P-7600	DMSO	10000
Ruboxistaurin	PKCbeta inhibitor	B. Kinase inhibitor	Investigational (Ph 3)			Axon Medchem	Axon 1401-2	DMSO	10000
Volasertib	PLK1 inhibitor	B. Kinase inhibitor	Investigational (Ph 3)			ChemieTek	CT-BI6727	DMSO	1000
Neratinib	EGFR inhibitor	B. Kinase inhibitor	Investigational (Ph 3)			Selleck	S2150(2)	DMSO	1000
Linifanib	VEGFR, PDGFR, CSF-1R, FLT3 inhibitor	B. Kinase inhibitor	Investigational (Ph 3)	AL-39324, RG3635		Selleck	S1003	DMSO	1000
Brivanib	VEGFR inhibitor	B. Kinase inhibitor	Investigational (Ph 3)	BMS-582664		Selleck	S1084	DMSO	1000
Dacomitinib	pan-HER inhibitor	B. Kinase inhibitor	Investigational (Ph 3)			ChemieTek	CT-DACO	DMSO	1000
Dinaciclib	CDK inhibitor	B. Kinase inhibitor	Investigational (Ph 3)			ChemieTek	CT-DINA	DMSO	1000
Apatinib	VEGFR inhibitor	B. Kinase inhibitor	Investigational (Ph 3)			Selleck	S2221	DMSO	10000

Semaxanib	VEGFR inhibitor	B. Kinase inhibitor	Investigational (Ph 3)		Selleck	S2845	DMSO	10000
NVP-LEE011	CDK4/6 inhibitor	B. Kinase inhibitor	Investigational (Ph 3)	LEE011	Selleck	S7440	DMSO	10000
Pacritinib	FLT3/JAK2	B. Kinase inhibitor	Investigational (Ph 3)		Selleck	S8057	DMSO	10000
Cobimetinib	MEK1/2 inhibitor	B. Kinase inhibitor	Investigational (Ph 3)	XL-518	Medchemexpress	HY-13064	DMSO	1000
Osimertinib	EGFR(L858R/T790M) inhibitor	B. Kinase inhibitor	Investigational (Ph 3)		Selleck	S7297	DMSO	2500
Losmapimod	p38MAPK inhibitor	B. Kinase inhibitor	Investigational (Ph 3)		Selleck	S7215	DMSO	10000
Rociletinib	EGFR(L858R/T790M) inhibitor	B. Kinase inhibitor	Investigational (Ph 3)	AVL-301, CNX-419	ChemieTek	CT-CO1686	DMSO	10000
Taselisib	PI3K alpha, gamma selective inhibitor	B. Kinase inhibitor	Investigational (Ph 3)	RG7604	Medchemexpress	HY-13898-2	DMSO	1000
Pexidartinib	KIT, CSF1R, FLT3 inhibitor	B. Kinase inhibitor	Investigational (Ph 3)		Medchemexpress	HY-16749	DMSO	10000
AZ 3146	Mps1 kinase (TTK) inhibitor	B. Kinase inhibitor	Probe		Tocris Biosciences	3994	DMSO	10000
PF-04708671	p70S6K inhibitor	B. Kinase inhibitor	Probe		Sigma-Aldrich	PZ0143	DMSO	10000
TGX-221	PI3K beta selective inhibitor	B. Kinase inhibitor	Probe		ChemieTek	CT-TGX221	DMSO	10000
VX-11E	ERK1 & 2 inhibitor	B. Kinase inhibitor	Probe		ChemieTek	CT-VX11e	DMSO	2500
PF-4800567	CK1epsilon inhibitor	B. Kinase inhibitor	Probe		Tocris Biosciences	4281	DMSO	10000
PF-670462	CK1epsilon and CK1delta inhibitor	B. Kinase inhibitor	Probe		Tocris Biosciences	3316	DMSO	10000
(5Z)-7-Oxozeaenol	TAK1 inhibitor	B. Kinase inhibitor	Probe	(5Z)-7-Oxozeaenol	Tocris Biosciences	3604-03-01	DMSO	10000
GSK269962	ROCK1 and ROCK2 inhibitor	B. Kinase inhibitor	Probe		Tocris Biosciences	4009	DMSO	10000
PF 431396	FAK/PYK2 inhibitor	B. Kinase inhibitor	Probe		Tocris Biosciences	4278	DMSO	10000
GSK650394	SGK1 & 2 inhibitor	B. Kinase inhibitor	Probe		Tocris Biosciences	3572	DMSO	10000

(continued)

Table 1 (continued)

DRUG_ NAME	Mechanism targets	Class explained	High phase approval status	Alias	Trade names	Supplier	Supplier Ref	Solvent	High Conc. (nM)
AZ-23	Trk inhibitor	B. Kinase inhibitor	Probe			Axon Medchem	Axon 1610	DMSO	1000
GSK-1838705A	IGF1R, INSR, ALK inhib	B. Kinase inhibitor	Probe			ChemieTek	CT-GSK183	DMSO	2500
GSK-1904529A	IGF1R, INSR inhib	B. Kinase inhibitor	Probe			ChemieTek	CT-GSK190	DMSO	10000
SP600125	pan-JNK inhibitor	B. Kinase inhibitor	Probe	Pyrazolanthrone, Anthrapyrazolone		LC Laboratories	S-7979	DMSO	100000
BX-912	PDK1 inhib	B. Kinase inhibitor	Probe			Selleck	S1275	DMSO	10000
GSK-2334470	PDK1 inhibitor	B. Kinase inhibitor	Probe			ChemieTek	CT-GSK233	DMSO	10000
SCH772984	ERK1 & 2 inhibitor	B. Kinase inhibitor	Probe			ChemieTek	CT-SCH772	DMSO	10000
PKI-402	PI3K/mTOR inhibitor	B. Kinase inhibitor	Probe			Selleck	S2739	DMSO	2500
PHA 408	IKK-2 inhibitor	B. Kinase inhibitor	Probe			Axon Medchem	Axon 1651	DMSO	10000
PS-1145	IKK-2 inhibitor	B. Kinase inhibitor	Probe			Axon Medchem	Axon 1568	DMSO	25000
KU-60019	ATM inhibitor	B. Kinase inhibitor	Probe			Selleck	S1570	DMSO	25000
TPCA-1	IKK-2 inhibitor	B. Kinase inhibitor	Probe			Selleck	S2824	DMSO	25000
IRAK1/4 inhibitor	IRAK1/4 inhibitor	B. Kinase inhibitor	Probe	IRAK1/4 inhibitor		Merck Millipore	407601	DMSO	25000
MK-8745	Aurora A inhibitor	B. Kinase inhibitor	Probe			Selleck	S7065	DMSO	2500
AZ191	DYRK1A inhibitor	B. Kinase inhibitor	probe			Selleck	S7338	DMSO	10000
VE-821	ATR inihibitor	B. Kinase inhibitor	Probe			Selleck	S8007	DMSO	10000
AZD7545	PDHK inhibitor	B. Kinase inhibitor	probe			Selleck	S7517	DMSO	10000
GNE-0877	LRRK2 inhibitor	B. Kinase inhibitor	probe			Selleck	S7367	DMSO	1000
OTSSP167	MELK inhibitor	B. Kinase inhibitor	probe			Selleck	S7159	DMSO	1000
UNC2881	MER inhibitor	B. Kinase inhibitor	probe			Selleck	S7325	DMSO	2500
AMG-925	FLT-3, CDK4 inhibitor	B. Kinase inhibitor	Probe			ChemieTek	CT-AMG925	DMSO	1000
FRAX486	PAK1, 2, 3 inhibitor	B. Kinase inhibitor	probe			ChemieTek	CT-F486	DMSO	5000

GNE-7915	LRRK2 inhibitor	B. Kinase inhibitor	probe			ChemieTek	CT-GNE79	DMSO	1000
TAK-632	pan-RAF inhibitor	B. Kinase inhibitor	probe			ChemieTek	CT-TAK632	DMSO	10000
GSK2656157	PERK inhibitor	B. Kinase inhibitor	probe			Medchemexpress	HY-13820	DMSO	2500
Tacrolimus	Binds FKBP12, causes inhibition of calcineurin	C. Rapalog	Approved	Fujimycin	Prograf, Advagraf, Protopic	Tocris Biosciences	3631	DMSO	10000
Everolimus	binds FKBP12, causes inhibition of mTORC1	C. Rapalog	Approved	SDZ-RAD	Afinitor, Certican, Zortress	LC Laboratories	E-4040	DMSO	100
Temsirolimus	binds FKBP12, causes inhibition of mTORC1	C. Rapalog	Approved		Torisel	LC Laboratories	T-8040	DMSO	100
Sirolimus	binds FKBP12, causes inhibition of mTORC1	C. Rapalog	Approved	Rapamycin	Rapamune	LC Laboratories	R-5000	DMSO	100
Ridaforolimus	binds FKBP12, causes inhibition of mTORC1	C. Rapalog	Investigational (Ph 3)	AP 23573, Deforolimus		Active Biochem	A-1004	DMSO	100
Dexamethasone	Immunosuppresant; glucocorticoid	D. Immunomodulatory	Approved		Decadron, Dexpak	Selleck	S1322	DMSO	10000
Thalidomide	Immunosuppresant	D. Immunomodulatory	Approved			National Cancer Institute	NSC 66847-R/5	DMSO	10000
Imiquimod	Immunomodulatory agent, TLR7 agonist	D. Immunomodulatory	Approved			National Cancer Institute	NSC 369100-F/4	DMSO	2500
Levamisole	Immunomodulatory agent	D. Immunomodulatory	Approved	Tetramisole		Sigma-Aldrich	L9756	DMSO	10000
Methylprednisolone	Immunosuppressant	D. Immunomodulatory	Approved			Santa Cruz Biotechnology	sc-205749	DMSO	10000
Prednisolone	Immunomodulatory agent	D. Immunomodulatory	Approved			Santa Cruz Biotechnology	sc-205815	DMSO	10000
Prednisone	Immunomodulatory agent	D. Immunomodulatory	Approved			Santa Cruz Biotechnology	sc-205816	DMSO	10000
Bimatoprost	Prostaglandin analog	D. Immunomodulatory	Approved		Latisse, Lumigan, Prostamide	Selleck	S1407	DMSO	5500
Lenalidomide	Immunomodulatory	D. Immunomodulatory	Approved		Revlimid	LC Laboratories	L-5499	DMSO	100000

(continued)

Table 1 (continued)

DRUG_ NAME	Mechanism targets	Class explained	High phase approval status	Alias	Trade names	Supplier	Supplier Ref	Solvent	High Conc. (nM)
Pomalidomide	Immunomodulatory agent, anti-angiogenic	D. Immunomodulatory	Approved	3-amino-thalidomide	Pomalyst	Sigma-Aldrich	P0018	DMSO	10000
VGX-1027	Immunomodulator	D. Immunomodulatory	Investigational (Ph 1)			Selleck	S7515	DMSO	10000
NLG919	IDO inhibitor	D. Immunomodulatory	Investigational (Ph 1)			Selleck	S7111	DMSO	10000
Epacadostat	IDO inhibitor	D. Immunomodulatory	Investigational (Ph 2)			Medchemexpress	HY-15689	DMSO	10000
Tasquinimod	S100A9, immunomodulatory, anti-angiogenic	D. Immunomodulatory	Investigational (Ph 3)			Medchemexpress	HY-10528	DMSO	10000
Tretinoin	Retinoic acid receptor agonist	E. Differentiating/ epigenetic modifier	Approved	ATRA, retinoic acid, vitamin A		National Cancer Institute	NSC 122758-Q/6	DMSO	10000
Bexarotene	Antineoplastic agent; retinoid specifically selective for retinoid X receptors	E. Differentiating/ epigenetic modifier	Approved			Santa Cruz Biotechnology	sc-217753	DMSO	10000
Decitabine	Nucleoside analog DNA methyl transferase inhibitor	E. Differentiating/ epigenetic modifier	Approved	5-aza-2'-deoxycytidine		Selleck	S1200	DMSO	10000
Vorinostat	HDAC inhibitor	E. Differentiating/ epigenetic modifier	Approved	SAHA	Zolinza	LC Laboratories	V-8477	DMSO	10000
Olaparib	PARP inhibitor	E. Differentiating/ epigenetic modifier	Approved	KU-0059436	Lynparza	LC Laboratories	O-9201	DMSO	10000
Arsenic(III) oxide	Thioredoxin reductase inhibitor; cytotoxic chemotherapeutic	E. Differentiating/ epigenetic modifier	Approved			Sigma-Aldrich	202673	AQ	2500
Azacitidine	Nucleoside analog DNA methyl transferase inhibitor	E. Differentiating/ epigenetic modifier	Approved	5-azacytidine, 5-AzaC	Vidaza	National Cancer Institute	NSC 102816-P/21	DMSO	10000
Valproic acid	HDAC inhibitor	E. Differentiating/ epigenetic modifier	Approved			Sigma-Aldrich	P4543	AQ	1E+06

Romidepsin	HDAC inhibitor	E. Differentiating/ epigenetic modifier	Approved	FR901228	Istodax	Medchemexpress	HY-15149	DMSO	1000
Belinostat	HDAC inhibitor	E. Differentiating/ epigenetic modifier	Approved (US)	PX105684		ChemieTek	CT-BELI	DMSO	10000
Panobinostat	HDAC inhibitor	E. Differentiating/ epigenetic modifier	Approved (US)		Farydak, Faridak	LC Laboratories	P-3703	DMSO	1000
Niraparib	PARP inhibitor	E. Differentiating/ epigenetic modifier	Investigational (Ph 1)			ChemieTek	CT-MK4827	DMSO	10000
Rocilinostat	HDAC-6 selective inhibitor	E. Differentiating/ epigenetic modifier	Investigational (Ph 1)			ChemieTek	CT-ACY12	DMSO	10000
AR-42	HDAC inhibitor	E. Differentiating/ epigenetic modifier	Investigational (Ph 1)	OSU-HDAC42		Selleck	S2244	DMSO	10000
E7438	EZH2 inhibitor	E. Differentiating/ epigenetic modifier	Investigational (Ph 1)	EPZ-6438		Chemietek	CT-EPZ438	DMSO	10000
EPZ-5676	DOT1L inhibitor	E. Differentiating/ epigenetic modifier	Investigational (Ph 1)			Selleck	S7062	DMSO	1000
OTX015	BET family inhibitor	E. Differentiating/ epigenetic modifier	Investigational (Ph 1)			Selleck	S7360	DMSO	10000
CUDC-907	HDAC1/2/3/10, PI3Kalpha inhibitor	E. Differentiating/ epigenetic modifier	Investigational (Ph 1)			Selleck	S2759	DMSO	10000
GSK525762	BET family inhibitor	E. Differentiating/ epigenetic modifier	Investigational (Ph 1)	I-BET762		ChemieTek	CT-BET762	DMSO	10000
BAY 87-2243	HIF1alpha inhibitor	E. Differentiating/ epigenetic modifier	Investigational (Ph 1)			Selleck	S7309	DMSO	1000
GSK2879552	LSD1 inhibitor	E. Differentiating/ epigenetic modifier	Investigational (Ph 1)			Chemietek	CT-GSK287	DMSO	10000
Rucaparib	PARP inhibitor	E. Differentiating/ epigenetic modifier	Investigational (Ph 2)	AG-014447, PF-01367338		Axon Medchem	Axon 1529	DMSO	10000
Entinostat	HDAC inhibitor	E. Differentiating/ epigenetic modifier	Investigational (Ph 2)	SNDX-275		LC Laboratories	E-3866	DMSO	10000
Mocetinostat	HDAC inhibitor (HDAC1 & 2-selective)	E. Differentiating/ epigenetic modifier	Investigational (Ph 2)			Selleck	S1122	DMSO	10000
Quisinostat	HDAC inhibitor	E. Differentiating/ epigenetic modifier	Investigational (Ph 2)			Active Biochem	A-1162	DMSO	1000
FG-4592	HIF prolyl hydroxylase inhibitor	E. Differentiating/ epigenetic modifier	Investigational (Ph 2)			Selleck	S1007	DMSO	10000

(continued)

Table 1 (continued)

DRUG_ NAME	Mechanism targets	Class explained	High phase approval status	Alias	Trade names	Supplier	Supplier Ref	Solvent	High Conc. (nM)
Lomeguatrib	O6-methylguanine-DNA methyltransferase inhibitor	E. Differentiating/ epigenetic modifier	Investigational (Ph 2)			Tocris Biosciences	4359	DMSO	10000
Pracinostat	HDAC inhibitor	E. Differentiating/ epigenetic modifier	Investigational (Ph 2)			Selleck	S1515	DMSO	10000
Resminostat	HDAC1, 3, 6 inhibitor	E. Differentiating/ epigenetic modifier	Investigational (Ph 2)			Medchemexpress	HY-14718A	DMSO	10000
Givinostat	HDAC inhibitor	E. Differentiating/ epigenetic modifier	Investigational (Ph 2)			Selleck	S2170	DMSO	1000
Abexinostat	HDAC1-selective inhibitor	E. Differentiating/ epigenetic modifier	Investigational (Ph 2)			Medchemexpress	HY-10990	DMSO	10000
Veliparib	PARP inhibitor	E. Differentiating/ epigenetic modifier	Investigational (Ph 3)			Selleck	S1004	DMSO	10000
Tipifarnib	Farnesyltransferase inhibitor	E. Differentiating/ epigenetic modifier	Investigational (Ph 3)	Zarnestra, IND 58359		Selleck	S1453	DMSO	10000
Tacedinaline	HDAC inhibitor	E. Differentiating/ epigenetic modifier	Investigational (Ph 3)	Acetyldinaline, Gö 5549, PD 123654,		LC Laboratories	C-2606	DMSO	1000
Iniparib	PARP inhibitor	E. Differentiating/ epigenetic modifier	Investigational (Ph 3)	IND-71677		Axon Medchem	Axon 1566	DMSO	10000
Lonafarnib	Farnesyl transferase inhibitor	E. Differentiating/ epigenetic modifier	Investigational (Ph 3)	Sarasar		Selleck	S2797	DMSO	10000
Talazoparib	PARP1/2 inhibitor	E. Differentiating/ epigenetic modifier	Investigational (Ph 3)			Medchemexpress	HY-16106	DMSO	1000
XAV-939	Tankyrase-1 and -2	E. Differentiating/ epigenetic modifier	Probe			Selleck	S1180	DMSO	10000
(+)JQ1	BET family inhibitor	E. Differentiating/ epigenetic modifier	Probe	(+)JQ1, SGCBD01(+)		SGC	SGCBD01(+)	DMSO	10000
Tubacin	HDAC6 inhibitor	E. Differentiating/ epigenetic modifier	Probe	Tubacin		Selleck	S2239	DMSO	10000
Tubastatin A	HDAC6 inhibitor	E. Differentiating/ epigenetic modifier	Probe	Tubastatin A		ChemieTek	CT-TUBA	DMSO	10000

StemRegenin 1	AHR antagonist, stem cell regenerating	E. Differentiating/ epigenetic modifier	Probe	StemRegenin 1, SR1	ChemieTek	CT-SR1	DMSO	10000
PFI-1	Selective chemical probe for BET Bromodomains	E. Differentiating/ epigenetic modifier	Probe		SGC	SGCPFI	DMSO	10000
I-BET151	BET family inhibitor	E. Differentiating/ epigenetic modifier	Probe	GSK1210151A	ChemieTek	CT-BET151	DMSO	10000
IOX-2	PHD2 inhibitor	E. Differentiating/ epigenetic modifier	Probe		Tocris Biosciences	4451	DMSO	50000
GSK-J4	JMJD3 (histone demethylase) inhibitor	E. Differentiating/ epigenetic modifier	Probe		Selleck	S7070-2	DMSO	100000
UNC1215	L3MBTL3 inhibitor	E. Differentiating/ epigenetic modifier	Probe		Tocris Biosciences	4666	DMSO	30000
SGC0946	DOT1L inhibitor	E. Differentiating/ epigenetic modifier	Probe		Selleck	S7079	DMSO	10000
UNC0642	G9a/GLP inhibitor	E. Differentiating/ epigenetic modifier	Probe		Tocris Biosciences	5132	DMSO	10000
GSK343	EZH2 inhibitor	E. Differentiating/ epigenetic modifier	Probe		SGC	SGCGSK343	DMSO	10000
UNC0638	G9a/GLP inhibitor	E. Differentiating/ epigenetic modifier	Probe		Tocris Biosciences	4343	DMSO	10000
C646	p300/CREB-binding protein (CBP) inhibitor	E. Differentiating/ epigenetic modifier	Probe		Axon Medchem	Axon 1781	DMSO	25000
EPZ-5687	EZH2 inhibitor	E. Differentiating/ epigenetic modifier	Probe		ChemieTek	CT-EPZ687	DMSO	10000
IOX-1	2-Oxoglutarate Oxygenase Inhibitor	E. Differentiating/ epigenetic modifier	Probe		Selleck	S7234	DMSO	100000
GSK2801	BAZ2B/A bromodomain inhibitor	E. Differentiating/ epigenetic modifier	Probe		SGC	SGCGSK2801	DMSO	10000
SGC-CBP30	CREBBP/EP300 bromodomain inhibitor	E. Differentiating/ epigenetic modifier	Probe		Medchemexpress	HY-15826	DMSO	25000
RGFP966	HDAC3 inhibitor	E. Differentiating/ epigenetic modifier	Probe		Selleck	S7229	DMSO	10000
PTC-209	BMI-1 inhibitor	E. Differentiating/ epigenetic modifier	probe		Selleck	S7539	DMSO	10000

(continued)

Table 1 (continued)

DRUG_ NAME	Mechanism targets	Class explained	High phase approval status	Alias	Trade names	Supplier	Supplier Ref	Solvent	High Conc. (nM)
UM729	Enhancer of aryl hydrocarbon receptor (AhR) antagonists	E. Differentiating/ epigenetic modifier	probe			Medchemexpress	HY-15972	DMSO	10000
PCI-34051	HDAC8 inhibitor	E. Differentiating/ epigenetic modifier	probe			Medchemexpress	HY-15224	DMSO	10000
EPZ015666	PRMT5 inhibitor	E. Differentiating/ epigenetic modifier	Probe			Selleck	S7748	DMSO	10000
Goserelin	Gonadotropin releasing hormone superagonist	F. Hormone therapy	Approved			Tocris Biosciences	3592	DMSO	10000
Raloxifene	Selective estrogen receptor modulator	F. Hormone therapy	Approved			National Cancer Institute	NSC 747974-X/1	DMSO	10000
Letrozole	Aromatase inhibitor	F. Hormone therapy	Approved		Femara	National Cancer Institute	NSC 719345-G/2	DMSO	10000
Anastrozole	Aromatase inhibitor	F. Hormone therapy	Approved			National Cancer Institute	NSC 719344-F/2	DMSO	10000
Bicalutamide	Nonsteriodal antiandrogen	F. Hormone therapy	Approved		Casodex, Cosudex, Calutide, Kalumid	ChemieTek	CT-BIC	DMSO	10000
Aminoglutethimide	Anti-steroid, aromatase inhibitor	F. Hormone therapy	Approved			Sigma-Aldrich	A9657	DMSO	10000
Clomifene	Selective estrogen receptor modulator	F. Hormone therapy	Approved	Clomid, Serophene, Milophene		Selleck	S2561	DMSO	10000
Finasteride	type II 5-alpha reductase inhibitor	F. Hormone therapy	Approved			Tocris Biosciences	3293	DMSO	10000
Flutamide	Nonsteroidal antiandrogen	F. Hormone therapy	Approved			Tocris Biosciences	4094	DMSO	10000
Fulvestrant	Estrogen receptor antagonist	F. Hormone therapy	Approved			Selleck	S1191	DMSO	1000
Megestrol	Progestogen	F. Hormone therapy	Approved		Megace	National Cancer Institute	NSC 71423-Q/12	DMSO	10000

Tamoxifen	Estrogen receptor antagonist	F. Hormone therapy	Approved			National Cancer Institute	NSC 180973-S/203	DMSO	10000
Nilutamide	Nonsteroidal antiandrogen	F. Hormone therapy	Approved			Santa Cruz Biotechnology	sc-203644	DMSO	10000
Exemestane	Aromatase inhibitor	F. Hormone therapy	Approved			National Cancer Institute	NSC 713563-U/2	DMSO	10000
Abiraterone	P450 17alpha-hydroxylase-17,20-lyase inhibitor	F. Hormone therapy	Approved			Selleck	S1123	DMSO	5000
Toremifene	selective estrogen receptor modulator	F. Hormone therapy	Approved		Fareston, Acapodene	Santa Cruz Biotechnology	sc-253712	DMSO	10000
Lasofoxifene	Selective estrogen receptor modulator	F. Hormone therapy	Approved		Oporia	Santa Cruz Biotechnology	sc-211721	DMSO	1000
Enzalutamide	AR antagonist	F. Hormone therapy	Approved		Xtandi	Axon Medchem	Axon 1613	DMSO	10000
ARN 509	AR antagonist	F. Hormone therapy	Investigational (Ph 2)			Axon Medchem	Axon 1979	DMSO	10000
Orteronel	CYP17A1, androgen synth inhib.	F. Hormone therapy	Investigational (Ph 3)			Selleck	S1195	DMSO	10000
4-hydroxy-tamoxifen	Selective estrogen receptor modulator	F. Hormone therapy	Investigational as a gel preparation			Santa Cruz Biotechnology	sc-3542	DMSO	10000
RD162	AR antagonist	F. Hormone therapy	Probe			Axon Medchem	Axon 1532	DMSO	10000
Serdemetan	HDM2-p53 antagonist	G. Apoptotic modulator	Investigational (Ph 1)			Selleck	S1172	DMSO	10000
APR-246	p53 activator, thioredoxin reductase 1 inhibitor	G. Apoptotic modulator	Investigational (Ph 1)	Prima-1 Met		Tocris Biosciences	3710	DMSO	10000
PAC-1	procaspase-3 activator	G. Apoptotic modulator	Investigational (Ph 1)			Selleck	S2738	DMSO	10000
AT-406	XIAP, cIAP1, cIAP2 inhibitor	G. Apoptotic modulator	Investigational (Ph 1)			Selleck	S2754	DMSO	10000
Venetoclax	Bcl-2-selective inhibitor	G. Apoptotic modulator	Investigational (Ph 1)	GDC-0199		ChemieTek	CT-A199	DMSO	1000
Verdinexor	XPO1/CRM1 inhibitor	G. Apoptotic modulator	Investigational (Ph 1)			Medchemexpress	HY-15970	DMSO	1000
SAR405838	MDM2 inhibitor	G. Apoptotic modulator	Investigational (Ph 1)	MI-773		Selleck	S7649	DMSO	10000

(continued)

Table 1 (continued)

DRUG_ NAME	Mechanism targets	Class explained	High phase approval status	Alias	Trade names	Supplier	Supplier Ref	Solvent	High Conc. (nM)
AT 101	Bcl-2 family inhibitor	G. Apoptotic modulator	Investigational (Ph 2)	R-(-).gossypol		Selleck	S2812	DMSO	100000
Navitoclax	Bcl-2/Bcl-xL inhibitor	G. Apoptotic modulator	Investigational (Ph 2)			Selleck	S1001-2	DMSO	10000
YM155	Survivin inhibitor	G. Apoptotic modulator	Investigational (Ph 2)			ChemieTek	CT-YM155	DMSO	10000
Birinapant	IAPs, SMAC mimetic	G. Apoptotic modulator	Investigational (Ph 2)			Chemietek	CT-BIRI	DMSO	1000
LCL161	IAPs, SMAC mimetic	G. Apoptotic modulator	Investigational (Ph 2)			Chemietek	CT-LCL161	DMSO	25000
Selinexor	CRM1 inhibitor	G. Apoptotic modulator	Investigational (Ph 2)			Selleck	S7252	DMSO	10000
AMG-232	MDM2 inhibitor	G. Apoptotic modulator	Investigational (Ph 2)			ChemieTek	CT-AMG232	DMSO	1000
Obatoclax	Bcl-2 family inhibitor	G. Apoptotic modulator	Investigational (Ph 3)			Selleck	S1057	DMSO	10000
Nutlin-3	MDM2 inhibitor	G. Apoptotic modulator	Probe	Nutlin-3		Selleck	S1061	DMSO	10000
RG7388	MDM2 inhibitor	G. Apoptotic modulator	Probe			Medchemexpress	HY-15676	DMSO	10000
WEHI-539	Bcl-XL inhibitor	G. Apoptotic modulator	Probe			Medchemexpress	HY-15607A	DMSO	2500
UMI-77	MCL1 inhibitor	G. Apoptotic modulator	probe			Selleck	S7531	DMSO	100000
Sabutoclax	pan-Bcl-2 family inhibitor	G. Apoptotic modulator	Probe	ONT-701		Selleck	S8061	DMSO	25000
Pyridoclax	MCL-1 inhibitor	G. Apoptotic modulator	probe			Medchemexpress	HY-12527	DMSO	25000
A-1210477	MCL-1 inhibitor	G. Apoptotic modulator	probe			Active Biochem	A-9036	DMSO	50000

Methotrexate	Antimetabolite; Anti-folate agent	H. Metabolic modifier	Approved			Selleck	S1210	DMSO	5000
Pemetrexed	Dihydrofolate reductase inhibitor	H. Metabolic modifier	Approved		Alimta	LC Laboratories	P-7177	AQ	10000
Atorvastatin	HMG CoA reductase inhibitor	H. Metabolic modifier	Approved		Lipitor	ChemieTek	CT-ATOR	DMSO	10000
Lovastatin	HMG-CoA reductase inhibitor	H. Metabolic modifier	approved (non-oncology)			Selleck	S2061	DMSO	10000
Metformin	AMPK activator	H. Metabolic modifier	Approved (non-oncology)			Tocris Biosciences	2864	AQ	100000
Simvastatin	HMG CoA reductase inhibitor	H. Metabolic modifier	Approved (non-oncology)		Zocor	Sigma-Aldrich	S6196	DMSO	10000
Disulfiram	alcohol dehydrogenase inhibitor	H. Metabolic modifier	Approved (non-oncology)		Antabuse	Selleck	S1680	DMSO	100000
Pravastatin	HMG CoA reductase inhibitor	H.Metabolic modifier	Approved (non-oncology)	Eptastatin		Tocris Biosciences	2318	AQ	10000
Pevonedistat	NAE inhibitor	H. Metabolic modifier	Investigational (Ph 1)			ChemieTek	CT-M4924	DMSO	10000
URB597	FAAH inhibitor	H. Metabolic modifier	Investigational (Ph 1)			Selleck	S2631	DMSO	1000
CPI-613	pyruvate dehydrogenase, alpha-ketoglutarate dehydrogenase inhibitor	H. Metabolic modifier	Investigational (Ph 2)			Selleck	S2776	DMSO	10000
AVN944	IMPDH inhibitor	H. Metabolic modifier	Investigational (Ph 2)	VX-944		ChemieTek	CT-AVN944	DMSO	10000
Triapine	ribonucleotide reductase inhibitor	H. Metabolic modifier	Investigational (Ph 2)	3-AP		Selleck	S7470	DMSO	10000
Daporinad	NAMPT inhibitor	H. Metabolic modifier	Probe	DGB		Axon Medchem	Axon 1546	DMSO	1000
AGI-5198	IDH1 R132H/R132C inhibitor	H. Metabolic modifier	Probe	IDH-C35, AGI 5198		Selleck	S7185	DMSO	10000
TH588	MTH1 inhibitor	H. Metabolic modifier	Probe			KI/Helleday	TH588	DMSO	25000

(continued)

Table 1 (continued)

DRUG_ NAME	Mechanism targets	Class explained	High phase approval status	Alias	Trade names	Supplier	Supplier Ref	Solvent	High Conc. (nM)
AGI-6780	IDH2-R140Q inhibitor	H. Metabolic modifier	Probe			Medchemexpress	HY-15734	DMSO	10000
GSK923295	CENP-E inhibitor	I. Kinesin inhibitor	Investigational (Ph 1)			Medchemexpress	HY-10299	DMSO	10000
SB 743921	Mitotic inhibitor. Eg5/KSP inhibitor	I. Kinesin inhibitor	Investigational (Ph 2)			Selleck	S2182	DMSO	100
ARRY-520	KSP/Eg5 inhibitor	I. Kinesin inhibitor	Investigational (Ph 2)			Medchemexpress	HY-15187	DMSO	1000
Rofecoxib	COX-2 inhibitor	J. NSAID	Approved		Vioxx	ChemieTek	CT-RX001	DMSO	10000
Celecoxib	Selective COX-2 inhibitor	J. NSAID	Approved			National Cancer Institute	NSC 719627-M/1	DMSO	10000
CUDC-305	HSP90 inhibitor	K. HSP inhibitor	Investigational (Ph 1)	DEBIO-0932		ChemieTek	CT-CU305	DMSO	10000
Tanespimycin	HSP90 inhibitor	K. HSP inhibitor	Investigational (Ph 2)	17-AAG		Selleck	S1141	DMSO	10000
Alvespimycin	HSP90 inhibitor	K. HSP inhibitor	Investigational (Ph 2)	17-DMAG		Selleck	S1142	DMSO	1000
BIIB021	HSP90 inhibitor	K. HSP inhibitor	Investigational (Ph 2)			Selleck	S1175	DMSO	10000
Luminespib	HSP90 inhibitor	K. HSP inhibitor	Investigational (Ph 2)	AUY922		ChemieTek	CT-AUY922	DMSO	1000
Onalespib	HSP90 inhibitor	K. HSP inhibitor	Investigational (Ph 2)			Medchemexpress	HY-14463	DMSO	2500
Ganetespib	HSP90 inhibitor	K. HSP inhibitor	Investigational (Ph 3)			Selleck	S1159	DMSO	1000
VER 155008	HSP70 inhibitor	K. HSP inhibitor	Probe			Axon Medchem	Axon 1608	DMSO	10000
Radicicol	HSP90 inhibitor	K. HSP inhibitor	Probe	Monorden		Tocris Biosciences	2/1/1589	DMSO	10000
Pilocarpine	Non-selective muscarinic receptor agonist	X. Other	Approved		Salagen	Tocris Biosciences	694	DMSO	40000

Anagrelide	PDE-3, PLA2 inhibitor	X. Other	Approved		Agrylin, Xagrid	Tocris Biosciences	2432	DMSO	10000
Mepacrine	Unclear. PLA2 inhibitor. NF-kB inhibitor, p53 activator	X. Other	Approved	Quinacrine, Achricrine		Sigma-Aldrich	Q3251	AQ	50000
Plerixafor	CXCR4 antagonist	X. Other	Approved	JM 3100, AMD 3100	Mozobil	Cayman Chemical Company	10011332-2	AQ	10000
Fingolimod	S1PR antagonist	X. Other	Approved		Gilenya	LC Laboratories	F-4633	DMSO	10000
Vismodegib	Smothened (Hh) inhibitor	X. Other	Approved	HhAntag691	Erivedge	LC Laboratories	V-4050	DMSO	10000
Deferoxamine	Iron chelator	X. Other	Approved (non-oncology)	desferrioxamine, DFOM		Sigma-Aldrich	D9533	AQ	10000
Itraconazole	antifungal, hedgehog signaling inhibitor	X. Other	Approved (non-oncology)			Selleck	S2476	DMSO	5000
NVP-LGK974	PORCN inhibitor	X. Other	Investigational (Ph 1)	LGK974		Selleck	S7143	DMSO	10000
Sonidegib	Smothened (Hh) inhib	X. Other	Investigational (Ph 2)	LDE225, erismodegib (USAN)		ChemieTek	CT-LDE225	DMSO	10000
2-methoxyestradiol	Angiogenesis inhibitor	X. Other	Investigational (Ph 2)	2ME2		Cayman Chemical Company	13021	DMSO	10000
MK-0752	gamma-secretase/notch inhibitor	X. Other	Investigational (Ph 2)			Selleck	S2660	DMSO	1000
Varespladib	Secretory phospholipase A2 inhibitor	X. Other	Investigational (Ph 2)	A-001		ChemieTek	CT-VARE	DMSO	10000
1-methyl-D-tryptophan	Indolamine 2,3-dioxygenase 1 and 2 inhibitor	X. Other	Investigational (Ph 2)	1-methyl-D-tryptophan		Sigma-Aldrich	452483	AQ	5000
Glasdegib	Smo inhibitor	X. Other	Investigational (Ph 2)			Medchemexpress	HY-16391	DMSO	1000
Tarenflurbil	Gamma-secretase inhibitor	X. Other	Investigational (Ph 3)	(R)-Flurbiprofen		Cayman Chemical Company	70255	DMSO	10000
Tosedostat	Aminopeptidase inhibitor	X. Other	Investigational (Ph 3)			Tocris Biosciences	3595	DMSO	10000

(continued)

Table 1 (continued)

DRUG_ NAME	Mechanism targets	Class explained	High phase approval status	Alias	Trade names	Supplier	Supplier Ref	Solvent	High Conc. (nM)
Cilengitide	alphaVbeta3 integrin inhibitor	X. Other	Investigational (Ph 3)	NSC 707544		Selleck	S7077	DMSO	10000
Darapladib	lipoprotein-associated phospholipase A2 inhibitor	X. Other	Investigational (Ph 3)			Selleck	S7520	DMSO	1000
Marimastat	MMP-9, MMP-1, MMP-2, MMP-14, MMP-7 inhibitor	X. Other	Investigational (Ph 3)			Selleck	S7156	DMSO	10000
Galiellalactone	STAT3-DNA interaction inhibitor	X. Other	Probe			Santa Cruz Biotechnology	sc-202165-6	DMSO	25000
PF-3845	FAAH inhibitor	X. Other	Probe			Selleck	S2666	DMSO	10000
15D-PGJ2	Endogenous PPARγ ligand, prostaglandin, NFkB signaling inhibitor	X. Other	Probe	15D-PGJ2, 15-deoxy delta(12,14) prostaglandin J2		Merck	538927-2	DMSO	3000
Stattic	STAT3 SH2 domain inhibitor	X. Other	Probe	Stattic		Tocris Biosciences	2798-03-01	DMSO	50000
TRAM-34	intermediate-conductance Ca2+-activated K+ channel inh.	X. Other	Probe			Selleck	S1160	DMSO	1000
deltarasin	Ras-PDEdelta inhibitor	X. Other	Probe	deltarasin		Chemietek	CT-DELT	DMSO	10000
NSC348884	NPM1 oligomerization inhibitor	X. Other	Probe			Axon Medchem	Axon 1402	DMSO	50000
ONX-0914	LMP7 (immuno-proteasome)	X.Other	Probe	PR-957		Selleck	S7172	DMSO	10000
NMS-873	p97/VCP inhibitor	X. Other	Probe			Selleck	S7285	DMSO	10000
ML323	USP1-UAF1 inhibitor	X. Other	probe			Selleck	S7529	DMSO	10000
GSK2830371	Wip1 inhibitor	X. Other	probe			ChemieTek	CT-GSK283	DMSO	5000
BCI	Dusp6 inhibitor	X. Other	probe	BCI, NSC 150117		Sigma	B4313	DMSO	50000
MST-312	Telomerase inhibitor	X. Other	probe	Telomerase Inhibitor IX		Sigma	M3949	DMSO	10000
SH-4-54	STAT3 inhibitor	X. Other	probe			Selleck	S7337	DMSO	25000

Compound name, classification, mechanism of action, clinical phase and supplier information are displayed

9. We provide an R-package SynergyFinder to calculate the drug synergy scores using four different reference models, acknowledging the fact that the optimal method for standardization of drug combination data analysis remains an open question (28). The users are therefore advised to apply all the models for their data and report a drug combination that can show a detectable level of synergy scores irrespective of the model in selection.
10. A strong synergy in a drug combination, as revealed using the synergy landscape analysis, might not be sufficient to warrant the next level confirmatory analysis if the synergy does not lead to sufficient overall responses. Therefore, the synergy scoring is always advised to be combined with the raw dose-response matrix data visualized in Fig. 4 to provide an overview of the extra benefits of drug combinations compared to single drugs.

Acknowledgments

This work was supported by the Academy of Finland (grants 272437, 269862, 279163, 295504, and 292611 for TA, 272577 and 277293 for KW); the Integrative Life Science Doctoral Program at the University of Helsinki (LH), the Sigrid Jusélius Foundation (KW) and the Cancer Society of Finland (JT, TA, and KW). This project has received funding from the European Union's Horizon 2020 research and innovation program 2014–2020 under Grant Agreement No 634143 (MedBioinformatics).

References

1. Vogelstein B, Papadopoulos N, Velculescu VE et al (2013) Cancer genome landscapes. Science 339:1546–1558
2. Pemovska T, Kontro M, Yadav B et al (2013) Individualized systems medicine strategy to tailor treatments for patients with chemorefractory acute myeloid leukemia. Cancer Discov 3:1416–1429
3. Yang W, Soares J, Greninger P et al (2013) Genomics of drug sensitivity in cancer (GDSC): a resource for therapeutic biomarker discovery in cancer cells. Nucleic Acids Res D41:D955–D961
4. Seashore-Ludlow B, Rees MG, Cheah JH et al (2015) Harnessing connectivity in a large-scale small-molecule sensitivity dataset. Cancer Discov 5:1210–1223
5. Tang J, Aittokallio T (2014) Network pharmacology strategies toward multi-target anticancer therapies: from computational models to experimental design principles. Curr Pharm Des 20:20–36
6. Gillies RJ, Verduzco D, Gatenby RA (2012) Evolutionary dynamics of carcinogenesis and why targeted therapy does not work. Nat Rev Cancer 12:487–493
7. Mathews Griner LA, Guha R, Shinn P et al (2014) High-throughput combinatorial screening identifies drugs that cooperate with ibrutinib to kill activated B-cell-like diffuse large B-cell lymphoma cells. Proc Natl Acad Sci U S A 111:2349–2354
8. Crystal AS, Shaw TA, Sequist VL et al (2014) Patient-derived models of acquired resistance can identify effective drug combinations for cancer. Science 346:1480–1486
9. Pemovska T, Johnson E, Kontro M et al (2015) Axitinib effectively inhibits BCR-ABL1 (T315I) with a distinct binding conformation. Nature 519:102–105
10. Kulesskiy E, Saarela J, Turunen L et al (2016) Precision cancer medicine in the acoustic dispensing era: ex vivo primary cell drug sensitivity testing. J Lab Autom 21:27–36

11. Haltia UM, Andersson N, Yadav B et al (2017) Systematic drug sensitivity testing reveals synergistic growth inhibition by dasatinib or mTOR inhibitors with paclitaxel in ovarian granulosa cell tumor cells. Gynecol Oncol 144:621
12. Saeed K, Rahkama V, Eldfors S et al (2017) Comprehensive drug testing of patient-derived conditionally reprogrammed cells from castration-resistant prostate cancer. Eur Urol 71:319. https://doi.org/10.1016/j.eururo.2016.04.019
13. Berenbaum MC (1989) What is synergy. Pharmacol Rev 41:93–141
14. Loewe S (1953) The problem of synergism and antagonism of combined drugs. Arzneimittelforschung 3:285–290
15. Bliss CI (1939) The toxicity of poisons applied jointly. Ann Appl Biol 26:585–615
16. Chou TC (2006) Theoretical basis, experimental design, and computerized simulation of synergism and antagonism in drug combination studies. Pharmacol Rev 58:621–681
17. Boik JC, Narasimhan B (2010) An R package for assessing drug synergism/antagonism. J Stat Softw 34:6
18. Ritz C, Baty F, Streibig JC (2005) Bioassay analysis using R. J Stat Softw 12:5
19. Greco WR, Bravo G, Parsons JC (1995) The search for synergy: a critical review from a response surface perspective. Pharmacol Rev 47:331–385
20. Zhao W, Sachsenmeier K, Zhang L et al (2014) A new bliss independence model to analyze drug combination data. J Biomol Screen 19:817–821
21. Yadav B, Wennerberg K, Aittokallio T et al (2015) Searching for drug synergy in complex dose-response landscapes using an interaction potency model. Comput Struct Biotechnol J 13:504–513
22. Szwajda A, Gautam P, Karhinen L et al (2015) Systematic mapping of kinase addiction combinations in breast cancer cells by integrating drug sensitivity and selectivity profiles. Chem Biol 22:1144–1155
23. Gautam P, Karhinen L, Szwajda A et al (2016) Identification of selective cytotoxic and synthetic lethal drug responses in triple negative breast cancer cells. Mol Cancer 15:34
24. Karjalainen R, Pemovska T, Majumder M et al (2017) JAK1/2 and BCL2 inhibitors synergize to counteract bone marrow stromal cell-induced protection of AML. Blood 130:789

INDEX

Louise von Stechow (ed.), *Cancer Systems Biology: Methods and Protocols*, Methods in Molecular Biology, vol. 1711, https://doi.org/10.1007/978-1-4939-7493-1, © Springer Science+Business Media, LLC 2018

T

U

W

MIX
Papier aus verantwortungsvollen Quellen
Paper from responsible sources
FSC® C105338

If you have any concerns about our products,
you can contact us on
ProductSafety@springernature.com

In case Publisher is established outside the EU,
the EU authorized representative is:
Springer Nature Customer Service Center GmbH
Europaplatz 3, 69115 Heidelberg, Germany

Printed by Libri Plureos GmbH
in Hamburg, Germany